HANDBOOK ON THE PHYSICS AND CHEMISTRY OF RARE EARTHS

VOLUME 13

HANDBOOK ON THE PHYSICS AND CHEMISTRY OF RARE EARTHS

HANDBOOK ON THE PHYSICS AND CHEMISTRY OF

RARE EARTHS

VOLUME 13

EDITORS:

Karl A. GSCHNEIDNER, Jr.

Ames Laboratory–US DOE, and
Dept. of Materials Science and Engineering
Iowa State University
Ames, Iowa 50011
USA

LeRoy EYRING

Department of Chemistry, and
Center for Solid State Science
Arizona State University
Tempe, Arizona 85287
USA

NORTH-HOLLAND
1990 AMSTERDAM, OXFORD, NEW YORK, TOKYO

ISBN: 0 444 88547 1

Published by:

North-Holland

Elsevier Science Publishers B.V.

P.O. Box 211
1000 AE Amsterdam
The Netherlands

Sole distributors for the USA and Canada:

Elsevier Science Publishing Company, Inc.

655 Avenue of the Americas
New York, NY 10010
USA

Library of Congress Cataloging in Publication Data

(Revised for vol. 13)
Handbook on the physics and chemistry of rare earths.

Vols. 5– distributed by: Elsevier Science Pub. Co.
Errata [for vols. 1–6]: v. 8, p. 375–378.
Includes bibliographies and indexes.
contents: v. 1. Metals. – v. 2. Alloys and intermetallics. – [etc.] – v. 13. [without special title].
1. Earths, Rare. I. Gschneidner, Karl A. II. Eyring, LeRoy. III. Title: Rare earths.
QD172.R2H26 546'.41 78-12371
ISBN 0-444-85022-8 (set)
ISBN 0-444-88547-1 (Vol. 13)

printed on acid-free paper.

Printed in The Netherlands

Handbook on the Physics and Chemistry of Rare Earths, Vol. 13
edited by K.A. Gschneidner, Jr. and L. Eyring

PREFACE

Karl A. GSCHNEIDNER, Jr., and LeRoy EYRING

These elements perplex us in our rearches [sic], *baffle us in our speculations, and haunt us in our very dreams. They stretch like an unknown sea before us – mocking, mystifying, and murmuring strange revelations and possibilities.*

Sir William Crookes (February 16, 1887)

One of the primary purposes of the handbook is to put archival material on record at a time and in a form readily available for the promotion of current research utilizing the rare earth elements. The chapters in this volume fulfill this aim admirably while at the same time intercollating new ideas, new results and systematics.

Gladyshevsky, Bodak and Pecharsky have collected together summaries of the literature of the last quarter century on phase equilibria and crystal chemistry of ternary metallic systems containing a rare earth element with a metallic member from the group III, IIIb, or IV and a transition metal. This literature contains studies of a surprising fraction of the enormous number of compounds possible (i.e. crystal structure data for over 1600 ternary intermetallic compounds are presented and the phase relationships of nearly 200 ternary systems are illustrated, generally as isothermal sections). The authors have been major contributors to these investigations.

A.A. Eliseev (recently deceased) and G.M. Kuzmichyeva have collected information on more than 2000 ternary chalcogenide compounds formed by the rare earths and elements of the main subgroups I to VIII of the periodic table. Members of this collection of compounds possess properties, not only of scientific importance, but of wide commercial value as well. The compositions, structure and thermodynamic conditions for their preparation are given together with information on representative phase diagrams and other physico-chemical characteristics. The authors have attempted to systematize this extensive literature.

In chapter 90 of this volume Kimizuka, Takayama-Muromachi and Siratori consider ternary oxides with the general formula R_2O_3–M_2O_3–$M'O$. In this formula "R" represents a rare earth cation, "M" a different trivalent cation, and "M'" a divalent cation. This study is singularly comprehensive and systematic covering the phase diagrams, thermochemistry, crystal chemistry, and physical characteristics, including especially electric and magnetic properties.

Elemental analysis by atomic emission and mass spectrometry with inductively coupled plasmas is shown by Houk to be the techniques of choice for the analysis of rare earth materials. In this chapter the instrumentation and principles of use of inductively coupled plasmas themselves are outlined as well as their function as a source for atomic emission spectroscopy and mass spectrometry. A number of important applications for which these techniques are eminently suited are also discussed.

The relatively high abundance of the rare earths in the earth's crust suggests a likely role for them in living systems. It is just this subject that Brown, Rathjen, Graham, and Tribe discuss in the final chapter of this volume. The incorporation and effects of the rare earth elements in plant and animal life is considered.

CONTENTS

CONTENTS OF VOLUMES 1–12

VOLUME 1: METALS

VOLUME 2: ALLOYS AND INTERMETALLICS

VOLUME 3: NON-METALLIC COMPOUNDS – I

VOLUME 4: NON-METALLIC COMPOUNDS – II

VOLUME 5

VOLUME 6

VOLUME 7

VOLUME 8

VOLUME 9

VOLUME 10: HIGH ENERGY SPECTROSCOPY

VOLUME 11: TWO-HUNDRED-YEAR IMPACT OF RARE EARTHS ON SCIENCE

VOLUME 12

Handbook on the Physics and Chemistry of Rare Earths, Vol. 13
edited by K.A. Gschneidner Jr. and L. Eyring

Chapter 88

PHASE EQUILIBRIA AND CRYSTAL CHEMISTRY IN TERNARY RARE EARTH SYSTEMS WITH METALLIC ELEMENTS

E.I. GLADYSHEVSKY, O.I BODAK and V.K. PECHARSKY
Department of Inorganic Chemistry, Lvov State University, Lomonosova, 6, 290005, Lvov, USSR

Contents

Symbols and abbreviations

a/c	composition (or contents) in atomic percent
a, b, c	unit cell dimensions (in Å)
α, β, γ	unit cell angles (in degrees)
x, y, z	atom coordinates
x	indicate variable component contents, when it is used as index in compound compositional

	formula. Expressions like: $N - x$ (where N is a number) means N minus x, the x value is given in the text; like: N–M (where M also is a number) means from N to M.
ST	structure type
DED	deformation electron density
DEDD	deformation electron density distribution
R	rare earth metal: Y, Sc, and the lanthanides
M, M′	transition metal
X	non-transition metal (Al, Ga, In, Tl, Ge, Sn, Pb)
...	is used to indicate unknown data for compounds listed in tables from 1–6
~	is used before the compositional formula of a compound to indicate that the composition is given approximately

1. Introduction

This survey is concerned with the results of phase equilibria and crystal structures found in ternary rare earth systems. We have taken into account only the metallic elements, such as Al, Ga, In, Tl, Ge, Sn, Pb, and a transition metal, because experimental investigations were prevailed for them. A few known ternary systems with actinides are also described.

An imposing number of experimental investigations (754 different ternary systems completely, or partially known; 1857 ternary existing compounds) does not allow us to give any experimental details. The results are described in an abbreviated manner.

The order of phase equilibria description is determined by the Mendeleev's Periodic System; only in a one case (ternary systems with two transition metals), an alphabetical order is used.

2. Ternary rare earth systems phase equilibria

2.1. *R–M–Al systems*

Phase equilibria have been investigated partially or for the full range of concentrations in 89 ternary R–M–Al systems up to the present time. It was established, that a total of 247 new ternary compounds are formed in these systems, of these 168 crystal structures were determined. The results (phase diagrams, compositions and crystallographic data of ternary compounds) are listed below. Also another 74 ternary systems were investigated without studying the phase diagram; the existence of 113 more ternary compounds has been reported in literature. The list of compositions and crystallographic data for these 113 are given in table 1.

Sc–Fe–Al

Two ternary compounds were found by Zarechnyuk et al. (1970a) during partial phase diagram study (fig. 1). They are: (1) ~ $ScFe_{7.1-4.0}Al_{4.9-8.0}$ ($ThMn_{12}$

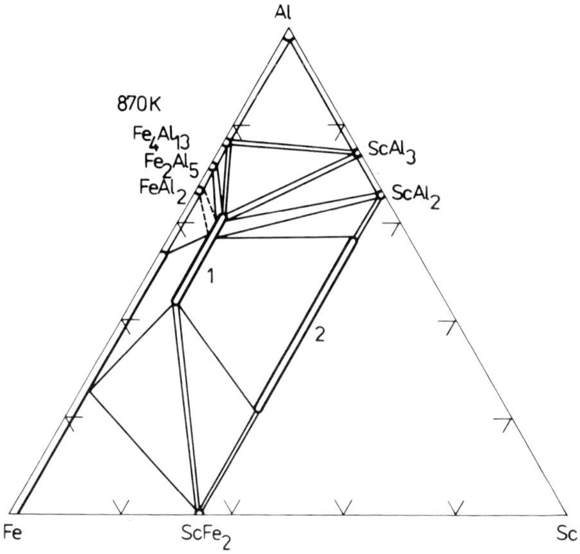

Fig. 1. Sc–Fe–Al, partial (0–33 a/c Sc) isothermal section at 870 K.

ST, $a = 8.68–8.70$, $c = 4.77–4.81$) and (2) ~ $ScFe_{1.55–0.30}Al_{0.45–1.70}$ ($MgZn_2$ ST, $a = 5.108–5.301$, $c = 8.319–8.587$). Practically no mutual solid solubility of binary phases was observed.

Sc–Ni–Al

Two partial isothermal sections are known for the system Sc–Ni–Al. The first one was derived by Goebel and Rosen (1968) in the region 100–40 a/c Ni at 1270 K, and the second by Zarechnyuk and Rykhal (1977) in the region 0–33a/c Sc at 870 K (both are shown in fig. 2). Crystal structures were determined for: (1) ~ $ScNi_2Al$ ($MnCu_2Al$ ST, $a = 5.99$) as reported by Zarechnyuk and Rykhal (1977), or (1) ~ $Sc_6Ni_{16}Al_7$ ($Mg_6Cu_{16}Si_7$ ST, $a = 12.16$) according to Markiv and Burnashova (1969); according to Teslyuk and Protasov (1965a), compound (2) ~ $ScNi_{1.0–0.6}Al_{1.0–1.4}$ has $MgZn_2$ structure type with $a = 5.12$, $c = 8.20$ for the 1 : 1 : 1 composition. The homogeneity regions for binary solid solutions and ternary compounds are different at 870 and 1270 K.

Sc–Cu–Al

The partial phase diagram at 770 K is shown in the fig. 3 and was studied by Prevarsky et al. (1976). There is no appreciable solubility of the third component in binary compounds. According to Prevarsky et al. (1976), compound (1) ~ $ScCu_{6.6–4.0}Al_{5.4–8.0}$ belong to $ThMn_{12}$ ST with $a = 8.63–8.66$, $c = 5.10–4.43$. Teslyuk and Protasov (1965b) reported for: (6) ~ $ScCu_2Al_2$ an orthorhombic unit cell with $a = 8.46$, $b = 8.86$, $c = 4.65$; (7) ~ $ScCu_2Al$ a cubic unit cell with $a = 3.10$, and (9) ~ ScCuAl the $MgZn_2$ type structure with $a = 5.04$, $c = 8.24$.

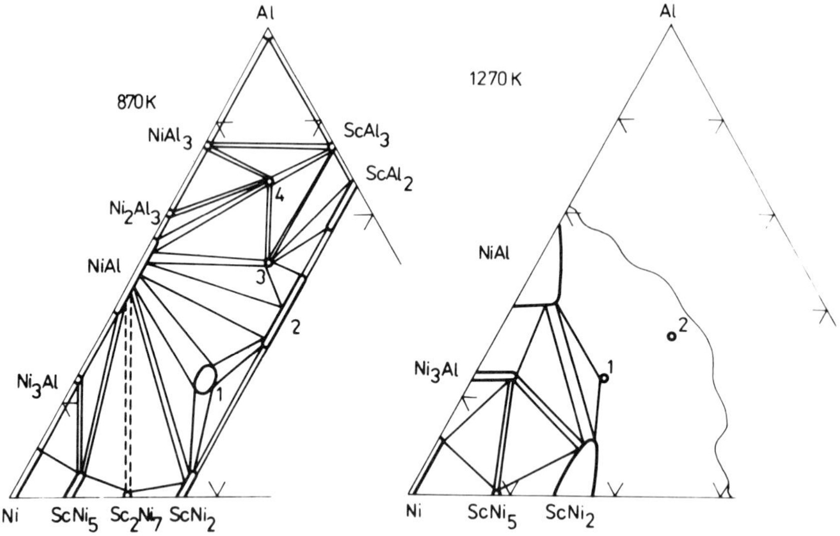

Fig. 2. Sc–Ni–Al, partial isothermal sections at 1270 K (100–40 a/c Ni) and at 870 K (0–33a/c Sc). Compounds with approximate compositions and unknown crystal structures: (3) ~ $ScNiAl_2$; (4) ~ $ScNiAl_4$.

Sc–Ru–Al

The system Sc–Ru–Al was investigated by Raevskaya and Sokolovskaya (1979) at 870 K and Karatygina et al. (1974a) at 1070 K. The results obtained are shown in fig. 4. Solid solution configuration is strongly dependent on temperature. Raevskaya and Sokolovskaya (1979) found for (6) ~ $ScRu_{0.7-0.3}Al_{1.3-1.7}$ a hexa-

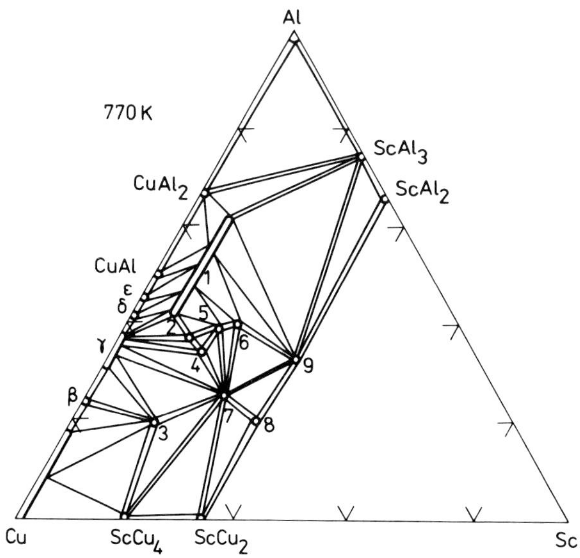

Fig. 3. Sc–Cu–Al, partial isothermal section at 770 K (0–33 a/c Sc). Compounds with undetermined crystal structure: (2) ~ $ScCu_4Al_4$; (3) ~ $Sc_3Cu_{13}Al_4$; (4) ~ $ScCu_3Al_2$; (5) ~ $Sc_3Cu_8Al_7$; (8) ~ $ScCu_{1.4}$-$Al_{0.6}$.

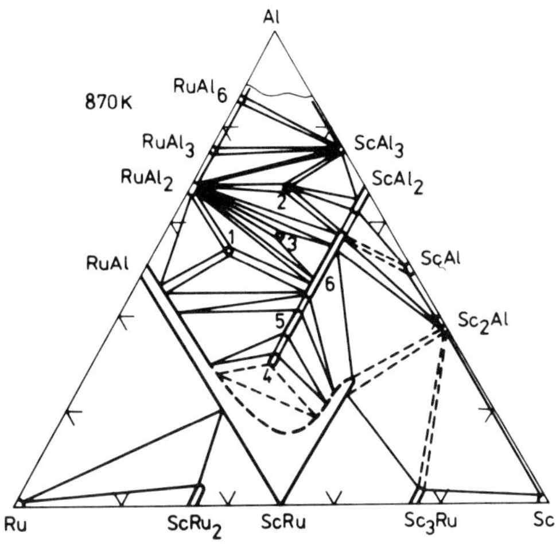

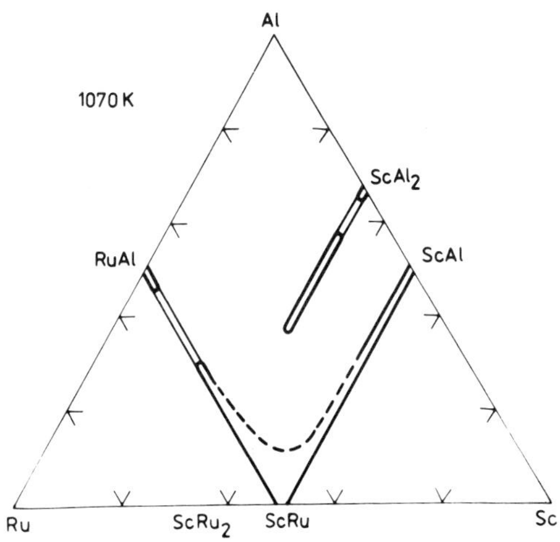

Fig. 4. Sc–Ru–Al, isothermal section at 870 K and partial isothermal section at 1070 K. Compounds with unknown crystal structure: (1) ~ $Sc_3Ru_6Al_{11}$; (2) ~ Sc_5Ru_4-Al_{18}; (4) ~ $ScRu_{1.07}Al_{0.93}$; (5) ~ $ScRu_{0.9-0.8}$-$Al_{1.1-1.2}$.

gonal unit cell with $a = 5.27$, $c = 8.57$ and for (3) ~ $ScRuAl_2$ a $MnCu_2Al$ ST with $a = 6.175$. On the other hand, Markiv and Storoghenko (1971) found for (3) ~ $ScRuAl_2$ a $Mg_6Cu_{16}Si_7$ ST with unknown cell dimensions.

Sc–Pd–Al

The isothermal section at 1070 K, derived by Karatygina et al. (1974b), is shown in fig. 5. It is characterized by the presence of continuous solid solution between PdAl and ScPd, and an up to 30 a/c solubility of ScAl in the ScPd. Crystal structure is determined for: (3) ~ $ScPdAl_2$ [$MnCu_2Al$ ST, $a = 6.188$ according to Karatygina et al. (1974b), or $Mg_6Cu_{16}Si_7$ ST after Markiv and Storoghenko (1971) with unknown unit cell]; (4) ~ $ScPd_{0.7-0.3}Al_{1.3-1.7}$ has $MgZn_2$ structure type with $a = 5.512$, $c = 8.621$ for unknown composition.

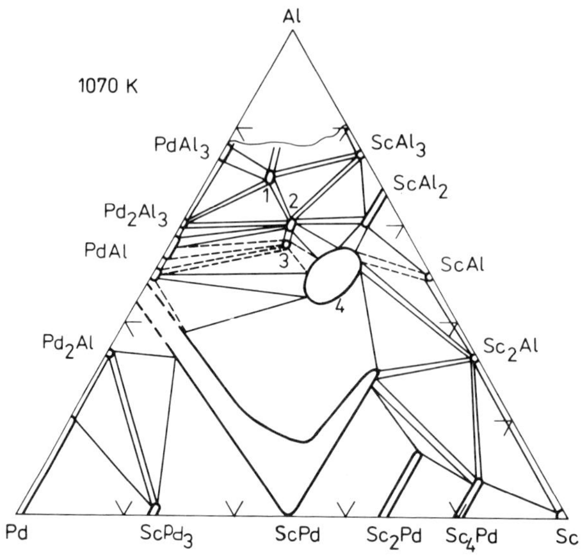

Fig. 5. Sc–Pd–Al, isothermal section at 1070 K. Compounds with unknown crystal structure: (1) ~ $Sc_3Pd_5Al_{19}$; (2) ~ $ScPdAl_3$.

Y–Be–Al

The partial isothermal section at 770 K (fig. 6) was established by Zarechnyuk et al. (1969a). Solubility of Al in YBe_{13} is near 3 a/c. Only one ternary compound, $YBe_{1.5-0.75}Al_{2.5-3.25}$, was found in the investigated region. It has La_3Al_{11} type structure. The unit cell could not be determined.

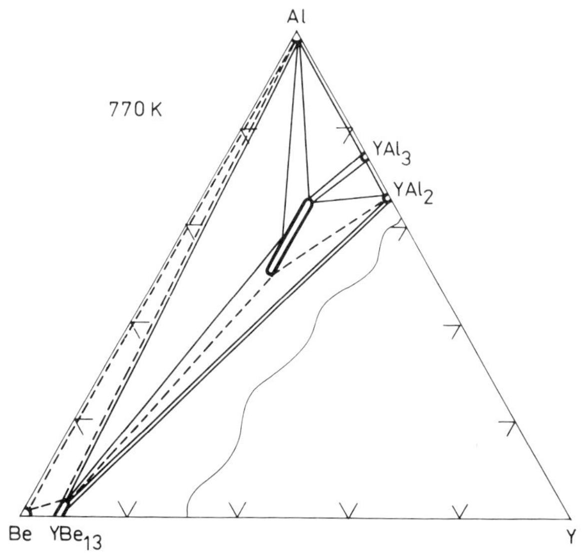

Fig. 6. Y–Be–Al, partial isothermal section at 770 K (0–33 a/c Y).

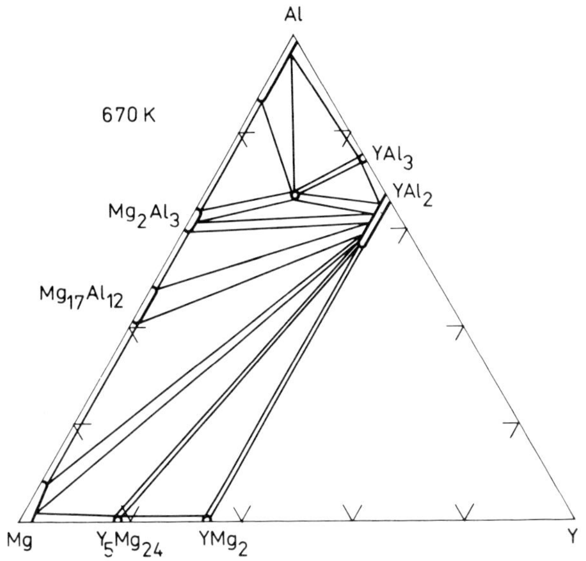

Fig. 7. Y–Mg–Al, partial isothermal section at 670 K (0–33 a/c Y).

Y–Mg–Al

The partial isothermal section at 670 K is shown in fig. 7 according to Zarechnyuk et al. (1980a). Solubility of Mg in YAl_2 is near 10 a/c. The only ternary compound $YMgAl_4$ has $MgZn_2$ ST with $a = 5.33$, $c = 8.57$.

Y–Ti–Al

No ternary compounds or solid solubilities were found in a partial study of the phase equilibria of the system (fig. 8) Rykhal (1972a).

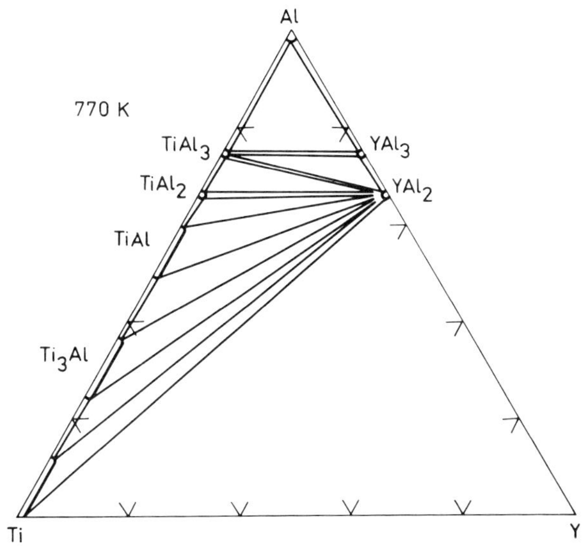

Fig. 8. Y–Ti–Al, partial isothermal section at 770 K (0–33 a/c Y).

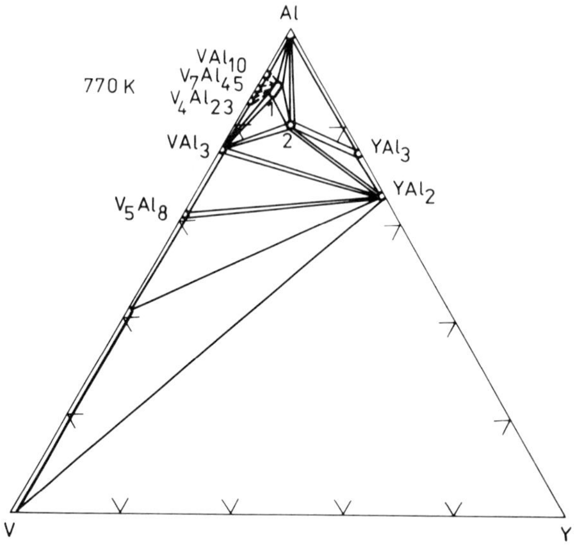

Fig. 9. Y–V–Al, partial isothermal section at 770 K (0–33 a/c Y).

Y–V–Al

The partial isothermal section at 770 K was derived by Zarechnyuk et al. (1971) and is shown in fig. 9. The existence of two ternary compounds was found by them. They are (1) ~ YV_2Al_{20} with $CeCr_2Al_{20}$ ST, $a = 14.51$, according to Krypyakevich and Zarechnyuk (1968) and (2) ~ $YVAl_8$ with hexagonal unit cell, $a = 10.98$, $c = 17.81$. The last is, possibly, isotypical with $YCrAl_8$.

Y–Cr–Al

Three ternary compounds were found in partial phase diagram studies (fig. 10): (1) ~ YCr_2Al_{20} ($CeCr_2Al_{20}$ ST, $a = 14.37$) according to Krypyakevich and

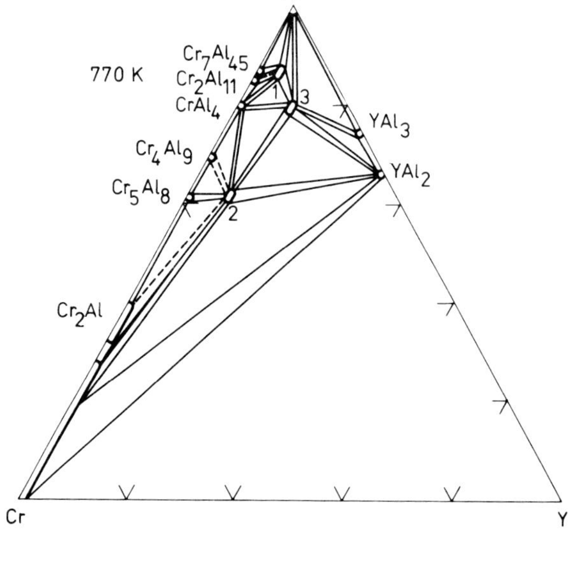

Fig. 10. Y–Cr–Al, partial isothermal section at 770 K (0–33 a/c Y).

Zarechnyuk (1968); (2) ~ YCr_4Al_8 ($CeMn_4Al_8$ ST, $a = 8.94$, $c = 5.12$) according to Zarechnyuk (1966); (3) ~ $YCrAl_8$ (hexagonal lattice, $a = 10.94$, $c = 17.74$) were found by Zarechnyuk et al. (1971).

Y–Mn–Al

The partial isothermal section (fig. 11) at 770 K was established by Rykhal et al. (1971). Ternary compound (2) ~ $Y_2Mn_9Al_8$ (Th_2Zn_{17} ST with $a = 8.90$, $c = 12.97$) was found by them, and compound (1) ~ YMn_4Al_8 ($CeMn_4Al_8$ ST, $a = 8.86$, $c = 5.09$), earlier found by Zarechnyuk (1966), was confirmed.

Y–Fe–Al

The partial isothermal section at 770 K was derived by Zarechnyuk et al. (1972b). Solubility of Al in Y_2Fe_{17} (Th_2Ni_{17} ST) reaches up to 25 a/c and in Y_2Fe_{17-x} (Th_2Zn_{17} ST) up to 45 a/c (fig. 12). It was reported by Zarechnyuk (1966), that compound (1) ~ $YFe_{5.8-4.0}Al_{6.2-8.0}$ has $ThMn_{12}$ ST with $a = 8.76–8.72$, $c = 4.91–5.04$; and by Rykhal (1972b), that compound (3), with $MgZn_2$ type structure, has composition $YFe_{1.2-1.0}Al_{0.8-1.0}$ and $a = 5.36–5.41$, $c = 8.74–8.81$.

Y–Co–Al

The partial isothermal section at 770 K is shown in fig. 13 and was established by Rykhal and Zarechnyuk (1971). According to Rykhal et al. (1972a) ternary compound (3) ~ $YCoAl_2$ has $MgCuAl_2$ ST and $a = 4.87$, $b = 10.15$, $c = 7.06$; Rykhal (1972b) reported for (4) ~ $YCo_{1.35-1}Al_{0.65-1}$ the $MgZn_2$ type structure with $a = 5.36–5.39$, $c = 8.63–8.67$; Rykhal and Zarechnyuk (1971) found for

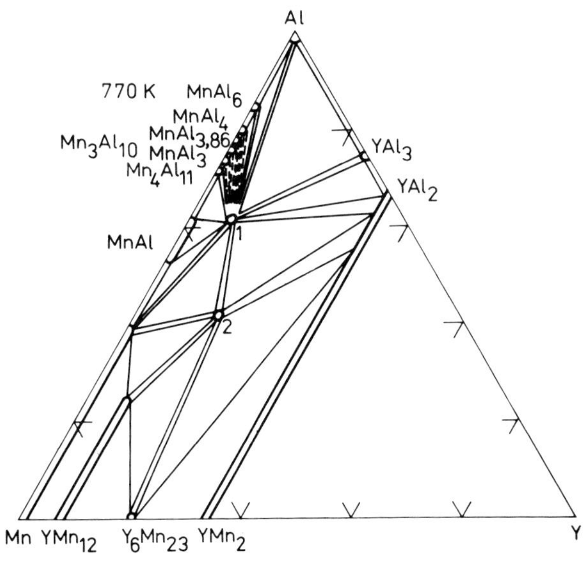

Fig. 11. Y–Mn–Al, partial isothermal section at 770 K (0–33 a/c Y).

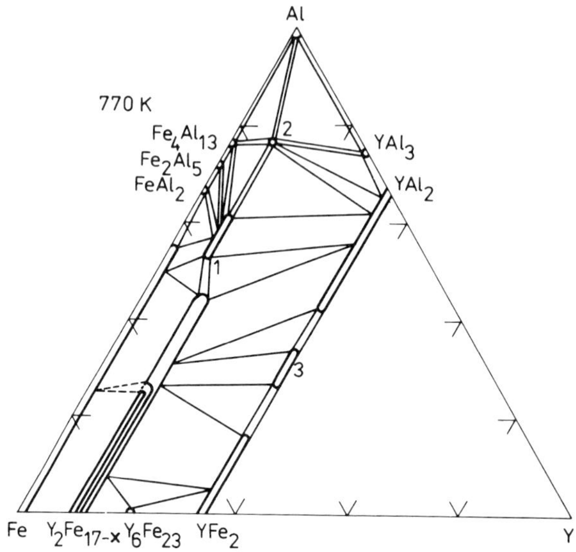

Fig. 12. Y–Fe–Al, partial isothermal section at 770 K (0–33 a/c Y). Compound with approximate composition: (2) ~ YFe_2Al_{10}.

(1) ~ YCo_2Al_7 and (2) ~ $YCoAl_4$ orthorhombic lattices with $a = 11.95$, $b = 16.90$, $c = 4.10$ and $a = 7.39$, $b = 12.76$, $c = 9.39$, respectively.

Y–Ni–Al

The partial isothermal sections were derived twice: 0–33 a/c Y at 1070 K by Rykhal and Zarechnyuk (1977) and at 1270 K by Rosen and Goebel (1968); both of them are shown in fig. 14. The biggest differences are observed for Al-rich

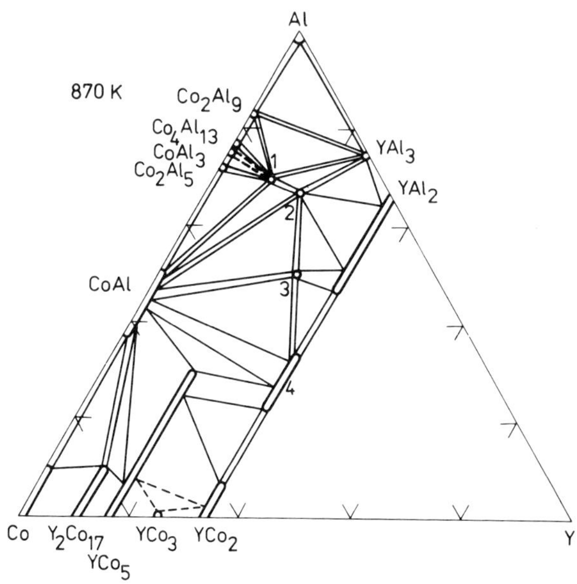

Fig. 13. Y–Co–Al, partial isothermal section at 870 K (0–33 a/c Y).

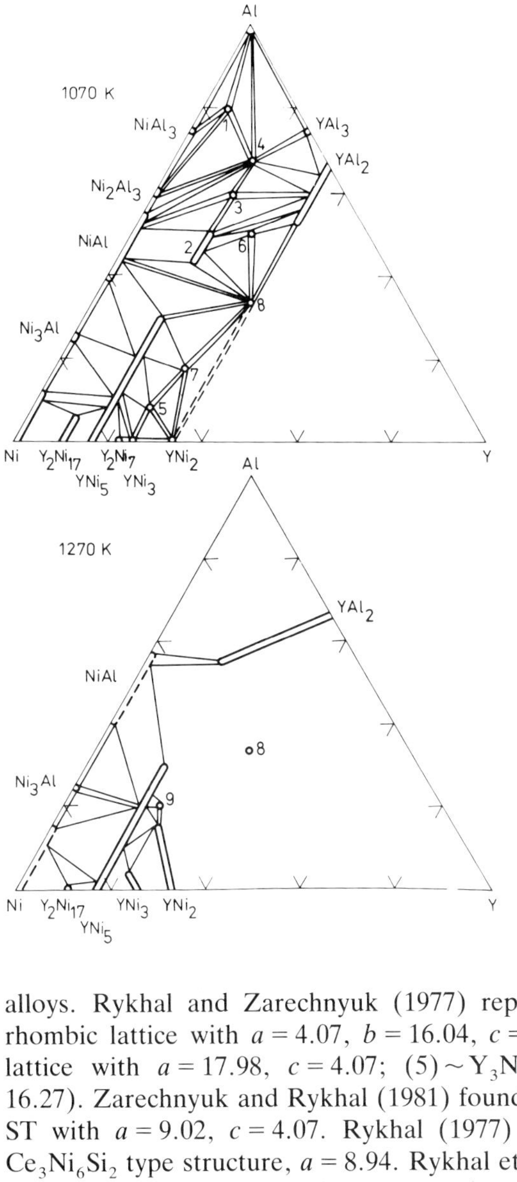

Fig. 14. Y–Ni–Al, partial isothermal sections at 1070 K and at 1270 K (both in the 0–33 a/c Y range).

alloys. Rykhal and Zarechnyuk (1977) reported for (1) ~ YNi_3Al_{16} an orthorhombic lattice with $a = 4.07$, $b = 16.04$, $c = 27.29$; (3) ~ $Y_2Ni_3Al_7$ a hexagonal lattice with $a = 17.98$, $c = 4.07$; (5) ~ Y_3Ni_8Al (Ce_3Co_8Si ST, $a = 5.07$, $c = 16.27$). Zarechnyuk and Rykhal (1981) found for (2) ~ YNi_2Al_3 the $HoNi_{2.6}Ga_{2.4}$ ST with $a = 9.02$, $c = 4.07$. Rykhal (1977) established for (7) ~ $Y_3Ni_6Al_2$ the $Ce_3Ni_6Si_2$ type structure, $a = 8.94$. Rykhal et al. (1972b) determined composition and crystal structure of (4) ~$YNiAl_4$ (a new ST, Cmcm, $a = 4.08$, $b = 15.44$, $c = 6.62$) and (6) ~ $YNiAl_2$ ($MgCuAl_2$ ST, $a = 4.072$, $b = 10.13$, $c = 7.06$). Rosen and Goebel (1968) proposed for (8) ~ YNiAl the ZrNiAl ST with $a = 7.04$, $c = 3.84$.

Y–Cu–Al

The partial isothermal section at 670 K (fig. 15) is based on the work of Zarechnyuk and Kolobnev (1968). The compositions and crystal structures for the

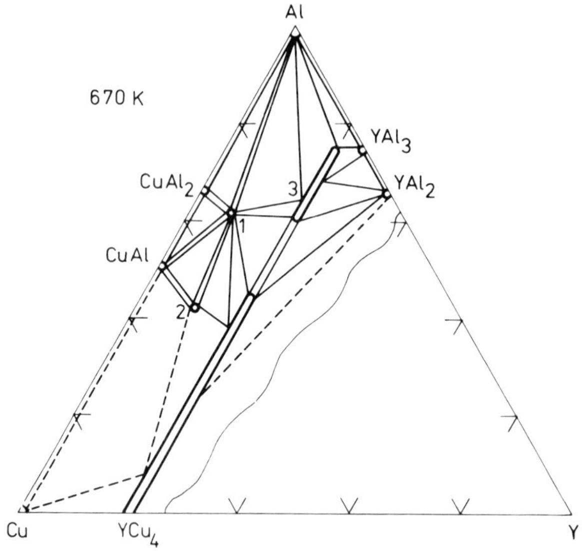

Fig. 15. Y–Cu–Al, partial isothermal section at 670 K (0–33 a/c Y).

three ternary compounds were determined and are as follows: (1) ~ YCu_4Al_8 ($CeMn_4Al_8$ ST, $a = 8.72$, $c = 5.16$); (2) ~ $Y_2Cu_9Al_8$ (Th_2Zn_{17} ST, $a = 8.79$, $c = 12.84$) and (3) ~ $YCu_{1-0.25}Al_{3-3.75}$ ($BaAl_4$ ST, $a = 4.22–4.25$, $c = 9.83–9.89$). Only one of the binary compounds, YCu_4, dissolve up to 45 a/c Al.

Y–Ga–Al

The partial isothermal section at 670 K is shown in fig. 16 after Dzyana and Zarechnyuk (1969). The crystal structure of only one ternary compound is

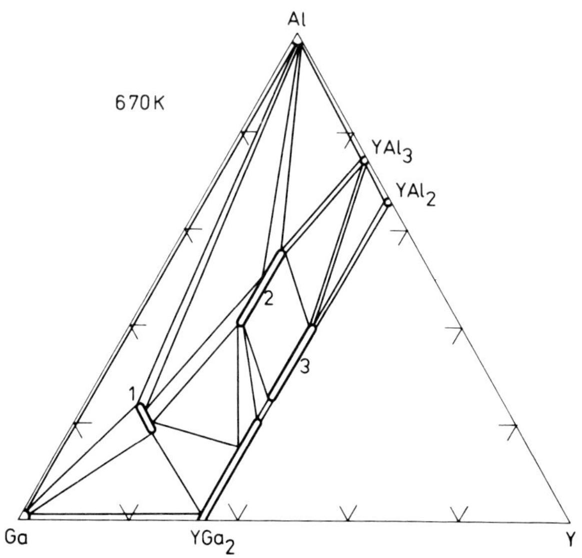

Fig. 16. Y–Ga–Al, partial isothermal section at 670 K (0–33 a/c Y). Compounds with unknown crystal structure: (1) ~ $Y_{0.30-0.45}Ga_2$-$Al_{0.70-0.55}$; (3) ~ $YGa_{1.3-0.8}Al_{0.7-1.2}$.

known: (2) ~ $YGa_{2-1.25}Al_{2-2.75}$ belongs to the La_3Al_{11} type structure. Unit cell was not determined. YCo_2 dissolves up to 20 a/c Al.

Y–Ge–Al

The partial isothermal section at 770 K was derived by Muravyeva et al. (1971) and is presented in fig. 17. The crystal structure of (1) ~ $YGe_{2.3-1.8}Al_{1.7-2.2}$ was determined in the work mentioned and belongs to the anti-La_2O_2S type with $a = 4.196–4.196$, $c = 6.702–6.775$. Yanson (1975) found that (3) ~ $YGe_{1.1-0.7}Al_{0.9-1.3}$ has DyGeAl ST and $a = 4.200$, $b = 10.62$, $c = 5.908$ for the 1 : 1 : 1 composition, and that (2) ~ Y_3Ge_6Al has hexagonal lattice with unknown unit cell dimensions.

Y–Ru–Al

Only one cross-section (YRu_2–YAl_2) at 770 K was investigated by Karatygina et al. (1974a). The existence of compounds: ~$YRu_{1.37}Al_{0.63}$; ~$YRu_{1.22}Al_{0.78}$; ~$YRu_{1.07}Al_{0.93}$; YRuAl; ~$YRu_{0.90}Al_{1.10}$ and ~$YRu_{0.8-0.4}Al_{1.2-1.6}$ was found. Crystal structures were not determined, but the authors made the assumption that multi-layer Laves phases can be realized in one or more of the described compounds. Both YRu_2 and YAl_2 dissolves up to 10 a/c Al and Ru, respectively.

La–Be–Al

Zarechnyuk et al. (1969a) described the partial isothermal section at 770 K. $LaBe_{13}$ dissolves 3 a/c Al. The only compound (fig. 18) has variable composition $LaBe_{1.65-1}Al_{2.35-3}$ and belongs to the $BaAl_4$ type structure, $a = 4.20–4.28$, $c = 10.71–10.58$.

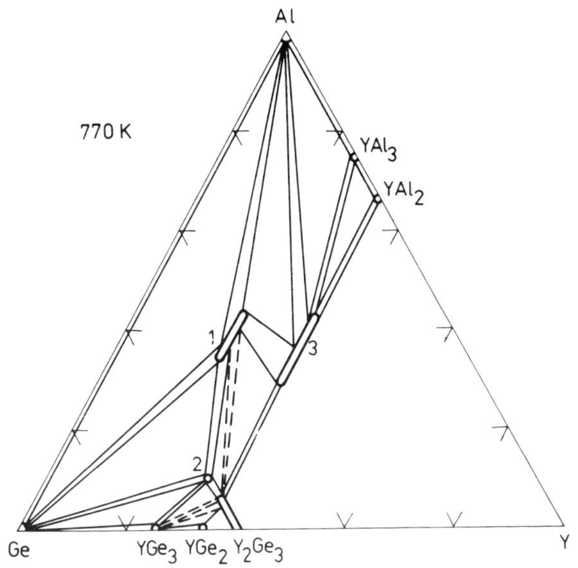

Fig. 17. Y–Ge–Al, partial isothermal section at 770 K (0–33 a/c Y).

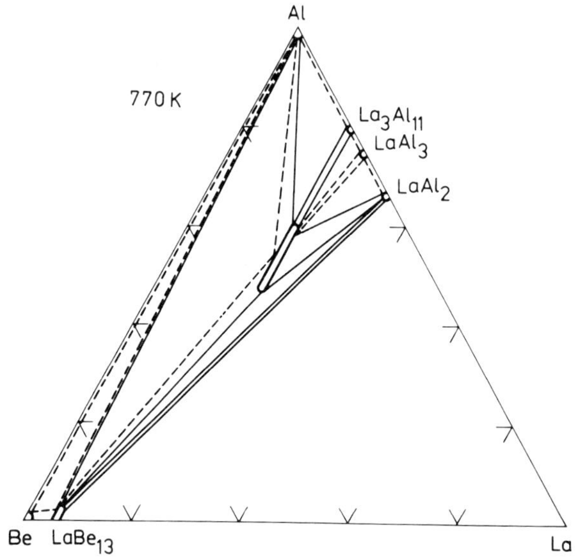

Fig. 18. La–Be–Al, partial isothermal section at 770 K (0–33 a/c La).

La–V–Al

The partial isothermal section was established by Zarechnyuk and Emes-Mysenko (1969) and is shown in fig. 19. The ternary compound LaV_2Al_{20} crystallized in the $ZrZn_{22}$ ST, $a = 14.61–14.69$ for variable Al contents from 80 to 85 a/c.

La–Cr–Al

The partial phase diagram at 770 K was derived by Emes-Mysenko (1971a). According to fig. 20, practically no mutual solid solubility in the binary com-

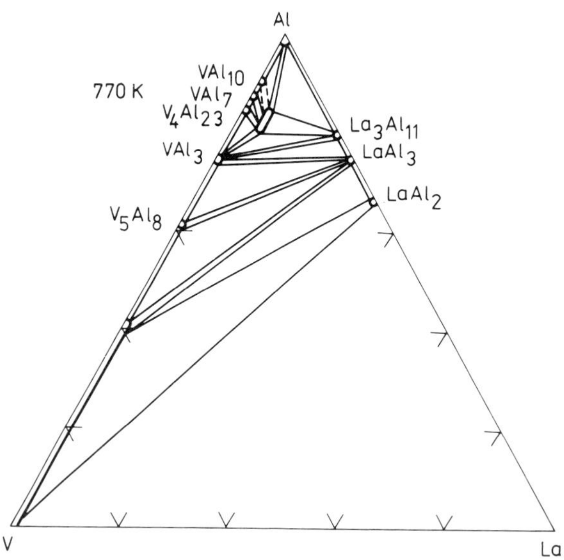

Fig. 19. La–V–Al, partial isothermal section at 770 K (0–33 a/c La).

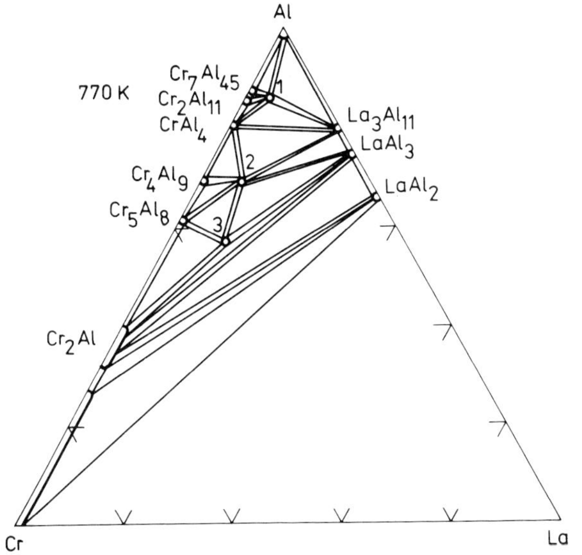

Fig. 20. La–Cr–Al, partial isothermal section at 770 K (0–33 a/c La).

pounds was found. The ternary compounds are: (1) ~ $LaCr_2Al_{20}$ ($CeCr_2Al_{20}$ ST, $a = 14.56$); (2) ~ $LaCr_3Al_9$ ($ThMn_{12}$ ST, $a = 9.138$, $c = 5.110$) and (3) ~ $La_2Cr_6Al_{11}$ (Th_2Zn_{17} ST, $a = 9.154$, $c = 13.39$). Crystal structure for (1) was reported by Krypyakevich and Zarechnyuk (1968).

La–Mn–Al

The partial isothermal section at 770 K (fig. 21) was investigated by Emes-Mysenko (1971a). (1) ~ $LaMn_{4-3.5}Al_{8-8.5}$ has the $ThMn_{12}$ type structure with

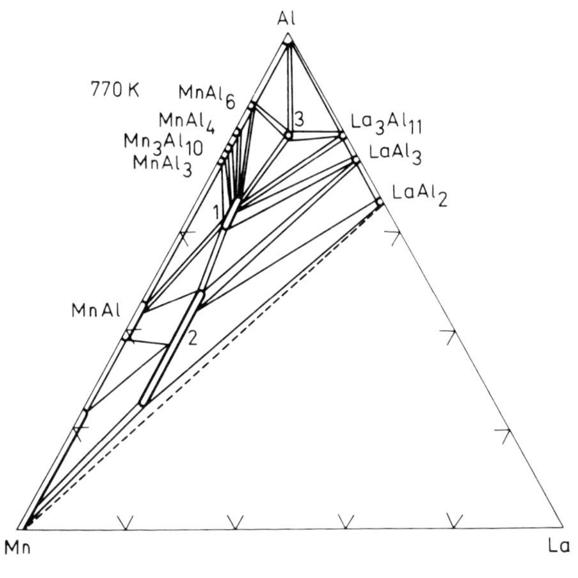

Fig. 21. La–Mn–Al, partial isothermal section at 770 K (0–33 a/c La). Compound with unknown crystal structure: (3) ~ $LaMnAl_8$.

$a = 9.02$, $c = 5.16$ (for the 1 : 4 : 8 composition), and (2) ~ $LaMn_{15-7}Al_{2-10}$ belong to the Th_2Zn_{17} ST with $a = 8.90–9.14$, $c = 13.09–13.28$.

La–Fe–Al

The partial isothermal section at 770 K was studied by Zarechnyuk et al. (1968) and is shown in fig. 22. The ternary compounds are: (1) ~ $LaFe_{7-6}Al_{6-7}$ ($NaZn_{13}$ ST, $a = 11.83–11.97$); (2) ~ $LaFe_4Al_8$ ($CeMn_4Al_8$ ST, $a = 8.82$, $c = 5.19$); (4) ~ $LaFe_2Al_8$ ($CeFe_2Al_8$ ST, $a = 12.65$, $b = 14.36$, $c = 4.27$); (5) ~ $La_2Fe_{7-6}Al_{10-11}$ (Th_2Zn_{17} ST, $a = 8.99–9.05$, $c = 13.04–13.13$). Crystal structure for (4) was reported by Yarmolyuk et al. (1974).

La–Ga–Al

The partial isothermal section at 670 K, shown in fig. 23, was derived by Dzyana and Zarechnyuk (1969). They found the existence of two compounds: (1) ~ $La_{0.15-0.30}Ga_2Al_{0.85-0.70}$ with unknown crystal structure and (2) ~ $LaGa_{2.5-2.0}Al_{1.5-2.0}$ with the $BaAl_4$ type structure, $a = 4.290–4.276$, $c = 11.09–11.16$.

La–Ge–Al

The partial isothermal section at 770 K was described by Muravyeva and Zarechnyuk (1970) and is shown in fig. 24. Raman and Steinfink (1967) reported for (5) ~ $LaGe_{0.5-0.2}Al_{1.5-1.8}$ the AlB_2 ST, $a = 4.35–4.40$, $c = 4.42–4.40$. This fact was confirmed by Muravyeva and Zarechnyuk (1970). They also proposed for (2) ~ $LaGe_2Al_2$ The La_2O_2S anti-type of crystal structure with $a = 4.28$, $c = 7.00$.

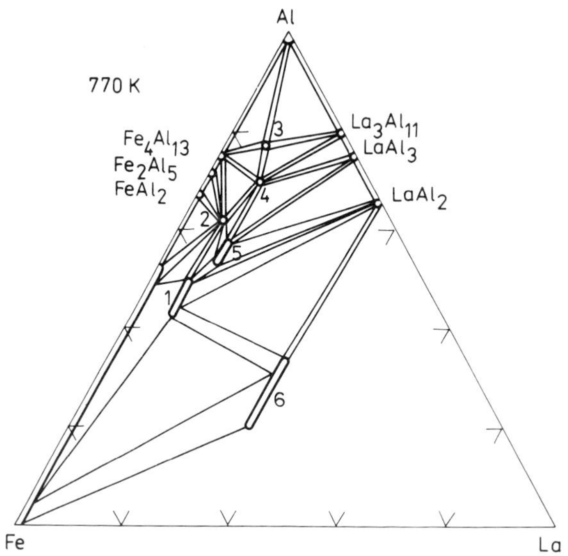

Fig. 22. La–Fe–Al, partial isothermal section at 770 K (0–33 a/c La). Compounds with approximate compositions (3) ~ $LaFe_2Al_{10}$; (6) ~ $LaFe_{1.4-1}Al_{0.6-1}$.

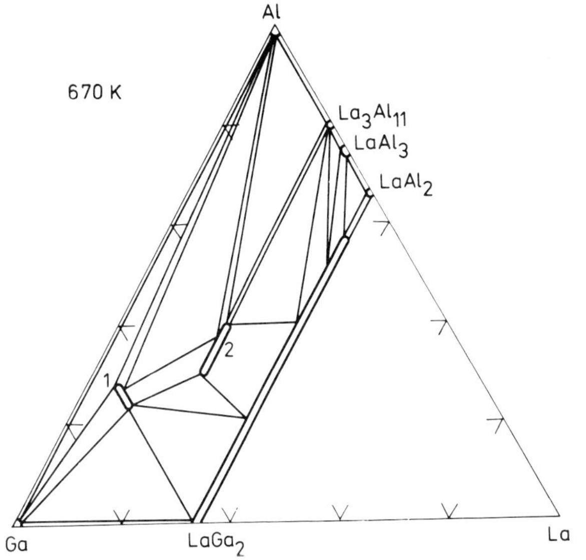

Fig. 23. La–Ga–Al, partial isothermal section at 670 K (0–33 a/c La).

{La, Ce, Pr, Nd}–{Zr, Nb, Hf}–Al

These twelve ternary systems were partially investigated by Tyvanchuk (1976) for the Nb, by Tyvanchuk (1975) for the Zr and by Tyvanchuk (1979) for the Hf. Figure 25 shows only the schemes of partial isothermal sections at 770 K, because of their close similarity. No ternary compounds were found.

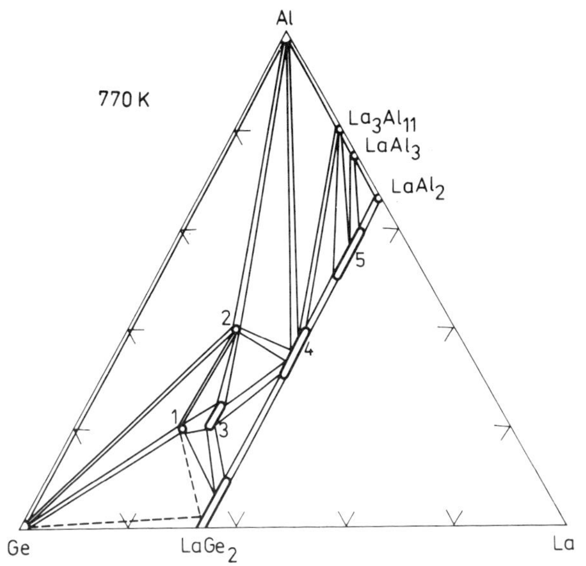

Fig. 24. La–Ge–Al, partial isothermal section at 770 K (0–33 a/c La). Compounds with unknown crystal structure: (1) ~ $LaGe_3Al$; (3) ~ $LaGe_{2.2-2}Al_{0.8-1}$; (4) ~ $LaGe_{1.1-0.8}Al_{0.9-1.2}$.

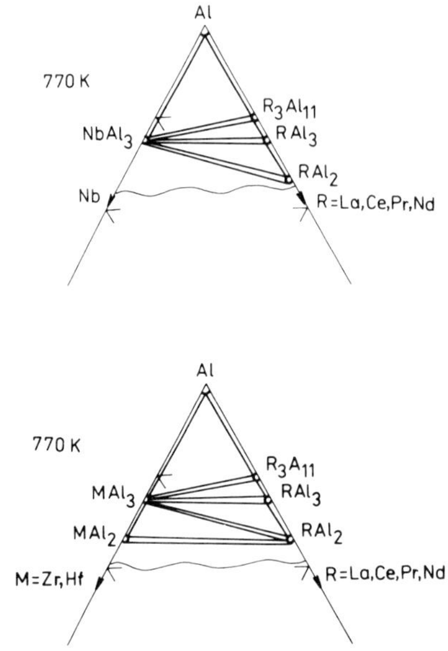

Fig. 25. {La, Ce, Pr, Nd}–{Zr, Nb, Hf}–Al, schemes of partial isothermal sections at 770 K (100–66 a/c Al).

Ce–Be–Al

Phase equilibria of the system Ce–Be–Al at 770 K have partially been studied by Zarechnyuk et al. (1966) and then by Zarechnyuk et al. (1969a). The derived isothermal section is presented in fig. 26. The only ternary compound found is $CeBe_{1.4-1.2}Al_{2.6-2.8}$ with $BaAl_4$ ST ($a = 4.16$–4.18, $c = 10.59$–10.57), as reported by Zarechnyuk et al. (1964).

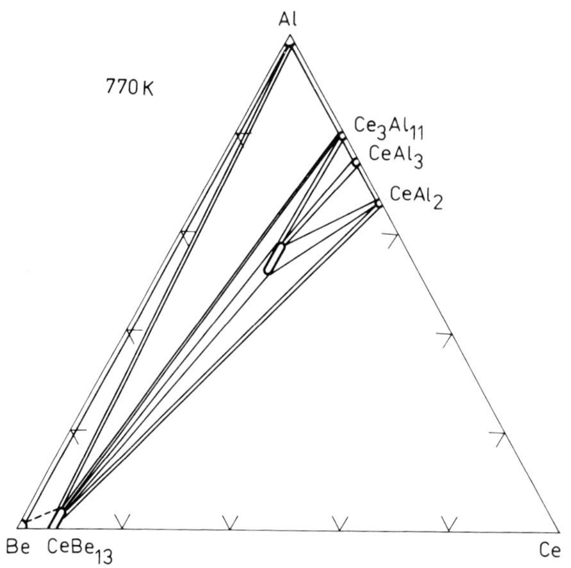

Fig. 26. Ce–Be–Al, partial isothermal section at 770 K (0–33 a/c Ce).

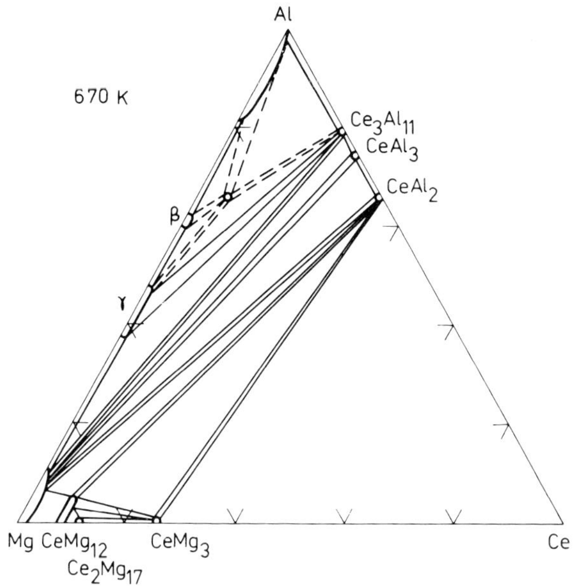

Fig. 27. Ce–Mg–Al, partial isothermal section at 670 K (0–33 a/c Ce).

Ce–Mg–Al

The partial isothermal section at 670 K, based on the results of Zarechnyuk and Krypyakevich (1967) is shown in fig. 27. The ternary compound $Ce_{0.15}Mg_{0.85}Al_2$ has the $MgZn_2$ type of structure with $a = 5.52$, $c = 8.89$. The atoms of Ce and Mg are placed statistically at the Mg position.

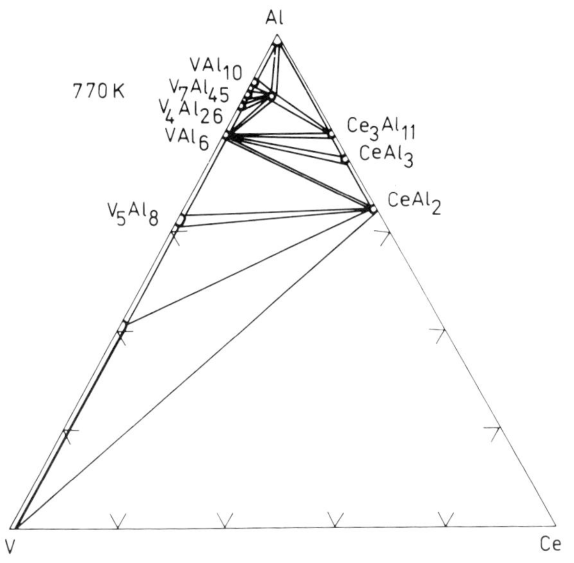

Fig. 28. Ce–V–Al, partial isothermal section at 770 K (0–33 a/c Ce).

Ce–V–Al

The phase equilibria in the system Ce–V–Al (fig. 28) were studied by Zarechnyuk and Rykhal (1974). The only ternary compound CeV_2Al_{20} with $CeCr_2Al_{20}$ type structure, $a = 14.55$, was earlier found by Krypyakevich and Zarechnyuk (1968). There is no solid solubility for the binary compounds.

Ce–Cr–Al

Two ternary compounds: (1) ~ $CeCr_2Al_{20}$ (an ordered superstructure of the type $ZrZn_{22}$, $a = 14.44$) and (3) ~ $Ce_2Cr_7Al_{10}$ (Th_2Zn_{17} ST, $a = 9.07$, $c = 13.01$) were found by Krypyakevich and Zarechnyuk (1968) and Zarechnyuk and Krypyakevich (1962), respectively. The third ternary compound (2) ~ $CeCr_{4-2.8}Al_{8-9.2}$ ($ThMn_{12}$ ST, $a = 8.98–9.11$, $c = 5.17–5.28$) was reported by Zarechnyuk and Rykhal (1974) together with the partial phase diagram (fig. 29).

Ce–Mn–Al

The two ternary compounds: (1) ~ $CeMn_4Al_8$ (an ordered superstructure of the $ThMn_{12}$ type ST, $a = 9.02$, $c = 5.11$) and (2) ~ $CeMn_{7-5}Al_{10-12}$ (Th_2Zn_{17} ST, $a = 8.85$, $c = 12.95$ for the 2 : 7 : 10 composition) were found by Zarechnyuk et al. (1963) during a partial study of the isothermal section at 770 K (fig. 30) of the Al-corner in the Ce–Mn–Al system.

Ce–Fe–Al

Zarechnyuk et al. (1969b) found a large solid solubility of Al in the Ce_2Fe_{17} (up to 60 a/c Al). They also derived a partial isothermal section at 770 K (fig. 31). According to Zarechnyuk and Krypyakevich (1962), compound (1) ~ $CeFe_4Al_8$ has $CeMn_4Al_8$ ST, $a = 8.86$, $c = 5.08$; (3) ~ $CeFe_2Al_8$ has a new type of crystal

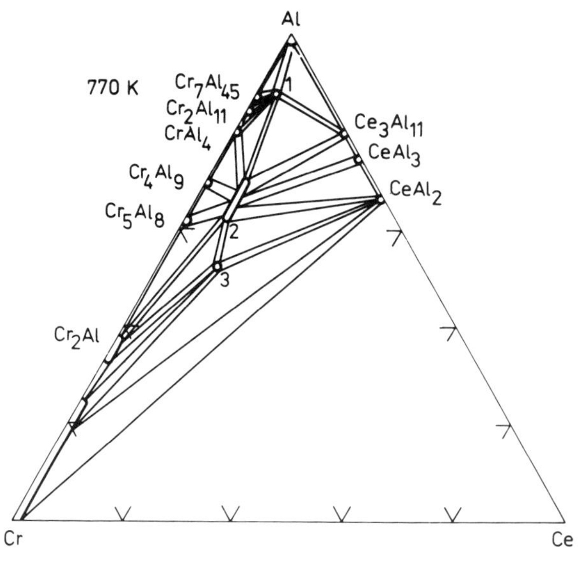

Fig. 29. Ce–Cr–Al, partial isothermal section at 770 K (0–33 a/c Ce).

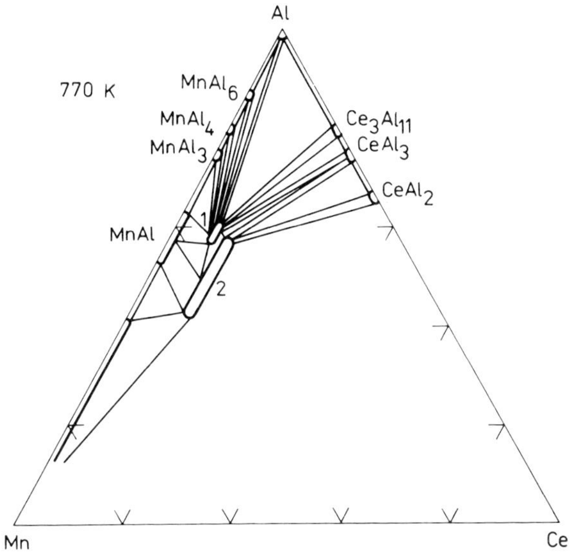

Fig. 30. Ce–Mn–Al, partial isothermal section at 770 K (0–33 a/c Ce).

structure with $a = 12.51$, $b = 14.48$, $c = 4.07$, space group is Pbam, as reported by Yarmolyuk et al. (1974).

Ce–Ni–Al

The two partial isothermal sections at 770 and 1070 K of the system Ce–Ni–Al (fig. 32) were reported by Zarechnyuk et al. (1983). The ternary compounds with a determined crystal structure are as follows: (2) ~ $CeNi_2Al_5$ ($PrNi_2Al_5$ ST,

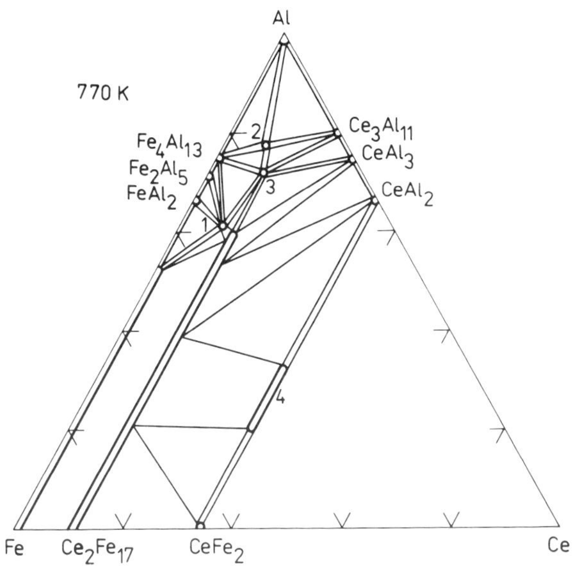

Fig. 31. Ce–Fe–Al, partial isothermal section at 770 K (0–33 a/c Ce). Compounds with unknown crystal structure: (2) ~ $CeFe_2Al_{10}$; (4) ~ $CeFe_{1.4-1.0}Al_{0.6-1.0}$.

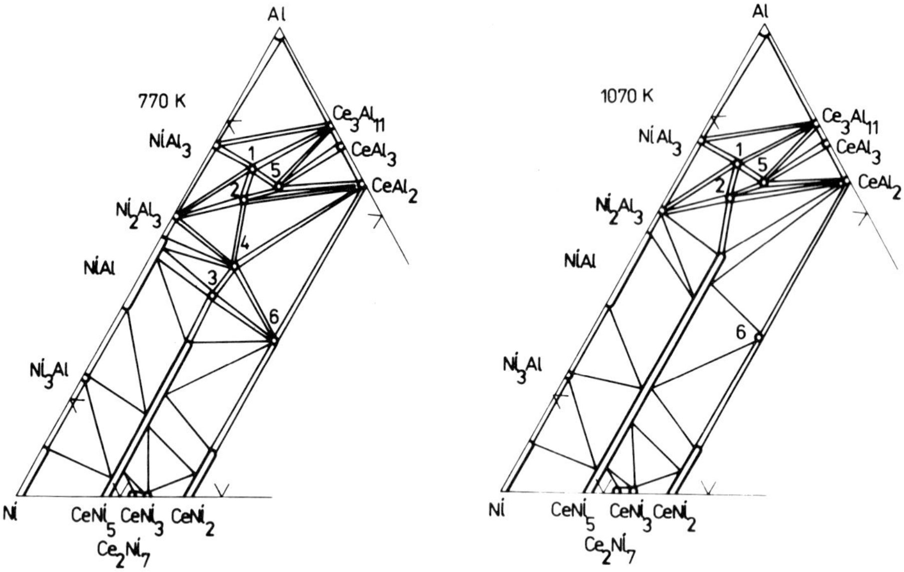

Fig. 32. Ce–Ni–Al, partial isothermal sections at 770 and at 1070 K (0–33 a/c Ce). Ternary compounds with undetermined crystal structure: (1) ~ $CeNi_2Al_7$; (3) ~ Ce_2Ni_5Al.

$a = 7.08$, $b = 9.67$, $c = 4.01$); (4) ~ $CeNi_2Al_3$ ($CaCu_5$ type superstructure, $a = 5.28$, $c = 4.04$); (5) ~ $CeNiAl_4$ ($YNiAl_4$ ST, $a = 4.097$, $b = 15.47$, $c = 6.643$) and (6) ~ CeNiAl (ZrNiAl ST, $a = 6.98$, $c = 4.02$). The most differences between the 770 and 1070 K isothermal sections are observed at less than 60 a/c Al.

Ce–Cu–Al

Gladyshevsky et al. (1961) partially investigated the isothermal section at 670 K (fig. 33). Up to 40 a/c Al solubility in $CeCu_4$ was found. Zarechnyuk and Krypyakevich (1962) reported for (1) ~ $CeCu_4Al_8$ the $CeMn_4Al_8$ type of crystal structure, $a = 8.84$, $c = 5.17$; (2) ~ $Ce_2Cu_{7.3-6.5}Al_{9.7-10.5}$ the Th_2Zn_{17} ST, $a = 8.96$–8.98, $c = 13.04$–13.07. The crystal structure for (3) ~ $CeCu_{1-0.75}Al_{3-3.25}$ was determined by Zarechnyuk et al. (1964) and belong to the $BaAl_4$ ST, $a = 4.25$, $c = 10.65$ for the 1 : 1 : 3 composition.

Ce–Ga–Al

Partial isothermal section (fig. 34) was studied by Dzyana and Zarechnyuk (1969). The ternary compound (1) ~ $Ce_{0.1-0.4}Ga_2Al_{0.9-0.6}$ has unknown crystal structure. Crystal structure of (2) ~ $GeGa_{2.5-2.0}Al_{1.5-2.0}$ is the $BaAl_4$ type with $a = 4.228$–4.229, $c = 11.08$–11.04, as reported by Zarechnyuk et al. (1964).

Ce–Ge–Al

Raman and Steinfink (1967) reported for (5) ~ $CeGe_{0.5-0.4}Al_{0.5-0.6}$ the AlB_2 type of structure with $a = 4.33$–4.36, $c = 4.32$–4.29. Muravyeva and Zarechnyuk

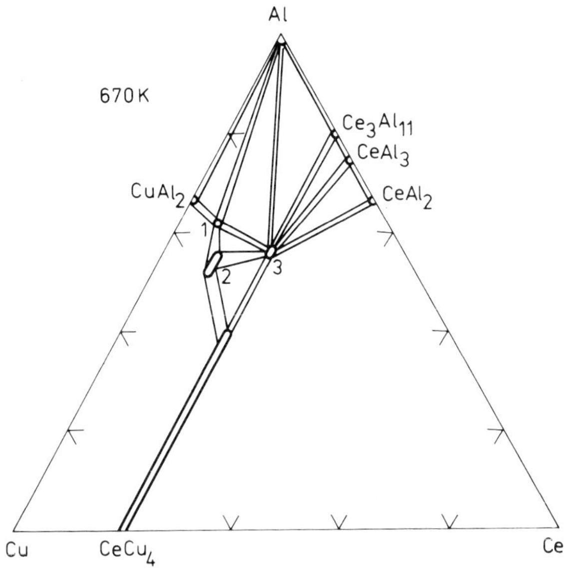

Fig. 33. Ce–Cu–Al, partial isothermal section at 670 K (0–33 a/c Ce).

(1970) derived a partial isothermal section at 770 K (fig. 35) and proposed the anti-La_2O_2S type of crystal structure for (2) ~ $CeGe_2Al_2$, $a = 4.27$, $c = 6.95$.

Ce–Th–Al

Isothermal section at 770 K was derived by Van Vucht (1961). A large binary phase solubility was found (fig. 36).

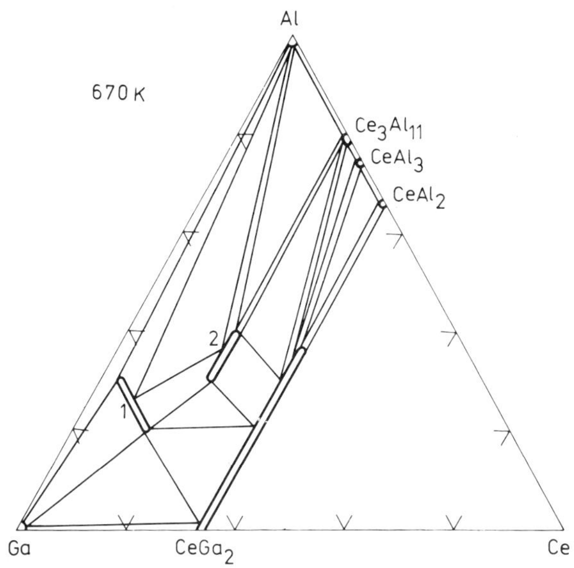

Fig. 34. Ce–Ga–Al, partial isothermal section at 670 K (0–33 a/c Ce).

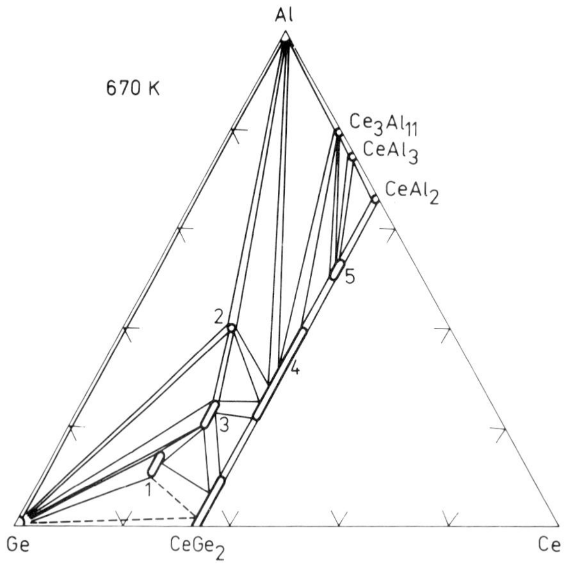

Fig. 35. Ce–Ge–Al, partial isothermal section at 670 K (0–33 a/c Ce). Compounds with unknown crystal structure: (1) ~ $CeGe_{3.5-3.2}Al_{0.5-0.8}$; (3) ~ $CeGe_{2.2-2}Al_{0.8-1}$; (4) ~ $CeGe_{1.3-0.8}Al_{0.7-1.2}$.

Pr–Ni–Al

The partial isothermal section at 770 K is shown in fig. 37, according to Rykhal et al. (1978a). (2) ~ $PrNi_2Al_5$ has its own type of crystal structure: Immm, $a = 7.073$, $b = 9.60$, $c = 4.013$; (3) ~ $PrNi_2Al_3$ belongs to the $CaCu_5$ ST, $a = 5.293$, $c = 4.064$; (4) ~ $PrNiAl_4$ to the $YNiAl_4$ ST, $a = 4.112$, $b = 15.49$, $c = 6.672$; (5) ~ Pr_3Ni_8Al to the Ce_3Co_8Si ST, $a = 5.124$, $c = 16.33$, (8) ~ Pr_2Ni_2Al to the

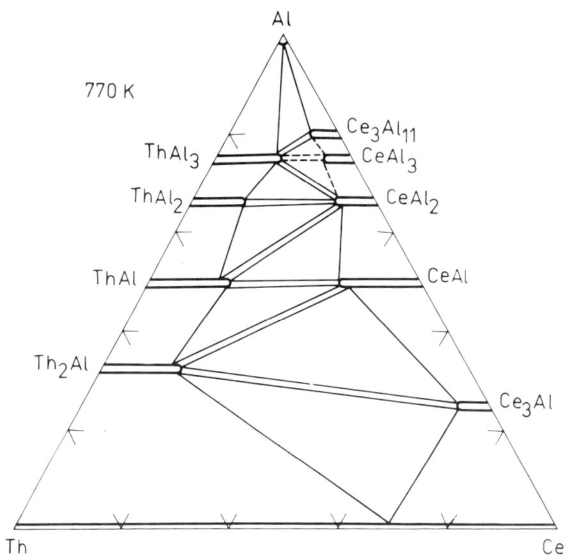

Fig. 36. Ce–Th–Al, isothermal section at 770 K.

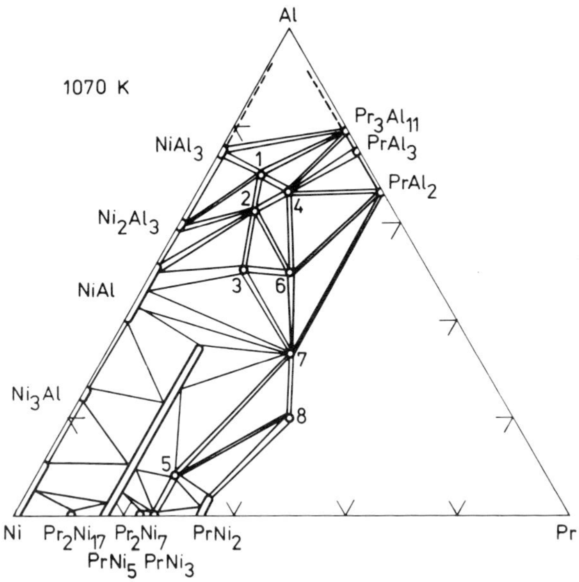

Fig. 37. Pr–Ni–Al, partial isothermal section at 1070 K (0–40 a/c Pr). Compounds with undetermined crystal structure: (1) ~ $PrNi_2Al_7$; (6) ~ $PrNiAl_2$.

Mo_2NiB_2 ST, $a = 5.523$, $b = 8.527$, $c = 4.273$. Leon and Wallace (1970) found the ZrNiAl type of crystal structure for (7) ~ PrNiAl, $a = 7.042$, $c = 4.032$.

Pr–Cu–Al

The partial phase diagram at 770 K, shown in fig. 38, was derived by Zarechnyuk and Rykhal (1978). Ternary compounds are as follows: (1) ~ $PrCu_7Al_6$

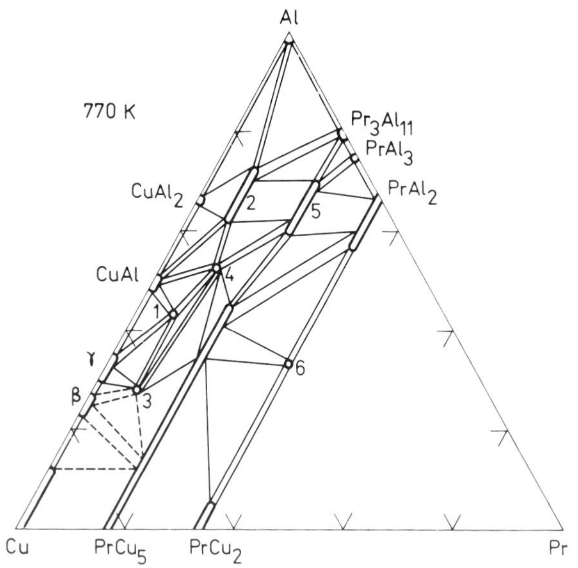

Fig. 38. Pr–Cu–Al, partial isothermal section at 770 K (0–33 a/c Pr).

($NaZn_{13}$ ST, $a = 11.83$); (2) ~ $PrCu_{4-2.6}Al_{8-9.4}$ ($ThMn_{12}$ ST, $a = 8.83–8.77$, $c = 5.17–5.16$); (3) ~ $PrCu_{7.6}Al_{3.4}$ ($BaCd_{11}$ ST, $a = 10.29$, $c = 6.57$); (4) ~ $Pr_2Cu_7Al_{10}$ (Th_2Zn_{17} ST, $a = 9.94$, $c = 13.02$); (5) ~ $PrCu_{1-0.5}Al_{3-3.5}$ ($BaAl_4$ ST, $a = 4.19–4.15$, $c = 11.09–10.95$). Compound (6) ~ PrCuAl has the ZrNiAl ST, $a = 7.173$, $c = 4.155$, according to Dwight et al. (1968).

Nd–Fe–Al

Phase equilibria in the system Nd–Fe–Al (fig. 39) have been studied by Vivchar et al. (1970). The crystal structure was determined for only one of the three existing ternary compounds: (1) ~ $NdFe_{4-3.3}Al_{8-8.7}$ ($ThMn_{12}$ ST, $a = 8.78–8.81$, $c = 5.04–5.05$).

Nd–Ni–Al

The partial isothermal section at 870 K, shown in fig. 40 was described by Rykhal et al. (1979b). Ternary compounds: (3) ~ $NdNi_2Al_3$ ($CaCu_5$ ST, $a = 5.282$, $c = 4.059$) and (4) ~ $NdNiAl_4$ ($YNiAl_4$ ST, $a = 4.110$, $b = 15.48$, $c = 6.669$) were also investigated by Rykhal et al. (1979b). Earlier, the existence and crystal structure of (5) ~ Nd_3Ni_8Al (Ce_3Co_8Si ST, $a = 5.107$, $c = 16.28$) and (8) ~ Nd_2Ni_2Al (Mo_2NiB_2 ST, $a = 5.448$, $b = 8.490$, $c = 4.269$) was established by Rykhal (1978), and (7) ~ NdNiAl (ZrNiAl ST, $a = 6.990$, $c = 4.010$) by Dwight et al. (1968).

Nd–Cu–Al

The system Nd–Cu–Al was studied by Zarechnyuk et al. (1975). The partial isothermal section at 770 K is shown in fig. 41. It was found, that (1) ~ $NdCu_7Al_6$

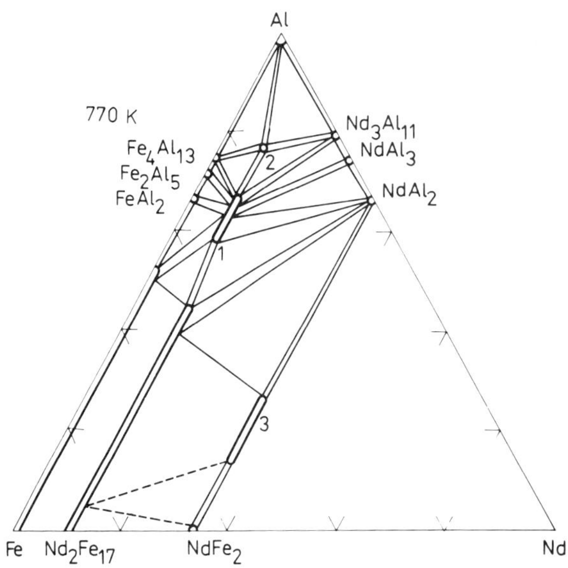

Fig. 39. Nd–Fe–Al, partial isothermal section at 770 K (0–33 a/c Nd). Compounds with undetermined crystal structure: (2) ~ $NdFe_2Al_{10}$; (3) ~ $NdFe_{1.65-1.2}Al_{0.35-0.8}$.

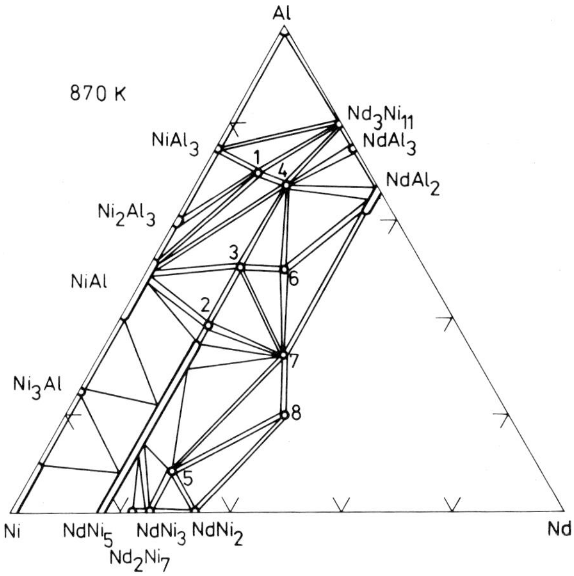

Fig. 40. Nd–Ni–Al, partial isothermal section at 870 K (0–33 a/c Nd). Compounds with unknown crystal structure: (1) ~ $NdNi_2Al_7$; (2) ~ $NdNi_{2.7}Al_{2.3}$; (6) ~ $NdNiAl_2$.

belongs to the $NaZn_{13}$ ST, $a = 11.72$; (2) ~ $NdCu_4Al_8$ to the $CeMn_4Al_8$ ST, $a = 8.84$, $c = 5.19$; (4) ~ $Nd_2Cu_8Al_9$ to the Th_2Zn_{17} ST, $a = 8.91$, $c = 12.97$, and (5) ~ $NdCuAl_3$ to the $BaAl_4$ ST, $a = 4.22$, $c = 10.70$. Bodak and Gladyshevsky (1969) investigated (3) ~ $NdCu_{8.6}Al_{2.4}$ and established it had the $BaCd_{11}$ type structure with $a = 10.27$, $c = 6.57$. (6) ~ NdCuAl, after Dwight et al. (1968), has the ZrNiAl type structure with $a = 7.17$, $c = 4.12$.

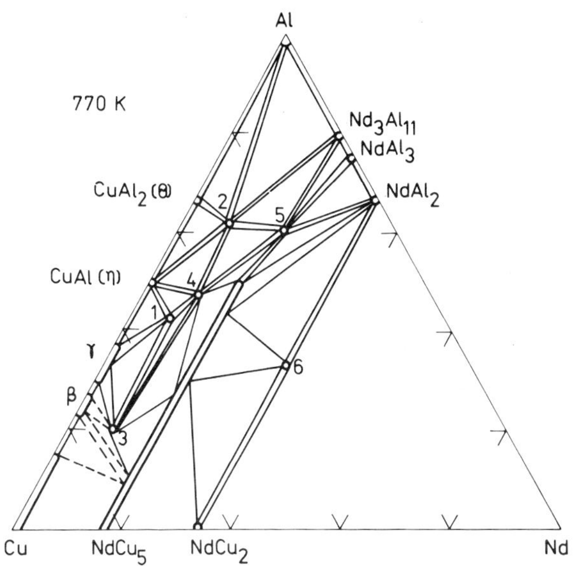

Fig. 41. Nd–Cu–Al, partial isothermal section at 770 K (0–33 a/c Nd).

Nd–{Sr, Ba}–Al

No ternary compounds were found by Eshonov et al. (1977) and Zukhuritdinov et al. (1976) in the systems Nd–Sr–Al and Nd–Ba–Al, respectiveley. Nd–Sr and Nd–Ba immiscibility reaches 50 and 58 a/c Al in the appropriate ternary systems (fig. 42).

Sm–Fe–Al

Figure 43 represents the partial isothermal section at 770 K for the system Sm–Fe–Al as derived by Vivchar et al. (1974). Ternary compounds are as follows: (1) ~ $SmFe_{5.6-3.3}Al_{6.4-8.7}$ ($ThMn_{12}$ ST, $a = 8.71$–8.81, $c = 5.01$–5.05); (2) ~ $SmFe_2Al_{10}$ with unknown crystal structure; (3) ~ $SmFe_{1.4-1.1}Al_{0.6-0.9}$ ($MgZn_2$ ST, $a = 5.36$–5.40, $c = 8.68$–8.75), according to Oesterreicher (1971).

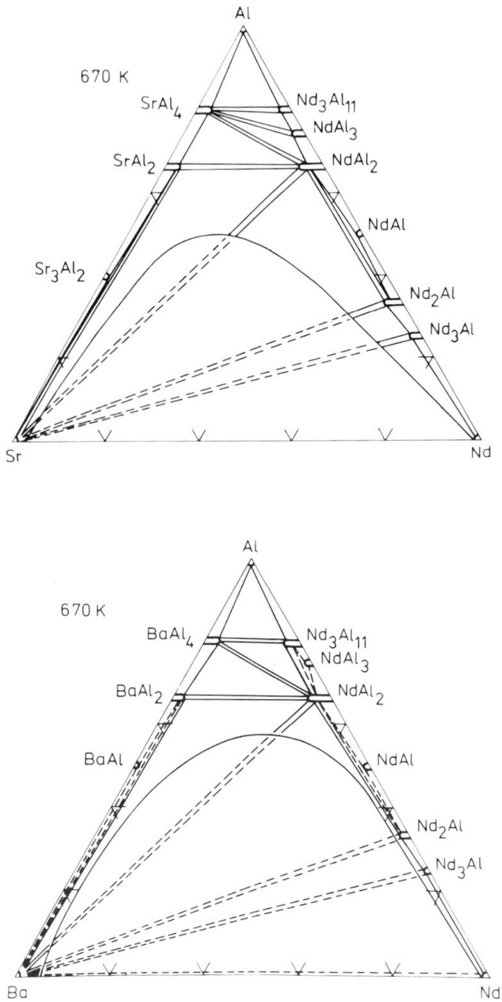

Fig. 42. Nd–{Sr, Ba}–Al, isothermal sections at 670 K.

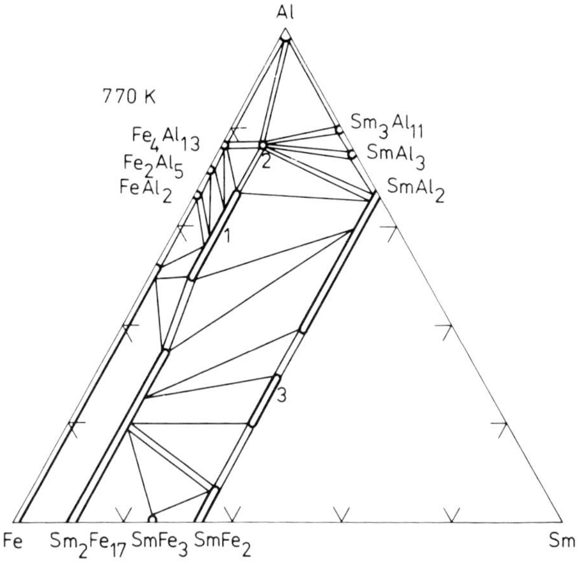

Fig. 43. Sm–Fe–Al, partial isothermal section at 770 K (0–33 a/c Sm).

Sm–Cu–Al

The phase diagram was described by Prevarsky et al. (1983) and is shown in fig. 44. Ternary compounds: (1) ~ $SmCu_4Al_8$ ($CeMn_4Al_8$ ST, $a = 8.774$, $c = 5.149$); (2) ~ $SmCu_7Al_6$ ($NaZn_{13}$ ST, $a = 11.829$); (3) ~ $SmCu_6Al_5$ (unknown crystal structure); (4) ~ $Sm_2Cu_7Al_{10}$ (Th_2Zn_{17} ST, $a = 8.855$, $c = 12.899$); (5) ~ $SmCuAl_3$ ($BaAl_4$ ST, $a = 4.173$, $c = 10.602$) and (6) ~ SmCuAl (ZrNiAl ST, $a = 7.123$, $c = 4.0828$) were found or confirmed. (6) was described earlier by Dwight et al. (1968).

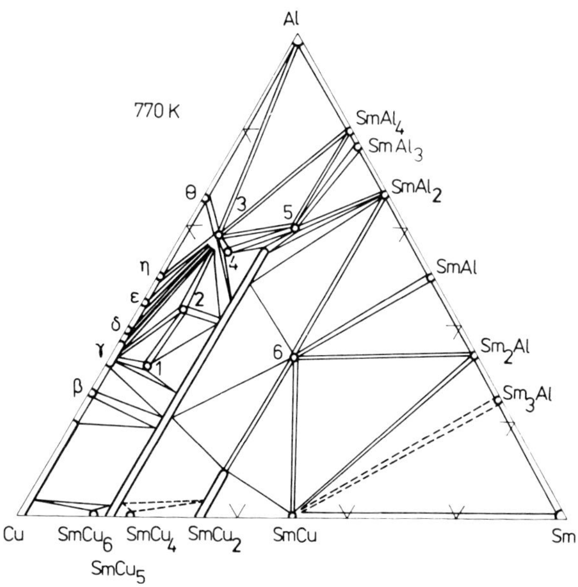

Fig. 44. Sm–Cu–Al, isothermal section at 770 K.

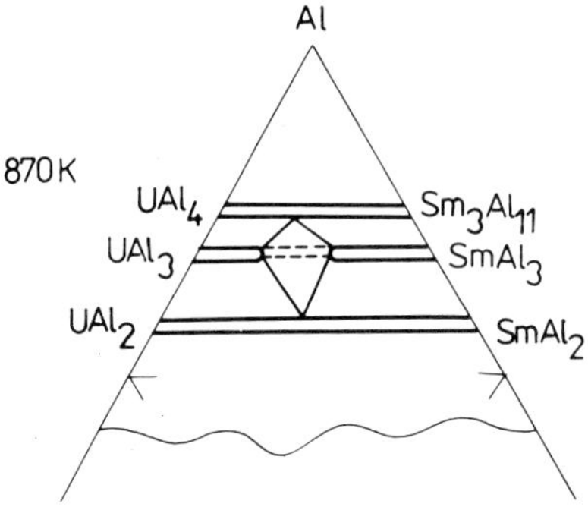

Fig. 45. Sm–U–Al, partial isothermal section at 870 K (100–66 a/c Al).

Sm–U–Al

Phase equilibria in the Al-corner of the system Sm–U–Al, according to Gasteels et al. (1967) are shown in fig. 45. The unit-cell volume shows positive deviation from Vegard's law in the region of continuous solid solution (Sm, U)Al_2.

Gd–Fe–Al

The partial isothermal section at 770 K (fig. 46) was investigated by Vivchar et al. (1973a). Ternary compounds are: (1) ~ $GdFe_4Al_8$ ($CeMn_4Al_8$ ST, $a = 8.57$, $c = 5.01$); (2) ~ $GdFe_2Al_{10}$ with undetermined crystal structure and (3) ~ $GdFe_{1.4-1}Al_{0.6-1}$ ($MgZn_2$ ST, $a = 5.35$–5.41, $c = 8.65$–8.81).

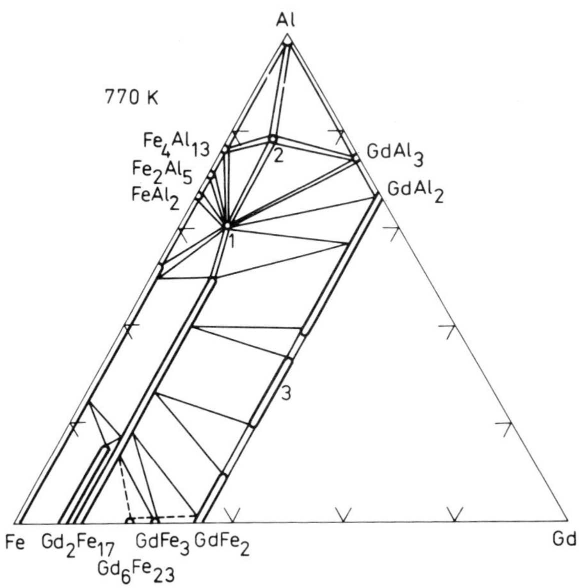

Fig. 46. Gd–Fe–Al, partial isothermal section at 770 K (0–33 a/c Gd).

Gd–Ni–Al

The phase diagram was derived in the region of 0–33 a/c Gd by Rykhal et al. (1978b) and is shown in fig. 47. Rykhal et al. (1978b) found and described the following compounds: (1) ~ $GdNi_3Al_{16}$ with orthorhombic unit cell, $a = 4.07$, $b = 16.01$, $c = 27.24$; (2) ~ $GdNi_2Al_3$ with the $HoNi_{2.6}Ga_{2.4}$ ST, $a = 9.04$, $c = 4.103$; (3) ~ $Gd_2Ni_3Al_7$ with hexagonal unit cell, $a = 17.94$, $c = 4.042$; (5) ~ Gd_3Ni_8Al with the Ce_3Co_8Si ST, $a = 5.07$, $c = 16.23$ and (9) ~ Gd_2Ni_2Al with the Mo_2NiB_2 ST, $a = 5.408$, $b = 8.416$, $c = 4.186$. (4) ~ $GdNiAl_4$ ($YNiAl_4$ ST, $a = 4.087$, $b = 15.30$, $c = 6.69$) and (6) ~ $GdNiAl_2$ ($MgCuAl_2$ ST, $a = 4.080$, $b = 10.14$, $c = 6.93$) were reported by Rykhal et al. (1973); (7) ~ $Gd_3NI_6Al_2$ ($Ce_3Ni_6Si_2$ ST, $a = 8.84$) by Rykhal (1977), and (8) ~ GdNiAl (ZrNiAl ST, $a = 7.02$, $c = 3.92$) by Dwight et al. (1968).

Gd–Ge–Al

The ternary compounds: (1) ~ $GdGe_2Al_2$ with the La_2O_2S anti-type structure, $a = 4.248$, $c = 6.714$ was found by Zarechnyuk et al. (1970b); (3) ~ GdGeAl with α-$ThSi_2$ structure type, $a = 4.152$, $c = 14.421$, reported by Ryabokon (1974). (2) ~ $GdGe_2Al$ with an undetermined crystal structure together with a partial phase diagram (fig. 48) was reported by Zarechnyuk et al. (1981).

Gd–{Zr, Hf}–Al and {Gd, Tb, Dy, Ho, Lu}–Nb–Al

These systems were partially investigated by Tyvanchuk (1979) and Tyvanchuk (1976). The isothermal sections presented in fig. 49 are schematic, because of close similarity in phase equilibria for the above-mentioned ternary systems. No ternary compounds or solid solutions were found.

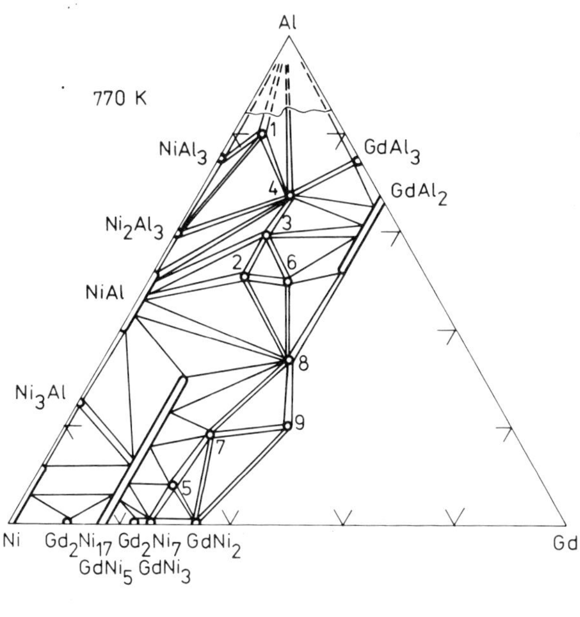

Fig. 47. Gd–Ni–Al, partial isothermal section at 770 K (0–33 a/c Gd).

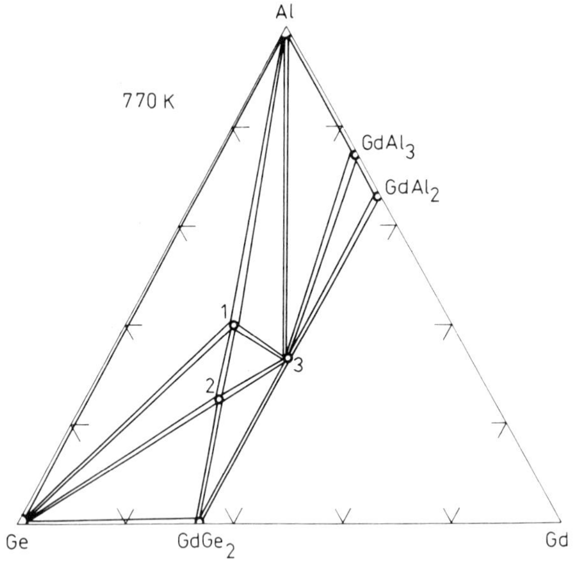

Fig. 48. Gd–Ge–Al, partial isothermal section at 770 K (0–33 a/c Gd).

Tb–Fe–Al

The system Tb–Fe–Al was investigated by Oesterreicher (1975). No phase equilibria were established (fig. 50). The ternary compounds are as follows: (1) ~ $Tb_2Fe_{12.2-9.4}Al_{4.8-7.6}$ (Th_2Zn_{17} ST, $a = 8.624–8.757$, $c = 12.614–12.743$); (2) ~ $Tb_2Fe_{5.6-3.7}Al_{11.4-13.3}$ (cubic unit cell with $a = 8.738–8.782$); (3) ~ $TbFe_{2.6}Al_{0.4}$ (trigonal or hexagonal lattice with $a = 5.168$, $c = 24.867$); (4) ~

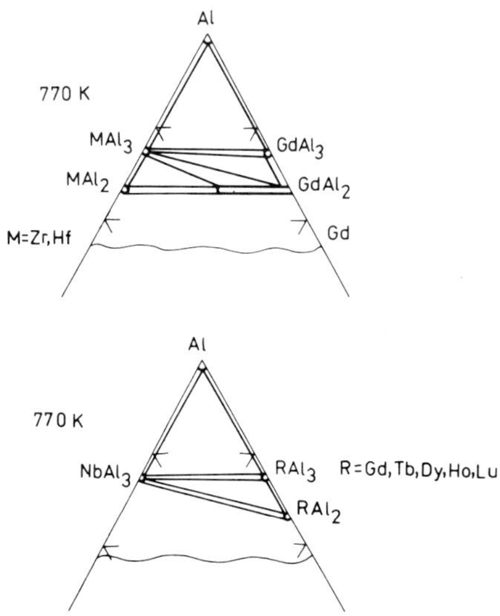

Fig. 49. Gd–M–Al (where M = Zr, Hf) and R–Nb–Al (where R = Gd, Tb, Dy, Ho, Lu), the partial isothermal sections are given at 770 K (100–66 a/c Al).

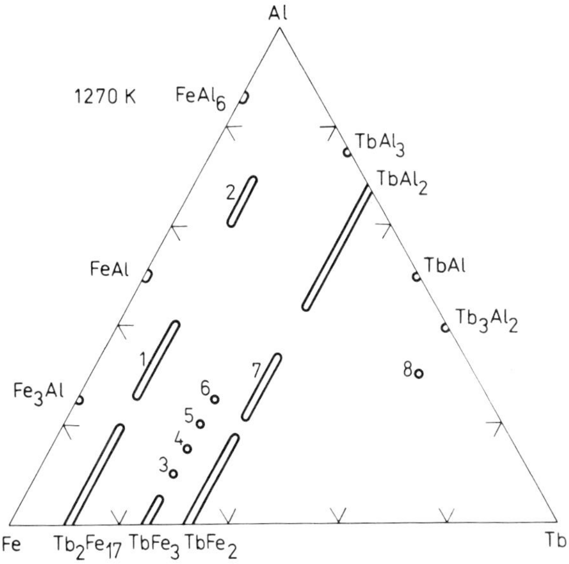

Fig. 50. Tb–Fe–Al, ternary phase compositions found at high temperatue, after annealing at 1270 K.

$TbFe_{2.4}Al_{0.6}$ (trigonal or hexagonal with $a = 5.189$, $c = 24.950$); (5) ~ $TbFe_{2.2}Al_{0.8}$ (trigonal or hexagonal with $a = 5.211$, $c = 75.352$); (6) ~ $TbFe_2Al$ ($CeNi_3$ ST, $a = 5.226$, $c = 16.794$); (7) ~ $TbFe_{1.4-1}Al_{0.6-1}$ ($MgZn_2$ ST, $a = 5.391$, $c = 8.702$ for the 1 : 1 : 1 composition); (8) ~ $Tb_3Fe_{0.5}Al_{1.5}$ with unknown lattice. Compounds (3), (4) and (5) were found only for arc-melted samples. Annealed at 1270 K, they have the $PuNi_3$ structure type. Vivchar and Zarechnyuk (1974) reported the existence of $TbFe_4Al_8$ with $ThMn_{12}$ ST, $a = 8.740$, $c = 5.018$.

{Tb, Dy}–Ni–Al

These two systems have completely identical phase diagrams at 1070 K. Figure 51 shows the phase relationships according to Rykhal et al. (1980). Ternary compounds found here are: (1) ~ RNi_3Al_{16} with orthorhombic lattice, $a = 4.035$ (4.021), $b = 15.91$ (15.86), $c = 27.08$ (27.01); (2) ~ RNi_2Al_3 with $HoNi_{2.6}Ga_{2.4}$ ST, $a = 8.93$ (8.92), $c = 3.972$ (3.936); (3) ~ $R_2Ni_3Al_7$ with hexagonal unit cell, $a = 17.86$ (17.82), $c = 4.015$ (3.988); (4) ~ $RNiAl_4$ with $YNiAl_4$ ST, $a = 4.075$ (4.056), $b = 15.24$ (15.13), $c = 6.61$ (6.63); (5) ~ R_3Ni_8Al with the Ce_3Co_8Si ST, $a = 5.042$ (5.037), $c = 16.14$ (16.10); (6) ~ $RNiAl_2$ with $MgCuAl_2$ ST, $a = 4.084$ (4.077), $b = 10.11$ (10.08), $c = 6.97$ (6.93); (7) ~ $R_3Ni_6Si_2$ with $Ce_3Ni_6Si_2$ ST, $a = 8.91$ (8.91); (8) ~ RNiAl with the ZrNiAl ST, $a = 7.003$ (6.994), $c = 3.877$ (3.821); (9) ~ R_2Ni_2Al with Mo_2NiB_2 ST, $a = 5.396$ (5.338), $b = 8.344$ (8.338), $c = 4.203$ (4.173). Here R = Tb or Dy; unit cell dimensions for R = Dy are always given in parentheses. The existence of compound (2) was also mentioned by Zarechnyuk and Rykhal (1981).

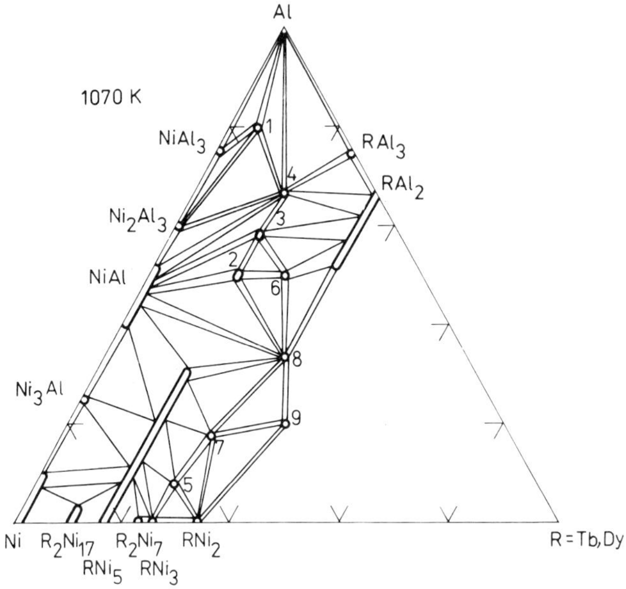

Fig. 51. {Tb, Dy}–Ni–Al, partial isothermal section at 1070 K (0–33 a/c R).

Tb–{Zr, Hf}–Al

Tyvanchuk (1979) reported the existence of two ternary compounds in both of the systems with Zr and Hf. They are: (1) ~ $Tb_{0.65}M_{0.35}Al_3$ with the $HoAl_3$ type structure, $a = 6.039$ (6.031), and (2) ~ $Tb_{0.8}M_{0.2}Al_3$ with the $AuCu_3$ structure type, $a = 4.203$ (4.196). Lattice parameters for M = Hf are given in the parentheses. A schematic partial isothermal section at 770 K after Tyvanshuk (1979) is shown in fig. 52.

Dy–V–Al

The partial phase diagram (fig. 53) was derived by Rykhal et al. (1979a). (1) ~ DyV_2Al_{20} has the $CeCr_2Al_{20}$ ST, $a = 14.54$, and (2) ~ $DyVAl_8$ has a hexagonal lattice with $a = 10.86$, $c = 17.71$.

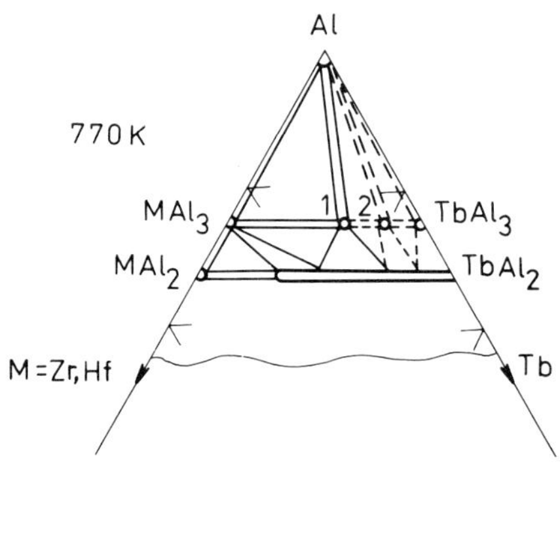

Fig. 52. Tb–{Zr, Hf}–Al, schematic partial isothermal section at 770 K (100–66 a/c Al).

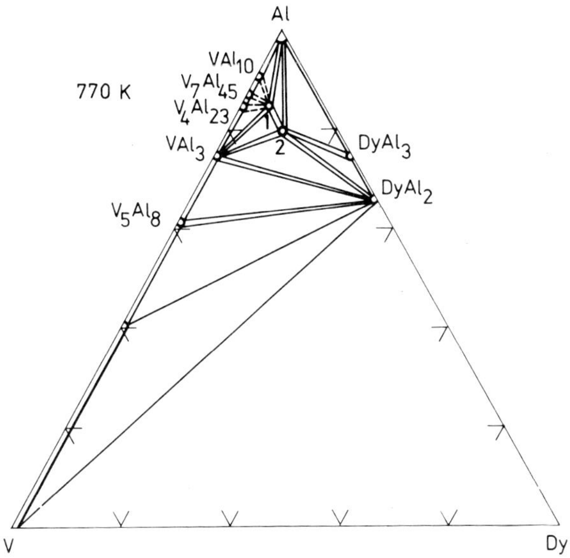

Fig. 53. Dy–V–Al, partial isothermal section at 770 K (0–33 a/c Dy).

Dy–Cr–Al

The system was investigated by Rykhal et al. (1979a). The ternary compounds that were found are: (1) ~ $DyCr_2Al_{20}$ with the $CeCr_2Al_{20}$ ST, $a = 14.39$; (2) ~ $DyCrAl_8$ with a hexagonal lattice, $a = 10.75$, $c = 17.60$, and (3) ~ $DyCr_4Al_8$ belongs to the $CeMn_4Al_8$ ST with $a = 8.87$, $c = 5.04$. The phase diagram is shown in fig. 54.

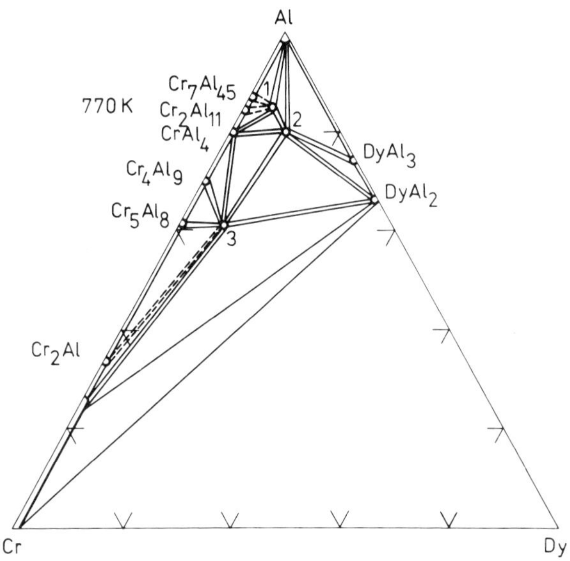

Fig. 54. Dy–Cr–Al, partial isothermal section at 770 K (0–33 a/c Dy).

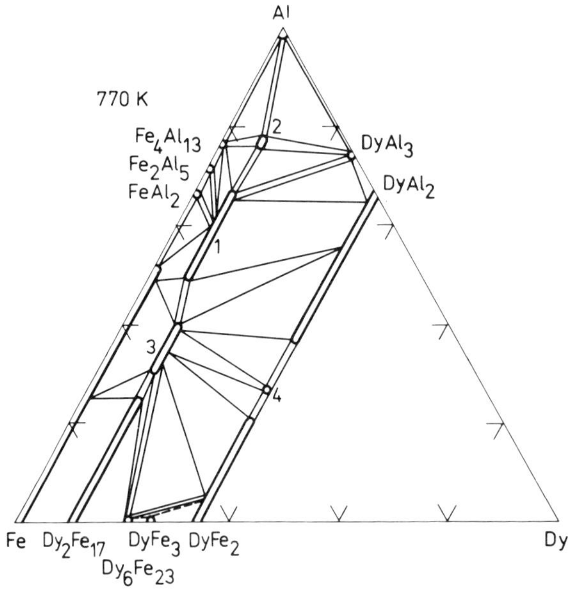

Fig. 55. Dy–Fe–Al, partial isothermal section at 770 K (0–33 a/c Dy).

Dy–Fe–Al

The partial isothermal section (fig. 55) was constructed by Vivchar et al. (1973b). Ternary compounds are as follows: (1) ~ $DyFe_{5.6-3.4}Al_{6.4-8.6}$ ($ThMn_{12}$ ST, $a = 8.93$–8.73, $c = 4.96$–5.02); (2) ~ $DyFe_2Al_{10}$ (unknown type of structure); (3) ~ $Dy_2Fe_{11-9}Al_{6-8}$ (Th_2Zn_{17} ST, $a = 8.64$–8.77, $c = 12.57$–12.64), and (4) ~ $DyFe_{1.2}Al_{0.8}$ ($MgZn_2$ ST, $a = 5.40$, $c = 8.78$). Earlier, the existence of (4) with the same crystal structure was reported by Wernick et al. (1962) and Oesterreicher (1971).

Dy–{Zr, Hf}–Al

The ternary compounds $Dy_{0.7}M_{0.3}Al_3$ (where M = Zr or Hf) belongs to the $AuCu_3$ type of structure with $a = 4.197$ (for M = Zr) and $a = 4.186$ (M = Hf), after Tyvanchuk (1979). A schematic representation of the partial isothermal section at 770 K is shown in fig. 56.

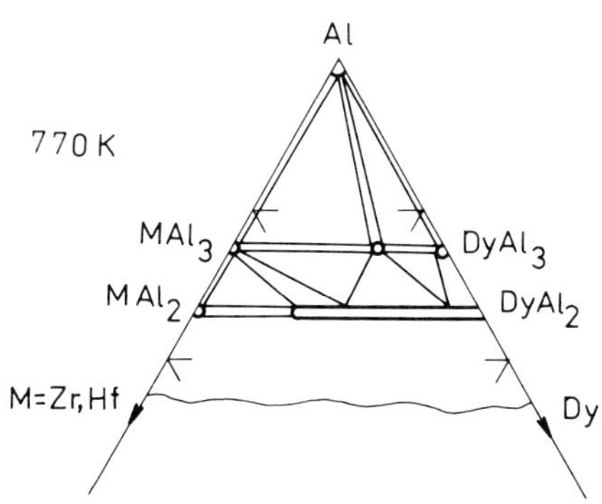

Fig. 56. Dy–M–Al (where M = Zr, Hf) schematic partial isothermal section at 770 K (100–66 a/c Al).

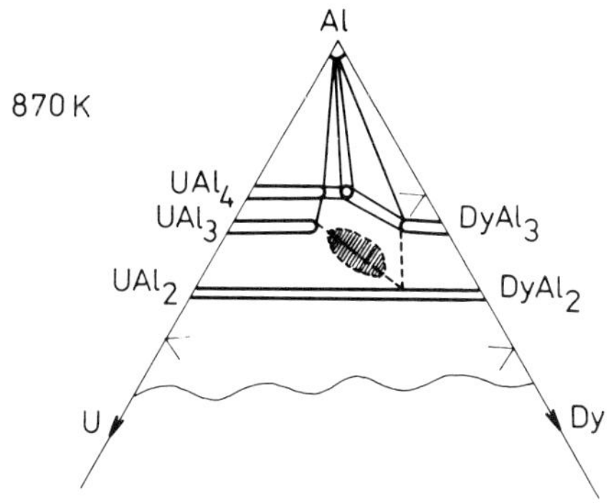

Fig. 57. Dy–U–Al, partial isothermal section at 870 K (100–66 a/c Al).

Dy–U–Al

The system Dy–U–Al was partially studied by Casteels et al. (1967). The derived partial isothermal section at 870 K is shown in fig. 57. The region of immiscibility is shaded. The only ternary compound $\sim Dy_3U_2Al_{15}$ has hexagonal lattice, $a = 6.05$, $c = 14.35$.

Ho–Fe–Al

Figure 58 shows the partial phase diagram as found by Ryabov et al. (1971). The $ThMn_{12}$ type of crystal structure was found for the compound (1) $\sim HoFe_{5.3-4}Al_{6.7-8}$, $a = 8.65$–8.759, $c = 4.965$–5.036; the Th_2Zn_{17} type was found for (3) $\sim Ho_2Fe_{11}Al_6$, $a = 8.714$, $c = 12.62$ and the $MgZn_2$ type of structure was found for (4) $\sim HoFe_{1.2}Al_{0.8}$, $a = 5.354$, $c = 8.695$.

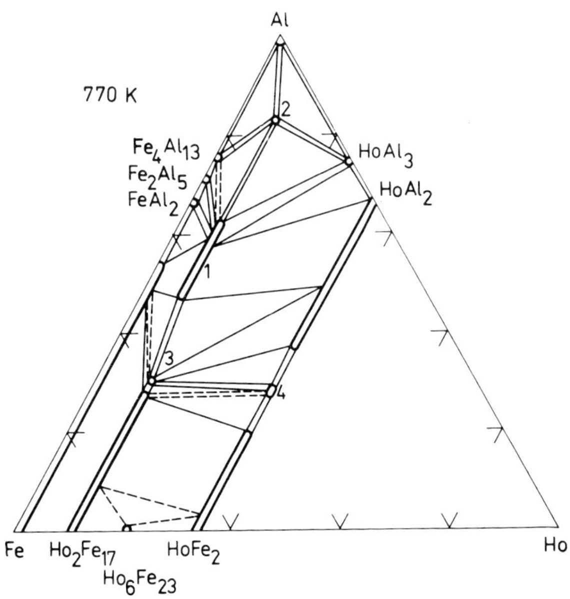

Fig. 58. Ho–Fe–Al, partial isothermal section at 770 K (0–33 a/c Ho). Compound with unknown crystal structure: (2) $\sim HoFe_2Al_{10}$.

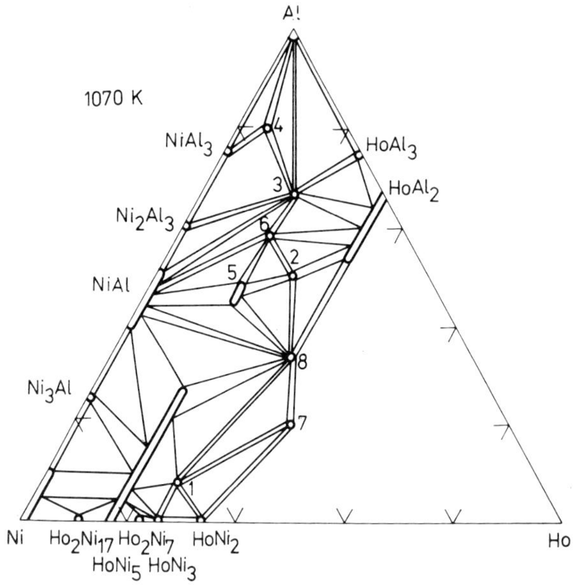

Fig. 59. Ho–Ni–Al, partial isothermal section at 1070 K (0–33 a/c Ho).

Ho–Ni–Al

The partial isothermal section at 1070 K (fig. 59) was investigated by Zarechnyuk et al. (1982). The crystal structures were determined for: (1) ~ Ho_3Ni_8Al (Ce_3Co_8Si ST, $a = 5.028$, $c = 16.07$); (2) ~ $HoNiAl_2$ ($MgCuAl_2$ ST, $a = 4.074$, $b = 10.07$, $c = 6.961$); (3) ~ $HoNiAl_4$ ($YNiAl_4$ ST, $a = 4.037$, $b = 15.27$, $c = 6.602$); (5) ~ $HoNi_2Al_3$ ($HoNi_{2.6}Ga_{2.4}$ ST, $a = 9.02$, $c = 4.052$); (7) ~ HoNiAl (ZrNiAl ST, $a = 6.991$, $c = 3.830$) and (8) ~ Ho_2Ni_2Al (Mo_2NiB_2 ST, $a = 5.388$, $b = 8.337$, $c = 4.172$). Only the lattice parameters were determined for (4) ~ $HoNi_3Al_{16}$ and (6) ~ $Ho_2Ni_3Al_7$. The first one is orthorhombic, $a = 3.991$, $b = 15.75$, $c = 26.92$ and the second is hexagonal with $a = 17.80$, $c = 3.974$.

Ho–{Zr, Hf}–Al

Tyvanchuk (1979) proposed a schematic phase diagram (fig. 60) and established that the compounds $Ho_{0.8}M_{0.2}Al_3$ had the $AuCu_3$ structure type with $a = 4.214$ (for Zr) and $a = 4.201$ (for Hf).

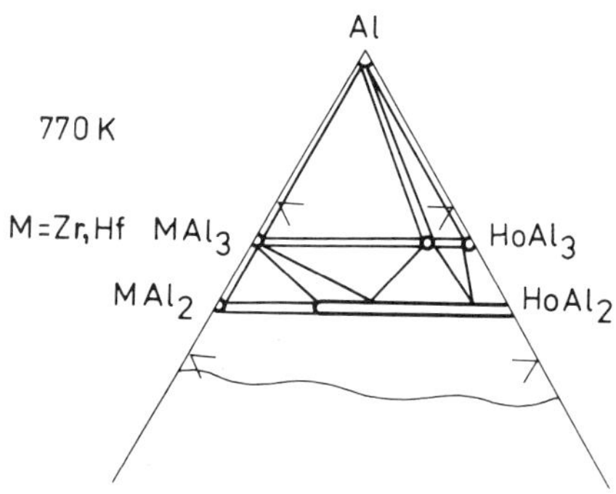

Fig. 60. Ho–M–Al (where M = Zr, Hf) schematic partial isothermal section at 770 K (100–66 a/c Al).

Er–Fe–Al

Zarechnyuk et al. (1972a) reported a partial phase diagram (fig. 61) and found the following ternary compounds: (1) ~ $ErFe_{5.6-3.4}Al_{6.4-8.6}$ ($ThMn_{12}$ ST, $a = 8.68–8.73$, $c = 4.99–5.04$); (2) ~ $ErFe_2Al_{10}$ (unknown crystal structure); (3) ~ $Er_2Fe_{11}Al_6$ (Th_2Zn_{17} ST, $a = 8.79$, $c = 12.68$), and (4) ~ $ErFe_{1.2}Al_{0.8}$ ($MgZn_2$ ST, $a = 5.40$, $c = 8.72$).

Er–Ni–Al

Phase equilibria (fig. 62) were derived by Zarechnyuk et al. (1982). Ternary compounds found in the system Er–Ni–Al are as follows: (1) ~ Er_3Ni_8Al (Ce_3Co_8Si ST, $a = 5.002$, $c = 15.99$), (2) ~ $ErNiAl_2$ ($MgCuAl_2$ ST, $a = 4.064$, $b = 10.06$, $c = 6.898$); (3) ~ $ErNiAl_4$ ($YNiAl_4$ ST, $a = 4.044$, $b = 15.08$, $c = 6.631$); (4) ~ $ErNi_3Al_{16}$ has a partially solved crystal structure (only the lattice parameters were established for an orthorhombic unit cell) $a = 3.960$, $b = 15.63$, $c = 26.81$; (5) ~ $ErNi_2Al_3$ ($HoNi_{2.6}Ga_{2.4}$ ST, $a = 9.01$, $c = 4.049$); (6) ~ $Er_2Ni_3Al_7$, $a = 17.77$, $c = 3.971$ for the hexagonal lattice, the crystal structure is unknown; (7) ~ $Er_3Ni_6Al_2$ ($Ce_3Ni_6Si_2$ ST, $a = 8.88$); (8) ~ ErNiAl (ZrNiAl ST, $a = 6.974$, $c = 3.801$) and (9) ~ Er_2Ni_2Al (Mo_2NiB_2 ST, $a = 5.347$, $b = 8.374$, $c = 4.157$).

Tm–Ni–Al

The system Tm–Ni–Al (fig. 63) was investigated by Rykhal et al. (1982) at 1070 K. Nine ternary compounds were found: (1) ~ Tm_3Ni_8Al (Ce_3Co_8Si ST, $a = 4.986$, $c = 15.96$); (2) ~ $TmNiAl_2$ ($MgCuAl_2$ ST, $a = 4.064$, $b = 10.030$, $c = 6.900$); (3) ~ $TmNiAl_4$ belongs to the $YNiAl_4$ ST, $a = 4.044$, $b = 15.04$, $c = 6.63$; (4) ~ $TmNi_3Al_{16}$ crystallizes in the orthorhombic lattice, $a = 3.927$, $b = 15.46$, $c = 26.59$; (5) ~ $TmNi_2Al_3$ has the $HoNi_{2.6}Ga_{2.4}$ ST, $a = 9.00$, $c = 4.037$; (6) ~ $Tm_2Ni_3Al_7$ with the hexagonal lattice, $a = 17.85$, $c = 4.048$; (7) ~ $Tm_3Ni_6Al_2$

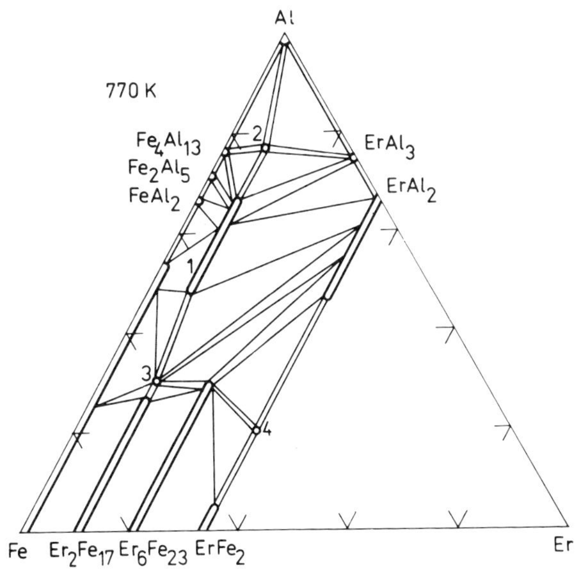

Fig. 61. Er–Fe–Al, partial isothermal section at 770 K (0–33 a/c Er).

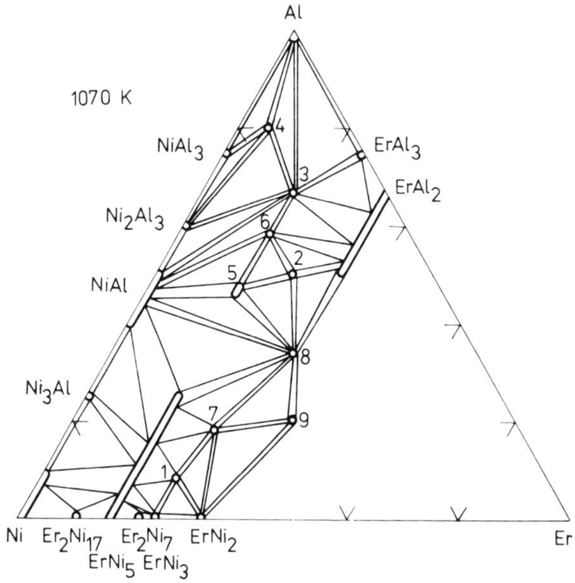

Fig. 62. Er–Ni–Al, partial isothermal section at 1070 K (0–33 a/c Er).

($Ce_3Ni_6Si_2$ ST, $a = 8.86$); (8) ~ TmNiAl reported as having ZrNiAl structure type, $a = 6.962$, $c = 3.775$, and (9) ~ Tm_2Ni_2Al with Mo_2NiB_2 ST, $a = 5.327$, $b = 8.308$, $c = 4.167$. Crystal structures of (4) and (6) are not solved.

Lu–Ni–Al

The partial isothermal section at 1070 K (fig. 64) was derived by Rykhal et al. (1982). Compositions and crystal structures for the eight existing ternary com-

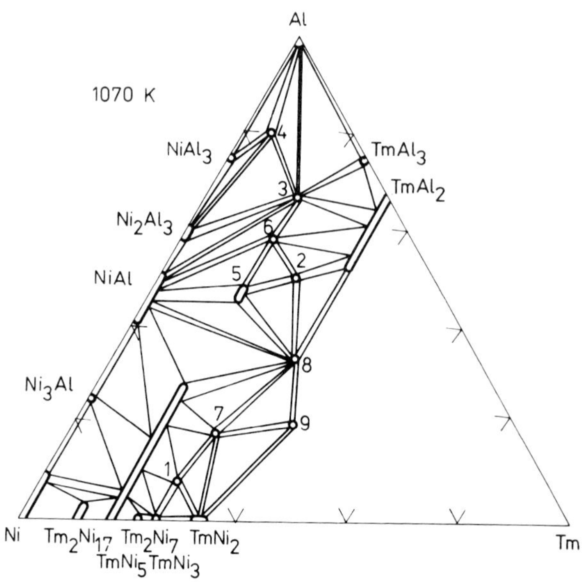

Fig. 63. Tm–Ni–Al, partial isothermal section at 1070 K (0–33 a/c Tm).

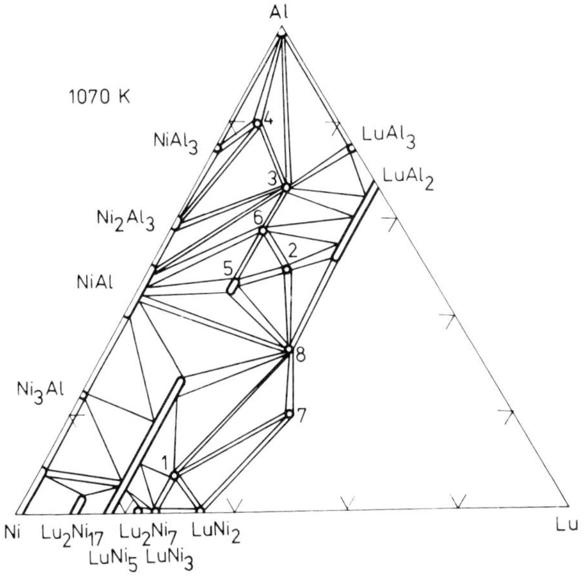

Fig. 64. Lu–Ni–Al, partial isothermal section at 1070 K (0–33 a/c Lu).

pounds (after Rykhal et al. 1982) are: (1) ~ Lu_3Ni_8Al (Ce_3Co_8Si ST, $a = 4.985$, $c = 15.94$); (2) ~ $LuNiAl_2$ ($MgCuAl_2$ ST, $a = 4.038$, $b = 9.875$, $c = 6.900$); (3) ~ $LuNiAl_4$ ($YNiAl_4$ ST, $a = 4.023$, $b = 14.97$, $c = 6.630$); (4) ~ $LuNi_3Al_{16}$ (unknown structure with the orthorhombic unit cell, $a = 3.924$, $b = 15.40$, $c = 26.54$); (5) ~ $LuNi_2Al_3$ ($HoNi_{2.6}Al_{2.4}$ ST, $a = 9.00$, $c = 4.032$); (6) ~ $Lu_3Ni_2Al_7$ (undetermined crystal structure with hexagonal lattice, $a = 17.85$, $c = 4.044$); (7) ~ LuNiAl (ZrNiAl ST, $a = 6.906$, $c = 3.748$), and (8) ~ Lu_2Ni_2Al (Mo_2NiB_2 ST $a = 5.623$, $b = 8.253$, $c = 4.124$).

Lu–{Zr, Hf}–Al

No ternary compounds (fig. 65) were found by Tyvanchuk (1979).

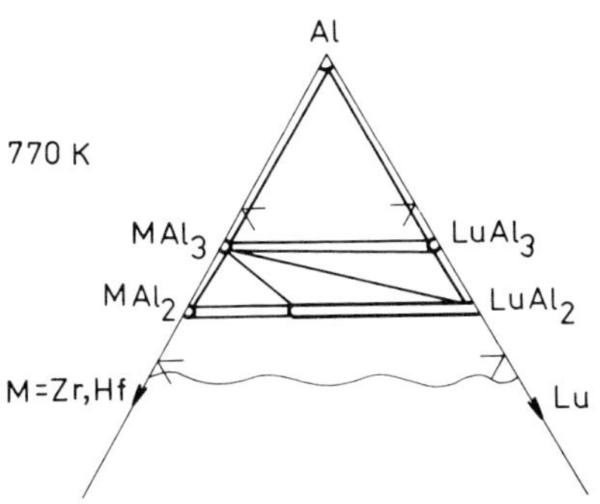

Fig. 65. Lu–M–Al, schematic partial isothermal section at 770 K (100–66 a/c Al), where M = Zr, Hf.

TABLE 1

Ternary compounds found in the partially investigated systems R–M–Al. In the tables, a short-hand notation for the references has been used: the first two letters of the surname of the first author have been taken followed by initial letter of the surname(s) of his co-author(s). The references can be found in the reference list accompanying the section.

Compound	Structure type	Lattice parameters			Ref.
		a	*b*	*c*	
ScCoAl	$MgZn_2$	5.10	–	8.20	TeP, 65a
$Sc_6Co_8Al_{15}$	Th_6Mn_{23}	12.09	–	–	MaB, 69
$Sc_6Rh_{16}Al_7$	$Mg_6Cu_{16}Si_7$	...	–	–	MaS, 71
$Sc_6Os_{16}Al_7$	$Mg_6Cu_{16}Si_7$	...	–	–	MaS, 71
$Sc_6Ir_{16}Al_7$	$Mg_6Cu_{16}Si_7$	...	–	–	MaS, 71
YMo_2Al_4	$YbMo_2Al_4$	6.748	–	5.331	GrR, 86
YRe_4Al_8	$CeMn_4Al_8$	9.02	–	5.22	RyZ, 74
LaCoAl	...	...	...	...	Oe, 71
$La_3Co_5Al_2$	...	...	...	...	Em, 71b
$La_3Co_4Al_3$	...	...	...	...	Em, 71b
$La_3Co_{15-18}Al_{8-5}$	...	...	...	...	Em, 71b
$LaCoAl_4$	$LaCoAl_4$	4.07	7.70	7.07	RyZY, 77
$LaCo_2Al_7$	...	...	...	...	Em, 71b
$LaNi_{1.5-1.2}Al_{0.5-0.8}$	...	...	...	...	Em, 71b
LaNiAl	...	...	...	...	Oe, 73
$LaNi_{0.9-0.76}Al_{1.1-1.24}$	...	...	...	...	Em, 71b
$LaNi_2Al_5$	$PrNi_2Al_5$	7.033	9.677	4.021	Ya RAZ, 81
$LaNi_{5-4}Al_{8-9}$	$NaZn_{13}$	11.72–11.78	–	–	Em, 71b
$La_2Ni_{2.25}Al_{0.75}$	Mo_2NiB_2	5.620	8.585	4.378	RoGYZS, 82b
La_4SnAl_7	AlB_2	4.474	–	4.397	RaS, 67
$LaRe_4Al_8$	$CeMn_4Al_8$	9.08	–	5.19	RyZ, 74
$Ce_2Ni_{2.15}Al_{0.85}$	Mo_2NiB_2	5.331	8.413	4.241	RoGYZS, 82b
$Ce(Zn, Al)_4$	$BaAl_4$	4.45	–	11.02	Za, 64
$Ce_8Sn_3Al_{13}$	AlB_2	4.454	–	4.284	RaS, 67
$CeRe_4Al_8$	$CeMn_4Al_8$	9.06	–	5.15	RyZ, 74
PrV_2Al_{20}	$CeCr_2Al_{20}$	14.58	–	–	KrZ, 68
$PrCr_4Al_8$	$CeMn_4Al_8$	9.033	–	5.128	BuV, 76
$PrCr_2Al_{20}$	$CeCr_2Al_{20}$	14.43	–	–	KrZ, 68
$PrMn_4Al_8$	$CeMn_4Al_8$	8.962	–	5.143	BuV, 76
$PrFe_4Al_8$	$CeMn_4Al_8$	8.796	–	5.056	BuV, 76
$PrCo_{1.6}Al_{0.4}$	...	...	...	...	OeW, 67
PrCoAl	...	...	...	...	Oe, 71
$PrCoAl_4$	$LaCoAl_4$	4.05	7.59	6.99	RyZY, 77
PrGeAl	α-$ThSi_2$	4.273	–	14.63	Rya, 74
$PrGe_2Al_2$	La_2O_2S	4.264	–	6.896	ZaMG, 70b
NdV_2Al_{20}	$CeCr_2Al_{20}$	14.57	–	–	KrZ, 68
$NdCr_2Al_{20}$	$CeCr_2Al_{20}$	14.41	–	–	KrZ, 68
NdCoAl	...	...	...	...	Oe, 71
NdGeAl	α-$ThSi_2$	4.272	–	14.66	Rya, 74
$NdGe_{1-0.75}Al_{1-1.25}$	AlB_2	4.298–4.308	–	4.210–4.204	RaS, 67
$NdGe_2Al_2$	La_2O_2S	4.265	–	6.827	ZaMG, 70b
$NdRe_4Al_8$	$CeMn_4Al_8$	9.03	–	5.14	RyZ, 74
SmCoAl	...	...	...	...	Oe, 71

TABLE 1 (cont'd)

Compound	Structure type	Lattice parameters a	b	c	Ref.
Sm_2Ni_2Al	Mo_2NiB_2	5.414	8.424	4.221	Ry, 78
SmNiAl	ZrNiAl	6.982	–	4.008	DwMCDK, 68
Sm_3Ni_8Al	Ce_3Co_8Si	5.082	–	16.19	Ry, 78
SmGeAl	α-$ThSi_2$	4.188	–	14.52	Rya, 74
$SmGe_2Al_2$	La_2O_2S	4.233	–	6.805	ZaMG, 70b
$EuMn_{1.3-3.6}Al_{10.7-8.4}$	$ThMn_{12}$	9.095–8.974	–	5.142–5.130	MaYZ, 86
$EuFe_4Al_8$	$CeMn_4Al_8$	8.846	–	5.084	ViZ, 74
$EuFe_2Al_8$	$CeFe_2Al_8$	12.530	14.503	4.036	MaSKZY, 83a
EuCuAl	...	...	...	...	Oe, 73
$EuCu_4Al_8$	$ThMn_{12}$	8.902	–	5.105	MaYZ, 83b
$EuCu_{6.2-9.1}Al_{6.8-3.9}$	$NaZn_{13}$	11.960–11.828	–	–	MaYZ, 83b
$EuGe_2Al_2$	La_2O_2S	4.22	–	7.31	ZaMG, 70b
GdV_2Al_{20}	$CeCr_2Al_{20}$	14.56	–	–	KrZ, 68
$GdCr_2Al_{20}$	$CeCr_2Al_{20}$	14.41	–	–	KrZ, 68
$GdCr_4Al_8$	$CeMn_4Al_8$	8.967	–	5.128	BuV, 76
$GdMn_4Al_8$	$CeMn_4Al_8$	8.887	–	5.119	BuV, 76
GdCoAl	$MgZn_2$	5.370	–	8.571	Oe, 71
GdCuAl	ZrNiAl	7.078	–	4.065	DwMCDK, 68
$GdCu_4Al_8$	$CeMn_4Al_8$	8.748	–	5.146	BuV, 76
$GdMo_2Al_4$	$YbMo_2Al_4$	6.780	–	5.331	FoP, 76
TbCoAl	$MgZn_2$	5.370	–	8.512	Oe, 71
TbCuAl	ZrNiAl	7.041	–	4.044	Oe, 73
TbGeAl	DyGeAl	4.121	10.52	5.831	Rya, 74
$TbGe_2Al_2$	La_2O_2S	4.24	–	6.63	ZaMG, 70b
$TbMo_2Al_4$	$YbMo_2Al_4$	6.759	–	5.319	GrR, 86
DyCoAl	$MgZn_2$	5.342	–	8.497	Oe, 71
DyCuAl	ZrNiAl	7.023	–	4.030	DwMCDK, 68
DyGeAl	DyGeAl	4.091	10.46	5.902	Rya, 74
$DyGe_2Al_2$	La_2O_2S	4.21	–	6.67	ZaMG, 70b
$DyMo_2Al_4$	$YbMo_2Al_4$	6.742	–	5.318	GrR, 86
$HoCr_2Al_{20}$	$CeCr_2Al_{20}$	14.39	–	–	KrZ, 68
HoCoAl	$MgZn_2$	5.336	–	8.487	Oe, 71
HoCuAl	ZrNiAl	6.997	–	4.015	DwMCDK, 68
HoGeAl	DyGeAl	4.123	10.40	5.827	Rya, 74
$HoGe_2Al_2$	La_2O_2S	4.196	–	6.668	ZaMG, 70b
$HoMo_2Al_4$	$YbMo_2Al_4$	6.734	–	5.317	GrR, 86
$ErCr_2Al_{20}$	$CeCr_2Al_{20}$	14.38	–	–	KrZ, 68
$ErCr_4Al_8$	$CeMn_4Al_8$	8.901	–	5.123	BuV, 76
$ErMn_4Al_8$	$CeMn_4Al_8$	8.892	–	5.117	BuV, 76
ErCoAl	$MgZn_2$	5.305	–	8.469	Oe, 71
ErCuAl	ZrNiAl	6.974	–	4.002	DwMCDK, 68
$ErCu_4Al_8$	$CeMn_4Al_8$	8.691	–	5.119	BuV, 76
ErGeAl	DyGeAl	4.101	10.34	5.840	Rya, 74
$ErGe_2Al_2$	La_2O_2S	4.184	–	6.662	ZaMG, 70b
$ErMo_2Al_4$	$YbMo_2Al_4$	6.722	–	5.317	FoP, 76
TmFeAl	$MgZn_2$	5.32	–	8.67	ZaRV, 73
$TmFe_4Al_8$	$CeMn_4Al_8$	8.695	–	5.015	BuV, 76

TABLE 1 (cont'd)

Compound	Structure type	Lattice parameters *a*	*b*	*c*	Ref.
TmCoAl	$MgZn_2$	5.276	–	8.435	Oe, 71
TmGeAl	DyGeAl	4.02	10.33	5.865	Rya, 74
$TmGe_2Al_2$	La_2O_2S	4.18	–	6.65	ZaMG, 70b
$TmMo_2Al_4$	$YbMo_2Al_4$	6.719	–	5.309	GrR, 86
$YbCr_4Al_8$	$CeMn_4Al_8$	8.710	–	5.110	BuV, 76
$YbMn_4Al_8$	$CeMn_4Al_8$	8.819	–	5.084	BuV, 76
YbFeAl	...	...	...	...	Oe, 71
$YbFe_4Al_8$	$CeMn_4Al_8$	8.706	–	5.004	BuV, 76
YbCoAl	$MgZn_2$	5.296	–	8.466	Oe, 71
YbNiAl	...	...	...	...	Oe, 73
$YbNiAl_2$	$MgCuAl_2$	4.065	10.03	6.97	RoZRYS, 82a
Yb_2Ni_2Al	Mo_2NiB_2	5.614	8.314	4.087	RoGYZS, 82b
YbCuAl	ZrNiAl	6.926	–	3.986	DwMCDK, 68
$YbCu_4Al_8$	$CeMn_4Al_8$	8.710	–	5.110	BuV, 76
YbGeAl	DyGeAl	4.307	10.57	5.935	Yan, 75
$YbGe_2Al_2$	La_2O_2S	4.118	–	7.05	ZaMG, 70b
$YbMo_2Al_4$	$YbMo_2Al_4$	6.717	–	5.312	FoP, 76
LuFeAl	ZrNiAl	6.926	–	3.896	DwMCDK, 68
$LuFe_4Al_8$	$CeMn_4Al_8$	8.687	–	4.982	BuV, 76
LuCoAl	$MgZn_2$	5.261	–	8.399	Oe, 71
LuCuAl	ZrNiAl	6.926	–	3.896	DwMCDK, 68
$LuGe_2Al_2$	La_2O_2S	4.17	–	6.63	ZaMG, 70b
$LuMo_2Al_4$	$YbMo_2Al_4$	6.695	–	5.311	GrR, 86

References

Bodak, O.I., and E.I. Gladyshevsky, 1969, Dopov. Akad Nauk Ukr. RSR, Ser. A, p. 452.

Buschow, K.H.J., and J.H.M. van Vucht, 1976, J. Less-Common Met. **50**, 145.

Casteels, E., P. Diels and A. Cools, 1967, J. Nucl. Mater. **24**, 87.

Dwight, A.E., M.H. Mueller, R.A. Conner, I.W. Donney and H. Knott, 1968, Trans. Met. Soc. AIME **242**, 2075.

Dzyana, D.I., and O.S. Zarechnyuk, 1969, Vestn. Lvovsk. Univ., Ser. Khim. **11**, 8.

Emes-Mysenko, E.I., 1971a, Vestn. Lvovsk. Univ., Ser. Khim. **13**, 12.

Emes-Mysenko, E.I., 1971b, Investigation of some of the ternary systems lanthanum–3d metal–aluminium in the regions of small lanthanum contents, Ph.D. Chemistry thesis, 1971 (Lvov State University, Lvov) pp. 2–16.

Eshonov, K.K., A.V. Vakhobov and T.D. Dghuraev, 1977, Izv. Akad. Nauk SSSR Met. (6), 196.

Fornasini, M.L., and A. Palenzona, 1976, J. Less-Common Met. **45**, 137.

Gladyshevsky, E.I., I.F. Kolobnev and O.S. Zarechnyuk, 1961, Zh. Neorg. Khim. **6**, 2104.

Goebel, J.A., and S. Rosen, 1968, J. Less-Common Met. **16**, 441.

Gryn, Yu.N., and P. Rogl, 1986, Vestn. Lvovsk. Univ., Ser. Khim. **27**, 38.

Karatygina, E.P., V.V. Burnashova, M.V. Raevskaya and E.M. Sokolovskaya, 1974a, Metallofizika **52**, 105.

Karatygina, E.P., V.V. Burnashova, M.V. Raevskaya and E.M. Sokolovskaya, 1974b, Physical chemistry investigation of the alloys in the system ruthenium–aluminium–scandium, palladium–aluminium–scandium, Manuscript reported to VINITI, Nr. 3021-741, Dep. 2.

Krypyakevich, P.I., and O.S. Zarechnyuk, 1968, Dopov. Akad. Nauk Ukr. RSR, Ser. A, p. 787.
Leon, B., and W.E. Wallace, 1970, J. Less-Common Met. **22**, 1.
Mansey, N.C., G.V. Raynor and I.R. Harris, 1968, J. Less-Common Met. **14**, 337.
Manyako, N.B., I.N. Stetz, J.V. Kavich, O.S. Zarechnyuk and T.I. Yanson, 1983a, Dopov. Akad. Nauk Ukr. RSR, Ser. A (7), 41.
Manyako, N.B., T.I. Yanson and O.S. Zarechnyuk, 1983b, Crystal structure of IMC in the systems Eu,Sr,Ba–Cu–Al, in: Chetvertaya Vsesoyuzn. Konf. Kristallokhim. Intermetal. Soed., Tezisy Dokl., Lvov, 1983 (Vyscha Shkola, Lvov) pp. 21, 22.
Manyako, N.B., T.I. Yanson and O.S. Zarechnyuk, 1986, Vestn. Lvovsk. Univ., Ser. Khim. **27**, 16.
Markiv, V.Ya., and V.V. Burnashova, 1969, Dopov. Akad. Nauk Ukr. RSR, Ser. A, p. 463.
Markiv, V.Ya., and A.I. Storoghenko, 1971, Ternary compounds with $Mg_6Cu_{16}Si_7$ type structure in some systems of two transition metals with aluminium and gallium, in: Vsesoyuzn. Konf. Kristallokhim. Intermetal. Soed., Tezisy Dokl., Lvov, 1971 (Vyscha Shkola, Lvov) pp. 18, 19.
Muravyeva, A.A., and O.S. Zarechnyuk, 1970, Izv. Akad Nauk SSSR Neorg. Mat. **6**, 1066.
Muravyeva, A.A., O.S. Zarechnyuk and E.I. Gladyshevsky, 1971, Izv. Akad Nauk SSSR, Neorg. Mat. **7**, 38.
Oesterreicher, H., 1971, J. Less-Common Met. **25**, 341.
Oesterreicher, H., 1973, J. Less-Common Met. **30**, 225.
Oesterreicher, H., 1975, J. Less-Common Met. **40**, 207.
Oesterreicher, H., and W.E. Wallace, 1967, J. Less-Common Met. **13**, 475.
Prevarsky, A.P., 1978, New compounds in the system samarium–copper–aluminium, in: Tretya Vsesoyuzn. Konf. Kristallokhim. Intermetal. Soed., Tezisy Dokl., Lvov, 1978 (Vyscha Shkola, Lvov) pp. 69, 70.
Prevarsky, A.P., O.S. Zarechnyuk, E.E. Cherkashin and V.I. Polikha, 1976, Vestn. Lvovsk. Univ., Ser. Khim. **18**, 14.
Prevarsky, A.P., Yu.B. Kuzma and O.S. Zarechnyuk, 1983, Dopov. Akad. Nauk Ukr. RSR, Ser. A (8), 53.
Raevskaya, M.V., and E.M. Sokolovskaya, 1979, Physical chemistry of ruthenium and its alloys (Moscow University Press, Moscow) p. 229.
Raman, A., and H. Steinfink, 1967, Inorg. Chem. **6**, 1789.
Romaka, V.A., O.S. Zarechnyuk, R.M. Rykhal, Ya.P. Yarmolyuk and R.V. Skolozdra, 1982a, Fiz. Met. Metalloved. **54**, 410.
Romaka, V.A., Yu.N. Gryn, Ya.P. Yarmolyuk, O.S. Zarechnyuk and R.V. Skolozdra, 1982b, Fiz. Met. Metalloved. **54**, 691.
Rosen, S., and J.A. Goebel, 1968, J. Less-Common Met. **16**, 285.
Ryabokon, T.I., 1974, Vestn. Lvovsk. Univ., Ser. Khim. **15**, 26.
Ryabov, V.R., O.S. Zarechnyuk, D.M. Rabkin and O.I. Vivchar, 1971, Avtom. Svarka (7), 76.
Rykhal, R.M., 1972a, Phase equilibria and crystal structure of compounds in the ternary systems yttrium–3d transition metal–aluminium, Ph.D. Chemistry thesis (Lvov State University, Lvov) pp. 2–15.
Rykhal, R.M., 1972b, Vestn. Lvovsk. Univ., Ser. Khim. **13**, 11.
Rykhal, R.M., 1977, Vestn. Lvovsk. Univ., Ser. Khim. **19**, 36.
Rykhal, R.M., 1978, The new representatives of structure types Ce_3Co_8Si and Mo_2NiB_2 in the systems rare earth–nickel–aluminium, in: Tretya Vsesoyuzn. Konf. Kristallokhim. Intermetal. Soed., Tezisy Dokl., Lvov, 1978 (Vyscha Shkola, Lvov) p. 17.
Rykhal, R.M., and O.S. Zarechnyuk, 1971, Dopov. Akad. Nauk Ukr. RSR, Ser. A, p. 854.
Rykhal, R.M., and O.S. Zarechnyuk, 1974, The new representatives of $CeMn_4Al_8$ structure type, in: Vtoraya Vsesoyuzn. Konf. Kristallokhim. Intermetal. Soed., Tezisy Dokl., Lvov, 1974 (Vyscha Shkola, Lvov) pp. 40, 41.
Rykhal, R.M., and O.S. Zarechnyuk, 1977, Dopov. Akad. Nauk Ukr. RSR, Ser. A, p. 376.
Rykhal, R.M., O.S. Zarechnyuk and N.V. German, 1971, Izv. Akad. Nauk SSSR, Met. (6), 205.
Rykhal, R.M., O.S. Zarechnyuk and G.V. Pyschyk, 1972a, Vestn. Lvovsk. Univ., Ser. Khim. **14**, 13.
Rykhal, R.M., O.S. Zarechnyuk and Ya.P. Yarmolyuk, 1972b, Kristallografiya **17**, 521.

Rykhal, R.M., O.S. Zarechnyuk and G.V. Pyschyk, 1973, Dopov. Akad. Nauk Ukr. RSR, Ser. A, p. 568.
Rykhal, R.M., O.S. Zarechnyuk and Ya.P. Yarmolyuk, 1977, Dopov. Akad. Nauk Ukr. RSR, Ser. A, p. 265.
Rykhal, R.M., O.S. Zarechnyuk and Ya.J. Kuten, 1978a, Dopov. Akad. Nauk Ukr. RSR, Ser. A, p. 1136.
Rykhal, R.M., O.S. Zarechnyuk and O.I. Marych, 1978b, Dopov. Akad. Nauk Ukr. RSR, Ser. A, p. 854.
Rykhal, R.M., O.S. Zarechnyuk and O.P. Matzkiv, 1979a, Vestn. Lvovsk. Univ., Ser. Khim. **21**, 46.
Rykhal, R.M., O.S. Zarechnyuk and T.I. Yanson, 1979b, Dopov. Akad. Nauk Ukr. RSR, Ser. A, p. 1057.
Rykhal, R.M., O.S. Zarechnyuk and V.M. Mandzyk, 1980, Dopov. Akad. Nauk Ukr. RSR, Ser. A, p. 77.
Rykhal, R.M., O.S. Zarechnyuk, V.S. Protasov and V.A. Romaka, 1982, Dopov. Akad. Nauk Ukr. RSR, Ser. A, (3), 43.
Teslyuk, M.Yu., and V.S. Protasov, 1965a, Dopov. Akad. Nauk Ukr. RSR, Ser. A, p. 599.
Teslyuk, M.Yu., and V.S. Protasov, 1965b, Kristallografiya **10**, 561.
Tyvanchuk, A.T., 1975, Dopov. Akad. Nauk Ukr. RSR, Ser. A, p. 660.
Tyvanchuk, A.T., 1976, Dopov. Akad. Nauk Ukr. RSR, Ser. A, p. 1046.
Tyvanchuk, A.T., 1979, Ternary systems Al–La,Ce,Pr,Nd,Gd,Tb,Dy,Ho,Lu–Zr,Nb,Hf at the region 66.7–100 at.% of Al, Ph.D. Chemistry thesis (Lvov State University, Lvov) pp. 1–16.
Van Vucht, J.H.N., 1961, Phil. Res. Rept. **16**, 1.
Vivchar, O.I., and O.S. Zarechnyuk, 1974, The compounds with crystal structure of $ThMn_{12}$ type in the systems R–Fe–Al, in: Vtoraya Vsesoyuzn. Konf. Kristallokhim. Intermetal. Soed., Tezisy Dokl., Lvov, 1974 (Vyscha Shkola, Lvov) p. 41.
Vivchar, O.I., O.S. Zarechnyuk and V.R. Ryabov, 1970, Izv. Akad. Nauk SSSR, Met. (1), 211.
Vivchar, O.I., O.S. Zarechnyuk and V.R. Ryabov, 1973a, Dopov. Akad. Nauk Ukr. RSR, Ser. A, p. 1040.
Vivchar, O.I., O.S. Zarechnyuk and V.R. Ryabov, 1973b, Dopov. Akad. Nauk Ukr. RSR, Ser. A, p. 159.
Vivchar, O.I., O.S. Zarechnyuk and V.R. Ryabov, 1974, Dopov. Akad. Nauk Ukr. RSR, Ser. A, p. 363.
Wernick, J.H., S.E.. Haszko and D. Dorsi, 1962, J. Chem. Solids **23**, 567.
Yanson, T.I., 1975, Crystal structures of the alumosilicides and alumogermanides of the rare earth elements, Ph.D. Chemistry thesis (Lvov State University, Lvov) pp. 2–18.
Yarmolyuk, Ya.P., R.M. Rykhal and O.S. Zarechnyuk, 1974, Crystal structure of the compounds $CeFe_2Al_8$ and $LaCoAl_4$, in: Vtoraya Vsesoyuzn. Konf. Kritallokhim. Intermetal. Soed., Tezisy Dokl., Lvov, 1974 (Vyscha Shkola, Lvov) pp. 39,40.
Yarmolyuk, Ya.P., R.M. Rykhal, L.G. Akselrud and O.S. Zarechnyuk, 1981, Dopov. Akad. Nauk Ukr. RSR, Ser. A (9), 86.
Zarechnyuk, O.S., 1964, Crystal structures of some aluminium intermetallic compounds, in: Nauk Konf. Prysv. 25-rich. Vozz'edn Ukr. Narodu Edyn. Ukr. Rad. Soc. Dergh., Tezy Dopov., Lvov (Lvov State University, Lvov) pp. 8, 9.
Zarechnyuk, O.S., 1966, Dopov. Akad. Nauk Ukr. RSR, Ser. A, p. 787.
Zarechnyuk, O.S., and E.I. Emes-Mysenko, 1969, Vestn. Lvovsk. Univ., Ser. Khim. **11**, 11.
Zarechnyuk, O.S., and I.F. Kolobnev, 1968, Izv. Akad. Nauk SSSR, Met. (5), 208.
Zarechnyuk, O.S., and P.I. Krypyakevich, 1962, Kristallografiya **7**, 543.
Zarechnyuk, O.S., and P.I. Krypyakevich, 1967, Izv. Akad. Nauk SSSR, Met. (4), 188.
Zarechnyuk, O.S., and R.M. Rykhal, 1974, Vestn. Lvovsk. Univ., Ser. Khim. **16**, 5.
Zarechnyuk, O.S., and R.M. Rykhal, 1977, Vestn. Lvovsk. Univ., Ser. Khim. **19**, 51.
Zarechnyuk, O.S., and R.M. Rykhal, 1978, Dopov. Akad. Nauk Ukr. RSR, Ser. A, p. 370.
Zarechnyuk, O.S., and R.M. Rykhal, 1981, Vestn. Lvovsk. Univ., Ser. Khim. **23**, 45.
Zarechnyuk, O.S., I.F. Kolobnev and M.Yu. Teslyuk, 1963, Zh. Neorg. Khim. **2**, 1668.

Zarechnyuk, O.S., P.I. Krypyakevich and E.I. Gladyshevsky, 1964, Kristallografiya **9**, 835.
Zarechnyuk, O.S., D.P. Frankevich and P.I. Krypyakevich, 1966, Izv. Akad. Nauk SSSR, Met. (1), 153.
Zarechnyuk, O.S., E.I. Emes-Mysenko, V.R. Ryabov and I.I. Dykij, 1968, Izv. Akad. Nauk SSSR, Met. (2), 219.
Zarechnyuk, O.S., O.I. Vivchar and D.P. Frankevich, 1969a, Izv. Vyssh. Uchebn. Zaved. Tsvetn. Metall., p. 122.
Zarechnyuk, O.S., M.G. Myskiv and V.R. Ryabov, 1969b, Izv. Akad. Nauk SSSR, Met. (2), 164.
Zarechnyuk, O.S., O.I. Vivchar and R.M. Rykhal, 1970a, Dopov. Akad. Nauk Ukr. RSR, Ser. A, p. 943.
Zarechnyuk, O.S., A.A. Muravyeva and E.I. Gladyshevsky, 1970b, Dopov. Akad. Nauk Ukr. RSR, Ser. A, p. 753.
Zarechnyuk, O.S., R.M. Rykhal and N.V. German, 1971, Vestn. Lvovsk. Univ., Ser. Khim. **12**, 10.
Zarechnyuk, O.S., O.I. Vivchar and V.R. Ryabov, 1972a, Vestn. Lvovsk. Univ., Ser. Khim. **14**, 16.
Zarechnyuk, O.S., R.M. Rykhal, V.R. Ryabov and O.I. Vivchar, 1972b, Izv. Akad. Nauk SSSR, Met. (1), 208.
Zarechnyuk, O.S., R.M. Rykhal and O.I. Vivchar, 1973, Metallofizika **46**, 92.
Zarechnyuk, O.S., R.M. Rykhal and M.Yu. Schtojko, 1975, Vestn. Lvovsk. Univ., Ser. Khim. **17**, 24.
Zarechnyuk, O.S., M.E. Dritz, R.M. Rykhal and V.V. Kinghybalo, 1980a, Izv. Akad. Nauk SSSR, Met. (5), 242.
Zarechnyuk, O.S., R.M. Rykhal and V.V. Koryn, 1980b, Dopov. Akad. Nauk Ukr. RSR, Ser. A, p. 86.
Zarechnyuk, O.S., T.I. Yanson and A.A. Muravyeva, 1981, Vestn. Lvovsk. Univ., Ser. Khim. **23**, 64.
Zarechnyuk, O.S., R.M. Rykhal, V.A. Romaka, O.K. Kovalskaya and A.I. Sharabura, 1982, Dopov. Akad. Nauk Ukr. RSR, Ser. A (1), 81.
Zarechnyuk, O.S., T.I. Yanson and R.M. Rykhal, 1983, Izv. Akad. Nauk SSSR, Met. (4), 192.
Zukhuritdinov, M.A., A.V. Vachobov and O.I. Bodak, 1976, Dokl. Akad. Nauk Tadgh. SSR **19**, 34.

2.2. *R–M–Ga systems*

Phase equilibria have been investigated in 30 ternary systems, and the existence of 210 ternary rare earth gallides was established. For 127 of them the crystal structure was solved. The investigation of another 59 ternary systems was carried out only partially – without deriving the phase diagrams. As the result of this study a total of 261 additional ternary compounds have been found, their compositions and crystallographic data for are given in table 2.

Sc–Ti–Ga

The phase diagram (fig. 66) was derived by Gavrylenko and Markiv (1978a). A total of three ternary compounds was found. The crystal structure was solved by Gryn et al. (1980a) only for the (1) ~ $ScTi_2Ga_4$ ($YbMo_2Al_4$ ST, $a = 6.578$, $c = 5.475$).

Sc–V–Ga

An investigation of the ternary system Sc–V–Ga was performed by Savitzky et al. (1978a). The phase diagram is shown in fig. 67. The only ternary compound, ScV_2Ga_4, belongs to the $YbMo_2Al_4$ ST, with $a = 6.482$, $c = 5.216$.

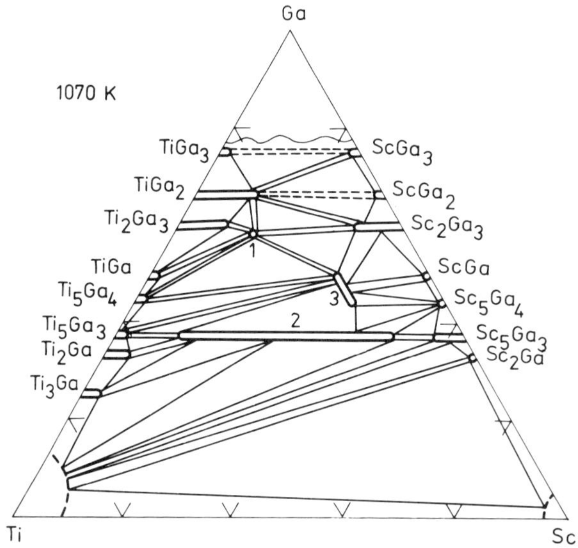

Fig. 66. Sc–Ti–Ga, isothermal section at 1070 K. Ternary compounds with unknown crystal structure: (2) ~ $Sc_{1-4}Ti_{4-1}Ga_3$; (3) ~ $Sc_{2-2.37}$-$TiGa_{3-2.63}$.

Sc–Cr–Ga

The only ternary compound, $Sc_{1.3}Cr_{0.7}Ga$, with unknown crystal structure was found by Gavrylenko and Markiv (1978a) during a phase diagram study of the system Sc–Cr–Ga (fig. 68).

Sc–Mn–Ga

Gavrylenko and Markiv (1978b) derived the isothermal section at 770 K, shown in fig. 69, and found three ternary compounds: (1) ~ $ScMnGa_2$; (2) ~ Sc_4MnGa_6, and (3) ~ $Sc_7Mn_{1.2}Ga_{0.8}$. Crystal structure of all of them are undetermined.

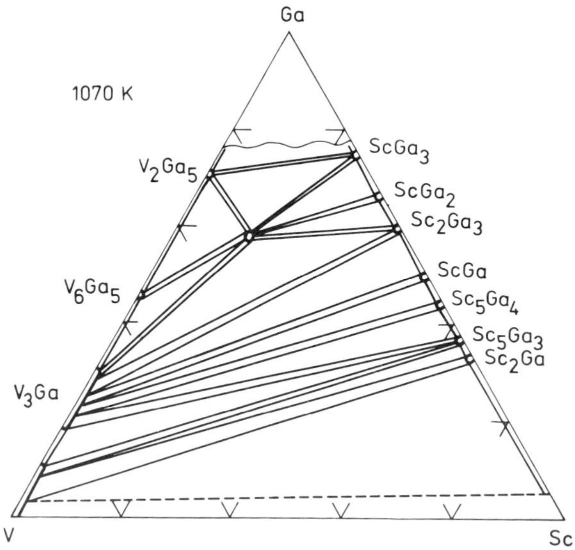

Fig. 67. Sc–V–Ga, isothermal section at 1070 K.

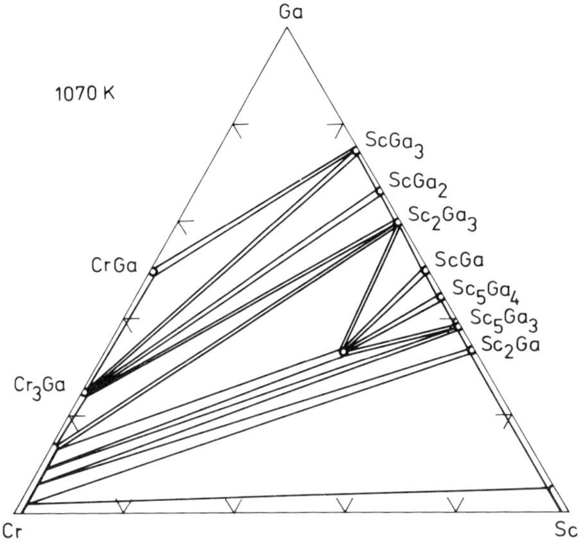

Fig. 68. Sc–Cr–Ga, isothermal section at 1070 K.

Sc–Fe–Ga

The isothermal section at 1070 K after Gavrylenko and Markiv (1978b) is shown in fig. 70. The crystal structures of ternary compounds are as follows: (1) ~ $ScFe_{6.8-5.0}Ga_{5.2-7.0}$ (the $ThMn_{12}$ ST, $a = 8.61–8.71$, $c = 4.73–4.81$); (2) ~ $ScFeGa_5$ ($HoCoGa_5$ ST, $a = 4.145$, $c = 6.623$); (4) ~ $ScFe_{1.82-1.44}Ga_{0.18-0.56}$ ($MgCu_2$ ST, $a = 7.03–7.15$), and (5) ~ $ScFe_{1.37-0.71}Ga_{0.63-1.29}$ ($MgZn_2$ ST, $a = 5.053–5.156$, $c = 8.289–8.386$).

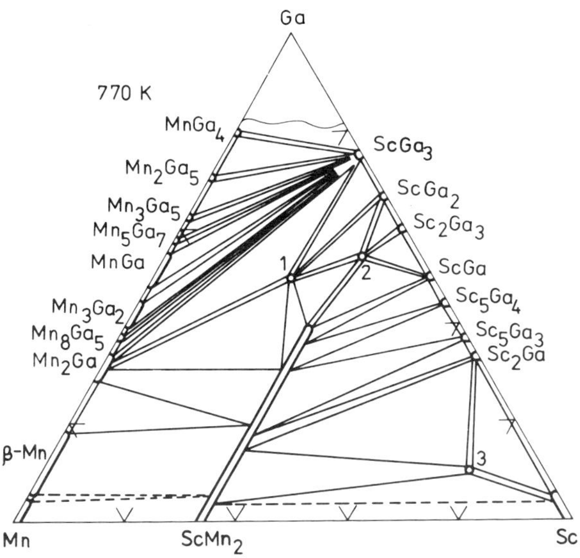

Fig. 69. Sc–Mn–Ga, isothermal section at 770 K.

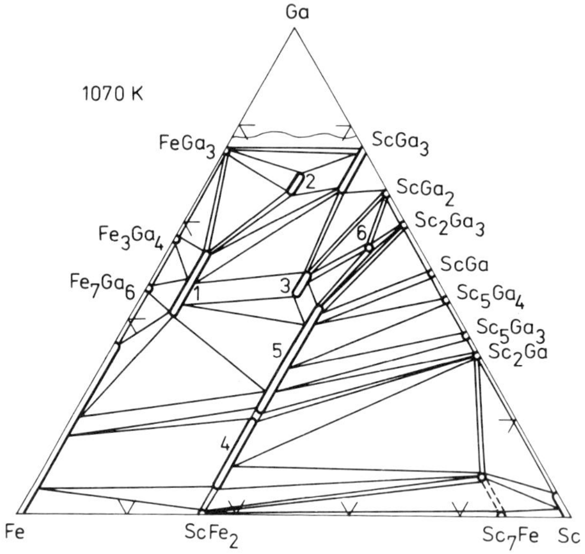

Fig. 70. Sc–Fe–Ga, isothermal section at 1070 K. Ternary compounds with unknown crystal structure: (3) ~ $Sc_4Fe_3Ga_7$; (6) ~ Sc_4FeGa_6; (7) ~ $Sc_4Fe_{0.6}Ga_{0.4}$.

Sc–Co–Ga

Figure 71 shows the phase diagram of the system Sc–Co–Ga after Gavrylenko and Markiv (1979). The existence of (7) ~ $Sc_6Co_8Ga_{15}$ with Th_6Mn_{23} ST, with $a = 12.14$, earlier found by Markiv and Burnasheva (1969), was confirmed. A total of 15 ternary compounds were synthesized by Gavrylenko and Markiv

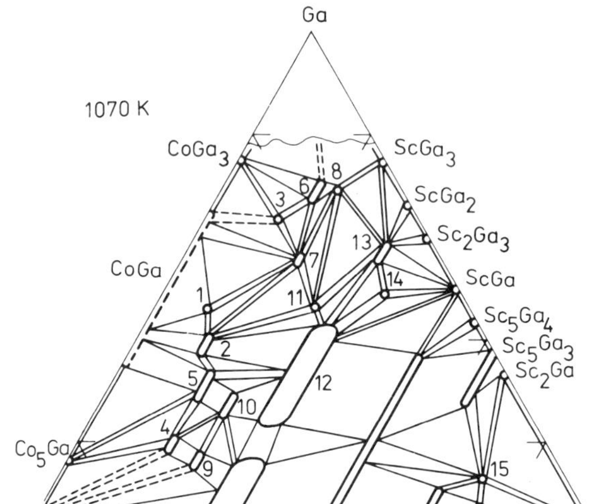

Fig. 71. Sc–Co–Ga, isothermal section at 1070 K. A total of 11 ternary compounds were reported with unknown crystal structures: (1) ~ $ScCo_5Ga_5$; (2) ~ $ScCo_4Ga_3$; (3) ~ $ScCo_2Ga_5$; (4) ~ $ScCo_{3.8}Ga_{1.2}$; (5) ~ $ScCo_3Ga_2$; (9) ~ $Sc_2Co_{5.5}Ga_{1.5}$; (10) ~ $Sc_2Co_{4.5}Ga_{2.5}$; (11) ~ $Sc_3Co_3Ga_5$; (13) ~ $ScCo_{0.35-0.2}Ga_{1.65-1.8}$; (14) ~ Sc_3CoGa_4; (15) ~ $Sc_7Co_{1.7}Ga_{1.5}$.

(1979). Crystal structure was established only for (12) ~ $Sc_{0.9-1.1}Co_{1.3-0.7}Ga_{0.7-1.3}$ ($MgZn_2$ type of crystal structure, $a = 5.021$–5.170, $c = 8.137$–8.175). Later, Belyavina and Markiv (1980) reported, that (6) ~ $ScCoGa_5$ has the $HoCoGa_5$ structure type, with $a = 4.135$, $c = 6.542$, and (8) ~ Sc_2CoGa_8 has Ho_2CoGa_8 structure type, with $a = 4.126$, $c = 10.68$.

Sc–Ni–Ga

The isothermal section at 1070 K is shown in fig. 72 according to Gavrylenko and Markiv (1979). Crystal structures were solved only for (8) ~ $ScNi_2Ga$ ($MnCu_2Al$ type of structure, $a = 6.001$); (11) ~ $Sc_{0.9-1.1}Ni_{1.54-1.24}Ga_{0.46-0.76}$ ($MgZn_2$ ST, $a = 5.054$, $c = 8.002$ for a Ga contents of 0.7), and (12) ~ $Sc_{0.88-1.12}Ni_{1.05-0.88}Ga_{0.95-1.32}$ ($CaIn_2$ ST, $a = 4.184$–4.240, $c = 6.387$–6.584). Belyavina and Markiv (1980) found the $HoCoGa_5$ type of crystal structure for (4) ~ $ScNiGa_5$ and the Ho_2CoGa_8 ST for (5) ~ Sc_2NiGa_8 with $a = 4.148$, $c = 6.521$ and $a = 4.127$, $c = 10.56$, respectively.

Sc–Cu–Ga

A total of 7 new ternary compounds were found by Gavrylenko and Markiv (1979) during a phase diagram (fig. 73) study. They are as follows: (1) ~ $ScCu_{5.5-4.9}Ga_{6.5-7.1}$ ($ThMn_{12}$ ST, $a = 8.62$, $c = 4.78$ for unknown composition); (2) ~ $Sc_4Cu_{16}Ga_9$; (3) ~ $ScCu_{3.6}Ga_{1.4}$; (4) ~ $ScCu_{3.4}Ga_{1.6}$; (5) ~ $Sc_2Cu_5Ga_2$; (6) ~ $ScCu_2Ga$ ($MnCu_2Al$ ST, $a = 6.158$) and (7) ~ $ScCu_{1.4-0.26}Ga_{0.6-1.74}$ ($CaIn_2$ ST, $a = 4.211$, $c = 6.549$ for a Ga contents of 0.6). Crystal structure of the compounds (2)–(5) are undetermined.

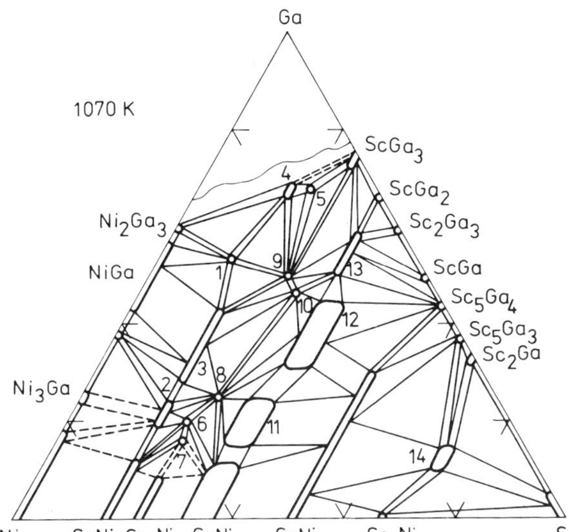

Fig. 72. Sc–Ni–Ga, isothermal section at 1070 K. Compounds with unknown crystal structure: (1) ~ $Sc_2Ni_5Ga_8$; (2) ~ $ScNi_{3.85-3.56}Ga_{1.15-1.44}$; (3) ~ $ScNi_{3.25-2.5}GA_{1.75-2.5}$; (6) ~ $Sc_3Ni_8Ga_3$; (7) ~ $Sc_3Ni_8Ga_2$; (9) ~ $ScNiGa_2$; (10) ~ $Sc_3Ni_3Ga_5$; (13) ~ $ScNi_{0.49-0.26}Ga_{1.51-1.74}$; (14) ~ $Sc_{6.95-7.05}Ni_{2-1.5}Ga_{1-1.5}$.

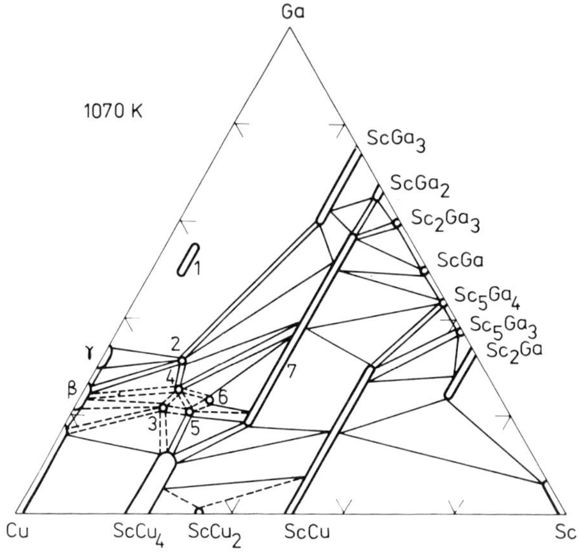

Fig. 73. Sc–Cu–Ga, isothermal section at 1070 K; compound (1) was found at 870 K, at 1070 K the alloys in this region become liquid.

Sc–Zr–Ga

Only one ternary compound was found by Markiv et al. (1979), who studied the isothermal section at 1070 K of the system Sc–Zr–Ga (fig. 74). Approximate composition of this compound is $Sc_{0.2-0.3}Zr_{0.35-0.25}Ga_{0.45}$. The authors made the assumption that it is isostructural with the Nb_5Ga_4 binary compound.

Sc–Nb–Ga

Savitzky et al. (1980) derived the isothermal section at 1070 K (fig. 75) and

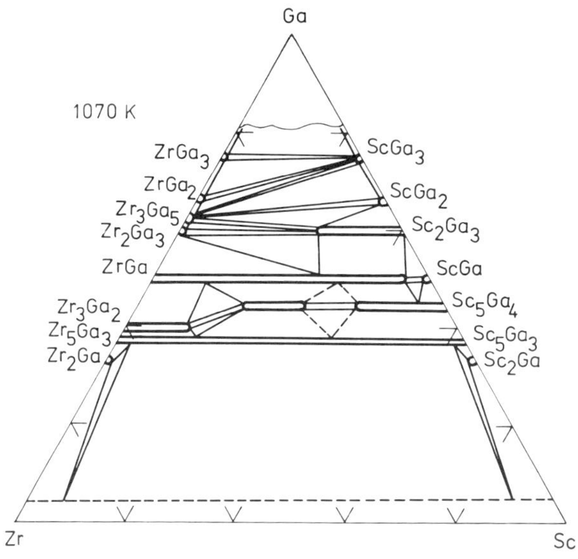

Fig. 74. Sc–Zr–Ga, isothermal section at 1070 K.

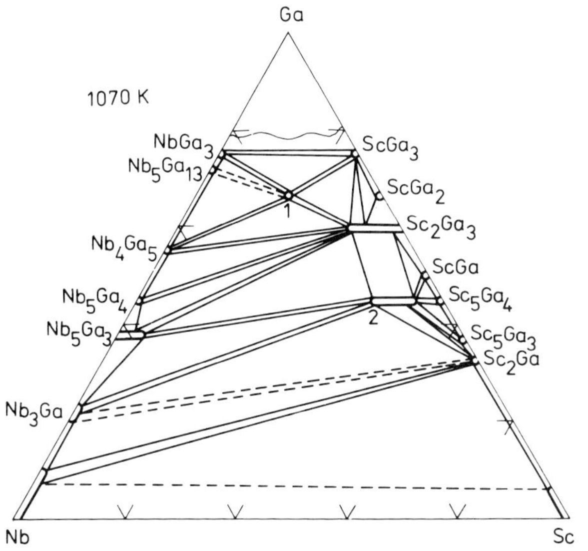

Fig. 75. Sc–Nb–Ga, isothermal section at 1070 K.

reported the existence of two ternary compounds: (1) ~ $ScNbGa_4$ and (2) ~ $Sc_{3.8-4.6}Nb_{1.2-0.4}Ga_4$. The crystal structure for both of them is undetermined.

Sc–Hf–Ga

Two ternary compounds: (1) ~ $Sc_{0.2}Hf_{0.8}Ga_3$ ($ZrAl_3$ ST, $a = 3.948$, $c = 17.35$) and (2) ~ $Sc_{0.2-0.9}Hf_{0.8-0.1}Ga$ (MoB ST, $a = 3.851$, $c = 20.27$ at a Hf content of 0.66) were found by Markiv and Belyavina (1981). They also derived the phase diagram (fig. 76).

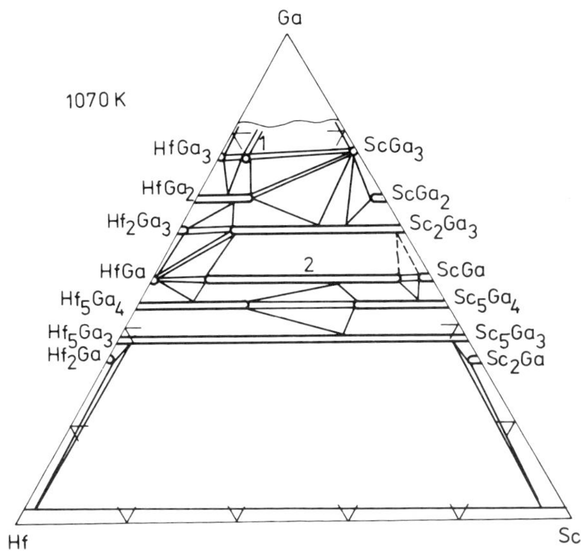

Fig. 76. Sc–Hf–Ga, isothermal section at 1070 K.

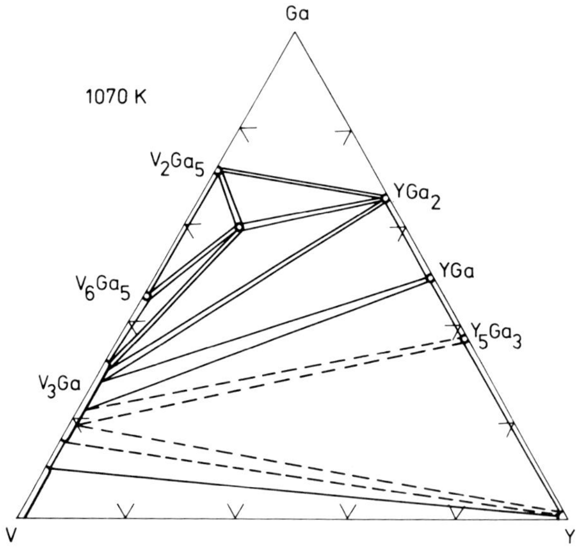

Fig. 77. Y–V–Ga, isothermal section at 1070 K.

Y–V–Ga

The isothermal section was described by Savitzky et al. (1977) and is shown in fig. 77. The only ternary compound, $\sim YV_3Ga_6$, has unknown crystal structure.

Y–Mn–Ga

The isothermal section at 970 K (fig. 78) has been investigated by Markiv et al. (1982b). The total number of compounds found is 6. They are as follows:

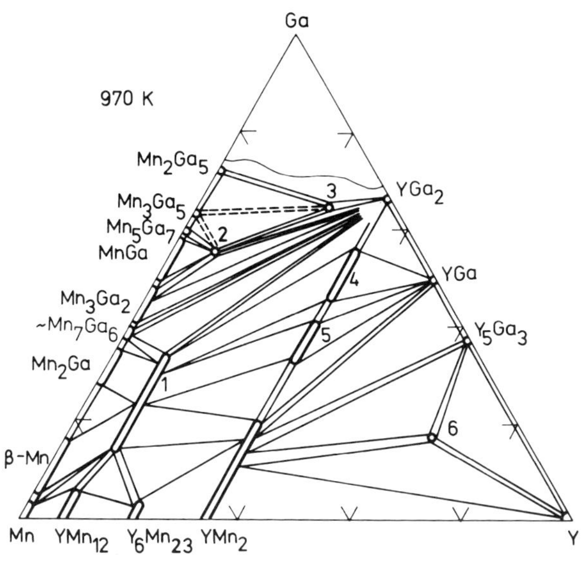

Fig. 78. Y–Mn–Ga, isothermal section at 970 K.

(1) ~ $Y_2(Mn_{0.85-0.63}Ga_{0.15-0.37})_{17}$ with Th_2Zn_{17} structure type, with $a = 8.712$, $c = 13.00$ at a 2 : 1 ratio between Mn and Ga; (2) ~ $Y(Mn_{0.4}Ga_{0.6})_{12}$ with $ThMn_{12}$ ST, with $a = 8.791$, $c = 5.242$; (3) ~ $YMnGa_5$ with $AuCu_3$ ST, $a = 4.296$; (4) ~ $Y(Mn_{0.33-0.16}Ga_{0.67-0.83})_2$ with $CeCu_2$ ST, with $a = 4.391$, $b = 7.090$, $c = 7.764$ at a 2 : 1 : 3 composition; (5) ~ YMnGa with orthorhombic lattice, $a = 4.296$, $b = 12.24$, $c = 7.067$, and (6) ~ Y_4MnGa for which the crystal structure was not determined.

Y–Cu–Ga

A partial phase diagram is shown in fig. 79 according to Markiv et al. (1982a). Ternary compounds are: (1) ~ $Y(Cu_{0.46}Ga_{0.54})_{12}$ with $ThMn_{12}$ ST, with $a = 8.600$, $c = 5.152$; (3) ~ $Y_2(Cu_{0.67}Ga_{0.33})_{17}$ with Th_2Zn_{17} ST, with $a = 8.675$, $c = 12.66$, and (7) ~ $YCu_{0.5}Ga_{1.5}$ with $CaIn_2$ ST, $a = 4.357$, $c = 7.205$. The compositions of the other 4 (a total of 7) ternary compounds are shown in fig. 79. Crystal structures are undetermined.

Y–Nb–Ga

No ternary compounds were found by Savitzky et al. (1978b). The isothermal section at 1070 K is shown in fig. 80.

{La, Ce}–Mg–Ga

These two systems were investigated by Kynghybalo and Grymak (1986). Figure 81 shows the phase diagrams. Crystal structure were undetermined for the ternary compounds ~$R_{10}Mg_3Ga_7$ (R = Ce or La).

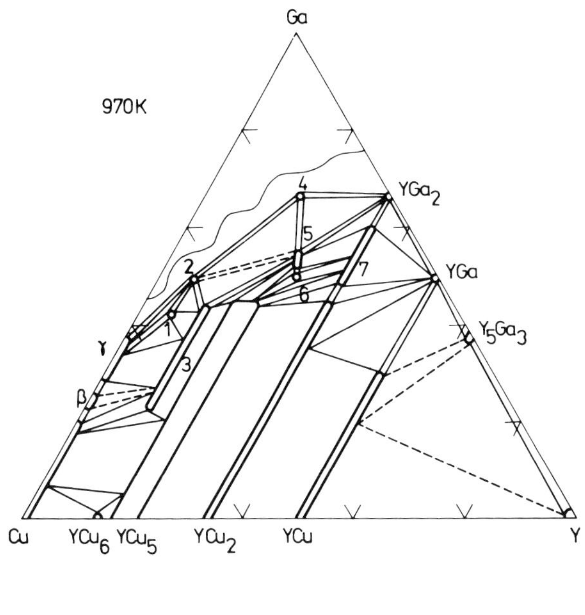

Fig. 79. Y–Cu–Ga, partial isothermal section at 970 K (0–66 a/c Ga).

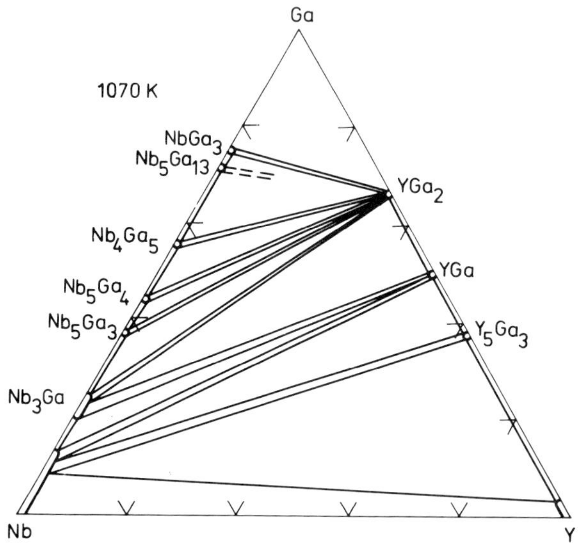

Fig. 80. Y–Nb–Ga, isothermal section at 1070 K.

Ce–Fe–Ga

Lapunova et al. (1983) found two ternary compounds in the system Ce–Fe–Ga at 870 K (fig. 82). They are as follows: (1) ~ $CeFe_2Ga_8$ ($CeFe_2Al_8$ ST, $a = 12.41$, $b = 14.39$, $c = 4.075$) and (2) ~ $CeFe_{5.74-4.7}Ga_{6.26-7.3}$ with $ThMn_{12}$ ST, and $a = 8.74$–8.70, $c = 5.010$–5.086.

Ce–Ni–Ga

Gryn (1980) reported the phase diagram of the system Ce–Ni–Ga (fig. 83) at 870 or 670 K. First, it was found that 19 ternary compounds exist here. According to Gryn (1980) they are as follows: (1) ~ $CeNi_{11-8.8}Ga_{2-4.2}$ ($CeNi_{8.5}Si_{4.5}$ ST, $a = 8.305$, $c = 11.49$ at $x_{Ga} = 4.2$); (5) ~ $CeNi_{5.7}Ga_{7.3}$ ($NaZn_{13}$ ST, $a = 11.742$); (8) ~ $CeNi_{3.0-2.8}Ga_{2.0-2.2}$ ($HoNi_{2.6}Ga_{2.4}$ ST, $a = 8.82$–8.83, $c = 4.19$–4.21); (9) ~ $CeNi_{2.2}Ga_{2.8}$ ($CaCu_5$ ST, $a = 5.263$, $c = 4.053$); (10) ~ $CeNi_{1.1-0.5}Ga_{2.9-3.5}$ ($BaAl_4$ ST, $a = 4.234$–4.297, $c = 10.19$–10.33); (12) ~ $CeNiGa_2$ ($NdNiGa_2$ ST, $a = 4.24$, $b = 17.59$, $c = 4.17$); (13) ~ CeNiGa (ZrNiAl ST, $a = 6.946$, $c = 3.987$), (15) ~ $Ce_2Ni_{2.2-2}Ga_{0.8-1}$ (Mo_2NiB_2 ST, $a = 5.410$, $b = 8.413$, $c = 4.306$ at the 2 : 2 : 1 composition) and (18) ~ $Ce_{26}(Ni_{0.65}Ga_{0.35})_{17}$ with $Sm_{26}(Co_{0.65}GA_{0.35})_{17}$ ST, $a = 12.09$, $c = 15.01$. Yarmolyuk et al. (1982) found that compound (7) ~ Ce_2NiGa_{11} has as true composition Ce_2NiGa_{10} with a new structure type: I4/mmm, $a = 4.262$, $b = 26.391$. Yarmolyuk et al. (1984b) established that $Ce_3Ni_2Ga_2$ belongs to a new structure type: Pbcm, $a = 5.746$, $b = 8.210$, $c = 13.748$. Gryn et al. (1983b) found that $Ce_4Ni_2Ga_{17}$ also has a new type of crystal structure: I4/mmm, $a = 4.283$, $c = 47.26$. Gryn et al. (1984) also reported on the crystal structure of $Ce_3Ni_2Ga_{15}$ (P4/nmm, $a = 4.278$, $c = 39.08$). The last three compounds were not found by Gryn (1980).

The total number of ternary compounds for this system, therefore, becomes 22.

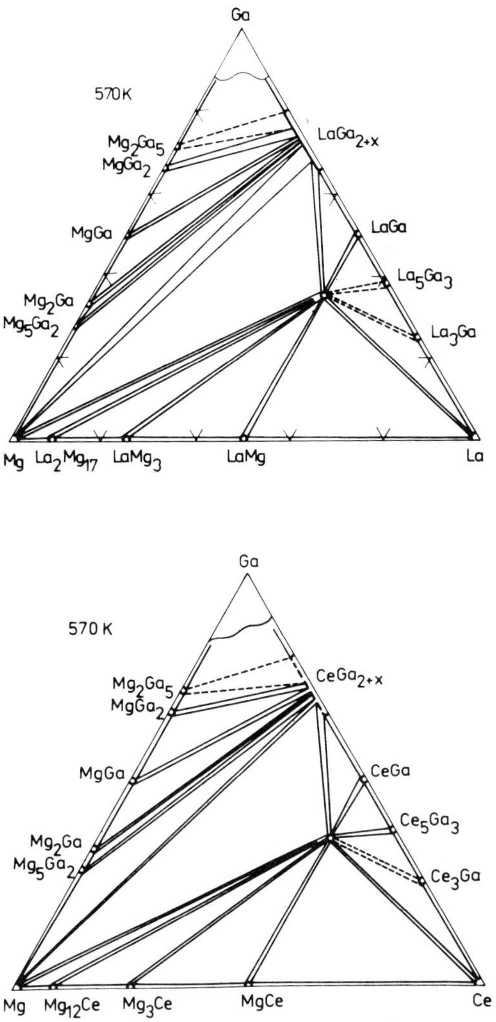

Fig. 81. {La, Ce}–Mg–Ga, isothermal sections at 570 K.

Gryn et al. (1985) found that compound (13) ~ CeNiGa with ZrNiAl ST represents the low-temperature modification. A high-temperature phase transition above 970 K was observed. The crystal structure of it was not determined.

Pr–Fe–Ga

The system Pr–Fe–Ga was investigated at 870 K by Lapunova et al. (1987b). The derived isothermal section is shown in fig. 84. Ternary compounds are: (1) ~ $PrFe_2Ga_8$ ($CeFe_2Al_8$ ST, $a = 12.47$, $b = 14.29$, $c = 4.080$); (2) ~ $PrFe_{6.1-5.6}Ga_{5.9-6.4}$ ($ThMn_{12}$ ST, $a = 8.715$–8.780, $c = 5.098$–5.084) and (3) ~ $Pr_6Fe_{11}Ga_3$ ($La_6Co_{11}Ga_3$ ST, $a = 8.052$, $c = 22.047$).

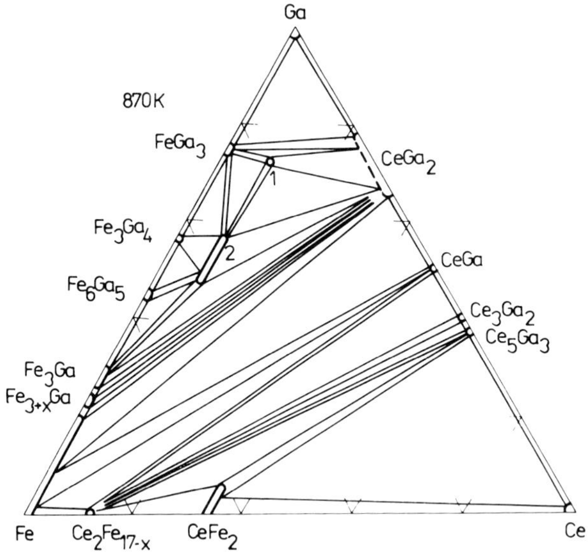

Fig. 82. Ce–Fe–Ga, isothermal section at 870 K.

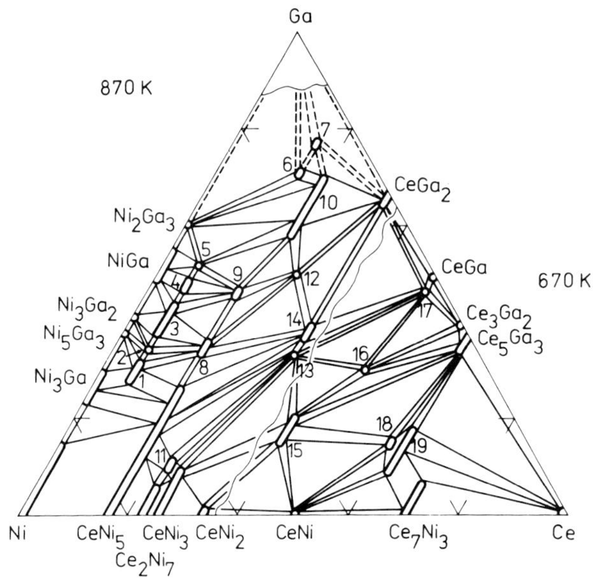

Fig. 83. Ce–Ni–Ga, isothermal section at 870 K (0–33 a/c Ce) and at 670 K (33–100 a/c Ce). Ternary compounds with unknown crystal structure: (2) ~ $CeNi_{8.1}Ga_{4.9}$; (3) ~ $CeNi_{7.7-6.8}Ga_{5.3-6.2}$; (4) ~ $CeNi_{6.5-6.0}Ga_{6.5-7.0}$; (6) ~ $Ce_2Ni_2Ga_9$; (11) ~ Ce_2Ni_6Ga; (14) ~ $CeNi_{0.9-0.8}Ga_{1.1-1.2}$; (16) ~ $Ce_6Ni_3Ga_4$; (17) ~ $CeNi_{0.95}Ga_{0.05}$; (19) ~ $Ce_5Ni_{2-1.9}Ga_{1-1.1}$.

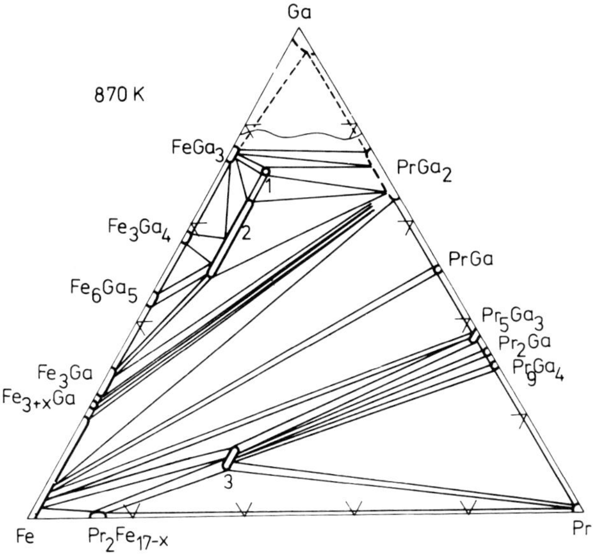

Fig. 84. Pr–Fe–Ga, isothermal section at 870 K.

Pr–Ni–Ga

The phase diagram shown in fig. 85 was derived by Gryn (1980). The existence of a total of 24 new ternary compounds was reported. The additional one, $Pr_3Ni_2Ga_2$, with $La_3Ni_2Ga_2$ ST, $a = 5.654$, $b = 8.110$, $c = 13.151$, was found by Yarmolyuk et al. (1984b). So, the full number of ternary compounds is now 25. Compound (1) ~ $PrNi_{11.2-8.8}Ga_{1.8-4.2}$ belong to the $CeNi_{8.5}Si_{4.5}$ ST with unreported lattice parameters; (5) ~ $PrNi_{5.7}Ga_{7.3}$ belongs to the $NaZn_{13}$ structure type, $a = 11.372$; (8) ~ $PrNi_{3.3-3}Ga_{1.7-2}$ crystallized in the $HoNi_{2.6}Ga_{2.4}$ ST, $a =$ 8.80–8.82, $c = 4.172$–4.152; (9) ~ $PrNi_{2.2}Ga_{2.8}$ in the $CaCu_5$ ST, $a = 5.365$, $c =$ 4.040; (12) ~ $PrNi_{1-0.5}Ga_{3-3.5}$ in the $BaAl_4$ ST, $a = 4.228$–4.275, $c = 10.18$–10.31; (16) ~ $PrNiGa_2$ in the $NdNiGa_2$ ST, $a = 4.20$, $b = 17.59$, $c = 4.15$; (17) ~ $Pr_3Ni_4Ga_2$ in the $La_3Ni_4Ga_2$ ST, $a = 9.870$, $b = 8.132$, $c = 4.212$, $\gamma = 99.2$); (18) ~ $Pr_3Ni_6Ga_2$ has the $Ce_3Ni_6Si_2$ structure type, $a = 9.046$; (19) ~ PrNiGa the TiNiSi ST, $a = 7.404$, $b = 4.518$, $c = 6.914$; (20) ~ $PrNi_{0.6}Ga_{1.4}$ the KHg_2 ST, $a = 7.24$, $b = 4.47$, $c = 7.72$; (21) ~ $Pr_2Ni_{2.7-1.9}Ga_{0.3-1.1}$ the Mo_2NiB_2 ST, $a =$ 5.514, $b = 8.516$, $c = 4.294$ at the 2 : 2 : 1 composition, and (23) ~ $Pr_{26}(Ni_{0.58}Ga_{0.42})_{17}$ the $Sm_{26}(Co_{0.65}Ga_{0.35})_{17}$ ST, $a = 12.06$, $c = 15.23$.

Sm–Fe–Ga

Lapunova et al. (1987b) found three ternary compounds during a study of the isothermal section at 870 K (fig. 86): (1) ~ $SmFe_2Ga_8$ with the $CeFe_2Al_8$ ST, $a = 12.29$, $b = 14.253$, $c = 4.079$; (2) ~ $SmFe_{5.9-4.4}Ga_{6.1-7.6}$ with the $ThMn_{12}$ ST, $a = 8.638$–8.738, $c = 5.089$–5.104, and (3) ~ $Sm_6Fe_{11}Ga_3$ with the $La_6Co_{11}Ga_3$ type of crystal structure, $a = 8.052$, $c = 23.047$.

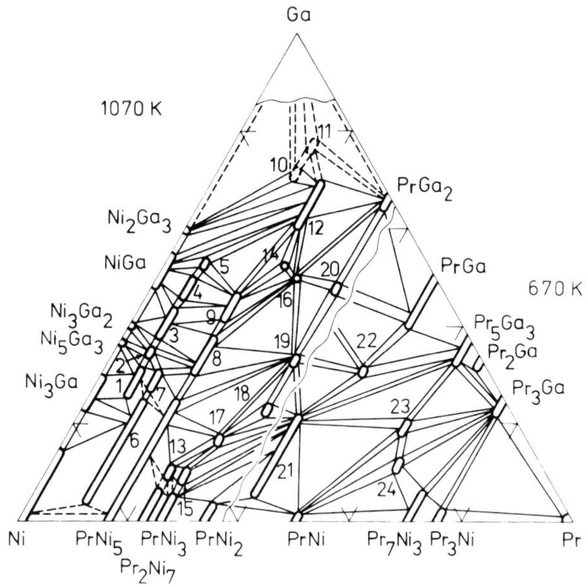

Fig. 85. Pr–Ni–Ga, isothermal section at 1070 K (0–33 a/c Pr) and at 670 K (33–100 a/c Pr). Compounds with unknown crystal structure: (2) ~ $PrNi_{8.1}Ga_{4.9}$; (3) ~ $PrNi_{7.7-6.8}Ga_{5.3-6.2}$; (4) ~ $PrNi_{6.5-6.0}Ga_{6.5-7.0}$; (6) ~ $Pr_2Ni_{16.5-12.4}Ga_{0.5-4.6}$; (7) ~ $Pr_2Ni_{12-11.3}Ga_{5-5.7}$; (10) ~ $Pr_2Ni_2Ga_9$; (11) ~ Pr_2NiGa_{11}; (13) ~ Pr_2Ni_6Ga; (14) ~ $Pr_2Ni_2Ga_5$; (15) ~ Pr_3Ni_8Ga; (22) ~ $Pr_6Ni_3Ga_4$; (24) ~ Pr_5Ni_2Ga.

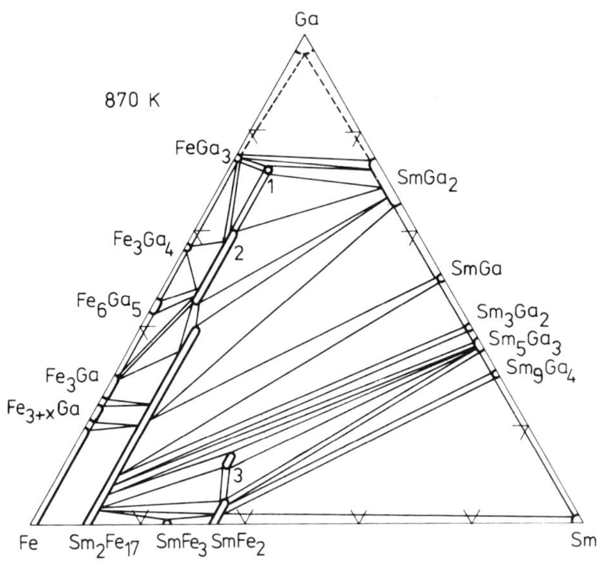

Fig. 86. Sm–Fe–Ga, isothermal section at 870 K.

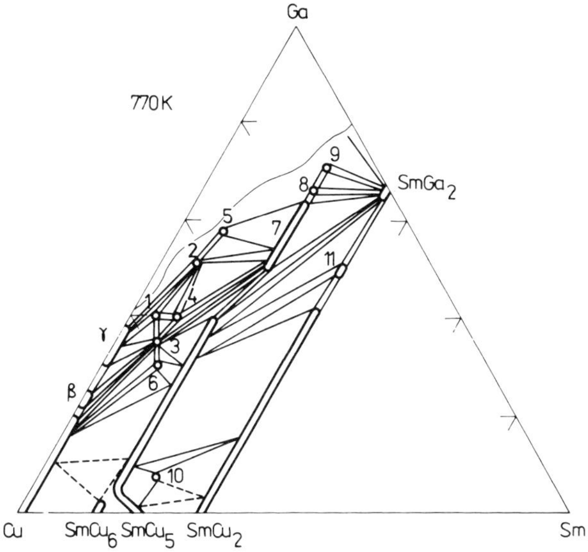

Fig. 87. Sm–Cu–Ga, partial isothermal section at 770 K (0–33 a/c Sm).

Sm–Cu–Ga

The partial isothermal section at 770 K is shown in fig. 87 after Markiv et al. (1986). A total of 11 ternary compounds were found: (1) ~ $SmCu_{7.4}Ga_{5.6}$ ($NaZn_{13}$ ST, $a = 11.74$); (2) ~ $SmCu_{5.5}Ga_{6.5}$ ($ThMn_{12}$ ST, $a = 8.644$, $c = 5.152$); (3) ~ $SmCu_{6.8}Ga_{4.2}$ ($BaCd_{11}$ ST, $a = 10.272$, $c = 6.611$); (4) ~ $SmCu_{6.2}Ga_{4.8}$ (a new structure type which is a monoclinic distortion of the $BaCd_{11}$ ST, $a = 10.331$, $b = 10.327$, $c = 6.591$, γ is not given); (5) ~ $SmCu_{4.1}Ga_{6.9}$ ($SmZn_{11}$ ST, $a = 8.874$, $c = 8.609$); (6) ~ $Sm_2Cu_{11.4}Ga_{5.6}$ (Th_2Zn_{17} ST, $a = 8.726$, $c = 12.74$); (7) ~ $SmCu_{1.4-0.8}Ga_{2.6-3.2}$ ($BaAl_4$ ST, $a = 4.157–4.168$, $c = 10.328–10.552$); (8) and (9) are compounds with approximate compositions $SmCu_{0.7}Ga_{3.3}$ and $SmCu_{0.5-0.3}Ga_{3.5-3.7}$, both of them are orthorhombic with $a = 4.178$, $b = 4.200$, $c = 10.525$ and $a = 4.234–4.239$, $b = 4.234–4.239$, $c = 10.018–9.933$, respectively; (10) ~ $Sm_{14}Cu_{46}Ga_5$ ($Gd_{14}Ag_{51}$ ST, $a = 11.717$, $c = 8.792$); (11) ~ $SmCu_{0.5}Ga_{1.5}$ (AlB_2 ST, $a = 4.386$, $c = 3.755$).

Tb–Fe–Ga

According to Lapunova et al. (1987a), a total of 6 new ternary compounds were found and the phase diagram was established (fig. 88). The compounds are as follows: (1) ~ $TbFe_{5.9-4.8}Ga_{6.1-7.2}$ ($ThMn_{12}$ ST, $a = 8.648–8.673$, $c = 5.077–5.087$); (2) ~ Tb_2FeGa_8 (Ho_2CoGa_8 ST, $a = 4.259$, $c = 11.33$); (3) ~ $TbFe_{0.1}Ga_{2.9}$ ($AuCu_3$ ST, $a = 4.283$); (4) ~ $TbFe_{2.4-2.1}Ga_{0.6-0.9}$ ($CeNi_3$ ST, $a = 5.185–5.195$, $c = 16.55–16.64$); (5) ~ $TbFe_{0.2-0.3}Ga_{1.8-1.7}$ (crystal structure unknown), and (6) ~ $TbFe_{0.5-0.6}Ga_{1.5-1.4}$ (KHg_2 ST, $a = 7.125–7.111$, $b = 4.380–4.389$, $c = 7.620–7.608$).

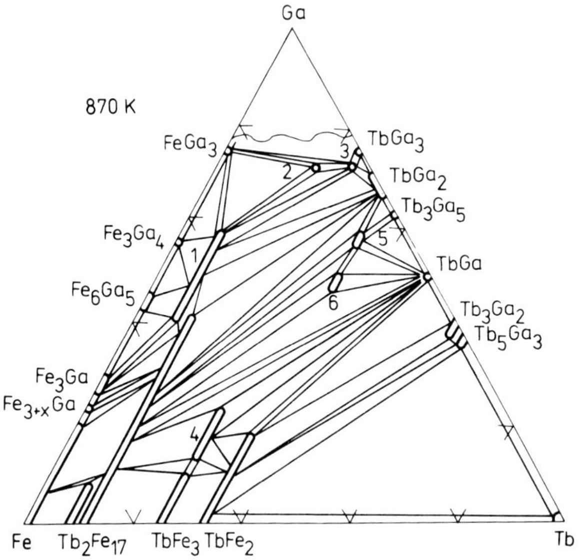

Fig. 88. Tb–Fe–Ga, isothermal section at 870 K.

Ho–Fe–Ga

Lapunova et al. (1987a) reported the phase diagram shown in fig. 89. The total number of ternary compounds found is 8: (1) ~ $HoFe_{6.1-5.1}Ga_{5.9-6.9}$ with the $ThMn_{12}$ ST, $a = 8.570–8.666$, $c = 5.053–5.082$; (2) ~ $Ho_2Fe_{13.6-8.5}Ga_{3.4-8.5}$ with the Th_2Zn_{17} ST, $a = 8.604–8.763$, $c = 12.558–12.590$; (3) ~ Ho_2FeGa_8 belongs to the Ho_2CoGa_8 ST, $a = 4.226$, $c = 11.088$; (4) ~ $HoFe_{0.1-0.2}Ga_{2.9-2.8}$ to the $AuCu_3$

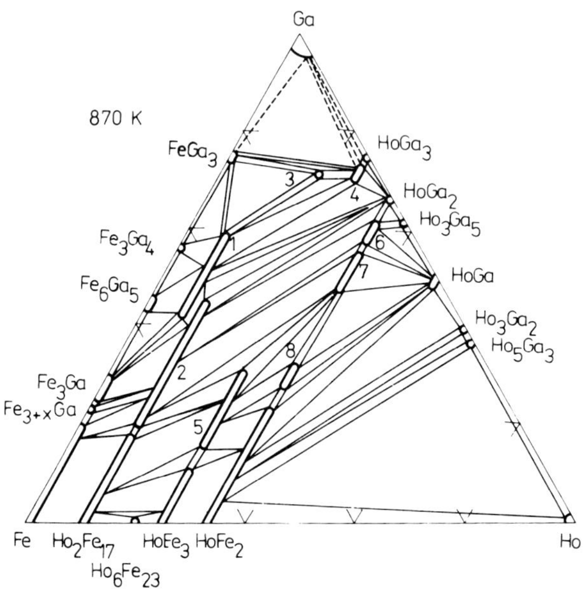

Fig. 89. Ho–Fe–Ga, isothermal section at 870 K.

ST, $a = 4.246–4.260$; (5) ~ $HoFe_{2.36-1.80}Ga_{0.64-1.20}$ to the $CeNi_3$ ST, $a = 5.154–5.148$, $c = 16.640–16.605$; (6) ~ $HoFe_{0.3-0.2}Ga_{1.7-1.8}$ with an undetermined structure; (7) ~ $HoFe_{0.56-0.35}Ga_{1.44-1.65}$ which belongs to the KHg_2 ST, $a = 7.069–7.001$, $b = 4.356–4.353$, $c = 7.601–7.611$, and (8) ~ $HoFe_{1.2-1.0}Ga_{0.8-1.0}$ with $MgZn_2$ type structure, and $a = 5.294–5.278$, $c = 8.645–8.635$.

Ho–Co–Ga

This system was investigated by Gladyshevsky et al. (1985). The phase equilibria are shown in fig. 90. The ternary compounds are as follows: (1) ~ $HoCo_{5.8-5.2}Ga_{6.2-6.8}$ ($ThMn_{12}$ ST, $a = 8.445$, $c = 5.072$ for the 1 : 5.4 : 6.6 composition); (2) ~ $Ho_2Co_{13.1-7.7}Ga_{3.9-9.9}$ (Th_2Zn_{17} ST, $a = 8.460–8.740$, $c = 12.358–12.159$); (3) ~ $Ho_2Co_3Ga_9$ ($Y_2Co_3Ga_9$ ST, $a = 12.721$, $b = 7.344$, $c = 9.353$); (4) ~ $HoCoGa_5$ (a new type of crystal structure: P4/mmm, $a = 4.207$, $c = 6.795$, after Gryn et al. (1979b)); (5) ~ $HoCo_{2.6}Ga_{2.4}$ (unknown crystal structure); (6) ~ $HoCo_{2.9}Ga_{2.1}$ ($HoNi_{2.6}Ga_{2.4}$ ST, $a = 8.801$, $c = 4.023$); (7) ~ $Ho_6Co_7Ga_{21}$ (a new type of structure: P4/mbm, $a = 16.687$, $c = 4.134$, according to Yarmolyuk et al. (1981)); (8) ~ Ho_2CoGa_8 (also a new type of crystal structure, after Gryn et al. (1979b): P4/mmm, $a = 4.217$, $c = 10.97$); (9) ~ $Ho_3Co_3Ga_8$ (La_3Al_{11} ST, $a = 4.186$, $b = 12.477$, $c = 9.228$); (10) ~ $HoCo_{2.4-1.8}Ga_{0.6-1.2}$ ($CeNi_3$ ST, $a = 5.053–5.101$, $c = 16.26–16.27$); (11) ~ $HoCo_{1.2}Ga_{0.8}$ ($MgZn_2$ ST, $a = 5.229$, $c = 8.438$); (12) ~ HoCoGa (TiNiSi ST, $a = 7.002$, $b = 4.361$, $c = 7.003$); (13) ~ $HoCo_{0.7-0.6}Ga_{1.3-1.4}$ (Lu_2CoGa_3 ST, $a = 6.941–6.966$, $b = 8.635–8.737$, $c = 4.479–4.639$); (14) ~ $HoCo_{0.5-0.2}Ga_{1.5-1.8}$ ($EuGe_2$ ST, $a = 4.351–4.327$, $c = 3.590–3.660$); (15) ~ Ho_2Co_2Ga (Mo_2NiB_2 ST, $a = 5.363$, $b = 8.345$, $c = 4.044$); (16) ~ $Ho_{26}Co_{13.7-6.7}Ga_{3.3-10.3}$

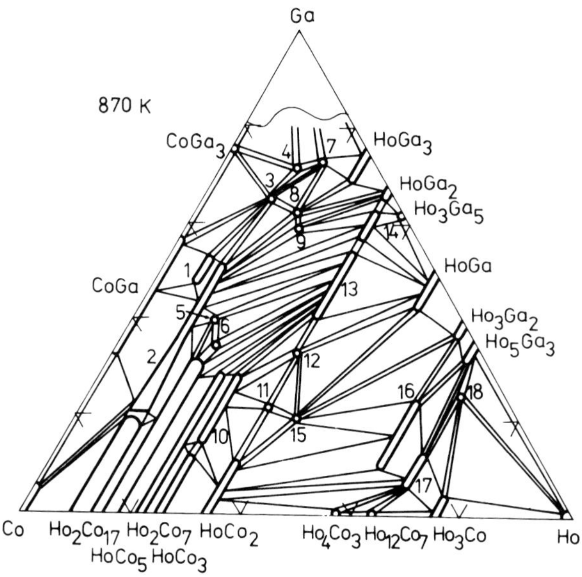

Fig. 90. Ho–Co–Ga, isothermal section at 870 K.

($Sm_{26}(Co_{0.65}Ga_{0.35})_{17}$ ST, $a = 11.38–11.44$, $c = 14.91–15.25$); (17) ~ Ho_6Co_2Ga (a new type of crystal structure, according to Gladyshevsky et al. 1983): Immm, $a = 9.294$, $b = 9.379$, $c = 9.804$), and (18) ~ Ho_8CoGa_3 (Pr_8CoGa_3 ST, $a = 9.969$, $c = 6.841$).

Ho–Ni–Ga

The isothermal section at 1070 K (fig. 91) was derived by Gryn (1980). A total of 24 new ternary compounds were found, for 14 of them their crystal structures were solved. They are as follows: (4) ~ $Ho_2Ni_{9.4-7.5}Ga_{7.6-9.5}$ (Th_2Zn_{17} ST, $a = 8.70–8.74$, $c = 11.98–12.27$); (5) ~ $Ho_4Ni_{10}Ga_{21}$ (a new ST, B2/m, $a = 20.802$, $b = 15.346$, $c = 4.074$, $\gamma = 125.80$); (7) ~ $HoNi_{2.9-2.5}Ga_{2.1-2.5}$ ($HoNi_{2.6}Ga_{2.4}$ ST, $a = 8.666–8.79$, $c = 4.117–4.09$); (9) ~ $HoNiGa_4$ ($YNiAl_4$ ST, $a = 4.053$, $b = 15.16$, $c = 6.545$); (11) ~ $Ho_3Ni_{3.4-1.8}Ga_{7.6-9.2}$ (La_3Al_{11} ST, $a = 4.142$, $b = 12.43$, $c = 9.608$ at $x_{Ga} = 9.2$); (14) ~ $HoNiGa_2$ ($MgCuAl_2$ structure type, $a = 4.119$, $b = 9.97$, $c = 6.596$); (15) ~ $Ho_3Ni_6Ga_2$ ($Ce_3Ni_6Si_2$ ST, $a = 8.849$); (16) ~ HoNiGa (TiNiSi ST, $a = 6.811$, $b = 4.274$, $c = 7.318$); (17) ~ $HoNi_{0.9-0.5}Ga_{1.1-1.5}$ (KHg_2 ST, $a = 6.866–6.990$, $b = 4.281–4.346$, $c = 7.365–7.527$); (18) ~ $HoNi_{0.4-0.3}Ga_{1.6-1.7}$ ($CaIn_2$ ST, $a = 4.301–4.308$, $c = 7.01–7.38$) (19) ~ Ho_2Ni_2Ga (Mo_2NiB_2 ST, $a = 5.321$, $b = 8.309$, $c = 4.152$); (20) ~ $HoNi_{0.8}Ga_{0.2}$ (α-ITl ST, $a = 3.705$, $b = 10.28$, $c = 4.241$); (21) ~ $Ho_6Ni_{2.3-2}Ga_{1.7-2}$ (Y_3Rh_2 ST, $a = 11.40–11.50$, $c = 24.42–42.61$); (22) ~ $Ho_5Ni_{0.4}Ga_{2.6}$ (Cr_5B_3 ST, $a = 7.557$, $c = 13.96$).

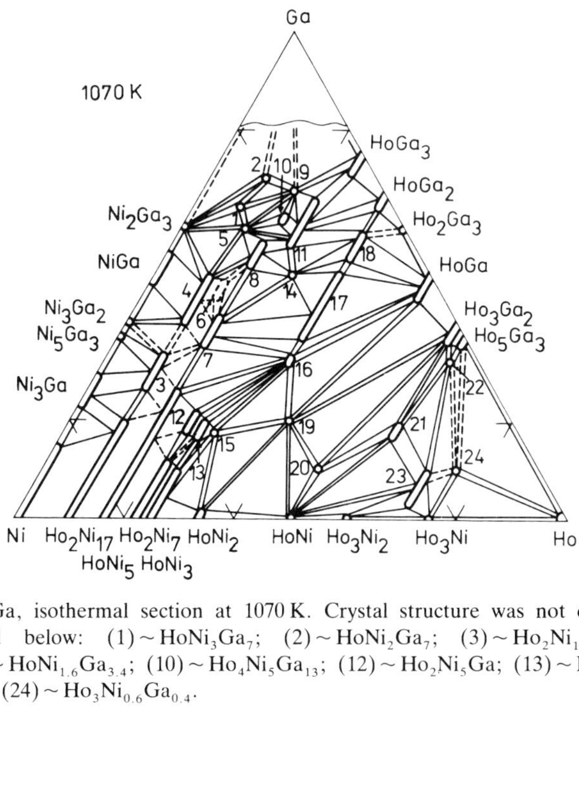

Fig. 91. Ho–Ni–Ga, isothermal section at 1070 K. Crystal structure was not determined for the compounds listed below: (1) ~ $HoNi_3Ga_7$; (2) ~ $HoNi_2Ga_7$; (3) ~ $Ho_2Ni_{12-9.6}Ga_{5-7.4}$; (6) ~ $Ho_3Ni_8Ga_{11}$; (8) ~ $HoNi_{1.6}Ga_{3.4}$; (10) ~ $Ho_4Ni_5Ga_{13}$; (12) ~ Ho_2Ni_5Ga; (13) ~ $Ho_3Ni_7Ga_2$; (23) ~ $Ho_7Ni_{2.8-2}Ga_{0.2-1}$; (24) ~ $Ho_3Ni_{0.6}Ga_{0.4}$.

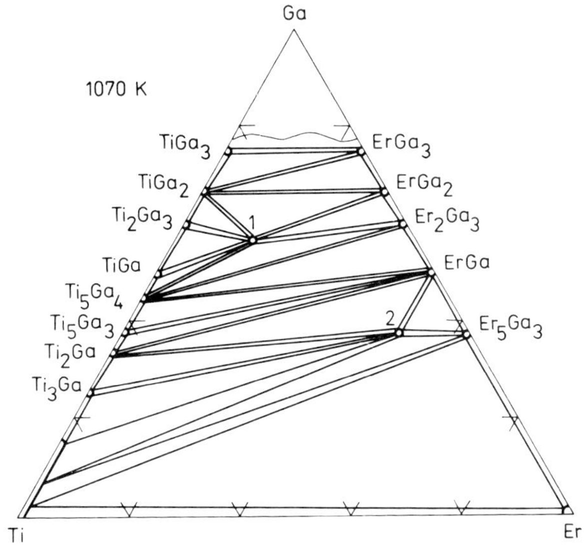

Fig. 92. Er–Ti–Ga, isothermal section at 1070 K.

Er–Ti–Ga

The isothermal section at 1070 K according to Markiv (1981) is shown in fig. 92. The two ternary compounds were found: (1) ~ $ErTi_2Ga_4$ with the $YbMo_2Al_4$ ST, $a = 6.706$, $c = 5.470$, solved by Gryn et al. (1980) and (2) ~ Er_4TiGa_3 with unknown crystal structure.

Er–V–Ga

The only ternary compound ErV_2Ga_4, belonging to the $YbMo_2Al_4$ type of crystal structure, was found by Gryn et al. (1980). The phase diagram reported by Markiv (1981) is shown in fig. 93.

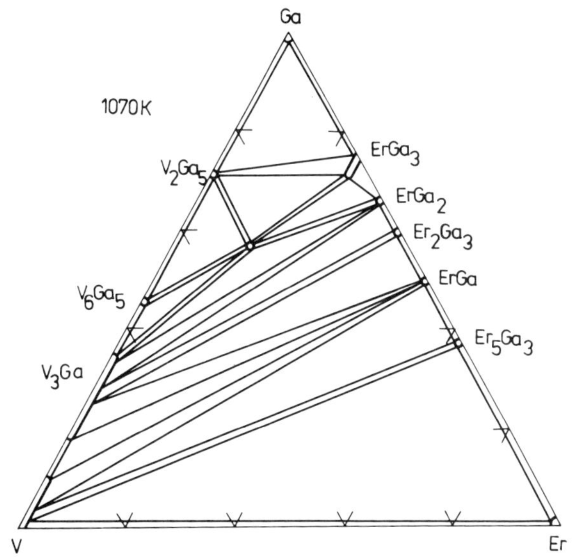

Fig. 93. Er–V–Ga, isothermal section at 1070 K.

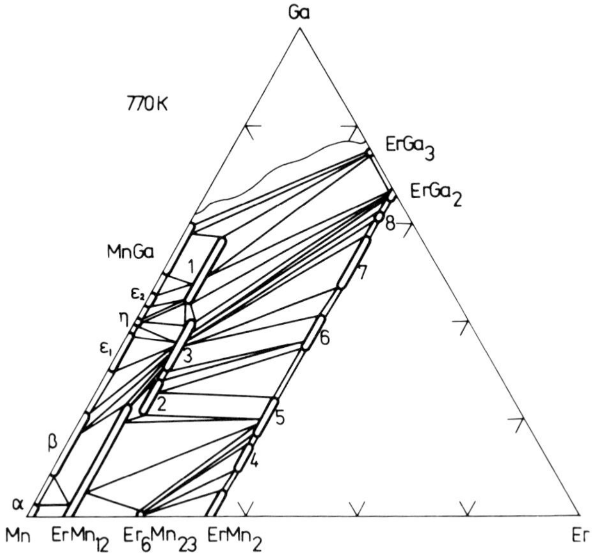

Fig. 94. Er–Mn–Ga, isothermal section at 770 K.

Er–Mn–Ga

The system Er–Mn–Ga (fig. 94) was partially investigated by Markiv et al. (1985). Compositions and crystal structures of the existing compounds are: (1) ~ $ErMn_{6\text{-}4.2}Ga_{6-7.8}$ ($ThMn_{12}$ ST, $a = 8.684$–8.764, $c = 5.200$–5.230); (2) ~ $Er_2Mn_{12.8-11.7}Ga_{4.2-5.3}$ (Th_2Ni_{17} ST, $a = 8.643$–8.719, $c = 8.604$–8.648); (3) ~ $Er_2Mn_{10.4-9.4}Ga_{6.6-7.6}$ (Th_2Zn_{17} ST, $a = 8.787$–8.802, $c = 12.93$–12.95); (4) ~

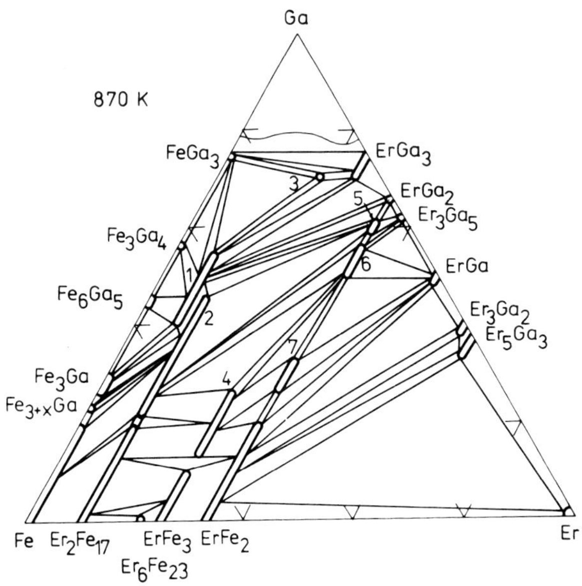

Fig. 95. Er–Fe–Ga, isothermal section at 870 K.

$ErMn_{1.7-1.6}Ga_{0.3-0.4}$ ($MgCu_2$ ST, $a = 7.550-7.545$); (5) ~ $ErMn_{1.5-1.3}Ga_{0.5-0.7}$ ($MgZn_2$ ST, $a = 5.347-5.361$, $c = 8.737-8.741$); (6) ~ $ErMn_{1-0.8}Ga_{1-1.2}$ (TiFeSi ST, $a = 8.508-8.526$, $b = 12.04-12.03$, $c = 6.955-6.943$); (7) ~ $ErMn_{0.5-0.3}Ga_{1.5-1.7}$ ($CeCu_2$ ST, $a = 4.366-4.380$, $b = 7.003-7.029$, $c = 7.694-7.622$), and (8) ~ $ErMn_{0.2}Ga_{1.8}$ ($CaIn_2$ ST, $a = 4.355$, $c = 7.193$).

Er–Fe–Ga

The isothermal section at 870 K (fig. 95) was proposed by Lapunova et al. (1986). The ternary compounds are listed below: (1) ~ $ErFe_{4.6-6.7}Ga_{7.1-5.3}$ belonging to the $ThMn_{12}$ ST, $a = 8.540-8.690$, $c = 5.048-5.065$; (2) ~ $Er_2Fe_{12.7-8.5}Ga_{4.3-8.5}$ to the Th_2Zn_{17} ST, $a = 8.583-8.759$, $c = 12.609-12.75$; (3) ~ Er_2FeGa_8 to the Ho_2CoGa_8 ST, $a = 4.221$, $c = 11.05$; (4) ~ $ErFe_{2-2.4}Ga_{1-0.6}$ to the $CeNi_3$ ST, $a = 5.133-5.157$, $c = 16.52-16.62$; for (5) ~ $ErFe_{0.2-0.3}Ga_{1.8-1.7}$ the crystal structure was not determined; (6) ~ $ErFe_{0.3-0.5}Ga_{1.7-1.5}$ belongs to the KHg_2 ST, $a = 6.975-6.990$, $b = 4.343-4.346$, $c = 7.596-7.598$, and (7) ~ $ErFe_{1-1.2}Ga_{1-0.8}$ to the $MgZn_2$ ST, $a = 5.258-5.296$, $c = 8.629-8.637$.

TABLE 2
Ternary compounds found in the partially investigated systems R–M–Ga.

Compound	Structure type	Lattice parameters			Refs.
		a	*b*	*c*	
YFe_xGa_{12-x} ($x = 6.1-4.8$)	$ThMn_{12}$	8.632–8.669	–	5.083–5.093	GrGLGY, 83
YCo_xGa_{3-x} ($x = 2.5-1.8$)	$CeNi_3$	5.131–5.163	–	16.31–16.32	Gl, 87
Y_4Co_4Ga	Y_4Co_4Ga	10.233 ($\gamma = 109.15°$)	8.811	4.115	Gl, 87
$Y_2Co_3Ga_9$	$Y_2Co_3Ga_9$	12.718	7.383	9.431	Gl, 87
Y_3Co_3Ga	W_3CoB_3	4.098	10.11	13.01	YaGG, 78b
Y_2Co_2Ga	Mo_2NiB_2	5.382	8.457	4.105	GlGY, 81
$Y_{10}Co_7Ga_3$	$Y_{10}Co_7Ga_3$	23.558	3.924	9.610	GrYG, 79a
Y_2CoGa_8	Ho_2CoGa_8	4.226	–	11.04	GrYG, 79b
$YCoGa_5$	$HoCoGa_5$	4.218	–	6.805	GrYG, 79b
Y_6Co_2Ga	Ho_6Co_2Ga	9.430	9.498	9.917	GlGY, 83
YCoGa	TiNiSi	7.042	4.367	7.078	SiGGY, 82
Y_5Co_5Ga	Y_5Co_5Ga	4.123	10.289	20.539	YaGBG, 84a
$Y_6Co_7Ga_{21}$	$Ho_6Co_7Ga_{21}$	. . .	–	. . .	YaGG, 81
YCo_xGa_{5-x} ($x = 2.9-2.6$)	$HoNi_{2.6}Ga_{2.4}$	8.798–8.817	–	4.055–4.060	SiGGY, 86a
Y_2Ni_2Ga	Mo_2NiB_2	5.381	8.329	4.177	GrY, 79
YNiGa	TiNiSi	6.891	4.297	7.307	YaGG, 79
$YNiGa_2$	$MgCuAl_2$	4.131	10.05	6.620	YaG, 81
$Y_4Ni_{10}Ga_{21}$	$Ho_4Ni_{10}Ga_{21}$	20.80	15.37	4.092	GrYG, 79c
$YNiGa_4$	$YNiAl_4$	4.075	15.21	6.558	RoGY, 83
$Y_3Ni_{3.3}Ga_{7.7}$	La_3Al_{11}	4.13	12.30	9.52	Gr, 82
$YPdGa_2$	$MgCuAl_2$	4.387	10.087	6.603	Gr, 84
$LaFe_2Ga_8$	$CeFe_2Al_8$	12.546	14.365	4.100	SiLGY, 85

TABLE 2 (cont'd)

Compound	Structure type	Lattice parameters a	b	c	Refs.
$LaCo_2Ga$	$PrCo_2Ga$	5.090	4.071	6.978	YaK, 76
$La_{26}(Co_xGa_{1-x})_{17}$ ($x = 0.6$)	$Sm_{26}(Co_xGa_{1-x})_{17}$ ($x = 0.65$)	12.28	–	15.38	YaGO, 80
$La_{10}Co_7Ga_3$	$Y_{10}Co_7Ga_3$	24.50	4.197	10.05	GrYG, 79a
$LaCo_3Ga_2$	$HoNi_{2.6}Ga_{2.4}$	8.892	–	4.176	FrGM, 85
$LaNi_{10.8}Ga_{2.2}$	$NaZn_{13}$	11.381	–	–	Gr, 80
$LaNi_xGa_{5-x}$ ($x = 3.2–2.9$)	$HoNi_{2.6}Ga_{2.4}$	8.81–8.91	–	4.20–4.22	GrYP, 83
$La_3Ni_2Ga_2$	$La_3Ni_2Ga_2$	5.746	8.210	13.478	YaGGF, 84b
$LaNiGa_2$	$NdNiGa_2$	4.29	17.83	4.273	GrY, 82
$LaNi_xGa_{4-x}$ ($x = 1.2–0.4$)	$BaAl_4$	4.267–4.371	–	10.20–10.40	Gr, 82
$La_2Ni_{2.3}Ga_{0.7}$	Mo_2NiB_2	5.445	8.671	4.366	RoGY, 82
LaNiGa	EuNiGe	6.679 ($\gamma = 106.35°$)	6.097	6.341	YaGVB, 86
$La_3Ni_4Ga_2$	$La_3Ni_4Ga_2$	10.273 ($\gamma = 99.32°$)	8.357	4.225	GrY, 80
La_2NiGa_{10}	Ce_2NiGa_{10}	4.292	–	26.41	YaGRUKBG, 82
$LaPt_xGa_{4-x}$ ($x = 0.25–1$)	$BaAl_4$	4.406–4.429	–	10.330–10.497	GrRh, 88
$LaAu_xGa_{4-x}$ ($x = 0.25–1$)	$BaAl_4$	4.406–4.379	–	10.231–10.632	GrRHWN, 87
$LaAu_{1.5}Ga_{2.5}$	$CaBe_2Ge_2$	4.379	–	10.632	GrRHWN, 87
CeCoGa	CeCoAl	10.987 ($\beta = 104.61°$)	4.367	4.840	Si, 85
Ce_8CoGa_3	Pr_8CoGa_3	10.663	–	6.913	Si, 83
$CeCo_xGa_{5-x}$ ($x = 3.1–2.7$)	$HoNi_{2.6}Ga_{2.4}$	8.741–8.731	–	4.158–4.155	SiGGY, 86a
$CeCo_2Ga_8$	$CeFe_2Al_8$	12.400	14.301	4.059	SiLGY, 85b
$Ce_{26}(Co_xGa_{1-x})_{17}$ ($x = 0.65$)	$Sm_{26}(Co_xGa_{1-x})_{17}$ ($x = 0.65$)	11.99	–	14.44	YaGO, 80
$CePd_xGa_{4-x}$ ($x = 0.2–0.9$)	$BaAl_4$	4.343–4.301	–	10.223–10.524	GrHR, 85
$CePt_xGa_{4-x}$ ($x = 0.25–0.9$)	$BaAl_4$	4.348–4.312	–	10.138–10.518	GrRH, 88
$CePt_{0.95}Ga_{3.05}$	$CePt_{0.95}Ga_{3.05}$	6.100	6.113	10.512	GrRH, 88
$CeAu_xGa_{4-x}$ ($x = 0.25–1.25$)	$BaAl_4$	4.364–4.330	–	10.134–10.726	GrRHWN, 87
$CeAu_{1.5}Ga_{2.5}$	$CaBe_2Ge_2$	4.349	–	10.671	GrRHWN, 87
$PrCoGa_4$	$YNiAl_4$	4.156	15.862	6.504	Si, 85
Pr_8CoGa_3	Pr_8CoGa_3	10.489	–	6.910	Si, 83
$PrCo_xGa_{5-x}$ ($x = 3–2.7$)	$HoNi_{2.6}Ga_{2.4}$	8.885–8.876	–	4.114–4.111	SiGGY, 86
$PrCo_2Ga$	$PrCo_2Ga$	5.021	4.043	6.860	YaK, 76
Pr_3Co_3Ga	W_3CoB_3	4.221	10.27	13.34	YaGG, 78b
$PrCo_2Ga_8$	$CeFe_2Al_8$	12.392	14.299	4.055	SiLGY, 85b
PrCoGa	TiNiSi	7.405	4.598	6.757	SiGGY, 82
$Pr_{26}(Co_xGa_{1-x})_{17}$ ($x = 0.65$)	$Sm_{26}(Co_xGa_{1-x})_{17}$ ($x = 0.65$)	11.87	–	15.22	YaGO, 80

TABLE 2 (cont'd)

Compound	Structure type	Lattice parameters a	b	c	Refs.
$PrPt_xGa_{4-x}$ ($x = 0.25$–0.9)	$BaAl_4$	4.322–4.296	–	10.107–10.527	GrRH, 88
$PrPt_{0.95}Ga_{3.05}$	$CePt_{0.95}Ga_{3.05}$	...	...	...	GrRH, 88
$PrAu_xGa_{4-x}$ ($x = 0.3$–1.25)	$BaAl_4$	4.334–4.309	–	10.096–10.739	GrRHWN, 87
$PrAu_{1.5}Ga_{2.5}$	$CaBe_2Ge_2$	4.324	–	10.682	GrRHWN, 87
$NdFe_xGa_{12-x}$ ($x = 5.5$–4.7)	$ThMn_{12}$	8.741–8.750	–	5.109–5.103	GrGLGY, 83b
$NdFe_2Ga_8$	$CeFe_2Al_8$	12.422	14.27	4.079	SiLGY, 85b
$Nd_6Fe_{11}Ga_3$	$La_6Co_{11}Ga_3$	8.043	–	23.023	SiLSGY, 85a
Nd_8CoGa_3	Pr_8CoGa_3	...	...	...	Si, 83
$NdCo_xGa_{5-x}$ ($x = 3$–2.7)	$HoNi_{2.6}Ga_{2.4}$	8.862–8.868	–	4.105–4.113	SiGGY, 86a
Nd_3Co_3Ga	W_3CoB_3	4.208	10.24	13.24	YaGG, 78b
NdCoGa	TiNiSi	7.358	4.334	6.803	SiGGY, 82
$Nd_{26}(Co_xGa_{1-x})_{17}$ ($x = 0.60$)	$Sm_{26}(Co_xGa_{1-x})_{17}$ ($x = 0.65$)	11.88	–	15.24	YaGO, 80
$NdCo_3Ga_9$	$NdCo_3Ga_9$	4.119	12.086	17.146	MoSSGY, 83
Nd_2Co_2Ga	Mo_2NiB_2	5.484	8.586	4.159	Si, 85
$Nd_2Co_3Ga_9$	$Y_2Co_3Ga_9$	12.896	7.444	9.543	Si, 85
$Nd_3Ni_6Ga_2$	$Ce_3Ni_6Ga_2$	8.978	–	–	YaGG, 78a
$NdNi_xGa_{5-x}$ ($x = 3.2$–2.8)	$HoNi_{2.6}Ga_{2.4}$	8.83–8.83	–	4.15–4.12	GrYP, 83
$Nd_{26}(Ni_xGa_{1-x})_{17}$ ($x = 0.65$)	$Sm_{26}(Co_xGa_{1-x})_{17}$ ($x = 0.65$)	12.03	–	15.21	YaGO, 80
$Nd_4Ni_2Ga_{17}$	$Ce_4Ni_2Ga_{17}$	4.261	–	45.55	GrYUKB, 83b
$NdNi_{0.6}Ga_{2.4}$	KHg_2	7.18	4.44	7.64	GrY, 83
$NdNiGa_4$	$YNiAl_4$	4.150	15.53	6.563	RoGY, 83
$NdNiGa_2$	$NdNiGa_2$	4.192	17.564	4.133	GrY, 82
$NdNi_xGa_{4-x}$ ($x = 1.1$–0.6)	$BaAl_4$	4.203–4.213	–	10.14–10.12	Gr, 82
Nd_2Ni_2Ga	Mo_2NiB_2	5.483	8.463	4.265	RoGY, 82
NdNiGa	TiNiSi	7.043	4.399	7.382	YaGG, 79
$NdPt_xGa_{4-x}$ ($x = 0.25$–0.9)	$BaAl_4$	4.303–4.269	–	10.102–10.511	GrRH, 88
$NdPt_{0.95}Ga_{3.05}$	...	...	...	...	GrRH, 88
$NdAu_xGa_{4-x}$ ($x = 0.5$–1.25)	$BaAl_4$	4.324–4.287	–	10.031–10.748	GrRHWN, 87
$NdAu_{1.5}Ga_{2.5}$	$CaBe_2Ge_2$	4.303	–	10.705	GrRHWN, 87
Sm_8CoGa_3	Pr_8CoGa_3	...	–	...	Si, 83
$SmCo_xGa_{5-x}$ ($x = 2.8$–2.5)	$HoNi_{2.6}Ga_{2.4}$	8.876–8.885	–	4.089–4.087	SiGGY, 86
Sm_3Co_3Ga	W_3CoB_3	4.183	10.19	13.19	YaGG, 78b
SmCoGa	TiNiSi	7.253	4.493	6.868	SiGGY, 82
$SmCoGa_2$	$NdNiGa_2$	4.132	17.862	4.048	SiGY, 83
$Sm_{26}(Co_xGa_{1-x})_{17}$ ($x = 0.65$)	$Sm_{26}(Co_xGa_{1-x})_{17}$ ($x = (0.65)$	11.713	–	15.171	YaGO, 80
$Sm_6Co_7Ga_{21}$	$Ho_6Co_7Ga_{21}$	16.838	–	4.208	YaGG, 81
$SmCo_3Ga_9$	$NdCo_3Ga_9$	4.116	12.079	17.111	MoSSGY, 83

TABLE 2 (cont'd)

Compound	Structure type	Lattice parameters			Refs.
		a	*b*	*c*	
Sm_2CoGa_8	Ho_2CoGa_8	4.264	–	11.14	GrYG, 79b
$Sm_2Co_3Ga_9$	$Y_2Co_3Ga_9$	12.819	7.401	9.491	Si, 85
Sm_6Co_2Ga	Ho_6Co_2Ga	9.420	9.636	10.031	Si, 85
Sm_2Co_2Ga	Mo_2NiB_2	5.462	8.552	4.151	Si, 85
$Sm_3Ni_6Ga_2$	$Ce_3Ni_6Si_2$	8.978	–	–	YaGG, 78a
$SmNi_xGa_{5-x}$	$HoNi_{2.6}Ga_{2.4}$	8.75–	–	4.14–	GrYP, 83
$(x=3.1–2.9)$		8.79	–	4.13	
$Sm_{26}(Ni_xGa_{1-x})_{17}$	$Sm_{26}(Co_xGa_{1-x})_{17}$	11.96	–	15.17	YaGO, 80
$(x=0.65)$	$(x=0.65)$				
$SmNiGa_4$	$YNiAl_4$	4.118	15.40	6.605	RoGY, 83
$SmNiGa_2$	$NdNiGa_2$	4.156	17.53	4.093	GrY, 82
$SmNi_xGa_{2-x}$	KHg_2	7.21–	4.41–	7.62–	GrY, 83
$(x=0.51–0.72)$		7.07	4.35	7.46	
$SmNi_xGa_{2-x}$	$CaIn_2$	4.341–	–	7.57–	GrY, 83
$(x=0.25–0.37)$		4.385		7.380	
$SmNi_xGa_{4-x}$	$BaAl_4$	4.180–	–	10.10–	Gr, 82
$(x=1–0.7)$		4.201		10.11	
Sm_2Ni_2Ga	Mo_2NiB_2	5.411	8.396	7.384	RoGY, 82
SmNiGa	TiNiSi	6.958	4.335	7.384	YaGG, 79
$SmPt_xGa_{4-x}$	$BaAl_4$	4.256–	–	9.965–	GrRH, 88
$(x=0.5–1.1)$		4.227		10.555	
$SmAu_xGa_{4-x}$	$BaAl_4$	4.280–	–	10.009–	GrRHWN, 87
$(x=0.5–1.25)$		4.247		10.818	
$SmAu_{1.5}Ga_{2.5}$	$CaBe_2Ge_2$	4.255	–	10.791	GrRHWN, 87
$EuCo_2Ga_3$	$BaZn_5$	10.126	7.691	5.191	SiKVGY, 86
$EuCo_2Ga_8$	$CeFe_2Al_8$	12.378	14.41	4.047	SiLGY, 85b
$Eu_3Ni_4Ga_4$	$Na_3Pt_4Ge_4$	7.501	–	–	GrR, 84
$GdFe_xGa_{12-x}$	$ThMn_{12}$	8.640–	–	5.074–	GrGLGY, 83b
$(x=6.1–4.2)$		8.720		5.083	
Gd_2Co_2Ga	Mo_2NiB_2	5.477	8.483	4.143	Si, 85
$GdCo_{2.2}Ga_{0.8}$	$CeNi_3$	5.145	–	16.388	Si, 85
$Gd_2Co_{2.9}Ga_{0.1}$	La_2NI_3	5.315	9.613	7.169	Si, 85
Gd_6Co_2Ga	Ho_6Co_2Ga	9.400	9.552	9.972	Si, 85
$Gd_2Co_3Ga_9$	$Y_2Co_3Ga_9$	12.777	7.377	9.460	Si, 85
$GdCo_xGa_{5-x}$	$HoNi_{2.6}Ga_{2.4}$	8.844–	–	4.083–	SiGGY, 86b
$(x=2.9–2.6)$		8.852		4.071	
Gd_3Co_3Ga	W_3CoB_3	4.119	10.08	13.14	YaGG, 78b
GdCoGa	TiNiSi	7.152	4.433	6.982	SiGGY, 82
$GdCoGa_2$	$NdNiGa_2$	4.118	17.876	4.001	SiGY, 83
$Gd_6Co_7Ga_{21}$	$Ho_6Co_7Ga_{21}$	16.793	–	4.163	YaGG, 81
Gd_2CoGa_8	Ho_2CoGa_8	4.259	–	11.11	GrYG, 79b
$GdCoGa_5$	$HoCoGa_5$	4.220	–	6.816	GrYG, 79b
$Gd_3Ni_6Ga_2$	$Ce_3Ni_6Si_2$	8.921	–	–	YaGG, 78a
$GdNi_xGa_{5-x}$	$HoNi_{2.6}Ga_{2.4}$	8.73–	–	4.13–	GrYP, 83a
$(x=3.2–2.8)$		8.77		4.13	
$GdNi_xGa_{2-x}$	$CaIn_2$	4.357–	–	7.43–	GrY, 83
$(x=0.28–0.37)$		4.361		7.32	
$GdNi_xGa_{2-x}$	KHg_2	7.19–	4.38–	7.57–	GrY, 83
$(x=0.52–0.75)$		7.05	4.34	7.43	

TABLE 2 (cont'd)

Compound	Structure type	Lattice parameters *a*	*b*	*c*	Refs.
$GdNiGa_4$	$YNiAl_4$	4.098	15.36	6.591	RoGY, 83
$GdNiGa_2$	$NdNiGa_2$	4.12	17.54	4.082	GrY, 82
$GdNi_xGa_{4-x}$ ($x = 1.1–0.6$)	$BaAl_4$	4.165–4.190	–	9.940–9.938	Gr, 82
Gd_2Ni_2Ga	Mo_2NiB_2	5.407	8.383	4.183	RoGY, 82
GdNiGa	TiNiSi	6.932	4.320	7.335	YaGG, 79
$GdNI_{0.8}Ga_{0.2}$	α-ITl	3.78	10.39	4.297	GrY, 81
$TbTi_2Ga_4$	$YbMo_2Al_4$	6.704	–	5.478	GrGMY, 80
$TbCo_xGa_{3-x}$ ($x = 2.4–1.8$)	$CeNi_3$	5.094–5.161	–	16.33–16.34	Gl, 87
Tb_2Co_2Ga	Mo_2NiB_2	5.410	8.402	4.091	Gl, 87
$Tb_2Co_3Ga_9$	$Y_2Co_3Ga_9$	12.762	7.369	9.434	Gl, 87
$TbCoGa_3$	Pr_8CoGa_3	10.065	–	6.888	Gl, 87
$TbCo_xGa_{5-x}$ ($x = 3.2–2.8$)	$HoNi_{2.6}Ga_{2.4}$	8.732–8.828	–	4.048–4.064	SiGGY, 86
Tb_3Co_3Ga	W_3CoB_3	4.109	10.07	13.07	YaGG, 78b
Tb_6Co_2Ga	Ho_6Co_2Ga	9.349	9.480	9.878	GlGY, 83
TbCoGa	TiNiSi	7.071	4.422	6.946	SiGGY, 82
$Tb_6Co_7Ga_{21}$	$Ho_6Co_7Ga_{21}$	16.734	–	4.154	YaGG, 81
Tb_2CoGa_8	Ho_2CoGa_8	4.230	–	11.01	GrYG, 79b
$TbCoGa_5$	$HoCoGa_5$	4.217	–	6.809	GrYG, 79b
$Tb_3Ni_6Ga_2$	$Ce_3Ni_6Si_2$	8.894	–	–	YaGG, 78a
$Tb_4Ni_{10}Ga_{21}$	$Ho_4Ni_{10}Ga_{21}$	20.84	15.44	4.094	GrYG, 79c
$TbNi_xGa_{5-x}$ ($x = 3.1–2.7$)	$HoNI_{2.6}Ga_{2.4}$	8.70–8.81	–	4.13–4.10	GrYP, 83a
$TbNi_xGa_{2-x}$ ($x = 0.2–0.37$)	$CaIn_2$	4.339–4.357	–	7.418–7.235	GrY, 83
$TbNi_xGa_{2-x}$ ($x = 0.6–0.75$)	KHg_2	7.094–7.038	4.351–4.328	7.53–7.48	GrY, 83
$TbNiGa_4$	$YNiAl_4$	4.075	15.29	6.506	RoGY, 83
$TbNi_xGa_{4-x}$ ($x = 1–0.6$)	$BaAl_4$	4.149–4.204	–	9.874–9.779	Gr, 82
Tb_2Ni_2Ga	Mo_2NiB_2	5.369	8.338	4.175	RoGY, 82
TbNiGa	TiNiSi	6.858	4.297	7.348	YaGG, 79
$TbNi_{0.75}Ga_{0.25}$	α-ITl	3.74	10.33	4.27	GrY, 81
$TbPdGa_2$	$MgCuAl_2$	4.422	10.115	6.609	Gr, 84
$DyTi_2Ga_4$	$YbMo_2Al_4$	6.717	–	5.482	GrGMY, 80
$DyFe_xGa_{12-x}$ ($x = 6.4–4.9$)	$ThMn_{12}$	8.591–8.659	–	5.056–5.091	GrGLGY, 83b
Dy_2FeGa_8	Ho_2CoGa_8	4.229	–	11.097	LaGY, 84
$DyCo_xGa_{5-x}$ ($x = 2.9–2.6$)	$HoNi_{2.6}Ga_{2.4}$	8.799–8.811	–	4.054–4.055	SiGGY, 86
Dy_6Co_2Ga	Ho_6Co_2Ga	9.272	9.478	9.843	GlGY, 83
DyCoGa	TiNiSi	7.037	4.398	6.981	SiGGY, 82a
$Dy_6Co_7Ga_{21}$	$Ho_6Co_7Ga_{21}$	16.710	–	4.144	YaGG, 81
Dy_2CoGa_8	Ho_2CoGa_8	4.222	–	10.99	GrYG, 79b
$DyCoGa_5$	$HoCoGa_5$	4.216	–	6.802	GrYG, 79b
Dy_8CoGa_3	Pr_8CoGa_3	10.014	–	6.871	Gl, 87
$DyCo_{5.5}Ga_{6.5}$	$ThMn_{12}$	8.434	–	5.072	Gl, 87

TABLE 2 (cont'd)

Compound	Structure type	Lattice parameters *a*	*b*	*c*	Refs.
$Dy_2Co_3Ga_9$	$Y_2Co_3Ga_9$	12.374	7.358	9.392	Gl, 87
$DyCo_xGa_{3-x}$ ($x = 2.5$–1.8)	$CeNi_3$	5.069–5.080	–	16.33–16.34	Gl, 87
$DyCo_{1.2}Ga_{0.8}$	$MgZn_2$	5.227	–	8.463	Gl, 87
Dy_2Co_2Ga	Mo_2NiB_2	5.363	8.389	4.069	Gl, 87
$Dy_3Ni_6Ga_2$	$Ce_3Ni_6Si_2$	8.862	–	–	YaGG, 78a
$Dy_4Ni_{10}Ga_{21}$	$Ho_4Ni_{10}Ga_{21}$	20.87	15.45	4.091	GrYG, 79c
$DyNi_xGa_{5-x}$ ($x = 3.1$–2.8)	$HoNi_{2.6}Ga_{2.4}$	8.67–8.79	–	4.13–4.09	GrYP, 83a
$DyNi_xGa_{2-x}$ ($x = 0.25$–0.37)	$CaIn_2$	4.318–4.344	–	7.401–7.160	GrY, 83
$DyNi_xGa_{2-x}$ ($x = 0.6$–0.8)	KHg_2	6.984–6.901	4.323–4.291	7.467–7.381	GrY, 83
$DyNiGa_4$	$YNiAl_4$	4.075	15.22	6.555	RoGY, 83
$DyNi_xGa_{4-x}$ ($x = 0.9$–0.7)	$BaAl_4$	4.143–4.156	–	9.78–9.71	Gr, 83
Dy_2Ni_2Ga	Mo_2NiB_2	5.336	8.272	4.167	RoGY, 82
$Dy_6Ni_xGa_{4-x}$ ($x = 2.2$–1.8)	Y_3Rh_2	11.47–11.66	–	24.56–24.93	GrYG, 80
$DyNiGa$	$TiNiSi$	6.832	4.285	7.323	YaGG, 79
$DyNi_{0.9}Ga_{0.1}$	α-ITl	3.72	10.29	4.258	GrY, 81
$DyPdGa_2$	$MgCuAl_2$	4.397	10.099	6.600	Gr, 84
$HoTi_2Ga_4$	$YbMo_2Al_4$	6.715	–	5.476	GrGMY, 80
$HoPdGa_2$	$MgCuAl_2$	4.369	10.055	6.595	Gr, 84
$ErCo_xGa_{3-x}$ ($x = 2.4$–1.8)	$CeNi_3$	5.042–5.100	–	16.23–16.26	Gl, 87
$ErCo_{1.25}Ga_{0.75}$	$MgZn_2$	5.206	–	8.428	Gl, 87
Er_2Co_2Ga	Mo_2NiB_2	5.294	8.330	4.026	Gl, 87
$Er_2Co_3Ga_9$	$Y_2Co_3Ga_9$	12.714	7.340	9.375	Gl, 87
Er_8CoGa_3	Pr_8CoGa_3	9.918	–	6.813	Gl, 87
Er_6Co_2Ga	Ho_6Co_2Ga	9.247	9.373	9.781	GlGY, 83
$ErCoGa$	$TiNiSi$	6.973	4.350	6.980	SiGGY, 82
$Er_6Co_7Ga_{21}$	$Ho_6Co_7Ga_{21}$	16.651	–	4.125	YaGG, 81
Er_2CoGa_8	Ho_2CoGa_8	4.195	–	10.96	GrYG, 79b
$ErCoGa_5$	$HoCoGa_5$	4.181	–	6.756	GrYG, 79b
$Er_4Ni_{10}Ga_{21}$	$Ho_4Ni_{10}Ga_{21}$	20.802	15.346	4.074	GrYG, 79c
$ErNi_xGa_{5-x}$ ($x = 3$–2.5)	$HoNi_{2.6}Ga_{2.4}$	8.65–8.82	–	4.12–4.07	GrYP, 83a
$ErNi_xGa_{2-x}$ ($x = 0.25$–0.43)	$CaIn_2$	4.321–4.336	–	7.231–6.978	GrY, 83
$ErNi_xGa_{2-x}$ ($x = 0.51$–0.9)	KHg_2	6.976–6.842	4.334–4.274	7.508–7.365	GrY, 83
$ErNiGa_4$	$YNiAl_4$	4.051	15.11	6.542	RoGY, 83
$Er_3Ni_{2.8}Ga_{8.2}$	La_3Al_{11}	4.13	12.25	9.41	Gr, 82
Er_2Ni_2Ga	Mo_2NiB_2	5.304	8.204	4.145	RoGY, 82
$Er_6Ni_xGa_{4-x}$ ($x = 2.6$–2.0)	Y_3Rh_2	11.42–11.48	–	24.47–24.61	GrYG, 80
$ErNiGa$	$TiNiSi$	6.771	4.254	7.307	YaGG, 79
$ErNi_{0.85}Ga_{0.15}$	α-ITl	3.69	10.22	4.23	GrY, 81

TABLE 2 (cont'd)

Compound	Structure type	Lattice parameters a	b	c	Refs.
$ErPdGa_2$	$MgCuAl_2$	4.356	10.020	6.588	Gr, 84
$TmTi_2Ga_4$	$YbMo_2Al_4$	6.657	–	5.467	GrGMY, 80
$TmFe_xGa_{12-x}$ ($x = 6.5$–6.1)	$ThMn_{12}$	8.576–8.604	–	5.054–5.070	GrGLGY, 83b
Tm_2FeGa_8	Ho_2CoGa_8	4.212	–	11.03	LaGY, 84
Tm_6Co_2Ga	Ho_6Co_2Ga	9.228	9.281	9.738	GlGY, 83
$Tm_6Co_7Ga_{21}$	$Ho_6Co_7Ga_{21}$	16.611	–	4.107	YaGG, 81
Tm_2CoGa_8	Ho_2CoGa_8	4.188	–	10.94	GrYG, 79b
$TmCoGa_5$	$HoCoGa_5$	4.172	–	6.737	GrYG, 79b
$Tm_2Co_3Ga_9$	$Y_2Co_3Ga_9$	12.688	7.326	9.359	Gl, 87
$Tm_2Ni_9Ga_8$	Th_2Zn_{17}	8.707	–	12.072	Gr, 80
$Tm_4Ni_{10}Ga_{21}$	$Ho_4Ni_{10}Ga_{21}$	20.79	15.41	4.057	GrYG, 79c
$TmNi_xGa_{5-x}$ ($x = 2.9$–2.7)	$HoNi_{2.6}Ga_{2.4}$	8.65–8.74	–	4.11–4.11	GrYP, 83
$TmNi_xGa_{2-x}$ ($x = 0.17$–0.37)	$CaIn_2$	4.312–4.333	–	7.243–7.061	GrY, 83
$TmNi_xGa_{2-x}$ ($x = 0.53$–0.9)	KHg_2	6.905–6.842	4.321–4.278	7.499–7.414	GrY, 83
$TmNiGa_4$	$YNiAl_4$	4.039	15.04	6.538	RoGY, 83
$Tm_3Ni_{2.2}Ga_{8.8}$	La_3Al_{11}	4.127	12.25	9.49	Gr, 82
Tm_2Ni_2Ga	Mo_2NiB_2	5.287	8.212	4.126	RoGY, 82
$Tm_6Ni_xGa_{4-x}$ ($x = 2.4$–1.8)	Y_3Rh_2	11.32–11.40	–	24.27–24.49	GrYG, 80
TmNiGa	TiNiSi	6.749	4.248	7.279	YaGG, 79
$TmNi_{0.9}Ga_{0.1}$	α-ITl	3.67	10.15	4.21	GrY, 81
$TmPdGa_2$	$MgCuAl_2$	4.354	9.979	6.574	Gr, 84
$YbFe_xGa_{12-x}$ ($x = 5.9$–5.5)	$ThMn_{12}$	8.555–8.580	–	5.048–5.065	GrGLGY, 83b
$YbCo_{5.6}Ga_{6.4}$	$ThMn_{12}$	8.411	–	5.070	Gl, 87
$YbCo_2Ga_8$	$CeFe_2Al_8$	12.402	14.422	3.985	Gl, 87
$Yb_2Co_3Ga_9$	$Y_2Co_3Ga_9$	12.691	7.327	9.353	Gl, 87
$YbCo_xGa_{3-x}$ ($x = 2.4$–1.8)	$CeNi_3$	5.033–5.075	–	16.19–16.21	Gl, 87
$Yb_4Ni_{10}Ga_{21}$	$Ho_4Ni_{10}Ga_{21}$	20.77	15.41	4.066	GrYG, 79c
$YbNi_xGa_{5-x}$ ($x = 3.1$–2.8)	$HoNi_{2.6}Ga_{2.4}$	8.78–8.79	–	4.06–4.08	GrYP, 83a
$YbNiGa_4$	$YNiAl_4$	4.086	15.58	6.615	RoGY, 83
Yb_2Ni_2Ga	Mo_2NiB_2	5.297	8.403	4.167	RoGY, 82
YbNiGa	TiNiSi	6.719	4.228	7.278	YaGG, 79
$YbPdGa_2$	$MgCuAl_2$	4.347	10.005	6.644	Gr, 84
$LuTi_2Ga_4$	$YbMo_2Al_4$	6.662	–	5.484	GrGMY, 80
Lu_2FeGa_8	Ho_2CoGa_8	4.195	–	11.00	LaGY, 84
$LuCo_{5.4}Ga_{6.6}$	$ThMn_{12}$	8.419	–	5.071	Gl, 87
$Lu_2Co_3Ga_9$	$Y_2Co_3Ga_9$	12.662	7.313	9.322	Gl, 87
$LuCo_xGa_{3-x}$ ($x = 2.4$–1.7)	$CeNi_3$	5.019–5.062	–	16.17–16.07	Gl, 87
$LuCo_{1.2}Ga_{0.8}$	$MgZn_2$	5.153	–	8.368	Gl, 87
Lu_6Co_2Ga	Ho_6Co_2Ga	9.197	9.155	9.661	GlGY, 83
$Lu_6Co_7Ga_{21}$	$Ho_6Co_7Ga_{21}$	16.549	–	4.099	YaGG, 81

TABLE 2 (cont'd)

Compound	Structure type	Lattice parameters *a*	*b*	*c*	Refs.
Lu_2CoGa_8	Ho_2CoGa_8	4.178	–	10.87	GrYG, 79b
$LuCoGa_5$	$HoCoGa_5$	4.146	–	6.708	GrYG, 79b
$Lu_4Ni_{10}Ga_{21}$	$Ho_4Ni_{10}Ga_{21}$	20.65	15.38	4.072	GrYG, 79c
$LuNi_xGa_{5-x}$ ($x = 3.1$–2.8)	$HoNi_{2.6}Ga_{2.4}$	8.66–8.75	–	4.09–4.06	GrYP, 83a
$LuNi_xGa_{2-x}$ ($x = 0.57$–0.9)	KHg_2	6.832–6.73	4.286–4.247	7.436–7.33	GrY, 83
$LuNiGa_4$	$YNiAl_4$	4.021	14.92	6.530	RoGY, 83
$Lu_3Ni_xGa_{11-x}$ ($x = 1.4$)	La_3Al_{11}	4.06	11.99	9.29	Gr, 82
Lu_2Ni_2Ga	Mo_2NiB_2	5.254	8.094	4.116	RoGY, 82
$Lu_6Ni_xGa_{4-x}$ ($x = 1.9$–1.3)	Y_3Rh_2	11.30–11.38	–	24.25–24.39	GrYG, 80
LuNiGa	TiNiSi	6.728	4.189	7.173	YaGG, 79
$LuPdGa_2$	$MgCuAl_2$	4.326	9.954	6.756	Gr, 84

References

Belyavina, N.N., and V.Ya. Markiv, 1980, Vestn. Kievsk. Univ., Ser. Fiz. **21**, 15.

Fremy, M.A., D. Grignoux and J.M. Moreau, 1985, J. Less-Common Met. **106**, 251.

Gavrylenko, I.S., and V.Ya. Markiv, 1978a, Vestn. Kievsk. Univ., Ser. Fiz. **19**, 18.

Gavrylenko, I.S., and V.Ya. Markiv, 1978b, Dopov. Akad. Nauk Ukr. RSR, Ser. A, p. 271.

Gavrylenko, I.S., and V.Ya. Markiv, 1979, Metallofizika **75**, 103.

Gladyshevsky, R.E., 1987, Phase equilibria and crystal structure of compounds in the systems R–Co–Ga, R – heavy rare earth, Ph.D. Chemistry thesis (Lvovsk. Politekhn. Inst., Moscow) pp. 1–20.

Gladyshevsky, R.E., Yu.N. Gryn and Ya.P. Yarmolyuk, 1981, Vestn. Lvovsk. Univ., Ser. Khim. **23**, 26.

Gladyshevsky, R.E., Yu.N. Gryn and Ya.P. Yarmolyuk, 1983, Dopov. Akad. Nauk Ukr. RSR, Ser. A (2), 67.

Gladyshevsky, R.E., Yu.N. Gryn and Ya.P. Yarmolyuk, 1985, Interaction in the system Ho–Co–Ga at 873K, in: Stable and metastable phase equilibria in metallic systems, ed. M.E. Dritz (Nauka, Moscow) pp. 49–54.

Gryn, Yu.N., 1980, Crystal chemistry of ternary rare earth gallides, Ph.D. Chemistry thesis, 1980, Lvov (Lvov State University, Lvov) pp. 1–16.

Gryn, Yu.N., 1982, Dopov. Akad. Nauk Ukr. RSR, Ser. A (2), 80.

Gryn, Yu.N., 1984, Dopov. Akad. Nauk Ukr. RSR, Ser. A (9), 33.

Gryn, Yu.N., and K.V. Remenyuk, 1984, Vestn. Lvovsk. Univ., Ser. Khim. **25**, 16.

Gryn, Yu.N., and Ya.P. Yarmolyuk, 1979, Vestn. Lvovsk. Univ., Ser. Khim. **21**, 13.

Gryn, Yu.N., and Ya.P. Yarmolyuk, 1980, Kristallografiya **25**, 613.

Gryn, Yu.N., and Ya.P. Yarmolyuk, 1981, Vestn. Lvovsk. Univ., Ser. Khim. **23**, 30.

Gryn, Yu.N., and Ya.P. Yarmolyuk, 1982, Dopov. Akad. Nauk Ukr. RSR, Ser. A (3), 69.

Gryn, Yu.N., and Ya.P. Yarmolyuk, 1983, Izv. Akad. Nauk SSSR, Met. (1), 192.

Gryn, Yu.N., Ya.P. Yarmolyuk and E.I. Gladyshevsky, 1979a, Kristallografiya **24**, 455.

Gryn, Yu.N., Ya.P. Yarmolyuk and E.I. Gladyshevsky, 1979b, Kristallografiya **24**, 242.

Gryn, Yu.N., Ya.P. Yarmolyuk and E.I. Gladyshevsky, 1979c, Dokl. Akad. Nauk SSSR **245**, 1102.
Gryn, Yu.N., I.S. Gavrylenko, V.Ya. Markiv and Ya.P. Yarmolyuk, 1980a, Dopov. Akad. Nauk Ukr. RSR, Ser. A (8), 75.
Gryn, Yu.N., Ya.P. Yarmolyuk and E.I. Gladyshevsky, 1980b, Dopov. Akad. Nauk Ukr. RSR, Ser. A (1), 80.
Gryn, Yu.N., Ya.P. Yarmolyuk and V.K. Pecharsky, 1983a, Izv. Akad. Nauk SSSR, Met. (3), 213.
Gryn, Yu.N., Ya.P. Yarmolyuk, O.A. Usov, A.M. Kuzmin and V.A. Bruskov, 1983b, Kristallografiya **28**, 1207.
Gryn, Yu.N., Ya.P. Yarmolyuk and V.E. Zavodnik, 1984, Kristallografiya **29**, 228.
Gryn, Yu.N., K. Hiebel and P. Rogl, 1985, J. Less-Common Met. **110**, 299.
Gryn, Yu.N., P. Rogl, K. Hiebl, F.E. Wagner and H. Noel, 1987, J. Solid State Chem. **70**, 168.
Gryn, Yu.N., P. Rogl and K. Hiebl, 1988, J. Less-Common Met. **136**, 329.
Gryniv, I.A., O.I. Godovanetz, R.V. Lapunova, Yu.N. Gryn and Ya.P. Yarmolyuk, 1983b, Dopov. Akad. Nauk Ukr. RSR, Ser. A (1), 75.
Kynghybalo, V.V., and M.I. Grymak, 1986, Izv. Akad. Nauk SSSR, Met. (5), 207.
Lapunova, R.V., Yu.N. Gryn and Ya.P. Yarmolyuk, 1983, Izv. Vyssh. Uchebn. Zaved. Tsvetn. Metall. (2), 116.
Lapunova, R.V., Yu.N. Gryn and Ya.P. Yarmolyuk, 1984, Dopov. Akad. Nauk Ukr. RSR, Ser. A (8), 45.
Lapunova, R.V., T.V. Snigur, Yu.N. Gryn and Ya.P. Yarmolyuk, 1986, Izv. Vyssh. Uchebn. Zaved. Tsvetn. Metall. (1), 111.
Lapunova, R.V., Yu.N. Gryn and Ya.P. Yarmolyuk, 1987a, Izv. Vyssh. Uchebn. Zaved. Tsvetn. Metall. (2), 74.
Lapunova, R.V., Yu.N. Gryn and Ya.P. Yarmolyuk, 1987b, Izv. Vyssh. Uchebn. Zaved. Tsvetn. Metall. (3), 75.
Markiv, V.Ya., 1981, Dopov. Akad. Nauk Ukr. RSR, Ser. A (4), 86.
Markiv, V.Ya., and N.N. Belyavina, 1981, Izv. Akad. Nauk SSSR, Met. (3), 213.
Markiv, V.Ya., and V.V. Burnashova, 1969, Dopov. Akad. Nauk Ukr. RSR, Ser. A, p. 463.
Markiv, V.Ya., A.I. Skrypka and L.S. Golubyak, 1979, Dopov. Akad. Nauk Ukr. RSR, Ser. A, p. 1054.
Markiv, V.Ya., N.N. Belyavina and T.I. Zhuntkovskaya, 1982a, Dopov. Akad. Nauk Ukr. RSR, Ser. A (2), 84.
Markiv, V.Ya., N.N. Belyavina, T.I. Zhuntkovskaya and A.A. Babenko, 1982b, Dopov. Akad. Nauk Ukr. RSR, Ser. A (4), 74.
Markiv, V.Ya., N.N. Belyavina, B.A. Karpenko and A.A. Karpenko, 1985, Dopov. Akad. Nauk Ukr. RSR, Ser. A (9), 77.
Markiv, V.Ya., I.P. Shevchenko, N.N. Belyavina and P.P. Kuzmenko, 1986, Dopov. Akad. Nauk Ukr. RSR, Ser. A (8), 78.
Moroz, O.I., G.E. Stetzkovych, O.M. Sichevych, Yu.N. Gryn and Ya.P. Yarmolyuk, 1983, Kristallografiya **28**, 795.
Romaka, V.A., Yu.N. Gryn and Ya.P. Yarmolyuk, 1982, Ukr. Fiz. Zh. **27**, 400.
Romaka, V.A., Yu.N. Gryn and Ya.P. Yarmolyuk, 1983, Ukr. Fiz. Zh. **28**, 1095.
Savitzky, E.M., Yu.V. Efimov and V.Ya. Markiv, 1977, Izv. Vyssh. Uchebn. Zaved. Tsvetn. Metall. (6), 77.
Savitzky, E.M., Yu.V. Efimov and V.Ya. Markiv, 1978a, Izv. Akad. Nauk SSSR, Met. (1), 211.
Savitzky, E.M., Yu.V. Efimov and V.Ya. Markiv, 1980, Dopov. Akad. Nauk Ukr. RSR, Ser. A (8), 81.
Savitzky, E.M., Yu.V. Efimov and V.Ya. Markiv, 1986, Dopov. Akad. Nauk Ukr. RSR, Ser. A, p. 1046.
Sichevych, O.M., 1983, Crystal structure of the compound Pr_8CoGa_3, in: Chetvertaya Vsesoyuzn. Konf. Kristallokhim. Intermetal. Soed. Tezisy Dokl., Lvov, 1983 (Vyscha Shkola, Lvov) p. 41.
Sichevych, O.M., 1985, Phase equilibria and crystal structures of compounds in the systems R–Co–Ga (R = La,Ce,Pr,Nd,Sm,Eu,Gd), Ph.D. Chemistry thesis, Lvov (Lvov State University, Lvov) pp. 1–24.

Sichevych, O.M., R.E. Gladyshevsky, Yu.N. Gryn and Ya.P. Yarmolyuk, 1982, Dopov. Akad. Nauk Ukr. RSR, Ser. A (11), 60.
Sichevych, O.M., Yu.N. Gryn and Ya.P. Yarmolyuk, 1983, Dopov. Akad. Nauk Ukr. RSR, Ser. A (7), 47.
Sichevych, O.M., R.V. Lapunova, A.N. Sobolev, Yu.N. Gryn and Ya.P. Yarmolyuk, 1985a, Kristallografiya **30**, 1077.
Sichevych, O.M., R.V. Lapunova, Yu.N. Gryn and Ya.P. Yarmolyuk, 1985b, Izv. Akad. Nauk SSSR, Met. (6), 117.
Sichevych, O.M., R.E. Gladyshevsky, Yu.N. Gryn and Ya.P. Yarmolyuk, 1986a, Vestn. Lvovsk. Univ., Ser. Khim. **27**, 20.
Sichevych, O.M., K.S. Gye, L.O. Vasylechko, Yu.N. Gryn and Ya.P. Yarmolyuk, 1986b, Izv. Akad. Nauk SSSR, Met. (4), 76.
Yarmolyuk, Ya.P., and Yu.N. Gryn, 1981, Izv. Akad. Nauk SSSR, Met. (5), 228.
Yarmolyuk, Ya.P., and P.I. Krypyakevich, 1976, Dopov. Akad. Nauk Ukr. RSR, Ser. A, p. 88.
Yarmolyuk, Ya.P., Yu.N. Gryn and E.I. Gladyshevsky, 1978a, Dopov. Akad. Nauk Ukr. RSR, Ser. A, p. 759.
Yarmolyuk, Ya.P., Yu.N. Gryn and E.I. Gladyshevsky, 1978b, Dopov. Akad. Nauk Ukr. RSR, Ser. A, p. 856.
Yarmolyuk, Ya.P., Yu.N. Gryn and R.E. Gladyshevsky, 1979, Dopov. Akad. Nauk Ukr. RSR, Ser. A, p. 772.
Yarmolyuk, Ya.P., Yu.N. Gryn and O.M. Olesh, 1980, Kristallografiya **25**, 248.
Yarmolyuk, Ya.P., Yu.N. Gryn and R.E. Gladyshevsky, 1981, Dokl. Akad. Nauk SSSR **260**, 1139.
Yarmolyuk, Ya.P., Yu.N. Gryn, I.V. Rozhdestvenskaya, O.A. Usov, A.M. Kuzmin, V.A. Bruskov and E.I. Gladyshevsky, 1982, Kristallografiya **27**, 999.
Yarmolyuk, Ya.P., R.E. Gladyshevsky, V.A. Bruskov and Yu.N. Gryn, 1984a, Dopov. Akad. Nauk Ukr. RSR, Ser. A (6), 50.
Yarmolyuk, Ya.P., Yu.N. Gryn, R.E. Gladyshevsky and V.S. Fundamensky, 1984b, Kristallografiya **29**, 697.
Yarmolyuk, Ya.P., Yu.N. Gryn, L.O. Vasylechko and V.K. Belsky, 1986, Kristallografiya **31**, 181.

2.3 *R–M–{In, Tl} systems*

No phase diagrams were reported for ternary rare earth systems with indium or thallium and another metallic element. The result of partial studying (without the establishment of phase equilibria) of 69 ternary systems with In, and 17 with Tl, the existence, compositions and crystal structures have been reported for 113 and 17 ternary compounds, respectively. The crystallographic data are listed in table 3.

TABLE 3
Ternary compounds found in the partially investigated systems R–M–In and R–M–Tl.

Compound	Structure type	Lattice parameters *a*	*b*	*c*	Refs.
$ScNi_4In$	$MgCu_4Sn$	6.872	–	–	ZaBK, 84
$ScCu_2In$	$MnCu_2Al$	6.362	–	–	KaDBO, 84
YNi_4In	$MgCu_4Sn$	7.034	–	–	ZaBK, 84
YNi_9In_2	YNi_9In_2	8.222	–	4.827	KaAZB, 84a
YNiIn	Fe_2P	7.486	–	3.784	FeMR, 74a
YCu_2In	$MnCu_2Al$	6.601	–	–	KaDBO, 84

TABLE 3 (cont'd)

Compound	Structure type	Lattice parameters a	b	c	Refs.
YPdIn	Fe_2P	7.649	–	3.819	FeMR, 74a
YAuIn	Fe_2P	7.691	–	3.909	RoFCM, 77
YAu_2In	CsCl	3.443	–	–	MaFR, 75
$LaNi_9In_2$	YNi_9In_2	8.339	–	4.877	KaAZB, 84a
$La_5Ni_6In_{11}$	$Pr_5Ni_6In_{11}$	14.640	14.674	4.439	KaZBDB, 87
$LaNi_3In_6$	$LaNi_3In_6$	4.388	7.574	12.110	KaGBDK, 85
LaNiIn	Fe_2P	7.613	–	4.035	FeMR, 74a
$LaCu_2In$	$MnCu_2Al$	6.845	–	–	KaDBO, 84b
$LaCu_{6.5}In_{6.5}$	$NaZn_{13}$	12.442	–	–	KaDBO, 84b
LaPdIn	Fe_2P	7.729	–	4.135	FeMR, 74a
LaCdIn	$CaIn_2$	4.936	–	7.652	RoMMF, 81
LaAuIn	Fe_2P	7.727	–	4.315	RoFCM, 77
$LaAu_2In$	α-Fe	3.560	–	–	MaFR, 75
$CeNi_9In_2$	YNi_9In_2	8.250	–	4.851	KaAZB, 84a
$Ce_5Ni_6In_{11}$	$Pr_5Ni_6In_{11}$	14.600	14.601	4.413	KaZBDB, 87
$CeNi_4In$	$MgCu_4Sn$	7.056	–	–	ZaBK, 84
$CeNi_3In_6$	$LaNi_3In_6$	4.385	7.537	12.06	KaGBDK, 85
CeNiIn	Fe_2P	7.523	–	3.973	FeMR, 74a
$CeCu_2In$	$MnCu_2Al$	6.785	–	–	KaDBO, 84
CePdIn	Fe_2P	7.698	–	4.079	FeMR, 74a
CeCdIn	$CaIn_2$	4.943	–	7.809	RoMMF, 81
CeAuIn	Fe_2P	7.698	–	4.256	RoFCM, 77
$Ce_{0.24}Au_{0.50}In_{0.26}$	α-Fe	3.544	–	–	MaFR, 75
$PrNi_9In_2$	YNi_9In_2	8.275	–	4.871	KaAZB, 84a
$Pr_5Ni_6In_{11}$	$Pr_5Ni_6In_{11}$	14.579	14.579	4.400	KaZBDE, 87
$PrNi_4In$	$MgCu_4Sn$	7.137	–	–	ZaBK, 84
$PrNi_3In_6$	$LaNi_3In_6$	4.336	7.538	12.043	KaGBDK, 85
PrNiIn	Fe_2P	7.541	–	3.950	FeMR, 74a
$PrCu_2In$	$MnCu_2Al$	6.757	–	–	KaDBO, 84
PrPdIn	Fe_2P	7.679	–	4.043	FeMR, 74a
PrCdIn	$CaIn_2$	4.907	–	7.609	RoMMF, 81
$PrAu_{1.4-1}In_{0.6-1}$	Fe_2P	7.700–7.735	–	4.195–4.137	RoFCM, 77
$PrAu_2In$	α-Fe	3.524	–	–	MaFR, 75
$NdNi_9In_2$	YNi_9In_2	8.262	–	4.847	KaAZB, 84a
$NdNi_4In$	$MgCu_4Sn$	7.118	–	–	ZaBK, 84
NdNiIn	Fe_2P	7.537	–	3.922	FeMR, 74a
$NdCu_2In$	$MnCu_2Al$	6.735	–	–	KaDBO, 84
NdPdIn	Fe_2P	7.683	–	3.997	FeMR, 74a
NdCdIn	$CaIn_2$	4.889	–	7.572	RoMMF, 81
NdAuIn	Fe_2P	7.719	–	4.124	RoFCM, 77
$NdAu_2In$	CsCl	3.516	–	–	MaFR, 75
$SmNi_9In_2$	YNi_9In_2	8.236	–	4.830	KaAZB, 84a
$SmNi_4In$	$MgCu_4Sn$	7.081	–	–	ZaBK, 84
SmNiIn	Fe_2P	7.503	–	3.880	FeMR, 74a
$SmCu_2In$	$MnCu_2Al$	6.732	–	–	KaDBO, 84
SmPdIn	Fe_2P	7.676	–	3.934	FeMR, 74a
SmCdIn	$CaIn_2$	4.86	–	7.34	RoMMF, 81
SmAuIn	Fe_2P	7.616	–	4.643	RoFCM, 77
$SmAu_2In$	α-Fe	3.488	–	–	MaFR, 75

TABLE 3 (cont'd)

Compound	Structure type	Lattice parameters a	b	c	Refs.
$EuNi_9In_2$	YNi_9In_2	8.246	–	4.840	KaAZB, 84a
$EuCu_{6.5}In_{6.5}$	$NaZn_{13}$	12.581	–	–	KaDBO, 84b
EuPdIn	Fe_2P	8.309	–	3.957	FeMR, 74a
$GdNi_9In_2$	YNi_9In_2	8.232	–	4.834	KaAZB, 84a
$GdNi_4In$	$MgCu_4Sn$	7.059	–	–	ZaBK, 84
GdNiIn	Fe_2P	7.467	–	3.845	FeMR, 74a
$GdCu_2In$	$MnCu_2Al$	6.641	–	–	KaDBO, 84
GdPdIn	Fe_2P	7.649	–	3.886	FeMR, 74a
GdCdIn	$CaIn_2$	4.845	–	7.398	RoMMF, 81
GdAuIn	Fe_2P	7.693	–	3.991	RoFCM, 77
$Gd_{0.27}Au_{0.49}In_{0.24}$	α-Fe	3.469	–	–	MaFR, 75
$TbNi_9In_2$	YNi_9In_2	8.212	–	4.821	KaAZB, 84a
$TbNi_4In$	$MgCu_4Sn$	7.034	–	–	ZaBK, 84
TbNiIn	Fe_2P	7.472	–	3.776	FeMR, 74a
$TbCu_2In$	$MnCu_2Al$	6.615	–	–	KaDBO, 84b
TbPdIn	Fe_2P	7.647	–	3.843	FeMR, 74a
TbAuIn	Fe_2P	7.591	–	3.951	RoFCM, 77
$TbAu_2In$	α-Fe	3.455	–	–	MaFR, 75
$DyNi_9In_2$	YNi_9In_2	8.200	–	4.815	KaAZB, 84
$DyNi_4In$	$MgCu_4Sn$	7.012	–	–	ZaBK, 84
DyNiIn	Fe_2P	7.462	–	3.718	FeMR, 74a
$DyCu_2In$	$MnCu_2Al$	6.593	–	–	KaDBO, 84b
DyPdIn	Fe_2P	7.638	–	3.770	FeMR, 74a
DyCdIn	$CaIn_2$	4.819	–	7.318	RoMMF, 81
DyAuIn	Fe_2P	7.686	–	3.902	RoFCM, 77
$DyAu_2In$	α-Fe	3.449	–	–	MaFR, 75
$HoNi_9In_2$	YNi_9In_2	8.197	–	4.810	KaAZB, 84a
$HoNi_4In$	$MgCu_4Sn$	7.004	–	–	ZaBK, 84
$Ho_{10}Ni_9In_{20}$	$Ho_{10}Ni_9In_{20}$	13.286	–	9.083	ZaBKPG, 87
HoNiIn	Fe_2P	7.485	–	3.749	FeMR, 74a
$HoCu_2In$	$MnCu_2Al$	6.511	–	–	KaDBO, 84
HoPdIn	Fe_2P	7.629	–	3.788	FeMR, 74a
HoCdIn	$CaIn_2$	4.805	–	7.301	RoMMF, 81
HoAuIn	Fe_2P	7.683	–	3.877	RoFCM, 77
$HoAu_2In$	α-Fe	3.437	–	–	MaFR, 75
$ErNi_9In_2$	YNi_9In_2	8.190	–	4.801	KaAZB, 84a
$ErNi_4In$	$MgCu_4Sn$	6.996	–	–	ZaBK, 84
$Er_{10}Ni_9In_{20}$	$Ho_{10}Ni_9In_{20}$	13.250	–	9.075	ZaBKPG, 87
ErNiIn	Fe_2P	7.441	–	3.723	FeMR, 74a
$ErCu_2In$	$MnCu_2Al$	6.581	–	–	KaDBO, 84b
ErPdIn	Fe_2P	7.632	–	3.754	FeMR, 74a
ErCdIn	$CaIn_2$	4.796	–	7.253	RoMMF, 81
ErAuIn	Fe_2P	7.674	–	3.847	RoFCM, 77
$Er_{0.22}Au_{0.54}In_{0.24}$	α-Fe	3.430	–	–	MaFR, 75
$TmNi_4In$	$MgCu_4Sn$	6.993	–	–	ZaBK, 84
$Tm_{10}Ni_9In_{20}$	$Ho_{10}Ni_9In_{20}$	13.251	–	9.060	ZaBKPG, 87
TmNiIn	Fe_2P	7.426	–	3.704	FeMR, 74a
$TmCu_2In$	$MnCu_2Al$	6.539	–	–	KaDBO, 84
TmPdIn	Fe_2P	7.628	–	3.726	FeMR, 74a
$YbNi_4In$	$MgCu_4Sn$	6.975	–	–	ZaBK, 84

TABLE 3 (cont'd)

Compound	Structure type	Lattice parameters *a*	*b*	*c*	Refs.
YbPdIn	Fe_2P	7.574	–	3.932	FeMR, 74a
YbCdIn	$CaIn_2$	4.948	–	7.302	RoMMF, 81
YbAuIn	Fe_2P	7.708	–	4.027	RoFCM, 77
$Yb_{0.25}Au_{0.53}In_{0.22}$	α-Fe	3.428	–	–	MaFR, 75
$LuNi_4In$	$MgCu_4Sn$	6.997	–	–	ZaBK, 84
$Lu_{10}Ni_9In_{20}$	$Ho_{10}Ni_9In_{20}$	13.155	–	8.993	ZaBKPG, 897
$LuCu_2In$	$MnCu_2Al$	6.509	–	–	KaDBO, 84b
LuPdIn	Fe_2P	7.629	–	3.684	FeMR, 74a
YPdTl	Fe_2P	7.748	–	3.751	FeMR, 74b
LaPdTl	Fe_2P	7.873	–	4.027	FeMR, 74b
CePdTl	Fe_2P	7.840	–	3.963	FeMR, 74b
PrPdTl	Fe_2P	7.844	–	3.930	FeMR, 74b
NdPdTl	Fe_2P	7.812	–	3.903	FeMR, 74b
SmPdTl	Fe_2P	7.787	–	3.842	FeMR, 74b
GdCuTl	$CaIn_2$	4.608	–	7.234	MaRMF, 81
GdPdTl	Fe_2P	7.766	–	3.809	FeMR, 74b
TbCuTl	$CaIn_2$	4.599	–	7.180	MaRMF, 81
TbPdTl	Fe_2P	7.751	–	3.771	FeMR, 74b
DyCuTl	$CaIn_2$	4.586	–	7.132	MaRMF, 81
DyPdTl	Fe_2P	7.72	–	3.73	FeMR, 74b
HoCuTl	$CaIn_2$	4.579	–	7.101	MaRMF, 81
HoPdTl	Fe_2P	7.727	–	3.722	FeMR, 74b
ErCuTl	$CaIn_2$	4.568	–	7.065	MaRMF, 81
ErPdTl	Fe_2P	7.722	–	3.702	FeMR, 74b
YbPdTl	Fe_2P	7.719	–	3.899	FeMR, 74b

References

Ferro, R., R. Marazza and G. Rombardi, 1974a, Z. Metallkd. **65**, 37.

Ferro, R., R. Marazza and G. Rombardi, 1974b, Z. Metallkd. **65**, 40.

Kalychak, Ya.M., L.G. Akselrud, V.I. Zaremba and V.M. Baranyak, 1984a, Dopov. Akad. Nauk Ukr. RSR, Ser. B (8), 35.

Kalychak, Ya.M., O.V. Dmytrakh, O.I. Bodak and M.M. Ogryzlo, 1984b, Dopov. Akad. Nauk Ukr. RSR, Ser. B (1), 33.

Kalychak, Ya.M., E.I. Gladyshevsky, O.I. Bodak, O.V. Dmytrakh and B.Ya. Kotur, 1985, Kristallografiya **30**, 591.

Kalychak, Ya.M., P.Yu. Zavalij, V.M. Baranyak, O.V. Dmytrakh and O.I. Bodak, 1987, Kristallografiya **32**, 1021.

Marazza, R., R. Ferro and D. Rossi, 1975, Z. Metallkd. **66**, 110.

Mazzone, D., D. Rossi, R. Marazza and R. Ferro, 1981, J. Less-Common Met. **80**, 47.

Rossi, D., R. Ferro, V. Contardi and R. Marazza, 1977, Z. Metallkd. **68**, 493.

Rossi, D., R. Marazza, D. Mazzone and R. Ferro, 1981, J. Less-Common Met. **78**, 1.

Zaremba, V.I., V.M. Baranyak and Ya.M. Kalychak, 1984, Vestn. Lvovsk. Univ., Ser. Khim. **25**, 18.

Zaremba, V.I., V.K. Belsky, Ya.M. Kalychak, V.K. Pecharsky and E.I. Gladyshevsky, 1987, Dopov. Akad. Nauk Ukr. RSR, Ser. B (3), 42.

2.4 *R–M–Ge systems*

A total of 117 ternary compounds were found in the 17 R–M–Ge ternary systems while deriving the phase diagrams. They are described below. Another 146 ternary systems of the rare earths, germanium and a metallic element have only been partially investigated. The existence of an additional 393 ternary compounds was reported in the literature. Crystallographic data for these are given in table 4. The crystal structure was solved for 83 of the first 117 ternary compounds.

Sc–V–Ge

Kotur (1987) reported the phase diagram (fig. 96) and crystal structure for the two existing ternary compounds: (1) ~ $Sc_{1-1.8}V_{5-4.2}Ge_5$ which crystallized in the V_6Si_5 ST, $a = 7.872–7.850$, $c = 5.091–5.157$; and (2) ~ $Sc_2V_3Ge_4$ belonging to the $Ce_2Sc_3Ge_4$ ST, $a = 6.614$, $b = 12.774$, $c = 6.776$.

Sc–Cu–Ge

Dwight et al. (1973) reported the existence of (2) ~ ScCuGe with ZrNiAl ST, $a = 6.515$, $c = 3.972$. The phase diagram, shown in fig. 97, was derived by Kotur and Andrusyak (1984a). They found one more ternary compound: (1) ~ $Sc_3Cu_4Ge_4$ and established the $Gd_6Cu_8Ge_8$ type of crystal structure for it with $a = 13.32$, $b = 6.523$, $c = 4.004$.

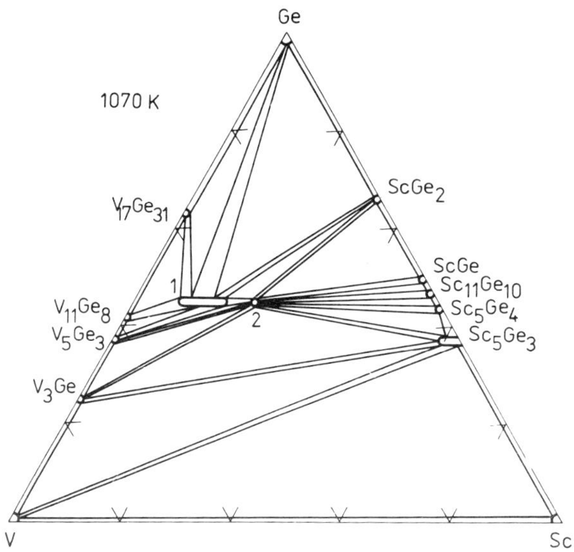

Fig. 96. Sc–V–Ge, isothermal section at 1070 K.

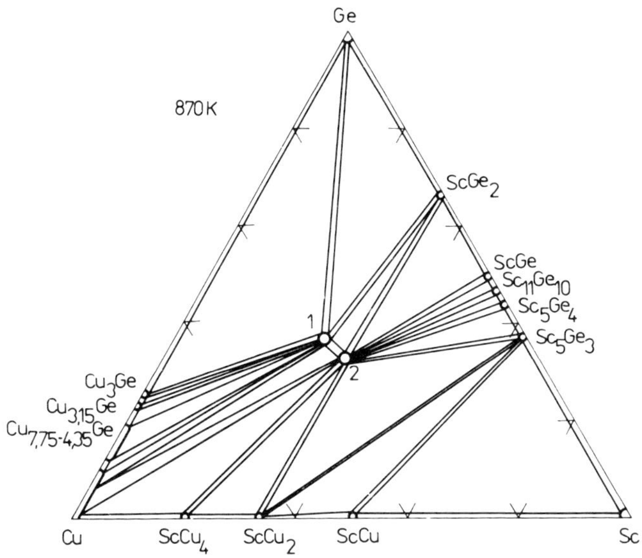

Fig. 97. Sc–Cu–Ge, isothermal section at 870 K.

Sc–Nb–Ge

Phase equilibria in the system Sc–Nb–Ge (fig. 98) were studied by Kotur (1986). The only one ternary compound $Sc_2Nb_3Ge_4$, which belongs to the $Ce_2Sc_3Ge_4$ ST, $a = 6.860$, $b = 13.39$, $c = 7.160$, was found.

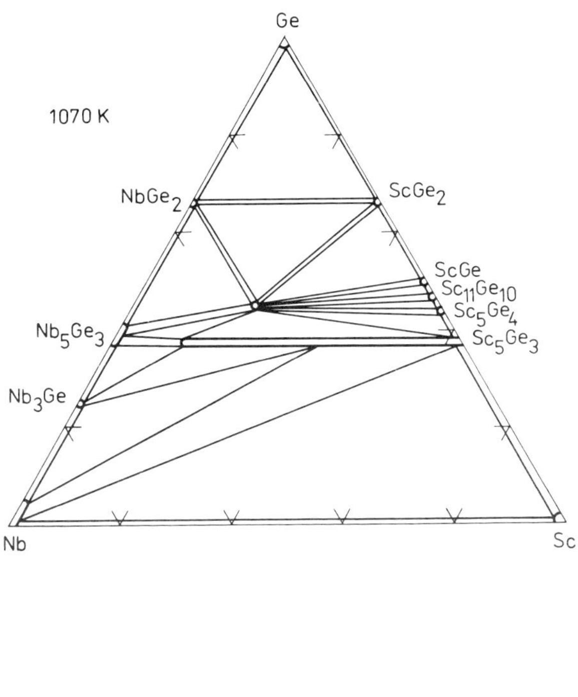

Fig. 98. Sc–Nb–Ge, isothermal section at 1070 K.

Pr–Fe–Ge

The phase diagram investigation was carried out by Fedyna (1988). The 870 K isothermal section is shown in fig. 99. The total number of ternary compounds found is 9. The existence of compounds (1) ~ $PrFe_6Ge_6$ (YCo_6Ge_6 ST, $a = 5.091$, $c = 4.036$), and (2) ~ $PrFe_2Ge_2$ ($CeAl_2Ga_2$ ST, $a = 4.055$, $c = 10.540$), found earlier by Starodub and Mruz (1983), and Rossi et al. (1978b), respectively, was confirmed. The compound (3) ~ $PrFeGe_3$ crystallized in the $BaNiSn_3$ ST, $a = 4.132$, $c = 9.932$; (5) ~ $PrFe_{0.55}Ge_2$ in the $CeNiSi_2$ ST, $a = 4.253$, $b = 16.478$, $c = 4.130$; (7) ~ $Pr_{0.32}Fe_{0.64}Ge_{0.04}$ has hexagonal unit cell, $a = 5.733$, $c = 11.514$; (8) ~ $Pr_4Fe_{0.85}Ge_7$ belong to the $Sm_4Co_{1-x}Ge_7$ ST, $a = 4.217$, $b = 30.520$, $c = 4.128$; and (9) ~ $Pr_{117}Fe_{52}Ge_{112}$ to the $Tb_{117}Fe_{52}Ge_{112}$ ST, $a = 29.428$. Ternary compounds (4) ~ $Pr_{0.23}Fe_{0.17}Ge_{0.60}$ and (6) ~ $Pr_{0.27}Fe_{0.13}Ge_{0.60}$ were found, but no structure details were given.

Pr–Co–Ge

The phase equilibria were studied by Fedyna (1988) and are shown in fig. 100. The ternary compounds are as follows: (1) ~ $Pr_{0.7}(Co, Ge)_{13}$ ($CeNi_{8.5}Si_{4.5}$ ST, $a = 7.932$, $c = 11.817$); (2) ~ $PrCo_2Ge_2$ ($CeAl_2Ga_2$ ST, $a = 4.048$, $c = 10.178$); (3) ~ $Pr_2Co_3Ge_5$ ($U_2Co_3Si_5$ ST, $a = 9.783$, $b = 11.896$, $c = 5.831$); (4) ~ $PrCo_{0.84}Ge_3$ ($BaNiSn_3$ ST, $a = 4.311$, $c = 9.833$). Crystal structures for the next 4 are undetermined: (5) ~ $Pr_{0.23}Co_{0.21}Ge_{0.56}$, (6) ~ $Pr_{0.24}Co_{0.18}Ge_{0.58}$, (7) ~ $Pr_{0.25}Co_{0.15}Ge_{0.60}$ and (8) ~ $Pr_{0.27}Co_{0.15}Ge_{0.58}$. The homogeneity range for (9) ~ $PrCo_{1-x}Ge_2$ reach the x-value from 0.52 to 0.10 with crystal structure of the $CeNiSi_2$ type, $a = 4.249$–4.262, $b = 16.650$–16.670, $c = 4.136$–4.230; (10) ~ PrCoGe belongs to the PbFCl ST, $a = 4.181$, $c = 6.842$; (11) ~ $PrCo_{0.5}Ge_{1.5}$ to the AlB_2 ST, $a = 4.163$, $c = 4.193$; (12) ~ $Pr_{117}Co_{52}Ge_{112}$ to the $Tb_{117}Fe_{52}Ge_{112}$ ST,

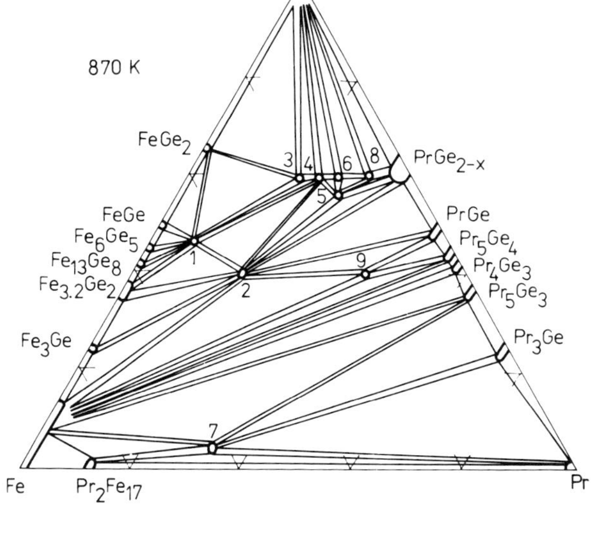

Fig. 99. Pr–Fe–Ge, isothermal section at 870 K.

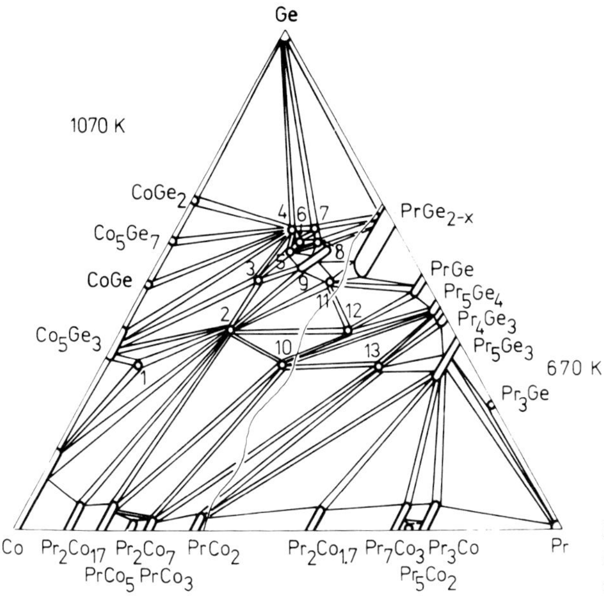

Fig. 100. Pr–Co–Ge, isothermal sections at 1070 K (0–33 a/c Pr) and at 670 K (33–100 a/c Pr).

$a = 29.318$, and (13) ~ Pr_3CoGe_2 to the La_3NiGe_2 type structure, $a = 11.840$, $b = 4.272$, $c = 11.519$. The results of crystal structure determination of (3) given above differ from those reported by Venturini et al. (1985), who reported the $Lu_2Co_3Si_5$ type of structure for $Pr_2Co_3Ge_5$, $a = 9.802$, $b = 11.94$, $c = 5.838$, $\beta = 90.7$).

Pr–Ni–Ge

The system Pr–Ni–Ge was studied by Fedyna et al. (1987). The phase diagram is shown in fig. 101. The ternary compounds are as follows: (1) ~ $Pr_{1-x}(Ni, Ge)_{13}$ ($CeNi_{8.5}Si_{4.5}$ ST, $a = 7.951$, $c = 11.739$); (2) ~ $PrNi_2Ge_2$ ($CeAl_2Ga_2$ ST, $a = 4.136$, $c = 9.844$, which is in a good agreement with the results of Rieger and Parthe (1969b)); (3) ~ $Pr_2Ni_3Ge_5$ (the $U_2Co_3Si_5$ ST, $a = 9.812$, $b = 11.832$, $c = 5.912$); (4) ~ $PrNiGe_3$ ($SmNiGe_3$ ST, $a = 21.779$, $b = 4.105$, $c = 4.144$); (5) ~ Pr_2NiGe_6 (Ce_2CuGe_6 ST, $a = 4.034$, $b = 21.722$, $c = 4.112$); (6) ~ $Pr_3Ni_4Ge_4$ ($U_3Ni_4Si_4$ ST, $a = 4.098$, $b = 4.182$, $c = 24.013$); (7) ~ $PrNi_{0.67}Ge_2$ ($CeNiSi_2$ ST, $a = 4.210$, $b = 16.68$, $c = 4.160$); (8) ~ PrNiGe (TiNiSi, ST, $a = 7.276$, $b = 4.343$, $c = 7.298$); (9) ~ $Pr(Ni_{0.42-0.28}\ Ge_{0.58-0.72})_{2-x}$ (the AlB_2 ST, $a = 4.158$–4.150, $c = 4.134$–4.220); (10) ~ Pr_3NiGe_2 (La_3NiGe_2 ST, $a = 11.810$, $b = 4.292$, $c = 11.611$, after Bodak et al. (1982)). Crystal structure for the last compound, (11) ~ Pr_5NiGe_4 is unknown.

Eu–Fe–Ge

No ternary compounds were found by Belan (1988) in a study of the phase equilibria (fig. 102).

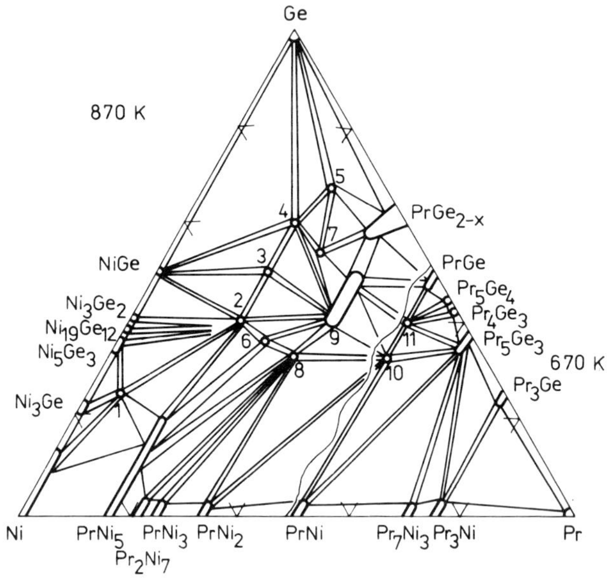

Fig. 101. Pr–Ni–Ge, isothermal sections at 870 K (0–50 a/c Pr) and at 670 K (50–100 a/c Pr).

Eu–Co–Ge

According to Felner and Nowik (1978), the compound $EuCo_2Ge_2$ with $CeAl_2Ga_2$ ST, has lattice parameters $a = 4.035$, $c = 10.46$. The isothermal section at 670 K, shown in fig. 103, has been derived by Oniskovetz et al. (1985). A total of 3 ternary compounds were found: (1) ~ $EuCo_{2.5}Ge_{1.5}$ ($BaAl_4$ type of crystal structure, $a = 4.035$, $c = 9.746$); (2) ~ $EuCoGe_3$ ($BaNiSn_3$ ST, $a = 4.345$, $c =$

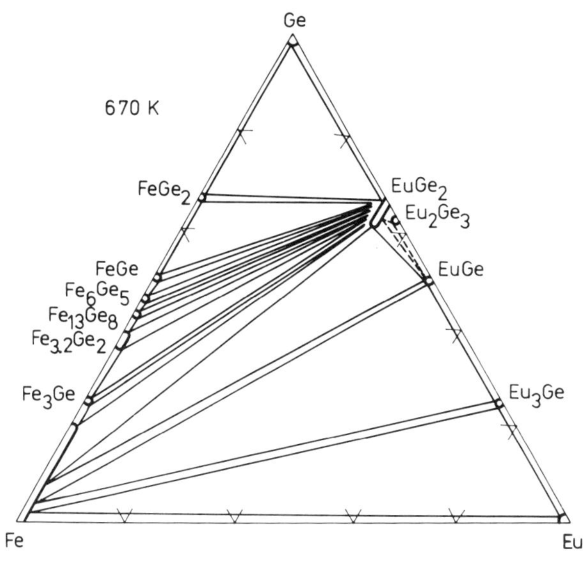

Fig. 102. Eu–Fe–Ge, isothermal section at 670 K.

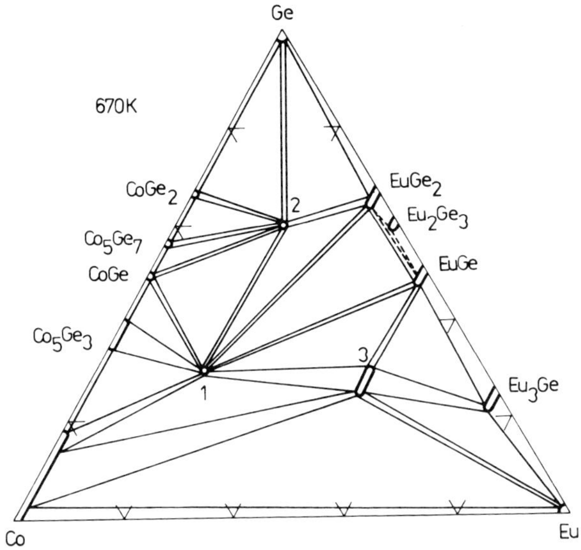

Fig. 103. Eu–Co–Ge, isothermal section at 670 K.

9.953), and (3) ~ $Eu_2Co_{1-1.2}Ge_{1-0.8}$ (crystal structure for this compound is unknown).

Eu–Ni–Ge

The phase diagram, shown in fig. 104, was derived by Belan (1988). The existence of (2) ~ $EuNi_2Ge_2$, found by Bodak et al. (1966), and (6) ~ $EuNi_{0.5-0.35}Ge_{1.5-1.65}$ has been confirmed. Crystal structures are determined for

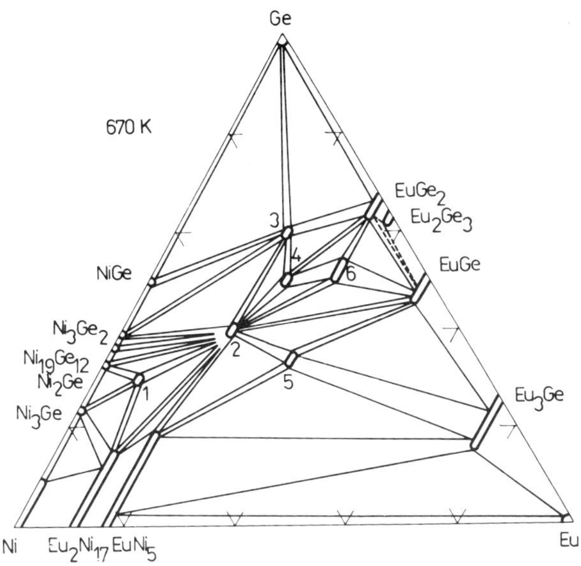

Fig. 104. Eu–Ni–Ge, isothermal section at 670 K.

all of the 6 existing ternary compounds: (1) ~ $EuNi_9Ge_4$ ($CeNi_{8.5}Si_{4.5}$ ST, a = 7.957, c = 11.793); (2) ~ $EuNi_2Ge_2$ ($CeAl_2Ga_2$ ST, a = 4.133, c = 10.126); (3) ~ $EuNiGe_3$ ($BaNiSn_3$, ST, a = 4.437, c = 9.891); (4) ~ $EuNiGe_2$ ($CeNiSi_2$ ST, a = 4.244, b = 16.91, c = 4.228); (5) ~ EuNiGe (according to Oniskovetz et al. (1987) a new type of crystal structure: $P2_1/b$, a = 6.996, b = 7.581, c = 6.187, γ = 130.22), and (6) ~ $EuNi_{0.5-0.35}Ge_{1.5-1.65}$ (AlB_2 ST, a = 4.184–4.169, c = 4.485–4.480).

Eu–Cu–Ge

The system Eu–Cu–Ge was investigated by Belan (1988), the phase diagram is shown in fig. 105. It was found, that (1) ~ $EuCu_{2.5-1.65}Ge_{1.5-2.35}$, earlier reported as $EuCu_2Ge_2$ by Felner and Nowik (1978), belongs to $CeAl_2Ga_2$ ST, a = 4.187–4.212, c = 10.116–10.336; (2) ~ EuCuGe to TiNiSi ST, a = 7.342, b = 4.372, c = 7.550; (3) ~ $EuCu_{0.8-0.5}Ge_{1.2-1.5}$ to AlB_2 ST, a = 4.217–4.218, c = 4.432–4.460. The crystal structure of the fourth one: (4) ~ Eu_2Cu_2Ge is undetermined.

Tb–Fe–Ge

The system Tb–Fe–Ge (fig. 106) was described by Starodub (1988). Six ternary compounds were found: (1) ~ $TbFe_6Ge_6$ (after Starodub and Mruz (1983) with YCo_6Ge_6 ST, a = 5.106, c = 4.055); (2) ~ $TbFe_4Ge_2$ with unknown crystal structure; (3) ~ $TbFe_2Ge_2$ (according to Malik et al (1975) $CeAl_2Ga_2$ ST, a = 3.968, c = 10.446); (4) ~ $Tb_2Fe_4Ge_3$ with unknown crystal structure; (5) ~ $TbFe_{0.33}Ge_2$ ($CeNiSi_2$ ST, a = 4.123, b = 15.830, c = 4.006), and (6) ~ $Tb_{117}Fe_{52}Ge_{112}$ (a new type of crystal structure with gigantic unit cell: Fm3m, a = 28.580, as reported by Pecharsky et al. (1987b)).

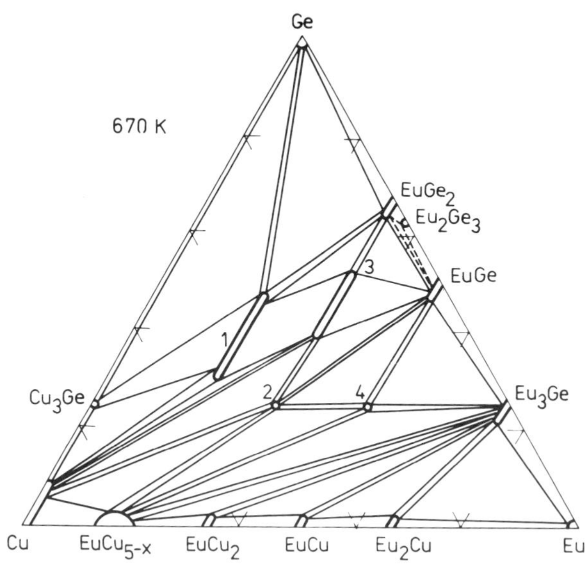

Fig. 105. Eu–Cu–Ge, isothermal section at 670 K.

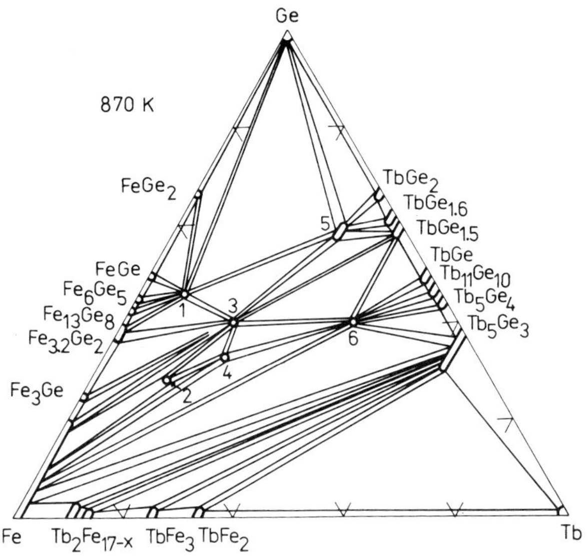

Fig. 106. Tb–Fe–Ge, isothermal section at 870 K.

Tb–Co–Ge

The phase diagram investigation was carried out by Starodub (1988). The existence of a total of 10 ternary compounds has been confirmed or established: (1) ~ $TbCo_6Ge_6$ (YCo_6Ge_6 ST, $a = 5.106$, $c = 4.055$, or, after Buchholz and Schuster (1981), $a = 5.071$, $c = 3.905$); (2) ~ $Tb_3Co_4Ge_{13}$ ($Y_3Co_4Ge_{13}$ ST, $a = 8.779$); (3) ~ $TbCo_2Ge_2$ ($CeAl_2Ge_2$ ST, $a = 3.984$, $c = 10.059$); (4) ~ $Tb_2Co_3Ge_5$ ($U_2Co_3Si_5$ ST with unknown lattice parameters, after Starodub (1988), or $Lu_2Co_3Si_5$ ST, $a = 9.649$, $b = 11.83$, $c = 5.715$, $\beta = 91.8$ after Venturini et al. (1985)); (5) ~ $TbCoGe_3$ and (6) ~ Tb_2CoGe_6, both with unknown crystal structure; (7) ~ $Tb_{0.27-0.30}Co_{0.1-0.2}Ge_{0.52-0.62}$ ($CeNiSi_2$ ST, $a = 4.122$–4.110, $b = 16.209$–15.991, $c = 4.050$–3.999); (8) ~ TbCoGe (TiNiSi ST, $a = 6.955$, $b = 4.242$, $c = 7.271$); (9) ~ $Tb_3Co_2Ge_4$ (after Starodub et al. (1986) a new type of crystal structure: B2/m, $a = 10.692$, $b = 8.067$, $c = 4.164$, $\gamma = 107.72$); (10) ~ Tb_2CoGe_2 (Sc_2CoSi_2 ST, $a = 10.569$, $b = 10.209$, $c = 4.212$, $\gamma = 118.18$ after Bodak et al. (1986)). The phase diagram is shown in fig. 107.

Tb–Ni–Ge

The isothermal section at 870 K is shown in fig. 108 after Starodub (1988). The known ternary compounds are: (1) ~ $TbNi_5Ge_3$ (YNi_5Si_3 ST, $a = 19.068$, $b = 3.871$, $c = 6.790$); (2) ~ $Tb_3Ni_{11}Ge_4$ ($Sc_3Ni_{11}Si_4$ ST, $a = 8.321$, $c = 8.832$); (3) ~ $TbNi_2Ge_2$ ($CeAl_2Ga_2$ ST, $a = 4.044$, $c = 9.782$, which is in the good agreement with the data of Rieger and Parthé (1969b)); (4) ~ $TbNiGe_3$ ($SmNiGe_3$ ST, $a = 21.496$, $b = 4.058$, $c = 4.054$); (5) and (6), both with unknown crystal structures and with approximate compositions of Tb_2NiGe_6 and $TbNiGe_2$, respectively; (7) ~ $TbNi_{0.6}Ge_2$ ($CeNiSi_2$ ST, $a = 4.105$, $b = 16.21$, $c = 3.993$); (8) ~ TbNiGe

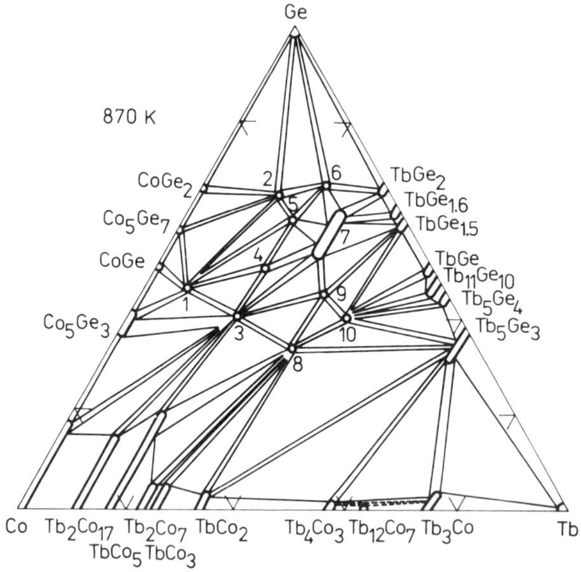

Fig. 107. Tb–Co–Ge, isothermal section at 870 K.

(TiNiSi ST, $a = 6.949$, $b = 4.196$, $c = 7.303$; (9) ~ $Tb_{0.47}Ni_{0.10}Ge_{0.43}$ with an unknown structure, and (10) ~ Tb_3NiGe_2 (La_3NiGe_2 ST, $a = 11.381$, $b = 4.222$, $c = 11.257$, as it was reported by Bodak et al. (1982)).

Tb–Cu–Ge

A total of 4 ternary compounds were found in the system Tb–Cu–Ge. The phase equilibria are shown in fig. 109, after Starodub (1988). Compound (1) ~

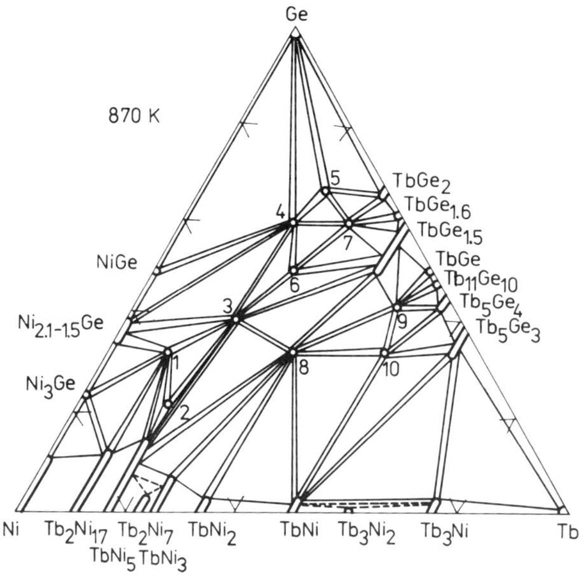

Fig. 108. Tb–Ni–Ge, isothermal section at 870 K.

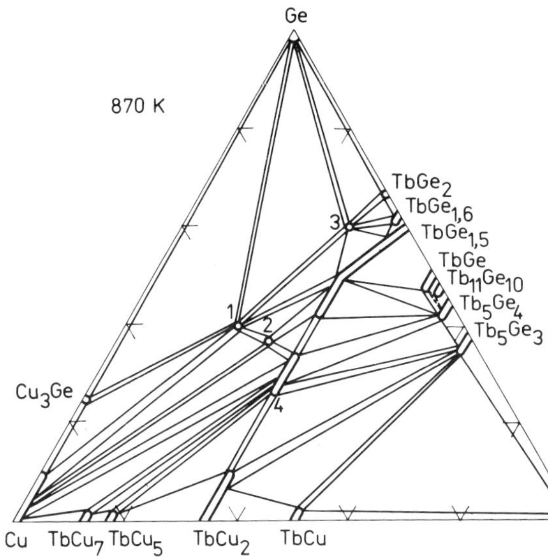

Fig. 109. Tb–Cu–Ge, isothermal section at 870 K.

$TbCu_2Ge_2$ belongs to the $CeAl_2Ga_2$ ST, $a = 4.044$, $c = 10.248$, as reported by Rieger and Parthé (1969b); (2) ~ $Tb_6Cu_8Ge_8$ to the $Gd_6Cu_8Ge_8$ ST, $a = 13.92$, $b = 6.64$, $c = 4.20$ after Hanel and Nowotny (1970), and (4) ~ $TbCu_{1-1.2}Ge_{1-0.8}$, earlier given as $TbCu_{1-0.67}Ge_{1-1.67}$ by Rieger and Parthé (1969a), belongs to the AlB_2 ST, $a = 4.234$–4.225, $c = 3.688$–3.658. The compound (3) is formed at approximate composition Tb_3CuGe_6, and its crystal structure was not determined.

Tm–Fe–Ge

The system Tm–Fe–Ge (fig. 110) was investigated by Fedyna (1988), at 870 K. The total number of compounds found is 10. They are: (1) ~ $TmFe_6Ge_6$, with YCo_6Ge_6 structure type according to Starodub and Mruz (1983), $a = 5.143$, $c = 4.098$; (2) ~ $TmFe_4Ge_2$ ($ZrFe_4Si_2$ ST, $a = 7.223$, $c = 3.746$); (3) ~ $Tm_{0.2}Fe_{0.53}Ge_{0.27}$ (unknown crystal structure); (4) ~ $TmFe_2Ge_2$ ($CeAl_2Ga_2$ ST, $a = 3.930$, $c = 10.418$); (5), (6), and (7) with unknown structure and approximate compositions $Tm_{0.225}Fe_{0.55}Ge_{0.225}$, $Tm_{0.275}Fe_{0.375}Ge_{0.35}$, and $Tm_3Fe_4Ge_3$, respectively; (8) ~ $Tm_9Fe_{10}Ge_{10}$ (a new type of crystal structure according to Pecharsky et al. (1987a): Immm, $a = 5.386$, $b = 13.306$, $c = 13.916$); (9) ~ $TmFe_{0.19}Ge_{0.95}$ ($CeNiSi_2$ ST, $a = 4.066$, $b = 15.705$, $c = 3.976$), and (10) ~ $Tm_{117}Fe_{52}Ge_{112}$ ($Tb_{117}Fe_{52}Ge_{112}$ST, $a = 27.964$).

Tm–Co–Ge

The phase diagram partially studied at 1070 or 670 K is shown in fig. 111, after Fedyna (1988). The crystal structure was established or confirmed for 9 of the total of 15 ternary compounds found. They are: (2) ~ $TmCo_6Ge_6$ (which agreed with Buchholz and Schuster (1981): YCo_6Ge_6 ST, $a = 5.098$, $c = 3.912$); (4) ~ $TmCo_4Ge_2$ ($ZrFe_4Si_2$ ST, $a = 7.229$, $c = 3.746$); (6) ~ $Tm_3Co_4Ge_{13}$ ($Y_3Co_4Ge_{13}$

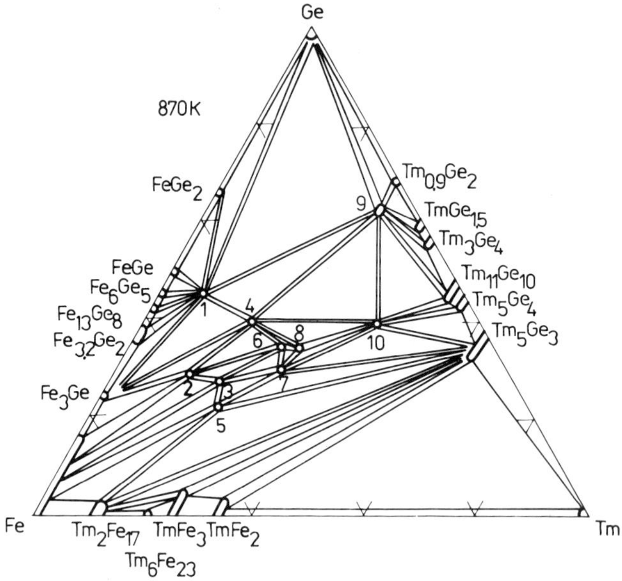

Fig. 110. Tm–Fe–Ge, isothermal section at 870 K.

ST, $a = 8.716$); (7) ~ $TmCo_2Ge_2$ ($CeAl_2Ga_2$ ST, $a = 4.259$, $c = 11.060$); (10) ~ $Tm_5Co_4Ge_{10}$ ($Sc_5Co_4Si_{10}$ ST, $a = 12.622$, $c = 4.139$, which confirms the results of Venturini et al (1984)); (11) ~ $TmCo_{1-x}Ge_{2-y}$, where x and y reach values ranging from 0.91 to 0.17 and 0.72 to 0, respectively, belongs to the $CeNiSi_2$ type of crystal structure, $a = 4.050$–4.065, $b = 15.690$–15.696, $c = 3.960$–3.966; (12) ~ TmCoGe (TiNiSi ST, $a = 6.775$, $b = 4.168$, $c = 7.224$, after Gorelenko et al. (1984)); (13) ~ $Tm_3Co_2Ge_4$ ($Tb_3Co_2Ge_4$ ST, $a = 10.410$, $b = 7.904$, $c = 4.143$,

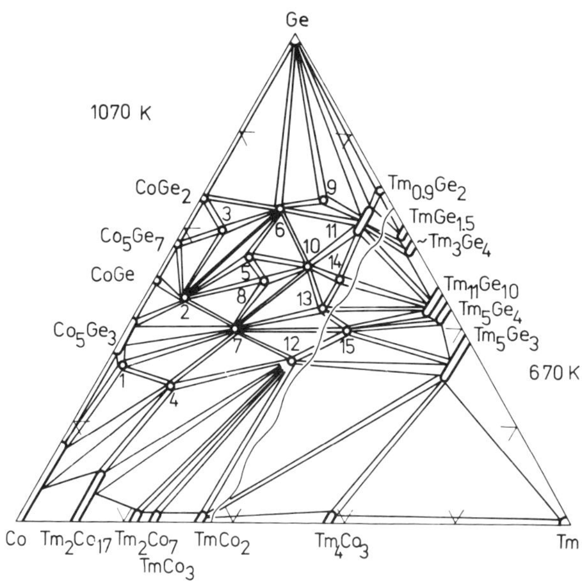

Fig. 111. Tm–Co–Ge, isothermal sections at 1070 K (0–33 a/c Tm) and at 670 K (33–100 a/c Tm). Compounds with unknown crystal structure: (1) ~ $Tm_{0.03}Co_{0.65}Ge_{0.32}$; (3) ~ $Tm_{0.07}Co_{0.33}Ge_{0.60}$; (5) ~ $Tm_{0.15}Co_{0.30}Ge_{0.55}$; (8) ~ $Tm_2Co_3Ge_5$; (9) ~ Tm_2CoGe_6; (14) ~ Tm_2CoGe_3.

$\gamma = 107.04$), and (15) ~ Tm_2CoGe_2 (Sc_2CoSi_2 ST, $a = 10.326$, $b = 10.125$, $c = 4.119$, $\gamma = 117.00$).

Tm–Ni–Ge

Fedyna (1988) proposed the partial isothermal sections at 870 and 670 K and a schematic representation of phase equilibria for arc-melted alloys. Both are shown in fig. 112. The crystal structures have been reported for only 6 of the 12 ternary compounds: (1) ~ $TmNi_5Ge_3$ (YNi_5Si_3 ST, $a = 19.018$, $b = 3.840$, $c = 6.787$); (2) ~ $Tm_3Ni_{11}Ge_4$ ($Sc_3Ni_{11}Si_4$ ST, $a = 8.279$, $c = 8.690$); (4) ~ $TmNi_{2-1.56}Ge_{2-2.20}$ ($CeAl_2Ga_2$ ST, $a = 4.003$–3.985, $c = 9.724$–9.596); (5) ~ $TmNiGe_3$ ($SmNiGe_3$ ST, $a = 21.372$, $b = 4.019$, $c = 4.003$); (9) ~ TmNiGe (TiNiSi ST, $a = 6.790$, $b = 4.183$, $c = 7.235$); (11) ~ $Tm(Ni_{0.25-0.165}Ge_{0.60-0.665})_{1.66}$ (AlB_2 ST, $a = 4.071$, $c = 4.008$ for an unknown composition).

The most differences for annealed and arc-melted alloys are observed in the central part of ternary phase diagram.

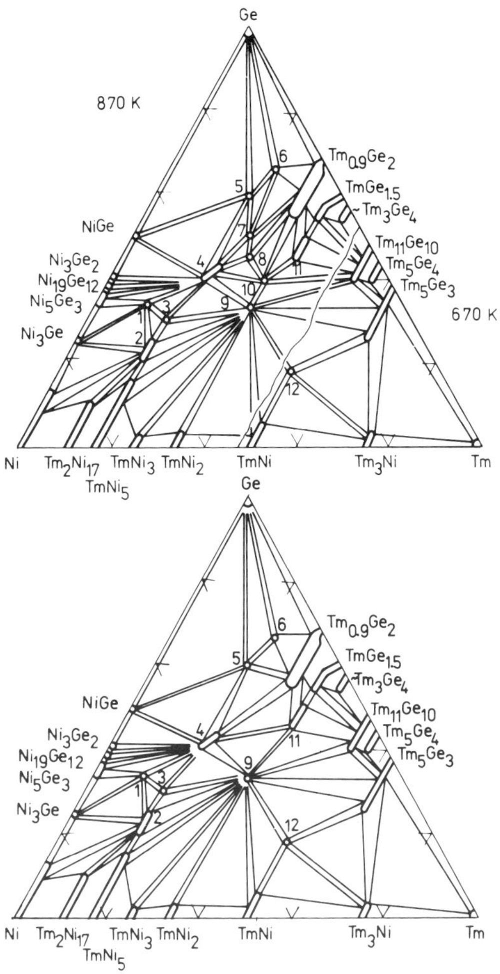

Fig. 112. Tm–Ni–Ge, isothermal sections at 870 K (0–50 a/c Tm) and 670 K (50–100 a/c Tm), and schematic of phase equilibria for arc-melted alloys. Compounds with unknown crystal structure: (3) ~ $Tm_{0.17}Ni_{0.53}Ge_{0.30}$; (6) ~ Tm_2NiGe_6; (7) ~ $TmNiGe_2$; (8) ~ $Tm_{0.27}Ni_{0.28}Ge_{0.45}$; (10) ~ $Tm_{0.33}Ni_{0.27}Ge_{0.40}$; (12) ~ $Tm_{0.50}Ni_{0.32}Ge_{0.18}$.

TABLE 4

Ternary compounds found in the partially investigated systems R–M–Ge.

Compound	Structure type	Lattice parameters a	b	c	Refs.
$ScCr_6Ge_6$	$HfFe_6Ge_6$	5.102	–	8.229	An, 86
$Sc_7Gr_{4+x}Ge_{10-x}$ (x unknown)	$Sc_7Gr_{4.8}Si_{9.2}$	10.150	–	14.352	AnKG, 86
ScMnGe	ZrNiAl	6.677	–	3.946	KoA, 84b
$ScMn_6Ge_6$	$HfFe_6Ge_6$	5.176	–	8.099	An, 86
$Sc_7Mn_{5.3}Ge_{8.7}$	$Sc_7Cr_{4.8}Si_{9.2}$	9.951	–	14.168	AnKG, 86
ScFeGe	ZrNiAl	6.547	–	3.510	KoA, 84b
$ScFe_6Ge_6$	$HfFe_6Ge_6$	5.069	–	8.077	OlAY, 81
$Sc_4Fe_4Ge_{6.6}$	$Zr_4Co_4Ge_7$	13.296	–	5.235	AnKG, 86
$Sc(Sc, Fe, Ge)_2$	$MgCu_2$	7.062	–	–	AnKG, 86
Sc_2Fe_3Ge	$MgZn_2$	5.000	–	8.147	AnKG, 86
$ScCo_6Ge_6$	$HfFe_6Ge_6$	5.056	–	7.799	BuS, 81
$ScCoGe_2$	$TiMnSi_2$	...	...	...	MeVMMMR, 85
ScCoGe	TiNiSi	6.495	4.000	7.072	KoA, 84b
$Sc_4Co_7Ge_6$	$U_4Re_7Si_6$	7.850	–	–	AnKG, 86
$Sc_4Co_4Ge_6$	$Zr_4Co_4Ge_7$	13.215	–	5.229	AnKG, 86
$Sc_3Co_2Ge_3$	$Hf_3Ni_2Si_3$	4.078	9.930	13.243	AnKG, 86
$Sc_3Ni_{11}Ge_4$	$Sc_3Ni_{11}Si_4$	8.130	–	8.505	AnKG, 86
$Sc_4Ni_4Ge_{6.2}$	$Zr_4Co_4Ge_7$	13.276	–	5.225	AnKG, 86
$Sc_6Ni_{18}Ge_{11}$	$Sc_6Ni_{18}Si_{11}$	17.949	12.42	8.23	AnK, 87
ScNiGe	TiNiSi	6.454	4.071	7.068	KoA, 84b
$ScNi_6Ge_6$	$ScNI_6Ge_6$	10.152	–	7.813	BuS, 81
$Sc_3Ni_4Ge_4$	$Gd_6Cu_8Ge_8$	12.910	6.598	3.908	KoS, 82
$ScRuGe_2$	$TiMnSi_2$	...	...	...	MeVMMMR, 85
$ScRhGe_2$	$TiMnSi_2$	...	...	...	MeVMMMR, 85
$ScOsGe_2$	$TiMnSi_2$	...	...	...	MeVMMMR, 85
YMn_2Ge_2	$CeAl_2Ga_2$	3.995	–	10.886	RoMMF, 78
YFe_6Ge_6	YCo_6Ge_6	5.110	–	4.049	StM, 83
$YCo_{0.55}Ge_2$	$CeNiSi_2$	4.106	16.12	4.026	MeVMR, 85
$Y_3Co_4Ge_{13}$	$Y_3Co_4Ge_{13}$	8.769	–	–	BrPB, 86b PeB, 83
$Y_2Co_3Ge_5$	$U_2Co_3Si_5$	9.642	11.79	5.707	VeMMMR, 85
YCo_2Ge_2	$CeAl_2Ga_2$	...	–	...	McNB, 73
YCo_6Ge_6	YCo_6Ge_6	5.074	–	3.908	BuS, 81
YCoGe	TiNiSi	6.880	4.212	7.258	GoSBSYBP, 84
$YNiGe_3$	$SmNiGe_3$	21.657	4.102	4.092	BoPMZVS, 85
YNi_2Ge_2	$CeAl_2Ga_2$	4.043	–	9.763	RiP, 69b
Y_3NiGe_2	La_3NiGe_2	11.304	4.215	11.299	BoBP, 82
YNiGe	TiNiSi	6.970	4.212	7.258	GoSBSYBP, 84
$YCu_{1-0.67}Ge_{1-1.33}$	AlB_2	4.217–4.04	–	3.696–4.05	Rip, 69a
$Y_6Cu_8Ge_8$	$Gd_6Cu_8Ge_8$	13.92	6.65	4.19	HaN, 70
YCu_2Ge_2	$CeAl_2Ga_2$	4.035	–	10.303	RiP, 69b
YRu_2Ge_2	$CeAl_2Ga_2$	4.225	–	9.837	FrVMMR, 85
$Y_2Ru_3Ge_5$	$U_2Co_3Si_5$	9.808	12.43	5.735	VeMMMR, 85
YRh_2Ge_2	$CeAl_2Ga_2$	4.103	–	10.207	FrVMMR, 85
$Y_2Rh_3Ge_5$	$Lu_2Co_3Si_5$	9.963 ($\beta = 91.9°$)	12.045	5.820	VeMMMR, 85
$Y_5Rh_4Ge_{10}$	$Sc_5Co_4Si_{10}$	...	...	...	MeVMMMR, 85

TABLE 4 (cont'd)

Compound	Structure type	Lattice parameters a	b	c	Refs.
YPd_2Ge	YPd_2Si	7.352	7.038	5.562	JoI, 83
$Y_5Os_4Ge_{10}$	$Sc_5Co_4Si_{10}$	13.006	–	4.229	BrYB, 80
$YIrGe_2$	$YIrGe_2$	4.264	15.978	8.813	FrVMMR, 87
$Y_2Ir_3Ge_5$	$U_2Co_3Si_5$	10.133	11.72	5.955	VeMMMR, 85
$Y_4Ir_7Ge_6$	$U_4Re_7Si_6$	8.324	–	–	FrVMMR, 85
$Y_5Ir_4Ge_{10}$	$Sc_5Co_4Si_{10}$	12.927	–	4.308	BrYB, 80
$YPtGe_2$	$YIrGe_2$	4.307	16.29	8.669	FrVMMR, 87
$LaLiGe_2$	$CaLiGe_2$	7.922	4.032	10.946	PaPB, 86
$LaMn_2Ge_2$	$CeAl_2Ga_2$	4.198	–	10.985	RoMMF, 78a
$LaFe_{0.67}Ge_{1.33}$	AlB_2	4.223	–	4.316	FeS, 73
$LaFe_2Ge_2$	$CeAl_2Ga_2$	4.110	–	10.581	RoMF, 78b
$LaCo_2Ge_2$	$CeAl_2Ga_2$	. . .	. . .	. . .	McNB, 73
$LaCo_{0.96}Ge_2$	$CeNiSi_2$	4.309	16.88	4.238	MeVMR, 85
$LaNi_{0.7-0.5}Ge_{1.3-1.5}$	AlB_2	4.202–4.193	–	4.306–4.336	RiP, 69a
$LaNi_2Ge_2$	$CeAl_2Ga_2$	4.187	–	9.918	RiP, 69b
$La_3Ni_4Ge_4$	$U_3Ni_4Si_4$	4.223	4.230	24.156	BrBP, 86a
$La_{11}Ni_4Ge_6$	$La_{11}Ni_4Ge_6$	18.637 ($\beta = 106.13°$)	4.384	14.191	PeBB, 86
La_3NiGe_2	La_3NiGe_2	12.041	4.358	11.871	BoBP, 82
La_8NiGe_5	La_8NiGe_5	15.586	18.384	11.351	BrPBG, 84
$La(Ni_{0.82}Ge_{0.18})_{13}$	$NaZn_{13}$	11.371	–	–	Br, 84
La_2NiGe	La_2NiGe	7.947	–	14.262	BoPG, 83
$LaCu_2Ge_2$	$CeAl_2Ga_2$	4.265	–	10.170	RiP, 69b
$LaCu_{1-0.67}Ge_{1-1.33}$	AlB_2	4.323–4.223	–	4.050–4.322	RiP, 69a
$LaRu_2Ge_2$	$CeAl_2Ga_2$	4.288	–	10.133	FrVMMR, 85
$La_2Ru_3Ge_5$	$U_2Co_3Si_5$	10.013	12.51	5.922	VeMMMR, 85
$LaRh_2Ge_2$	$CeAl_2Ga_2$	4.184	–	10.541	FrVMMR, 85
$La_2Rh_3Ge_5$	$U_2Co_3Si_5$	10.132	12.16	6.027	VeMMMR, 85
$LaPd_2Ge$	YPd_2Si	7.711	6.972	5.828	JoI, 83
$LaPd_2Ge_2$	$CeAl_2Ga_2$	4.338	–	10.063	RoMF, 79
$La_2Ir_3Ge_5$	$U_2Co_3Si_5$	10.218	12.06	6.091	VeMMMR, 85
$LaIr_2Ge_2$	$CaBe_2Ge_2$	4.273	–	10.134	FrVMMR, 85
$LaPt_2Ge_2$	$CeAl_2Ga_2$	4.412	–	9.923	RoMF, 79
Ce_2LiGe_6	Pr_2LiGe_6	4.140	21.054	4.363	PaPB, 88a
$CeLiGe_2$	$CaLiGe_2$	7.815	3.965	10.748	PaPB, 86b
$Ce_{26}Li_5Ge_{22+x}$	$Ce_{26}Li_5Ge_{22+x}$	15.402	–	10.990	PaPB, 87a
$Ce_2Li_2Ge_3$	$Ce_2Li_2Ge_3$	4.480	18.846	6.957	PaPBB, 88b
$GeMn_2Ge_2$	$CeAl_2Ga_2$	4.135	–	10.928	RoMMF, 78a
$CeFe_2Ge_2$	$CeAl_2Ga_2$	4.070	–	10.883	RoMF, 78b
$CeCo_{0.89}Ge_2$	$CeNiSi_2$	4.245	16.74	4.197	MeVMR, 87
$Ce_2Co_3Ge_5$	$U_2Co_3Si_5$	9.814	11.84	5.862	VeMMMR, 85
$Ce(Co, Ge)_2$	AlB_2	4.182	–	4.151	GlB, 65
$CeCo_2Ge_2$	$CeAl_2Ga_2$	4.071	–	10.170	RiP, 69b
$Ce_3Ni_2Ge_7$	$U_3Fe_2Si_7$	4.243	25.728	4.289	Kon, 86
$CeNiGe_2$	$CeNiSi_2$	4.244	16.647	4.199	BoG, 69
$CeNi_2Ge_2$	$CeAl_2Ga_2$	4.150	–	9.854	BoGK, 66
Ce_3NiGe_2	La_3NiGe_2	11.924	4.311	11.624	BoBP, 82
Ce_5NiGe_2	Ce_5NiGe_2	11.760	–	6.429	BeKPB, 87

TABLE 4 (cont'd)

Compound	Structure type	Lattice parameters a	b	c	Refs.
$CeCu_2Ge_2$	$CeAl_2Ga_2$	4.172	–	10.212	RiP, 69b
$CeCu_{1-0.7}Ge_{1-1.3}$	AlB_2	4.296–4.197	–	3.988–4.245	RiP, 69a
$CeRu_2Ge_2$	$CeAl_2Ga_2$	4.270	–	10.088	FrVMMR, 85
$Ce_2Ru_3Ge_5$	$U_2Co_3Si_5$	9.919	12.38	5.887	VeMMMR, 85
$CeRh_2Ge_2$	$CeAl_2Ga_2$	4.160	–	10.438	FrVMMR, 85
$Ce_2Rh_3Ge_5$	$U_2Co_3Si_5$	10.097	12.09	5.987	VeMMMR, 85
$CePd_2Ge$	YPd_2Si	7.668	6.972	5.757	JoI, 83
$CePd_2Ge_2$	$CeAl_2Ga_2$	4.369	–	10.055	RoMF, 79
$CeAg_2Ge_2$	$CeAl_2Ga_2$	4.296	–	10.955	PaPBPV, 86a
$Ce_2Ir_3Ge_5$	$U_2Co_3Si_5$	10.189	11.97	6.060	VeMMMR, 85
$CeIr_2Ge_2$	$CaBe_2Ge_2$	4.246	–	10.098	FrVMMR, 85
$CePtGe_2$	$YIrGe_2$	4.428	16.48	8.795	FrVMMR, 87
$CePt_2Ge_2$	$CeAl_2Ga_2$	4.40	–	9.87	RoMF, 79
$PrLiGe_2$	$CaLiGe_2$	7.813	3.948	10.695	PaPB, 86b
Pr_2LiGe_6	Pr_2LiGe_6	4.127	21.052	4.346	PaPB, 88a
$PrMn_2Ge_2$	$CeAl_2Ga_2$	4.117	–	10.902	RoMMF, 78a
$PrCu_{1-0.67}Ge_{1-1.33}$	AlB_2	4.374–4.183	–	3.971–4.197	RiP, 69a
$PrCu_2Ge_2$	$CeAl_2Ga_2$	4.154	–	10.206	RiP, 69b
$PrRu_2Ge_2$	$CeAl_2Ga_2$	4.263	–	10.022	FrVMMR, 85
$Pr_2Ru_3Ge_5$	$U_2Co_3Si_5$	9.933	12.47	5.846	VeMMMR, 85
$PrRh_2Ge_2$	$CeAl_2Ga_2$	4.147	–	10.436	FrVMMR, 85
$Pr_2Rh_3Ge_5$	$U_2Co_3Si_5$	10.073	12.09	5.969	VeMMMR, 85
$PrPd_2Ge$	YPd_2Si	7.617	6.993	5.735	JoI, 83
$PrIrGe_2$	$YIrGe_2$	4.393	16.17	8.908	FrVMMR, 87
$Pr_2Ir_3Ge_5$	$U_2Co_3Si_5$	10.185	11.90	6.045	VeMMMR, 85
$PrIr_2Ge_2$	$CaBe_2Ge_2$	4.235	–	10.057	FrVMMR, 85
$PrPtGe_2$	$YIrGe_2$	4.416	16.44	8.793	FrVMMR, 87
$NdLiGe_2$	$CaLiGe_2$	7.728	3.915	10.564	PaPB, 86b
$NdMn_2Ge_2$	$CeAl_2Ga_2$	4.105	–	10.912	RoMMF, 78a
$NdFe_6Ge_6$	YCo_6Ge_6	5.142	–	4.096	StM, 83
$NdFe_{0.67}Ge_{1.33}$	AlB_2	4.174	–	4.163	FeS, 73
$NdCo_{0.82}Ge_2$	$CeNiSi_2$	4.210	16.70	4.152	MeVMR, 85
$Nd_6Co_5Ge_{2.2}$	$Nd_6Co_5Ge_{2.2}$	9.272	–	4.188	SaPBB, 86
$Nd_2Co_3Ge_5$	$Lu_2Co_3Si_5$	9.765 ($\beta = 91.1°$)	11.93	5.812	VeMMMR, 85
$NdCo_2Ge_2$	$CeAl_2Ga_2$	4.038	–	10.510	RoMF, 78b
Nd_2CoGe_2	Sc_2CoSi_2	10.882	10.438	4.301	BoPSSMB, 86
$NdCo_{0.5}Ge_{1.5}$	AlB_2	4.144	–	4.151	Sa, 86
$NdNiGe_3$	$SmNiGe_3$	21.733	4.123	4.124	BoPMZVS, 85
$NdNi_{0.6}Ge_{1.4}$	AlB_2	4.157	–	4.145	GlB, 65
$NdNi_2Ge_2$	$CeAl_2Ga_2$	4.115	–	9.842	RiP, 69b
Nd_3NiGe_2	La_3NiGe_2	11.764	4.279	11.593	BoBP, 82
$NdCu_2Ge_2$	$CeAl_2Ga_2$	4.129	–	10.216	RiP, 69b
$NdCu_{0.7-0.6}Ge_{1.3-1.4}$	AlB_2	4.151–4.172	–	4.188–4.160	Sa, 86
$NdCuGe$	AlB_2	4.277	–	3.881	Sa, 86
$NdRu_2Ge_2$	$CeAl_2Ga_2$	4.250	–	9.979	FrVMMR, 85
$Nd_2Ru_3Ge_5$	$U_2Co_3Si_5$	9.897	12.43	5.818	VeMMMR, 85

TABLE 4 (cont'd)

Compound	Structure type	Lattice parameters *a*	*b*	*c*	Refs.
$NdRh_2Ge_2$	$CeAl_2Ga_2$	4.143	–	10.408	FrVMMR, 85
$Nd_2Rh_3Ge_5$	$U_2Co_3Si_5$	10.041	12.083	5.931	VeMMMR, 85
$NdPd_2Ge$	YPd_2Si	7.598	6.992	5.719	JoI, 83
$NdIrGe_2$	$YIrGe_2$	4.354	16.13	8.891	FrVMMR, 87
$Nd_2Ir_3Ge_5$	$U_2Co_3Si_5$	10.172	11.88	6.030	VeMMMR, 85
$NdIr_2Ge_2$	$CaBe_2Ge_2$	4.222	–	10.032	FrVMMR, 85
$NdPtGe_2$	$YIrGe_2$	4.395	16.41	8.764	FrVMMR, 87
$NdPt_2Ge_2$	$CeAl_2Ga_2$	4.38	–	9.865	RoMF, 79
$SmLiGe_2$	$CaLiGe_2$	7.741	3.873	10.488	PaPB, 86
$SmMn_2Ge_2$	$CeAl_2Ga_2$	4.062	–	10.896	RoMMF, 78a
$SmFe_{0.67}Ge_{1.33}$	AlB_2	...	–	...	FeMCS, 72
$Sm_4Fe_{1-x}Ge_7$	$Sm_4Co_{1-x}Ge_7$	4.067	30.21	4.161	MrPBB, 87
$SmFe_2Ge_2$	$CeAl_2Ga_2$	4.005	–	10.476	MrBBKZ, 83
$SmFe_6Ge_6$	YCo_6Ge_6	5.135	–	4.091	StM, 83
$SmCo_{0.71}Ge_2$	$CeNiSi_2$	4.169	16.49	4.097	MeVMR, 85
$SmCo_{0.75}Ge_3$	$BaNiSn_3$	4.277	–	9.799	Mr, 86
$Sm_3Co_{1.44}Ge_7$	$U_3Fe_2Si_7$	4.193	25.534	4.216	Mr, 86
$Sm_2Co_3Ge_5$	$Lu_2Co_3Si_5$	9.718 ($\beta = 91.3°$)	11.93	5.768	VeMMMR, 85
$SmCo_2Ge_2$	$CeAl_2Ga_2$	4.016	–	10.133	MrBBKZ, 83
Sm_2CoGe_2	Sc_2CoSi_2	10.818 ($\gamma = 118.53°$)	10.343	4.254	BoPSSMB, 86
$Sm_4Co_{1-x}Ge_7$	$Sm_4Co_{1-x}Ge_7$	4.083	30.18	4.166	MrPBB, 87
$SmNiGe_3$	$SmNiGe_3$	21.657	4.102	4.092	BoPMZVS, 85
$SmNi_{0.6}Ge_{1.4}$	AlB_2	4.135	–	4.059	GlB, 65
$SmNi_2Ge_2$	$CeAl_2Ga_2$	4.086	–	9.809	RiP, 69b
Sm_3NiGe_2	La_3NiGe_2	11.605	4.242	11.470	BoBP, 82
$SmCu_{1-0.67}Ge_{1-1.33}$	AlB_2	4.255–4.152	–	3.847–4.101	RiP, 69a
$SmCu_2Ge_2$	$CeAl_2Ga_2$	4.088	–	10.223	RiP, 69b
$Sm_6Cu_8Ge_8$	$Gd_6Cu_8Ge_8$	14.12	6.68	4.26	HaN, 70
$SmRu_2Ge_2$	$CeAl_2Ga_2$	4.238	–	9.953	FrVMMR, 85
$Sm_2Ru_3Ge_5$	$U_2Co_3Si_5$	9.867	12.44	5.790	VeMMMR, 85
$SmRh_2Ge_2$	$CeAl_2Ga_2$	4.125	–	10.305	FrVMMR, 85
$Sm_2Rh_3Ge_5$	$Lu_2Co_3Si_5$	10.005 ($\beta = 91.5°$)	12.084	5.894	VeMMMR, 85
$SmPd_2Ge$	YPd_2Si	7.501	6.996	5.673	JoI, 83
$SmIrGe_2$	$YIrGe_2$	4.314	16.03	8.871	FrVMMR, 87
$Sm_2Ir_3Ge_5$	$U_2Co_3Si_5$	10.151	11.82	6.005	VeMMMR, 85
$SmIr_2Ge_2$	P4/nmm	4.203	–	10.084	FrVMMR, 85
$EuLiGe_2$	$CaLiGe_2$	8.174	3.977	10.988	PaPB, 86
$EuMg_3Ge_3$	$EuMg_3Ge_3$	4.485	30.60	4.485	ZmGB, 73
$EuMn_2Ge_2$	$CeAl_2Ga_2$	4.244	–	10.75	FeN, 78
$EuRu_2Ge_2$	$CeAl_2Ga_2$	4.276	–	10.204	FrVMMR, 85
$EuRh_2Ge_2$	$CeAl_2Ga_2$	4.164	–	10.601	FrVMMR, 85
$EuPd_2Ge$	YPd_2Si	7.582	6.981	5.834	JoI, 83
EuPdGe	EuNiGe	6.132 ($\gamma = 132.64°$)	7.934	7.441	Be, 88
$EuIr_2Ge_2$	$CeAl_2Ga_2$	4.172	–	10.543	FrVMMR, 85

TABLE 4 (cont'd)

Compound	Structure type	Lattice parameters *a*	*b*	*c*	Refs.
$GdMn_2Ge_2$	$CeAl_2Ga_2$	4.031	–	10.900	RoMMF, 78a
$GdFe_6Ge_6$	YCo_6Ge_6	5.118	–	4.056	StM, 83
$GdFe_2Ge_2$	$CeAl_2Ga_2$	3.989	–	10.485	RoMF, 78b
$GdCo_{0.64}Ge_2$	$CeNiSi_2$	4.139	16.27	4.062	MeVMR, 85
$Gd_3Co_2Ge_4$	$Tb_3Co_2Ge_4$	10.472 ($\gamma = 107.84°$)	8.059	4.177	Fe, 88
$Gd_3Co_4Ge_{13}$	$Y_3Co_4Ge_{13}$	8.791	–	–	BoSFPG, 87
$GdCo_6Ge_6$	YCo_6Ge_6	5.016	–	3.931	BuS, 81
$Gd_2Co_3Ge_5$	$Lu_2Co_3Si_5$	9.683 $\beta = 91.65$	11.83	5.750	VeMMMR, 85
$GdCo_2Ge_2$	$CeAl_2Ga_2$	3.996	–	10.096	RiP, 69b
Gd_2CoGe_2	Sc_2CoSi_2	10.647 ($\gamma = 118.48°$)	10.259	4.224	BoPSSMB, 86
GdCoGe	TiNiSi	6.995	4.269	7.253	GoSBSYBP, 84
$GdNi_5Ge_3$	YNi_5Si_3	19.072	3.880	6.792	FePSKB, 87
$GdNiGe_3$	$SmNiGe_3$	21.545	4.079	4.073	BoPMZVS, 85
$GdNi_{0.7-0.6}Ge_{1.3-1.4}$	AlB_2	4.111–4.101	–	4.033–4.053	CoFMR, 77
$GdNi_2Ge_2$	$CeAl_2Ga_2$	4.063	–	9.783	RiP, 69b
Gd_3NiGe_2	La_3NiGe_2	11.450	4.220	11.355	BoBP, 82
GdNiGe	TiNiSi	7.009	4.280	7.283	GoSBSYBP, 84
$Gd_6Cu_8Ge_8$	$Gd_6Cu_8Ge_8$	14.000	6.655	4.223	Ri, 70
$GdCu_{1.5}Ge_{0.5}$	AlB_2	...	–	...	Ri, 70
$GdCu_{1-0.67}Ge_{1-1.33}$	AlB_2	4.228–4.136	–	3.796–4.041	RiP, 69a
$GdCu_2Ge_2$	$CeAl_2Ga_2$	4.057	–	10.248	RiP, 69b
$GdRu_2Ge_2$	$CeAl_2Ga_2$	4.231	–	9.895	FrVMMR, 85
$Gd_2Bu_3Ge_5$	$U_2Co_3Si_5$	9.840	12.41	5.767	VeMMMR, 85
$Gd_5Rh_4Ge_{10}$	$Sc_5Co_4Si_{10}$	...	–	...	MeVMMMR, 85
$GdRh_2Ge_2$	$CeAl_2Ga_2$	4.127	–	10.142	RoMF, 79
$Gd_2Rh_3Ge_5$	$Lu_2Co_3Si_5$	9.993 ($\beta = 91.7°$)	12.062	5.875	VeMMMR, 85
$GdPd_2Ge$	YPd_2Si	7.431	7.014	5.607	JoI, 83
$GdPd_2Ge_2$	$CeAl_2Ga_2$	4.255	–	10.035	RoMF, 79
$Gd_5Ir_4Ge_{10}$	$Sc_5Co_4Si_{10}$	...	–	...	MeVMMMR, 85
$GdIrGe_2$	$YIrGe_2$	4.287	16.00	8.845	FrVMMR, 87
$Gd_2Ir_3Ge_5$	$U_2Co_3Si_5$	10.127	11.78	5.980	VeMMMR, 85
$GdIr_2Ge_2$	Pnmm	4.113	4.207	10.167	FrVMMR, 85
$GdPt_2Ge_2$	$CeAl_2Ga_2$	4.36	–	9.77	RoMF, 79
$TbMn_2Ge_2$	$CeAl_2Ga_2$	9.999	–	10.850	RoMMF, 78
$TbRu_2Ge_2$	$CeAl_2Ga_2$	4.221	–	9.879	FrVMMR, 85
$Tb_2Ru_3Ge_5$	$U_2Co_3Si_5$	9.793	12.41	5.742	VeMMMR, 85
$TbRh_2Ge_2$	$CeAl_2Ga_2$	4.118	–	10.280	FrVMMR, 85
$Tb_2Rh_3Ge_5$	$Lu_2Co_3Si_5$	9.951 ($\beta = 91.9°$)	12.046	5.849	VeMMMR, 85
$Tb_5Rh_4Ge_{10}$	$Sc_5Co_4Si_{10}$	...	–	...	MeVMMMR, 85
$TbPd_2Ge$	YPd_2Si	7.390	7.028	5.598	JoI, 83
$TbIrGe_2$	$YIrGe_2$	4.266	15.95	8.829	FrVMMR, 87
$Tb_2Ir_3Ge_5$	$U_2Co_3Si_5$	10.122	12.76	5.790	VeMMMR, 85

TABLE 4 (cont'd)

Compound	Structure type	Lattice parameters a	b	c	Refs.
$Tb_4Ir_7Ge_6$	$U_4Re_7Si_6$	8.345	–	–	FrVMMR, 85
$TbIr_2Ge_2$	P4/nmm	4.144	–	10.222	FrVMMR, 85
$Tb_5Ir_4Ge_{10}$	$Sc_5Co_4Si_{10}$	...	–	...	MeVMMMR, 85
$DyMn_2Ge_2$	$CeAl_2Ga_2$	3.980	–	10.737	RoMMF, 78a
$DyFe_6Ge_6$	YCo_6Ge_6	5.108	–	4.044	StM, 83
$DyFe_2Ge_2$	$CeAl_2Ga_2$	3.957	–	10.446	RoMF, 78b
$DyFe_4Ge_2$	$ZrFe_4Si_2$	7.303	–	3.865	Fe, 88
$Dy_3Ni_{11}Ge_4$	$Sc_3Ni_{11}Si_4$	8.317	–	8.807	FePSKB, 87
$DyNi_5Ge_3$	YNi_5Si_3	19.065	3.864	6.787	FePSKB, 87
$DyNi_{0.6}Ge_{1.4}$	AlB_2	4.081	–	3.991	CoFMR, 77
$DyNi_2Ge_2$	$CeAl_2Ga_2$	4.035	–	9.758	RiP, 69b
$DyNiGe_3$	$SmNiGe_3$	21.430	4.054	4.042	BoPMZVS, 85
Dy_3NiGe_2	La_3NiGe_2	11.272	4.207	11.218	BoBP, 82
DyNiGe	TiNiSi	6.903	4.213	7.285	GoSBSYBP, 84
$DyCo_{0.56}Ge_2$	$CeNiSi_2$	4.098	16.07	4.020	MeVMR, 85
$Dy_3Co_2Ge_4$	$Tb_3Co_2Ge_4$	10.590 ($\gamma = 107.83°$)	8.114	4.135	Fe, 88
$Dy_3Co_4Ge_{13}$	$Y_3Co_4Ge_{13}$	8.750	–	–	BoSFPG, 87
Dy_2CoGe_2	Sc_2CoSi_2	10.467 ($\gamma = 117.33°$)	10.132	4.178	BoPSSMB, 86
$DyCo_6Ge_6$	YCo_6Ge_6	5.081	–	3.918	BuS, 81
$Dy_2Co_3Ge_5$	$Lu_2Co_3Si_5$	9.627 ($\beta = 92.0°$)	11.80	5.707	VeMMMR, 85
DyCoGe	TiNiSi	6.886	4.204	7.256	GoSBSYBP, 84
$Dy_6Cu_8Ge_8$	$Gd_6Cu_8Ge_8$	13.845	6.618	4.173	Ri, 70
$DyCu_{1-0.67}Ge_{1-1.33}$	AlB_2	4.221–4.038	–	3.664–3.637	RiP, 69a
$DyCu_2Ge_2$	$CeAl_2Ga_2$	4.029	–	10.281	RiP, 69b
$DyRu_2Ge_2$	$CeAl_2Ga_2$	4.219	–	9.856	FrVMMR, 85
$Dy_2Ru_3Ge_5$	$U_2Co_3Si_5$	9.783	12.41	5.723	VeMMMR, 85
$Dy_4Rh_7Ge_6$	$U_4Re_7Si_6$	8.326	–	–	FrVMMR, 85
$Dy_2Rh_3Ge_5$	$Lu_2Co_3Si_5$	9.940 ($\beta = 92.0°$)	12.035	5.827	VeMMMR, 85
$DyRh_2Ge_2$	$CeAl_2Ga_2$	4.104	–	10.253	FrVMMR, 85
$Dy_5Rh_4Ge_{10}$	$Sc_5Co_4Si_{10}$	...	–	...	MeVMMMR, 85
$DyPd_2Ge$	YPd_2Si	7.355	7.033	5.565	JoI, 83
$DyPd_2Ge_2$	$CeAl_2Ga_2$	4.035	–	9.758	RoMF, 79
$Dy_5Os_4Ge_{10}$	$Sc_5Co_4Si_{10}$	...	–	...	MeVMMMR, 85
$Dy_5Ir_4Ge_{10}$	$Sc_5Co_4Si_{10}$	...	–	...	MeVMMMR, 85
$DyIrGe_2$	$YIrGe_2$	4.256	15.93	8.815	FrVMMR, 87
$Dy_2Ir_3Ge_5$	$U_2Co_3Si_5$	10.122	11.71	5.950	VeMMMR, 85
$Dy_4Ir_7Ge_6$	$U_4Re_7Si_6$	8.322	–	–	FrVMMR, 85
$DyPt_2Ge_2$	$CeAl_2Ga_2$	4.34	–	9.75	RoMF, 79
$HoMn_2Ge_2$	$CeAl_2Ga_2$	3.977	–	10.845	RoMMF, 78a
$HoFe_6Ge_6$	YCo_6Ge_6	5.110	–	4.056	StM, 83
$HoFe_4Ge_2$	$ZrFe_4Si_2$	7.255	–	3.852	Fe, 88
$HoCo_{0.48}Ge_2$	$CeNiSi_2$	4.094	15.95	4.009	MeVMR, 85
$Ho_3Co_2Ge_4$	$Tb_3Co_2Ge_4$	10.528 ($\gamma = 107.36°$)	7.958	4.152	Fe, 88

TABLE 4 (cont'd)

Compound	Structure type	Lattice parameters			Refs.
		a	b	c	
$Ho_3Co_4Ge_{13}$	$Y_3Co_4Ge_{13}$	8.742	–	–	BoSFPG, 87
Ho_2CoGe_2	Sc_2CoSi_2	10.408	10.109	4.166	BoPSSMB, 86
		($\gamma = 117.45°$)			
$HoCo_6Ge_6$	YCo_6Ge_6	5.074	–	3.910	BuS, 81
$Ho_2Co_3Ge_5$	$Lu_2Co_3Ge_5$	9.618	11.805	5.691	VeMMMR, 85
		($\beta = 91.9°$)			
$HoCo_2Ge_2$	$CeAl_2Ga_2$	...	–	...	McNB, 73
HoCoGe	TiNiSi	6.851	4.198	7.253	GoSBYBP, 84
$Ho_3Ni_{11}Ge_4$	$Sc_3Ni_{11}Si_4$	8.292	–	8.682	FePSKB, 87
$HoNi_5Ge_3$	YNi_5Si_3	19.050	3.856	7.785	FePSKB, 87
$HoNiGe_3$	$SmNiGe_3$	21.434	4.038	4.047	BoPMZVS, 85
$HoNi_{0.4}Ge_{1.6}$	AlB_2	4.080	–	3.958	CoFMR, 77
$HoNi_2Ge_2$	$CeAl_2Ga_2$	4.021	–	9.757	RiP, 69b
Ho_3NiGe_2	La_3NiGe_2	11.195	4.197	11.192	BoGP, 82
HoNiGe	TiNiSi	6.856	4.204	7.256	GoSBSYBP, 84
$Ho_6Cu_8Ge_8$	$Gd_6Cu_8Ge_8$	13.84	6.62	4.17	HaN, 70
$HoCu_{1-0.67}Ge_{1-1.33}$	AlB_2	4.225– 4.00	–	3.608– 4.02	RiP, 69a
$HoCu_2Ge_2$	$CeAl_2Ga_2$	4.016	–	10.302	RiP, 69b
$HoRu_2Ge_2$	$CeAl_2Ga_2$	4.211	–	9.801	FrVMMR, 85
$Ho_2Ru_3Ge_5$	$U_2Co_3Si_5$	9.771	12.41	5.706	VeMMMR, 85
$Ho_4Rh_7Ge_6$7	$U_4Re_7Si_6$	8.305	–	–	FrVMMR, 85
$HoRh_2Ge_2$	$CeAl_2Ga_2$	4.101	–	10.196	FrVMMR, 85
$Ho_2Rh_3Ge_5$	$Lu_2Co_3Si_5$	9.930	11.976	5.813	VeMMMR, 85
		($\beta = 91.73°$)			
$Ho_5Rh_4Ge_{10}$	$Sc_5Co_4Si_{10}$	...	–	...	MeVMMMR, 85
$HoPd_2Ge$	YPd_2Si	7.318	7.036	5.553	JoI, 83
$Ho_5Os_4Ge_{10}$	$Sc_5Co_4Si_{10}$	...	–	...	MeVMMMR, 85
$Ho_5Ir_4Ge_{10}$	$Sc_5Co_4Si_{10}$	...	–	...	MeVMMMR, 85
$Ho_4Ir_7Ge_6$	$U_4Re_7Si_6$	8.311	–	–	FrVMMR, 85
$HoIrGe_2$	$YIrGe_2$	4.244	15.95	8.829	FrVMMR, 87
$ErMn_2Ge_2$	$CeAl_2Ga_2$	3.977	–	10.845	RoMMF, 78
$ErFe_6Ge_6$	YCo_6Ge_6	5.100	–	4.041	StM, 83
$ErFe_2Ge_2$	$CeAl_2Ga_2$	3.941	–	10.414	MaSRO, 75
$ErFe_4Ge_2$	$ZrFe_4Si_2$	7.229	–	3.842	Fe, 88
$ErCo_{0.47}Ge_2$	$CeNiSi_2$	4.078	15.86	3.985	MeVMR, 85
$Er_3Co_2Ge_4$	$Tb_3Co_2Ge_4$	10.459	7.938	4.148	Fe, 88
		($\gamma = 107.20°$)			
$Er_3Co_4Ge_{13}$	$Y_3Co_4Ge_{13}$	8.731	–	–	BoSFPG, 87
$ErCo_4Ge_2$	$ZrFe_4Si_2$	7.217	–	3.743	Fe, 88
Er_2CoGe_2	Sc_2CoSi_2	10.359	10.181	4.142	BoPSSMB, 86
		($\gamma = 116.89°$)			
$ErCo_6Ge_6$	YCo_6Ge_6	5.074	–	3.910	BuS, 81
$Er_2Co_3Ge_5$	$Lu_2Co_3Si_5$	9.601	11.78	5.680	VeMMMR,85
		($\beta = 91.9°$)			
$ErCo_2Ge_2$	$CeAl_2Ga_2$	...	–	...	McNB, 73
ErCoGe	TiNiSi	6.815	4.181	7.245	GoSBSYBP, 84
$Er_3Ni_{11}Ge_4$	$Sc_3Ni_{11}Si_4$	8.273	–	8.680	FePSKB, 87
$ErNi_5Ge_3$	YNi_5Si_3	19.030	3.848	6.784	FePSKB, 87

TABLE 4 (cont'd)

Compound	Structure type	Lattice parameters a	b	c	Refs.
$ErNiGe_3$	$SmNiGe_3$	21.396	4.031	4.031	BoPMZVS, 85
$ErNi_{0.6}Ge_{1.4}$	AlB_2	4.068	–	3.970	CoFMR, 77
$ErNi_2Ge_2$	$CeAl_2Ga_2$	4.016	–	9.733	RiP, 69b
Er_3NiGe_2	La_3NiGe_2	11.180	4.131	11.124	BoBP, 82
ErNiGe	TiNiSi	6.818	4.188	7.239	GoSBSYBP, 84
$Er_6Cu_8Ge_8$	$Gd_6Cu_8Ge_8$	13.790	6.617	4.153	Ri, 70
$ErCu_{0.67}Ge_{1.33}$	AlB_2	3.98	–	4.04	RiP, 69a
$ErCu_2Ge_2$	$CeAl_2Ga_2$	4.003	–	10.317	RiP, 69b
$ErRu_2Ge_2$	$CeAl_2Ga_2$	4.211	–	9.801	FrVMMR, 85
$Er_2Ru_3Ge_5$	$U_2Co_3Si_5$	9.752	12.40	5.696	VeMMMR, 85
$Er_4Rh_7Ge_6$	$U_4Re_7Si_6$	8.296	–	–	FrVMMR, 85
$ErRh_2Ge_2$	$CeAl_2Ga_2$	4.094	–	10.170	FeVMMR, 85
$Er_2Rh_3Ge_5$	$Lu_2Co_3Si_5$	9.933 ($\beta = 92.1°$)	12.025	5.808	VeMMMR, 85
$Er_5Rh_4Ge_{10}$	$Sc_5Co_4Si_{10}$	...	–	...	MeVMMMR, 85
$ErPd_2Ge$	YPd_2Si	7.296	7.038	5.551	JoI, 83
$ErOsGe_2$	$TiMnSi_2$	...	...	...	MeVMMMR, 85
$Er_4Os_7Ge_6$	$U_4Re_7Si_6$	8.314	–	–	FrVMMR, 85
$Er_5Os_4Ge_{10}$	$Sc_5Co_4Si_{10}$	...	–	...	MeVMMMR, 85
$Er_5Ir_4Ge_{10}$	$Sc_5Co_4Si_{10}$	...	–	...	MeVMMMR, 85
$Er_4Ir_7Ge_6$	$U_4Re_7Si_6$	8.295	–	–	FrVMMR, 85
$ErIrGe_2$	$YIrGe_2$	4.227	15.94	8.783	FrVMMR, 87
$TmCu_{1-0.67}Ge_{1-1.33}$	AlB_2	4.220–3.97	–	3.545–3.97	RiP, 69a
$Tm_6Cu_8Ge_8$	$Gd_6Cu_8Ge_8$	13.74	6.60	4.13	HaN, 70
$TmCu_2Ge_2$	$CeAl_2Ga_2$	3.994	–	10.307	RiP, 69b
$TmRu_2Ge_2$	$CeAl_2Ga_2$	4.202	–	9.797	FrVMMR, 85
$Tm_2Ru_3Ge_5$	$U_2Co_3Si_5$	9.725	12.39	5.676	VeMMMR, 85
$Tm_4Rh_7Ge_6$	$U_4Re_7Si_6$	8.276	–	–	FrVMMR, 85
$TmRh_2Ge_2$	$CeAl_2Ga_2$	4.062	–	10.129	FrVMMR, 85
$Tm_2Rh_3Ge_5$	$Lu_2Co_3Si_5$	9.903 ($\beta = 92.4°$)	12.009	5.781	VeMMMR, 85
$Tm_5Rh_4Ge_{10}$	$Sc_5Co_4Si_{10}$	...	–	...	MeVMMMR, 85
$TmPd_2Ge$	YPd_2Si	7.254	7.045	5.523	JoI, 83
$Tm_4Os_7Ge_6$	$U_4Re_7Si_6$	8.276	–	–	FrVMMR, 85
$Tm_5Os_4Ge_{10}$	$Sc_5Co_4Si_{10}$	...	–	...	MeVMMMR, 85
$Tm_5Ir_4Ge_{10}$	$Sc_5Co_4Si_{10}$	...	–	...	MeVMMMR, 85
$Tm_4Ir_7Ge_6$	$U_4Re_7Si_6$	8.273	–	–	FrVMMR, 85
$TmPtGe_2$	$YIrGe_2$	4.263	16.20	8.657	FrVMMR, 87
Yb_8LiGe_{13}	Yb_8LiGe_{13}	11.317 ($\gamma = 106.24°$)	15.577	10.817	PaBBP, 87b
$YbMn_2Ge_2$	$CeAl_2Ga_2$	4.067	–	10.871	RoMMF, 78a
$YbFe_6Ge_6$	YCo_6Ge_6	5.087	–	4.037	StM, 83
$YbFe_2Ge_2$	$CeAl_2Ga_2$	3.924	–	10.503	RoMF, 78b
$YbCo_{0.37}Ge_2$	$CeNiSi_2$	4.077	15.63	3.967	MeVMR, 85
$Yb_3Co_4Ge_{13}$	$Y_3Co_4Ge_{13}$	8.725	–	–	BoSFPG, 87
$YbCo_6Ge_6$	YCo_6Ge_6	5.078	–	3.910	BuS, 81
$YbCo_2Ge_2$	$CeAl_2Ga_2$	...	–	...	McNB, 73
$YbCo_4Ge_2$	$ZrFe_4Si_2$	7.236	–	3.761	Fe, 88

TABLE 4 (cont'd)

Compound	Structure type	Lattice parameters *a*	*b*	*c*	Refs.
$Yb_5Co_4Ge_{10}$	$Sc_5Co_4Si_{10}$	...	–	...	MeVMMMR, 85
$YbNi_{0.6}Ge_{1.4}$	AlB_2	4.193	–	3.965	CoFMR, 77
$YbNi_2Ge_2$	$CeAl_2Ga_2$	4.001	–	9.733	RiP, 69b
YbNiGe	TiNiSi	6.740	4.162	7.217	GoSBSYBP, 84
$YbNi_5Ge_3$	YNi_5Si_3	18.892	3.830	6.779	FePSKB, 87
$Yb_3Ni_{11}Ge_4$	$Sc_3Ni_{11}Si_4$	8.259	–	8.648	FePSKB, 87
$Yb_6Cu_8Ge_8$	$Gd_6Cu_8GE_8$	13.71	6.59	4.13	HaN, 70
$YbCu_2Ge_2$	$CeAl_2Ga_2$	4.045	–	10.228	RiP, 69b
$YbRu_2Ge_2$	$CeAl_2Ga_2$	4.203	–	9.763	FrVMMR, 85
$YbRh_2Ge_2$	$CeAl_2Ga_2$	4.062	–	10.065	FrVMMR, 85
$Yb_4Rh_7Ge_6$	$U_4Re_7Si_6$	8.254	–	–	FrVMMR, 85
$Yb_5Rh_4Ge_{10}$	$Sc_5Co_4Si_{10}$	...	–	...	MeVMMMR, 85
$YbPd_2Ge$	...	...	...	...	JoI, 83
$Yb_4Os_7Ge_6$	$U_4Re_7Si_6$	8.288	–	–	FrVMMR, 85
$Yb_4Ir_7Ge_6$	$U_4Re_7Si_6$	8.262	–	–	FeVMMR, 85
$Yb_5Ir_4Ge_{10}$	$Sc_5Co_4Si_{10}$	...	–	...	MeVMMMR, 85
$LuFe_6Ge_6$	YCo_6Ge_6	5.087	–	4.034	StM, 83
$LuFe_4Ge_2$	$ZrFe_4Si_2$	7.200	–	3.854	Fe, 88
$Lu_3Fe_2Ge_3$	$Hf_3Ni_2Si_3$	4.127	10.471	13.611	Fe, 88
$LuCo_{0.34}Ge_2$	$CeNiSi_2$	4.018	15.54	5.939	MeVMR, 85
$Lu_3Co_4Ge_{13}$	$Y_3Co_4Ge_{13}$	8.694	–	–	BoSFPG, 87
Lu_2CoGe_2	Sc_2CoSi_2	10.234 ($\gamma = 117.31°$)	10.049	4.109	BoPSSMB, 86
LuCoGe	TiNiSi	6.720	4.140	7.226	GoSBSYBP, 84
$LuCo_4Ge_2$	$ZrFe_4Si_2$	7.199	–	3.735	Fe, 88
$LuCoGe_2$	$TiMnSi_2$	...	...	...	MeVMMMR, 85
$LuNiGe_3$	$SmNiGe_3$	21.305	3.932	3.983	BoPMZVS, 85
LuNiGe	TiNiSi	6.727	4.147	7.229	GoSBSYBP, 84
$LuNi_5Ge_3$	YNi_5Si_3	18.973	3.826	6.775	FePSKB, 87
$Lu_3Ni_{11}Ge_4$	$Sc_3Ni_{11}Si_4$	8.249	–	8.641	FePSKB, 87
$Lu_6Cu_8Ge_8$	$Gd_6Cu_8Ge_8$	13.661	6.581	4.113	Ri, 70
$LuCu_{0.67}Ge_{1.33}$	AlB_2	3.83	–	4.05	RiP, 69a
$LuRuGe_2$	$TiMnSi_2$	...	...	...	MeVMMMR, 85
$Lu_4Ru_7Ge_6$	$U_4Re_7Si_6$	8.256	–	–	FrVMMR, 85
$Lu_4Rh_7Ge_6$	$U_4Re_7Si_6$	8.243	–	–	FrVMMR, 85
$Lu_5Rh_4Ge_{10}$	$Sc_5Co_4Si_{10}$	...	–	...	MeVMMMR, 85
$LuPd_2Ge$	YPd_2Si	7.189	7.049	5.512	JoI, 83
$Lu_4Os_7Ge_6$	$U_4Re_7Si_6$	8.276	–	–	FrVMMR, 85
$LuOsGe_2$	$TiMnSi_2$	...	...	...	MeVMMMR, 85
$Lu_4Ir_7Ge_6$	$U_4Re_7Si_6$	8.258	–	–	FrVMMR, 85
$Lu_5Ir_4Ge_{10}$	$Sc_5Co_4Si_{10}$	...	–	...	MeVMMMR, 85

References

Andrusyak, R.I., 1986, Vestn. Lvovsk. Univ., Ser. Khim. **27**, 24.

Andrusyak, R.I., and B.Ya. Kotur, 1987, Kristallografiya **32**, 1018.

Andrusyak, R.I., B.Ya. Kotur and E.I. Gladyshevsky, 1986, Crystal structure of the new ternary scandium germanides, in: XI Resp. Konf. Neorg. Khim., Tezisy Dokl., Kiev, 1968, p. 8.

Belan, B.D., 1988, Phase equilibria, crystal structures and properties of the compounds in the systems Eu–Fe,Co,Ni,Cu–Si,Ge, Ph.D. Chemistry thesis, Lvov, 1988 (Lvov State University) pp. 1–17. .

Belsky, V.K., M.B. Konyk, V.K. Pecharsky and O.I. Bodak, 1987, Kristallografiya **32**, 241.

Bodak, O.I., and E.I. Gladyshevsky, 1969, Kristallografiya **14**, 990.

Bodak, O.I., E.I. Gladyshevsky and P.I. Krypyakevich, 1966, Izv. Akad. Nauk SSSR, Neorg. Mater. **2**, 2151.

Bodak, O.I., V.A. Bruskov and V.K. Pecharsky, 1982, Kristallografiya **27**, 896.

Bodak, O.I., V.K. Pecharsky, O.Ya. Mruz, V.E. Zavodnik, G.M. Vytvitzkaya and P.S. Salamakha, 1985, Dopov. Akad. Nauk Ukr. RSR, Ser. B (2), 37.

Bodak, O.I., V.K. Pecharsky, P.K. Starodub, P.S. Salamakha, O.Ya. Mruz and V.A. Bruskov, 1986, Izv. Akad. Nauk SSSR, Met. (4), 214.

Bodak, O.I., R.V. Skolozdra, M.F. Fedyna, V.K. Pecharsky and Yu.K. Gorelenko, 1987, Anomalous electrokinetical and magnetic properties of the $R_3Co_4Ge_{13}$ germanides, in: IY Vsesoyuzn. Konf. Fiz. Khim. Redkozem. Poluprovodnikov. Tezisy Dokl., Novosibirsk, 1987 (Nauka, Novosibirsk) p. 29.

Braun, H.F., K. Yvon and R.M. Braun, 1980, Acta Crystallogr. B **36**, 2397.

Bruskov, V.A., 1984, New structure types of rare earth germanides, Ph.D. Chemistry thesis, Lvov, 1984 (Lvov State University, Lvov) pp. 1–17.

Bruskov, V.A., V.K. Pecharsky and E.I. Gladyshevsky, 1983, New ternary germanides of lanthanum and nickel, in: Chetvertaya Vsesoyuzn. Konf. Kristallokhim. Intermetal. Soed., Tezisy Dokl., Lvov, 1983 (Vyscha Shkola, Lvov) pp. 95,96.

Bruskov, V.A., V.K. Pecharsky, O.I. Bodak and E.I. Gladyshevsky, 1984, Kristallografiya **29**, 1071.

Bruskov, V.A., O.I. Bodak and V.K. Pecharsky, 1986a, Izv. Akad. Nauk SSSR, Neorg. Mater. **22**, 1573.

Bruskov, V.A., V.K. Pecharsky and O.I. Bodak, 1986b, Izv. Akad. Nauk SSSR, Neorg. Mater. **22**, 1471.

Buchholz, W., and H.-U. Schuster, 1981, Z. Anorg. Allg. Chem. **482**, 40.

Contardi, V., R. Ferro, R. Marazza and D. Rossi, 1977, J. Less-Common Met. **51**, 277.

Dwight, A.E., W.C. Harper and C.W. Kimball, 1973, J. Less-Common Met. **30**, 1.

Fedyna, M.F., 1986, Vestn. Lvovsk. Univ., Ser. Khim. **27**, 29.

Fedyna, M.F., 1988, Phase equilibria, crystal structures and physical properties of compounds in the systems Pr,Tm–Fe,Co,Ni–Ge, Ph.D. Chemistry thesis, Lvov, 1988 (Lvov State University, Lvov) pp. 1–17.

Fedyna, M.F., V.K. Pecharsky, P.K. Starodub, G.M. Koterlyn and O.I. Bodak, 1987, Ternary compounds with structure types YNi_5Si_3 and $Sc_3Ni_{11}Si_4$. Manuscript referred to Ukr NIITI, 07.09.1987, Nr2306–Uk87, pp. 1–15.

Fedyna, M.F., V.K. Pecharsky and O.I. Bodak, 19xx, Dopov. Akad. Nauk Ukr. RSR, Ser. B (2), 50.

Felner, I., and I. Nowik, 1978, J. Phys. Chem. Solids **39**, 763.

Felner, I., and M. Schieber, 1973, Solid State Commun. **13**, 457.

Felner, I., I. Mayer, H. Crill and M. Schieber, 1972, Solid State Commun. **11**, 1231.

Francois, M., G. Venturini, J.F. Mareche, B. Malaman and B. Roques, 1985, J. Less-Common Met. **113**, 231.

Francois, M., G. Venturini, E. McRae, B. Malaman and B. Roques, 1987, J. Less-Common Met. **128**, 249.

Gladyshevsky, E.I., and O.I. Bodak, 1965, Dopov. Akad. Nauk Ukr. RSR, p. 601.

Gorelenko, Yu.K., P.K. Starodub, V.A. Bruskov, R.V. Skolozdra, V.I. Yarovetz, O.I. Bodak and V.K. Pecharsky, 1984, Ukr. Fiz. Zh. **29**, 867.

Hanel, G., and H. Nowotny, 1970, Monatsh. Chem. **101**, 463.
Jorda, J.L., and M. Ishikawa, 1983, J. Less-Common Met. **92**, 155.
Konyk, M.B., 1986, Vestn. Lvovsk. Univ., Ser. Khim. **27**, 35.
Kotur, B.Ya., 1986, Dopov. Akad. Nauk Ukr. RSR, Ser. A (1), 79.
Kotur, B.Ya., 1987, Izv. Akad. Nauk SSSR, Neorg. Mater. **23**, 558.
Kotur, B.Ya., and R.I. Andrusyak, 1984a, Vestn. Lvovsk. Univ., Ser. Khim. **25**, 35.
Kotur, B.Ya., and R.I. Andrusyak, 1984b, Dopov. Akad. Nauk Ukr. RSR, Ser. B (12), 41.
Kotur, B.Ya., and M. Sikirica, 1982, Acta Crystallogr. B **38**, 917.
Malik, S.K., S.G. Saukar, V.U.S. Rao and R. Obermayer, 1975, Magnetic behaviour of some rare earth germanides of the type RFe_2Ge_2, in: Magnetism and Magnet. Mater., 21st Annu. Conf., 1975, Philadelphia, PA, 1976, pp. 585,586.
McCall, W.M., K.S.V.L. Narasimhan and R.A. Butera, 1973, J. Appl. Phys. **44**, 4724.
Meot, M., G. Venturini, E. McRae, J.F. Mareche, B. Malaman and B. Roques, 1985, New ternary rare earth–transition metal germanides with $TiMnSi_2$, $Sc_5Co_4Si_{10}$ or $U_2Co_3Si_5$ type structures. Superconductivity in this compounds, in: YIII Int. Conf. Solid Comp. Trans. Elements (Extendet Abstr., Vienna) pp. P3B9, 1–3.
Meot-Mayer, M., G. Venturini, B. Malaman and B. Roques, 1985, Mater. Res. Bull. **20**, 1515.
Mruz, O.Ya., 1986, Vestn. Lvovsk. Univ., Ser. Khim. **27**, 32.
Mruz, O.Ya., V.K. Belsky, O.I. Bodak, B.Ya. Kotur and V.E. Zavodnik, 1983, Dopov. Akad. Nauk Ukr. RSR, Ser. B (6), 41.
Mruz, O.Ya., V.K. Pecharsky, O.I. Bodak and V.A. Bruskov, 1987, Dopov. Akad. Nauk Ukr. RSR, Ser. B (6), 51.
Olenych, R.R., L.G. Akselrud and Ya.P. Yarmolyuk, 1981, Dopov. Akad. Nauk Ukr. RSR, Ser. A (2), 84.
Oniskovetz, B.D., 1984, Vestn. Lvovsk. Univ., Ser. Khim. **25**, 21.
Oniskovetz, B.D., V.V. Pavlyuk and O.I. Bodak, 1985, Dopov. Akad. Nauk Ukr. RSR, Ser. B (10), 40.
Oniskovetz, B.D., V.K. Belsky, V.K. Pecharsky and O.I. Bodak, 1987, Kristallografiya **32**, 888.
Pavlyuk, V.V., V.K. Pecharsky, O.I. Bodak, A.S. Protzyk and N.B. Vytvitzky, 1986a, Crystal structure of the compounds $Ce_2Li_2Ge_3$ and $CeAg_2Ge_2$, in: XI Ukr. Resp. Konf. Neorg. Khim., Tezisy Dokl., 1986, Kiev, p. 164.
Pavlyuk, V.V., V.K. Pecharsky and O.I. Bodak, 1986b, Dopov. Akad. Nauk Ukr. RSR, Ser. A (7), 76.
Pavlyuk, V.V., V.K. Pecharsky, O.I. Bodak and V.A. Bruskov, 1987a, Kristallografiya **32**, 70.
Pavlyuk, V.V., V.K. Belsky, O.I. Bodak and V.K. Pecharsky, 1987b, Dopov. Akad. Nauk Ukr. RSR, Ser. B (10), 44.
Pavlyuk, V.V., V.K. Pecharsky and O.I. Bodak, 1988a, Krystallografiya **33**, 43.
Pavlyuk, V.V., V.K. Pecharsky, O.I. Bodak and V.A. Bruskov, 1988b, Kristallografiya **33**, 45.
Pecharsky, V.K., and V.A. Bruskov, 1983, Crystal structure of the compound $Y_3Co_4Ge_{13}$, in: Chetvertaya Vsesoyuzn. Konf. Kristallokhim. Intermetal. Soed., Tezisy Dokl., Lvov, 1983 (Vyscha Shkola, Lvov) pp. 96, 97.
Pecharsky, V.K., O.I. Bodak and V.A. Bruskov, 1986, Kristallografiya **31**, 312.
Pecharsky, V.K., M.F. Fedyna, O.I. Bodak and V.E. Zavodnik, 1987a, Dopov. Akad. Nauk Ukr. RSR, Ser. A (5), 43.
Pecharsky, V.K., O.I. Bodak, V.K. Belsky, P.K. Starodub, I.R. Mokraya and E.I. Gladyshevsky, 1987b, Kristallografiya **32**, 334.
Rieger, W., 1970, Monatsh. Chem. **101**, 449.
Rieger, W., and E. Parthe, 1969a, Monat. Chem. **100**, 439.
Rieger, W., and E. Parthe, 1969b, Monatsh. Chem. **100**, 444.
Rossi, D., R. Marazza, D. Mazzone and R. Ferro, 1978a, J. Less-Common Met. **59**, 79.
Rossi, D., R. Marazza and R. Ferro, 1978b, J. Less-Common Met. **58**, 203.
Rossi, D., R. Marazza and R. Ferro, 1979, J. Less-Common Met. **66**, 15.
Salamakha, P.S., 1986, Vestn. Lvovsk. Univ., Ser. Khim. **27**, 27.

Salamakha, P.S., V.K. Pecharsky, V.A. Bruskov and O.I. Bodak, 1986, Kristallografiya **31**, 587.
Starodub, P.K., 1988, Phase equilibria, crystal structures and some physical properties of the ternary compounds in the systems Tb–Fe,Co,Ni,Gu–Ge, Ph.D. Chemistry thesis, Lvov, 1988 (Lvov State University, Lvov) pp. 1–17.
Starodub, P.K., and O.Ya. Mruz, 1983, The new representatives of YCo_6Ge_6 structure type, in: Chetvertaya Vsesoyuzn. Konf. Kristallokhim. Intermetal. Soed., Tezisy Dokl., Lvov, 1983 (Vyscha Shkola, Lvov) p. 91.
Starodub, P.K., I.R. Mokraya, O.I. Bodak, V.K. Pecharsky and V.A. Bruskov, 1986, Kristallografiya **31**, 394.
Venturini, G., M. Meot-Meyer and E. McRae, 1984, Mater. Res. Bull. **19**, 1647.
Venturini, G., M. Meot-Meyer, J.F. Mareche, B. Malaman and B. Roques, 1985, Mater. Res. Bull. **21**, 33.
Zmij, O.F., E.I. Gladyshevsky and V.S. Bulyo, 1973, Kristallografiya **18**, 277.

2.5. *R–M–{Sn, Pb} systems*

Phase equilibria were investigated for only 7 ternary R–M–Sn systems and a total of 61 ternary compounds were found. The crystal structures were solved for 29 of the them. The existence of 163 more ternary compounds (they are listed in table 5) was reported for another 108 partially investigated ternary R–M–Sn systems without any phase equilibria details. The crystal structure was solved for 131 of them.

Only two ternary compounds (see also table 5) are known with lead.

Y-Pd-Sn

The system Y–Pd–Sn was partially investigated by Jorda et al. (1985). Five ternary compounds have been found (fig. 113). Compound (2) ~ $Y_{13}Pd_{40}Sn_{31}$ represents a new structure type: $P6_3/mmc$, $a = 19.891$, $c = 9.246$, and (1) ~

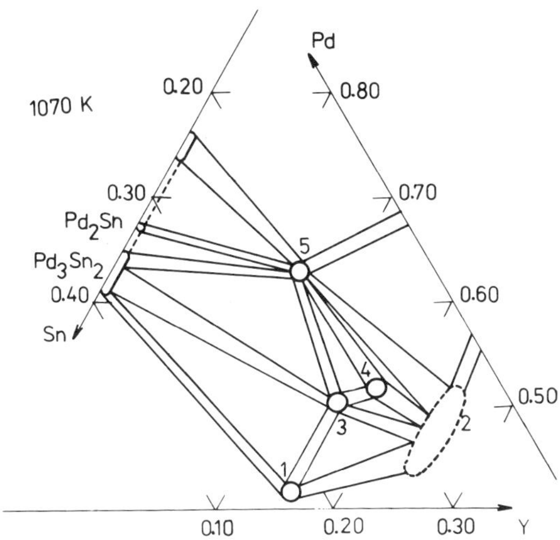

Fig. 113. Y–Pd–Sn, partial isothermal section at 1070 K.

YPd_2Sn belongs to the $MnCu_2Al$ ST, $a = 6.702$. The remaining three ternary compounds have the approximate compositions: (3) ~ $Y_{0.17}Pd_{0.54}Sn_{0.29}$, (4) ~ $Y_{0.19}Pd_{0.54}Sn_{0.27}$ and (5) ~ $P_{0.09}Pd_{0.64}Sn_{0.27}$. The crystal structures are unknown.

Ce–Ni–Sn

Komarovskaya (1984) and Skolozdra et al. (1982b) reported the phase diagram shown in fig. 114. The total number of ternary compounds is 10, for 6 of them the crystal structure was solved, and for 1 only the lattice parameters were established: (1) ~ $CeNi_5Sn$ (new type of crystal structure according to Skolozdra et al. (1981a): $P6_3/mmc$, $a = 4.905$, $c = 19.731$); (2) ~ $Ce_5Ni_{1.4}Sn_{1.6}$ (Cr_5B_3 ST, $a = 7.975$, $c = 13.59$); (5) ~ CeNiSn (TiNiSi ST, $a = 7.523$, $b = 4.592$, $c = 7.561$); (6) ~ $CeNi_2Sn_2$ (after Skolozdra et al. (1981b) belongs to the $CeAl_2Ga_2$ ST, $a = 4.436$, $c = 10.14$); (8) ~ $CeNiSn_2$ ($CeNiSi_2$ structure type, $a = 4.485$, $b = 17.74$, $c = 4.485$), as it was reported by Skolozdra (1977). The compound has variable composition from 50 to 60 a/c Sn, and 25 to 10 a/c Ni); (9) ~ $CeNi_{2.67}Sn_{5.44}$ ($GdNi_{2.67}Sn_{5.44}$ ST, $a = 11.934$, after Komarovskaya and Skolozdra (1985)). For (4) ~ $CeNi_4Sn_2$ a hexagonal unit cell with $a = 7.730$, $c = 7.869$ has been established.

Pr–Cu–Sn

Phase equilibria were established by Komarovskaya (1984) and are shown in fig. 115. Skolozdra and Komarovskaya (1982b) reported for (7) ~ $PrCu_2Sn_2$ $CaBe_2Ge_2$ type structure, $a = 4.435$, $c = 10.469$; Komarovskaya et al. (1983b) for (5) ~ PrCuSn the $CaIn_2$ ST, $a = 4.583$, $c = 7.759$. The remaining 4 (of a total of

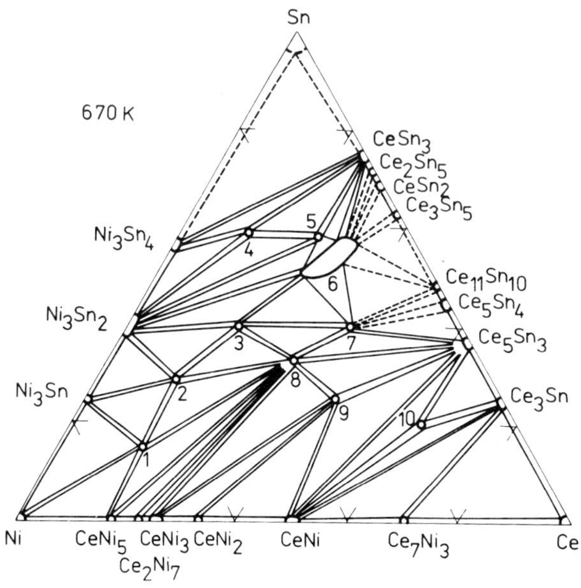

Fig. 114. Ce–Ni–Sn, isothermal section at 670 K. Compounds with unknown crystal structure: (3) ~ $Ce_9Ni_6Sn_5$; (7) ~ $Ce_{0.37}Ni_{0.20}Sn_{0.43}$; (10) ~ $Ce_3Ni_4Sn_{13}$.

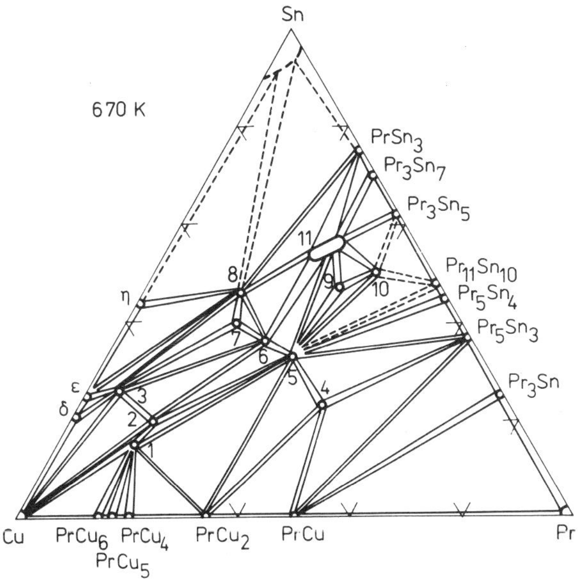

Fig. 115. Pr–Cu–Sn, isothermal section at 670 K. Compounds with unknown crystal structure: (1) ~ $PrCu_5Sn$; (2) ~ $Pr_3Cu_{13}Sn_4$; (4) ~ $Pr_{0.49}Cu_{0.28}Sn_{0.23}$; (8) ~ $Pr_{0.35}Cu_{0.18}Sn_{0.47}$; (11) ~ $Pr_{0.15}Cu_{0.27}Sn_{0.58}$.

11) ternary compounds for which crystallographic data exist are as follows: (3) ~ $PrCu_{9.4}Sn_{3.6}$ ($NaZn_{13}$ ST, $a = 12.162$); (6) ~ $Pr_6Cu_8Sn_8$ ($Gd_6Cu_8Ge_8$ ST, $a = 14.820$, $b = 6.990$, $c = 4.548$); (9) ~ $PrCu_{0.25}Sn_{1.25}$ (AlB_2 ST, $a = 4.393$, $c = 4.464$); and (10) ~ $PrCu_{1-x}Sn_{2-y}$ ($x = 0.28$–0.66, $y = 0.18$–0.22) belongs to the $CeNiSi_2$ ST, $a = 4.486$, $b = 17.770$, $c = 4.433$.

Gd–Ni–Sn

The phase diagram derived by Skolozdra et al. (1982b) and Komarovskaya (1984) is shown in fig. 116. Crystal structures were reported for 7 of a total of 15 existing ternary compounds: (2) ~ Gd_6Ni_2Sn (Ho_6Co_2Ga ST, $a = 9.480$, $b = 9.658$, $c = 10.023$); (5) ~ $Gd_3Ni_{12}Sn_5$ (orthorhombic lattice with $a = 24.3$, $b = 4.9$, $c = 10.4$); (6) ~ $GdNi_3Sn_2$ ($HoNi_{2.6}Ga_{2.4}$ ST, $a = 9.197$, $c = 4.271$); (7) ~ GdNiSn (TiNiSi ST, $a = 7.198$, $b = 4.469$, $c = 7.671$, which is in good agreement with the results of Dwight (1983)); (8) ~ $GdNi_2Sn_2$ (after Komarovskaya et al. (1983a) with $CaBe_2Ge_2$ ST, $a = 4.369$, $c = 9.709$); (12) ~ $GdNiSn_2$ (after Komarovskaya et al. (1983b), with $LuNiSn_2$ type of crystal structure, $a = 16.131$, $b = 4.434$, $c = 14.670$), and (13) ~ $GdNi_{2.67}Sn_{5.44}$ (according to Akselrud et al. (1983) a new type of crystal structure: Im3, $a = 11.855$).

Gd–Cu–Sn

Komarovskaya (1984) derived the phase diagram (fig. 117) and established the existence of a total of 6 ternary compounds. The crystal structure was determined for only 2 of them: (4) ~ GdCuSn ($CaIn_2$ ST, $a = 4.531$, $c = 7.364$) and (5) ~ $Gd_6Cu_8Sn_8$ ($Gd_6Cu_8Ge_8$ ST, $a = 14.737$, $b = 6.945$, $c = 4.474$).

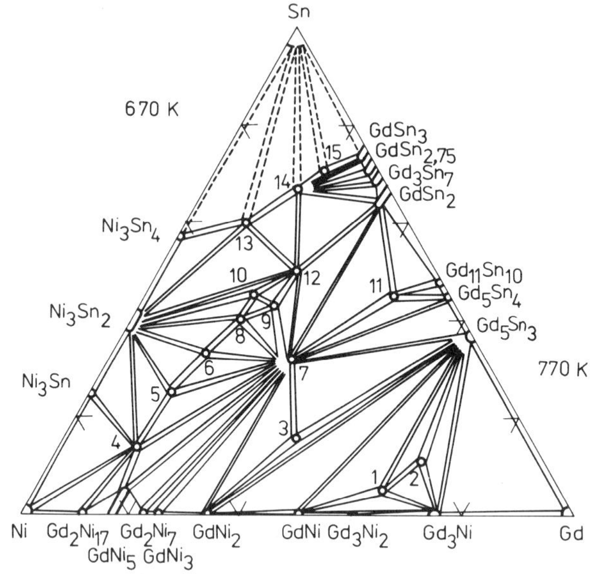

Fig. 116. Gd–Ni–Sn, isothermal sections at 770 K (0–50 a/c Sn) and at 670 K (55–75 a/c Sn). Compounds with unknown crystal structure: (1) ~ $Gd_{0.63}Ni_{0.32}Sn_{0.05}$; (3) ~ $Gd_{0.42}Ni_{0.42}Sn_{0.16}$; (4) ~ $GdNi_5Sn$; (9) ~ $Gd_{0.25}Ni_{0.32}Sn_{0.43}$; (10) ~ $Gd_2Ni_3Sn_4$; (11) ~ $Gd_{0.45}Ni_{0.10}Sn_{0.45}$; (14) ~ $Gd_3Ni_4Sn_{13}$; (15) ~ Gd_2NiSn_7.

Lu–Ni–Sn

The partial isothermal section at 770 or 670 K is shown in fig. 118 after Skolozdra et al. (1982b) and Komarovskaya (1984). The crystal structure of 5 from the 12 existing ternary compounds was reported: (2) ~ Lu_6Ni_2Sn (Ho_6Co_2Ga ST, $a = 9.141$, $b = 9.279$, $c = 9.710$); (6) ~ $LuNi_4Sn$ ($MgCu_4Sn$ ST, $a = 6.981$), and (7) ~ $LuNi_2Sn$ ($MnCu_2Al$ ST, $a = 6.345$). Both of them were found and described by Skolozdra and Komarovskaya (1983). For (8) ~ LuNiSn, Dwight (1983) found the TiNiSi ST, $a = 6.926$, $b = 4.386$, $c = 7.589$, and (9),

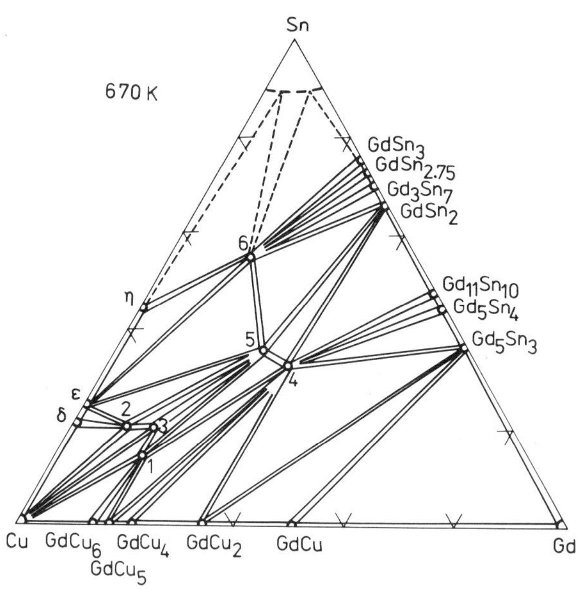

Fig. 117. Gd–Cu–Sn, isothermal section at 670 K. Compounds with unknown crystal structure are as follows: (1) ~ $Gd_{0.14}Cu_{0.72}Sn_{0.14}$; (2) ~ $Gd_3Cu_{13}Sn_4$; (3) ~ $GdCu_7Sn_2$; (6) ~ $Gd_3Cu_6Sn_{11}$.

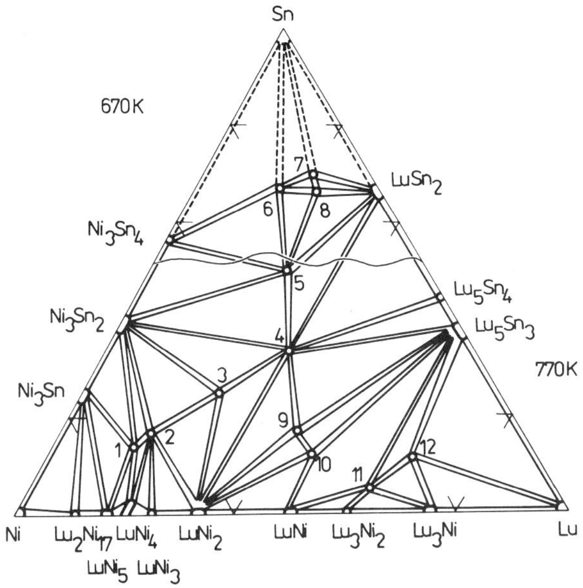

Fig. 118. Lu–Ni–Sn, isothermal sections at 770 K (0–50 a/c Sn) and at 670 K (50–75 a/c Sn). Compounds with unknown crystal structure: (1) ~ $Lu_{0.62}Ni_{0.33}Sn_{0.05}$; (3) ~ $Lu_{0.48}Ni_{0.40}Sn_{0.12}$; (4) ~ $Lu_{0.43}Ni_{0.40}Sn_{0.17}$; (5) ~ $LuNi_5Sn$; (10) ~ $Lu_3Ni_4Sn_{13}$; (11) ~ $Lu_{0.25}Ni_{0.12}Sn_{0.63}$; (12) ~ Lu_2NiSn_7.

according to Komarovskaya et al. (1983a), has the composition $LuNiSn_2$ and a new type of structure: Pnma, $a = 15.944$, $b = 4.361$, $c = 14.345$.

Lu–Cu–Sn

The phase diagram is shown in fig. 119 after Komarovskaya (1984). Ternary compounds are as follows: (1) ~ $Lu_{0.125}Cu_{0.675}Sn_{0.20}$, and (3) ~ $Lu_3Cu_4Sn_4$, both with unknown crystal structures, and (2) ~ LuCuSn, earlier reported by Komarovskaya et al. (1983b), belonging to $CaIn_2$ ST, $a = 4.457$, $c = 7.093$.

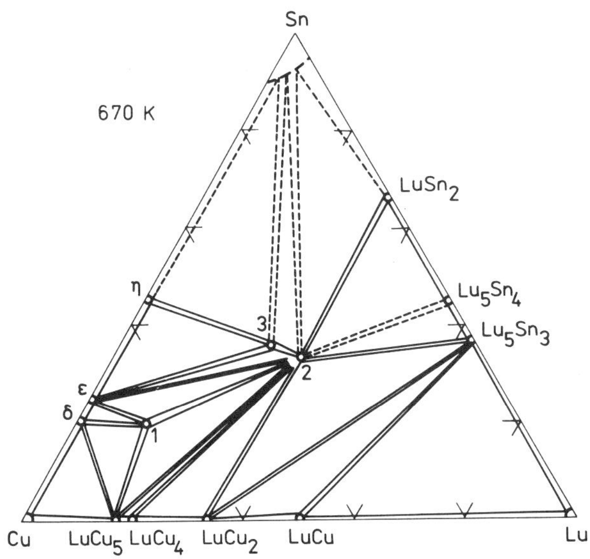

Fig. 119. Lu–Cu–Sn, isothermal section at 670 K.

TABLE 5
Ternary compounds found in the partially investigated systems R–M–{Sn, Pb}.

Compound	Structure type	Lattice parameters			Refs.
		a	*b*	*c*	
~$ScRh_{1.1}Sn_{3.6}$	tetragonal	13.565	–	9.498	Co, 80
ScPdSn	Fe_2P	7.471	–	3.620	DwHK, 73
~$ScIr_{1.1}Sn_{3.6}$	tetragonal	13.574	–	9.566	Co, 80
ScPtSn	Fe_2P	7.41	–	3.62	DwHK, 73
YFe_6Sn_6	YCo_6Ge_6	5.382	–	4.441	KoS, 86
YCoSn	TiNiSi	7.208	4.499	7.503	SkKG, 82a
YCo_3Sn	YCo_3Sn	11.026	–	8.891	Sk, 83
$YNi_{2.67}Sn_{5.44}$	$GdNi_{2.67}Sn_{5.44}$	11.837	–	–	KoS, 85
YNiSn	TiNiSi	7.115	4.490	7.665	Dw, 83
YCuSn	$CaIn_2$	4.505	–	7.286	KoSF, 83b
~$YRu_{1.1}Sn_{3.1}$	cubic	13.772	–	–	EsCBR, 80
YRhSn	Fe_2P	7.54	–	3.77	DwHK, 73
~$YRh_{1.1}Sn_{3.6}$	tetragonal	13.772	–	9.707	Co, 80
~$YOs_{1.1}Sn_{3.1}$	cubic	13.801	–	–	EsCBR, 80
~$YIr_{1.1}Sn_{3.1}$	cubic	13.773	–	–	EsCBR, 80
YPtSn	Fe_2P	7.433	–	3.992	DwHK, 73
$LaFeSn_2$	$CeNiSi_2$	4.511	17.81	4.511	Sk, 77
$LaCo_{1.33}Sn_{4.33}$	$PrRh_{1.33}Sn_{4.33}$	9.635	–	–	SkYKA, 83
$LaCoSn_2$	$CeNiSi_2$	4.529	17.86	4.529	Sk, 77
$La_3Co_2Sn_7$	$La_3Co_2Sn_7$	4.59	27.60	4.60	DoS, 80
$LaNi_{2.67}Sn_{5.44}$	$GdNi_{2.67}Sn_{5.44}$	11.965	–	–	KoS, 85
$LaNi_5Sn$	$CeNi_5Sn$	4.971	–	19.99	SkK, 82a
$LaNiSn_2$	$CeNiSi_2$	4.513	17.71	4.513	Sk, 77
$LaNi_2Sn_2$	$CeAl_2Ga_2$	4.461	–	10.19	SkMGT, 81b
LaNiSn	TiNiSi	7.552	4.661	7.636	Dw, 83
$LaNi_{0.74}Sn_2$	$CeNiSi_2$	4.52	17.79	4.51	DoSSS, 82
LaCuSn	AlB_2	4.582	–	4.082	KoSF, 83b
$LaCu_2Sn_2$	$CaBe_2Ge_2$	4.485	–	10.523	SkK, 82b
$LaCu_{0.56}Sn_2$	$CeNiSi_2$	4.53	18.34	4.46	DoSSS, 82
~$LaRu_{1.4}Sn_{4.6}$	cubic	9.772	–	–	EsCBR, 80
$LaRuSn_3$	$LaRuSn_3$	9.773	–	–	EiS, 86
$La_3Rh_4Sn_{13}$	$La_3Rh_4Sn_{13}$	9.745	–	–	MiHMLGE, 86
~$LaIr_{1.4}Sn_{4.6}$	cubic	9.775	–	–	Co, 80
$La_xPt_ySn_z$ (*x*, *y* and *z* unknown)	cubic	15.63	–	–	EsCBR, 80
$CeCo_{1.33}Sn_{4.33}$	$PrRh_{1.33}Sn_{4.33}$	9.594	–	–	SkYKA, 83
$CeCoSn_2$	$CeNiSi_2$	4.536	17.26	4.534	Sk, 77
Ce_5CuSn_3	Ti_5Ga_4	9.474	–	6.699	RiP, 68
CeCuSn	$CaIn_2$	4.570	–	7.903	KoSF, 83b
$CeCu_2Sn_2$	$CaBe_2Ge_2$	4.456	–	10.475	SkK, 82b
$CeRuSn_3$	$LaRuSn_3$	9.726	–	–	EiS, 86
$Ce_3Rh_4Sn_{13}$	$La_3Rh_4Sn_{13}$	9.708	–	–	MiHMLGE, 86
CePdSn	TiNiSi	7.541	4.705	7.984	RoMF, 85
Ce_5AgSn_3	Ti_5Ga_4	9.600	–	6.730	RiP, 68
~$CeIr_{1.4}Sn_{4.6}$	cubic	9.720	–	–	Co, 80
$PrCo_{1.33}Sn_{4.33}$	$PrRh_{1.33}Sn_{4.33}$	9.571	–	–	SkYKA, 83
$PrNi_{2.67}Sn_{5.44}$	$GdNi_{2.67}Sn_{5.44}$	11.903	–	–	KoS, 85
$PrNi_5Sn$	$CeNi_5Sn$	4.925	–	19.84	SkK, 82a
$PrNi_2Sn_2$	$CeAl_2Ga_2$	4.429	–	10.13	SkMGT, 81b

TABLE 5 (cont'd)

Compound	Structure type	Lattice parameters			Refs.
		a	*b*	*c*	
PrNiSn	TiNiSi	7.440	4.56	7.706	Dw, 83
$PrRuSn_3$	$LaRuSn_3$	9.708	–	–	EiS, 86
$PrRh_{1.33}Sn_{4.33}$	$PrRh_{1.33}Sn_{4.33}$	9.693	–	–	Va, 80
PrPdSn	TiNiSi	7.475	4.680	7.979	RoMF, 85
$\sim PrIr_{1.4}Sn_{4.6}$	cubic	9.704	–	–	Co, 80
$NdCo_{1.33}Sn_{4.33}$	$PrRh_{1.33}Sn_{4.33}$	9.553	–	–	SkYKA, 83
$NdNi_{2.67}Sn_{5.44}$	$GdNi_{2.67}Sn_{5.44}$	11.892	–	–	KoS, 85
$NdNi_5Sn$	$CeNi_5Sn$	4.923	–	19.73	SkK, 82a
$NdNi_2Sn_2$	$CeAl_2Ga_2$	4.411	–	10.12	SkMGT, 81b
NdNiSn	TiNiSi	7.395	4.541	7.69	Dw, 83
NdCuSn	$CaIn_2$	4.562	–	7.614	KoSF, 83b
$NdCu_2Sn_2$	$CaBe_2Ge_2$	4.432	–	10.264	SkK, 82b
$NdRuSn_3$	$LaRuSn_3$	9.687	–	–	EiS, 86
$Nd_3Rh_4Sn_{13}$	$La_3Rh_4Sn_{13}$	9.675	–	–	MiHMLGE, 86
NdPdSn	TiNiSi	7.424	4.665	7.951	Dw, 83
$\sim NdIr_{1.4}Sn_{4.6}$	cubic	9.691	–	–	Co, 80
$SmCo_{1.33}Sn_{4.33}$	$PrRh_{1.33}Sn_{4.33}$	9.524	–	–	SkYKA, 83
$SmNi_{2.67}Sn_{5.44}$	$GdNi_{2.67}Sn_{5.44}$	11.871	–	–	KoS, 85
$SmNi_2Sn_2$	$CeAl_2Ga_2$	4.391	–	10.09	SkMGT, 81b
SmNiSn	TiNiSi	7.304	4.509	7.68	Dw, 83
SmCuSn	$CaIn_2$	4.549	–	7.476	KoSF, 83b
$SmCu_2Sn_2$	$CaBe_2Ge_2$	4.401	–	10.095	SkK, 82b
SmRhSn	Fe_2P	7.46	–	3.98	DwHK, 73
$Sm_3Rh_4Sn_{13}$	$La_3Rh_4Sn_{13}$	9.656	–	–	MiHMLGE, 86
SmPdSn	TiNiSi	7.323	4.639	7.929	Dw, 83
$\sim SmIr_{1.4}Sn_{4.6}$	cubic	9.668	–	–	Co, 80
$Eu_3Rh_4Sn_{13}$	$La_3Rh_4Sn_{13}$	9.749	–	–	MiHMLGE, 86
$GdFe_6Sn_6$	YCo_6Ge_6	5.412	–	4.446	KoS, 86
$GdCo_{1.33}Sn_{4.33}$	$PrRh_{1.33}Sn_{4.33}$	9.518	–	–	SkYKA, 83
$GdCo_3Sn$	YCo_3Sn	...	–	...	Sk, 83
GdRhSn	Fe_2P	7.527	–	3.862	DwHK, 73
$Gd_3Rh_4Sn_{13}$	$La_3Rh_4Sn_{13}$	9.638	–	–	MiHMLGE, 86
GdPdSn	TiNiSi	7.254	4.618	7.912	Dw, 83
$\sim GdIr_{1.1}Sn_{3.6}$	cubic	13.811	–	–	Co, 80
GdPtSn	Fe_2P	7.462	–	4.034	DwHK, 73
GdAuSn	$CaIn_2$	...	–	...	Oe, 77
$TbFe_6Sn_6$	YCo_6Ge_6	5.397	–	4.477	KoS, 86
TbCoSn	TiNiSi	7.260	4.534	7.348	SkKG, 82a
$TbCo_{1.33}Sn_{4.33}$	$PrRh_{1.33}Sn_{4.33}$	9.501	–	–	SkYKA, 83
$TbNi_{2.67}Sn_{5.44}$	$GdNi_{2.67}Sn_{5.44}$	11.838	–	–	KoS, 85
$TbNiSn_2$	$LuNiSn_2$	16.062	4.425	14.605	KoAS, 83a
TbNiSn	TiNiSi	7.146	4.448	7.661	Dw, 83
TbCuSn	$CaIn_2$	4.509	–	7.291	KoSF, 83b
TbRhSn	Fe_2P	7.542	–	3.796	DwHK, 73
$Tb_5Rh_6Sn_{16}$	$Tb_5Rh_6Sn_{16}$	13.774	–	–	Va, 80
TbPdSn	TiNiSi	7.182	4.598	7.916	Dw, 83
$\sim TbIr_{1.1}Sn_{3.6}$	cubic	13.781	–	–	Co, 80
TbPtSn	Fe_2P	7.437	–	4.001	DwHK, 73
$DyFe_6Sn_6$	YCo_6Ge_6	5.407	–	4.462	KoS, 86
DyCoSn	TiNiSi	7.202	4.526	7.448	SkKG, 82a

TABLE 5 (cont'd)

Compound	Structure type	Lattice parameters *a*	*b*	*c*	Refs.
$DyNiSn_2$	$LuNiSn_2$	16.029	4.415	14.556	KoAS, 83a
DyNiSn	TiNiSi	7.112	4.439	7.656	Dw, 83
DyCuSn	$CaIn_2$	4.500	–	7.245	KoSF, 83b
DyRhSn	Fe_2P	7.529	–	3.771	DwHK, 73
$\sim DyRh_{1.1}Sn_{3.6}$	cubic	13.750	–	–	Co, 80
DyPdSn	TiNiSi	7.157	4.589	7.897	Dw, 83
$\sim DyIr_{1.1}Sn_{3.6}$	cubic	13.766	–	–	Co, 80
DyPtSn	Fe_2P	7.427	–	3.981	DwHK, 73
$HoFe_6Sn_6$	YCo_6Ge_6	5.398	–	4.459	KoS, 86
HoCoSn	TiNiSi	7.154	4.503	7.461	SkKG, 82a
$Ho_7Co_6Sn_{23}$	$Ho_7Co_6Sn_{23}$	9.617	–	9.821	SkAK, 85
$HoNiSn_2$	$LuNiSn_2$	16.000	4.404	14.503	KoAS, 83a
HoNiSn	TiNiSi	7.063	4.438	7.64	Dw, 83
HoCuSn	$CaIn_2$	4.499	–	7.229	KoSF, 83b
HoRhSn	Fe_2P	7.527	–	3.748	DwHK, 73
$\sim HoRh_{1.2}Sn_{3.9}$	cubic	13.750	–	–	Co, 80
$\sim HoRh_{1.2}Sn_{3.9}$	tetragonal	13.747	–	9.683	Co, 80
HoPdSn	Fe_2P	7.433	–	3.979	DwHK, 73
HoPdSn	TiNiSi	7.117	4.581	7.884	Dw, 83
HoIrSn	Fe_2P	7.456	–	3.848	DwHK, 73
$\sim HoIr_{1.1}Sn_{3.6}$	cubic	13.750	–	–	Co, 80
HoPtSn	Fe_2P	7.418	–	3.966	DwHK, 73
$ErFe_6Sn_6$	YCo_6Ge_6	5.383	–	4.441	KoS, 86
ErCoSn	TiNiSi	7.093	4.486	7.431	SkKG, 82a
$\sim ErCo_{1.1}Sn_{3.6}$	cubic	13.528	–	–	Co, 80
$ErNiSn_2$	$LuNiSn_2$	15.986	4.399	14.470	KoAS, 83a
ErNiSn	TiNiSi	7.016	4.425	7.629	Dw, 83
ErCuSn	$CaIn_2$	4.473	–	7.133	KoSF, 83b
$\sim ErRu_{1.1}Sn_{3.6}$	tetragonal	13.730	–	9.671	Co, 80
$\sim ErRh_{1.1}Sn_{3.6}$	cubic	13.725	–	–	Co, 80
$\sim ErRh_{1.1}Sn_{3.6}$	tetragonal	13.725	–	9.675	Co, 80
ErPdSn	Fe_2P	7.43	–	3.94	DwHK, 73
$\sim ErOs_{1.1}Sn_{3.6}$	cubic	13.760	–	–	EsCBR, 80
ErIrSn	Fe_2P	7.452	–	3.822	DwHK, 73
$\sim ErIr_{1.1}Sn_{3.6}$	tetragonal	13.756	–	9.677	Co, 80
$TmFe_6Sn_6$	YCo_6Ge_6	5.371	–	4.434	KoS, 86
TmCoSn	TiNiSi	7.082	4.456	7.445	SkKG, 82a
$TmNiSn_2$	$LuNiSn_2$	15.979	4.382	14.409	KoAS, 83a
TmNiSn	TiNiSi	7.000	4.414	7.621	Dw, 83
TmCuSn	$CaIn_2$	4.473	–	7.133	KoSF, 83b
TmRhSn	Fe_2P	7.534	–	3.670	DwHK, 73
$\sim TmRh_{1.3}Sn_{4.0}$	cubic	13.701	–	–	Co, 80
$\sim TmRh_{1.3}Sn_{4.0}$	tetragonal	13.706	–	9.675	Co, 80
TmPdSn	Fe_2P	7.42	–	3.92	DwHK, 73
TmIrSn	Fe_2P	7.46	–	3.77	DwHK, 73
$\sim YbCo_{1.4}Sn_{4.6}$	cubic	9.563	–	–	Co, 80
$YbNi_2Sn$	$MnCu_2Al$	6.355	–	–	SkK, 83
YbNiSn	TiNiSi	6.983	4.426	7.616	Dw, 83
Yb_5CuSn_3	Ti_5Ga_4	8.892	–	6.518	RiP, 68

TABLE 5 (cont'd)

Compound	Structure type	Lattice parameters *a*	*b*	*c*	Refs.
YbRhSn	Fe_2P	7.52	–	3.67	DwHK, 73
$YbRh_{1.33}Sn_{4.33}$	$PrRh_{1.33}Sn_{4.33}$	9.676	–	–	HoCMR, 80
~$YbRh_{1.1}Sn_{3.0}$	cubic	13.735	–	–	JaRE, 81
YbIrSn	Fe_2P	7.49	–	3.73	DwHK, 73
$YbIr_{1.1}Sn_{3.1}$	cubic	9.709	–	–	Co, 80
~$YbIr_{1.1}Sn_{3.6}$	cubic	13.751	–	–	Co, 80
~$YbIr_{1.1}Sn_{3.6}$	tetragonal	13.751	–	9.703	Co, 80
YbPtSn	Fe_2P	7.349	–	3.933	DwHK, 73
$LuFe_6Sn_6$	YCo_6Ge_6	5.370	–	4.435	KoS, 86
LuCoSn	TiNiSi	7.018	4.462	7.424	SkKG, 82a
~$LuRu_{1.1}Sn_{3.6}$	tetragonal	13.692	–	9.639	Co, 80
LuRhSn	Fe_2P	7.529	–	3.646	DwHK, 73
LuPdSn	Fe_2P	7.513	–	3.783	DwHK, 73
LuIrSn	Fe_2P	7.45	–	3.75	DwHK, 73
~$LuIr_{1.1}Sn_{3.6}$	tetragonal	13.708	–	9.666	Co, 80
LuPtSn	Fe_2P	7.382	–	3.902	DwHK, 73
Ce_5CuPb_3	Ti_5Ga_4	9.542	–	6.772	RiP, 68
Ce_5AgPb_3	Ti_5Ga_4	9.643	–	6.770	RiP, 68

References

Akselrud, L.G., L.P. Komarovskaya and R.V. Skolozdra, 1983, Dopov. Akad. Nauk Ukr. RSR, Ser. B (5), 35.

Cooper, A.S., 1980, Mater. Res. Bull. **15**, 799.

Dörrscheidt, A., and H. Schäfer, 1980, J. Less-Common Met. **70**, 1.

Dörrscheidt, A., G. Savelsberg, J. Stöhr and H. Schäfer, 1982, J. Less-Common Met. **83**, 269.

Dwight, A.E., 1983, J. Less-Common Met. **93**, 411.

Dwight, A.E., W.C. Harper and C.W. Kimball, 1973, J. Less-Common Met. **30**, 1.

Eisenmann, B., and H. Schäfer, 1986, J. Less-Common Met. **123**, 89.

Espinosa, G.P., A.S. Cooper, H. Barz and J.P. Remeika, 1980, Mater. Res. Bull. **15**, 1635.

Hodeau, J.L., J. Chenevas, M. Marezio and J.P. Remeika, 1980, Solid State Commun. **36**, 839.

Jayaraman, A., J.P. Remeika and G.P. Espinosa, 1981, Solid State Commun. **39**, 1049.

Jorda, J.L., M. Ishikawa and J. Muller, 1985, J. Less-Common Met. **107**, 321.

Komarovskaya, L.P., 1984, Phase equilibria, crystal structures and behaviours of compounds in the systems R–Ni,Cu–Sn, where R = Ce,Pr,Gd,Lu, Ph.D. Chemistry thesis, 1984, Lvov (Lvov State University, Lvov) pp. 1–17.

Komarovskaya, L.P., and R.V. Skolozdra, 1985, Dopov. Akad. Nauk Ukr. RSR, Ser. A (1), 81.

Komarovskaya, L.P., L.G. Akselrud and R.V. Skolozdra, 1983a, Kristallografiya **28**, 1201.

Komarovskaya, L.P., R.V. Skolozdra and I.V. Filatova, 1983b, Dopov. Akad. Nauk Ukr. RSR, Ser. A (1), 81.

Koretzkaya, O.E., and R.V. Skolozdra, 1986, Izv. Akad. Nauk SSSR, Neorg. Mater. **22**, 690.

Miraglia, S., J.L. Hodeau, M. Marezio, C. Laviran, M. Ghedira and G.P. Espinosa, 1986, J. Solid State Chem. **63**, 358.

Oesterreicher, H., 1977, J. Less-Common Met. **55**, 131.

Rieger, W., and E. Parthe, 1968, Monatsh. Chem. **99**, 291.

Rossi, D., R. Marazza and R. Ferro, 1985, J. Less-Common Met. **107**, 99.

Skolozdra, R.V., 1977, Vestn. Lvovsk. Univ., Ser. Khim. **19**, 40.
Skolozdra, R.V., 1983, Crystal structure of compounds RCo_3Sn and magnetic behaviours of $GdCo_3Sn$, in: Chetvertaya Vsesoyuzn. Konf. Kristallokhim. Intermetal. Soed., Tezisy Dokl., 1983, Lvov (Vyscha Shkola, Lvov) pp. 44,45.
Skolozdra, R.V., and L.P. Komarovskaya, 1982a, Dopov. Akad. Nauk Ukr. RSR, Ser. A (6), 83.
Skolozdra, R.V., and L.P. Komarovskaya, 1982b, Ukr. Fiz. Zh. **27**, 1834.
Skolozdra, R.V., and L.P. Komarovskaya, 1983, Ukr. Fiz. Zh. **28**, 1093.
Skolozdra, R.V., V.M. Mandzyk and L.G. Akselrud, 1981a, Kristallografiya **26**, 480.
Skolozdra, R.V., V.M. Mandzyk, Yu.K. Gorelenko and V.D. Tkachuk, 1981b, Fiz. Metal. Metalloved. **52**, 966.
Skolozdra, R.V., O.E. Koretzkaya and Yu.K. Gorelenko, 1982a, Ukr. Fiz. Zh. **27**, 263.
Skolozdra, R.V., L.P. Komarovskaya and O.E. Koretzkaya, 1982b, Phase diagrams in the ternary systems Y,Ce,Gd,Lu–Co,Ni–Sn, in: IY Vsesoyuzn. Sovesch. "Diagr. Sostoyaniya Metal. Sistem", Tezisy Dokl., Moscow, 1982 (Nauka, Moscow) p. 99.
Skolozdra, R.V., I.V. Yasnitzkaya, O.E. Koretzkaya and L.G. Akselrud, 1983, Dopov. Akad. Nauk Ukr. RSR, Ser. B (6), 43.
Skolozdra, R.V., L.G. Akselrud and O.E. Koretzkaya, 1985, Kristallografiya **30**, 1003.
Vandenberg, J.M., 1980, Mater. Res. Bull. **15**, 835.

2.6. *R–M–M′ systems*

A total of 72 ternary R–M–M′ systems were studied, and phase equilibria at different temperatures were derived. These systems are briefly described below. The existence of a total of 89 ternary compounds (49 of which with a determined crystal structure) was reported. Also in 64 more related ternary systems, the existence of another 71 ternary compounds, 54 of which with known crystal structures, were found and are listed in table 6.

In this subsection order of description is taken in the alphabetical order for the non-rare earth metals.

Sc–Cr–Ni

The system was partially investigasted by Maslennikov and Braslavskaya (1986), the phase diagram is shown in fig. 120. There are no ternary compounds.

Sc–Nb–V

The diagram was investigated at 1070 K by Savitzky et al. (1975). The continuous solid solutions between Nb and V is in equilibrium with Sc. Ternary compounds are not formed.

Y–Ag–Pd

Schematic representation of the phase equilibria in the Y–Ag–Pd system is shown in fig. 121 after Slavev et al. (1978).

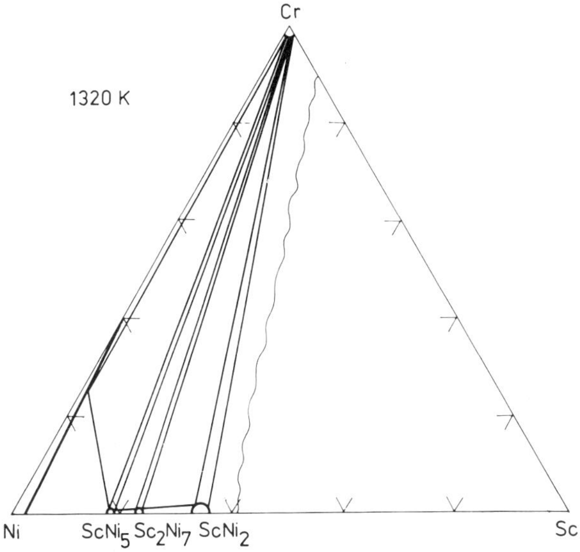

Fig. 120. Sc–Cr–Ni, partial isothermal section at 1320 K.

Y–Co–Cu

Kharchenko (1981) reported the phase diagram of the system Y–Co–Cu, shown in fig. 122. The ternary compounds are as follows: (1) ~ $YCo_{0.7}Cu_{0.3}$ (MoB ST, $a = 3.971$, $c = 21.08$), and (2) ~ $YCo_{0.26-0.64}Cu_{0.74-0.36}$ (YNi ST, $a = 4.103–4.203$, $b = 7.273–7.450$, $c = 5.495–5.496$, $\gamma = 90.5$).

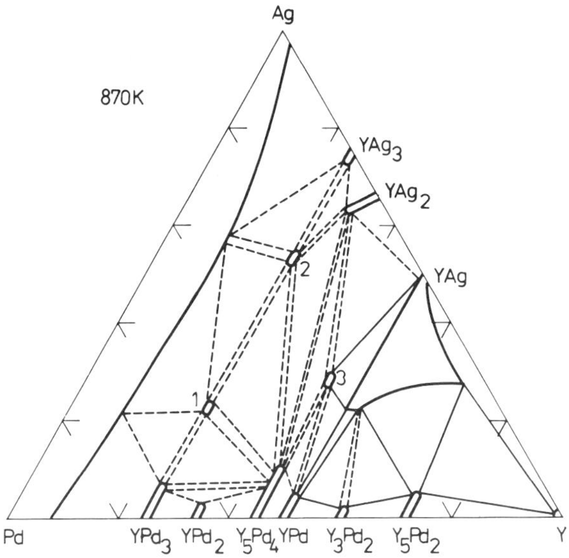

Fig. 121. Y–Ag–Pd, schematic isothermal section at 870 K. Compounds with unknown crystal structure: (1) ~ $YAgPd_2$; (2) ~ YAg_2Pd; (3) ~ $Y_5Ag_{2.5}Pd_{1.5}$.

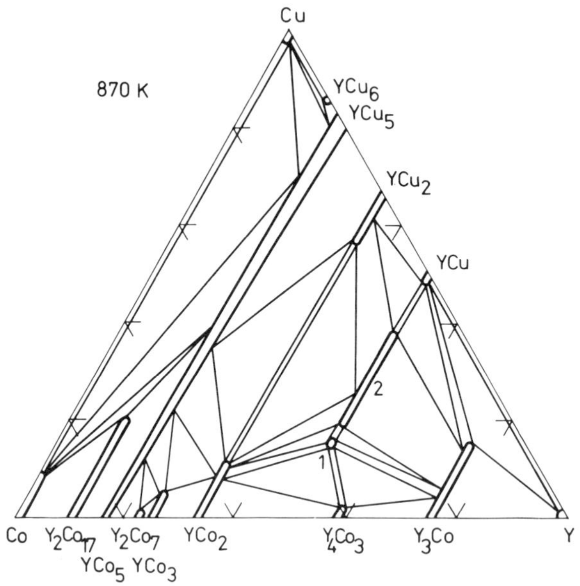

Fig. 122. Y–Co–Cu, isothermal section at 870 K.

Y–Co–Fe

No ternary compounds were reported by Kharchenko et al. (1977a). The two partial isothermal sections are shown in fig. 123.

Y–Co–Ni

Isothermal section at 870 K has been derived by Kharchenko et al. (1977a) and is shown in fig. 124. The only ternary compound has the composition

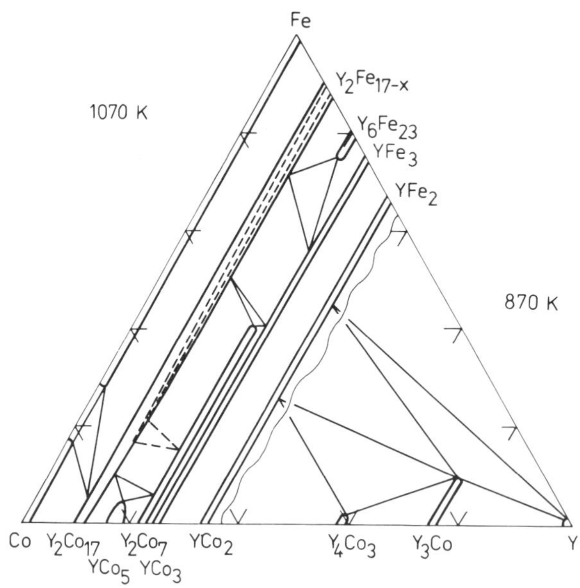

Fig. 123. Y–Co–Fe, isothermal sections at 1070 K (0–33 a/c Y) and at 870 K (33–100 a/c Y).

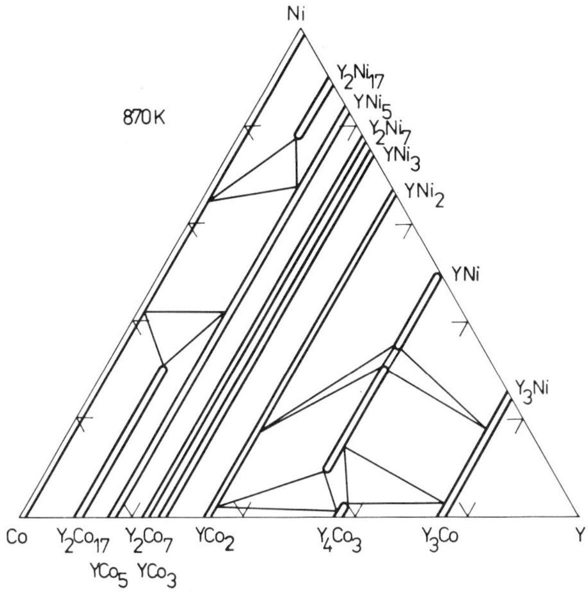

Fig. 124. Y–Co–Ni, isothermal section at 870 K.

$YCo_{0.8-0.44}Ni_{0.2-0.56}$ and the MoB type of crystal structure ($a = 3.953$–3.931, $c = 20.78$–20.98). Lattice parameters in the regions of continous solid solutions changes linearly for the $Y(Co, Ni)_5$, $Y_3(Co, Ni)$, and nonlinearly for the remaining ones.

Y–Co–Pd

The phase diagram is shown in fig. 125, after Sokolovskaya et al. (1985b). The dotted line shows the immiscibility range. No ternary compounds were found.

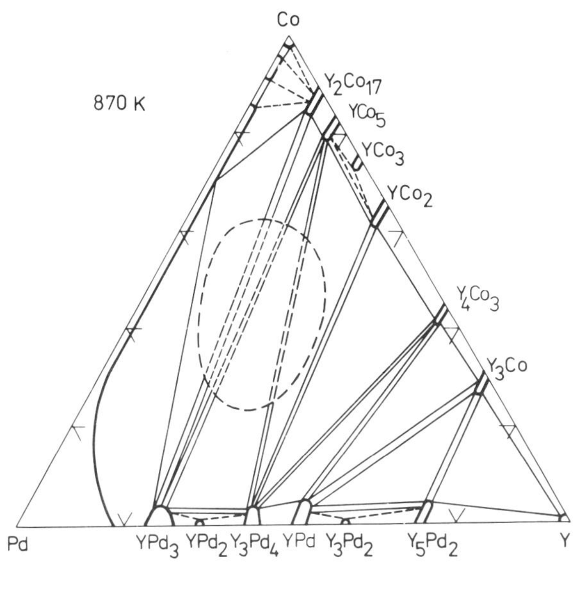

Fig. 125. Y–Co–Pd, isothermal section at 870 K.

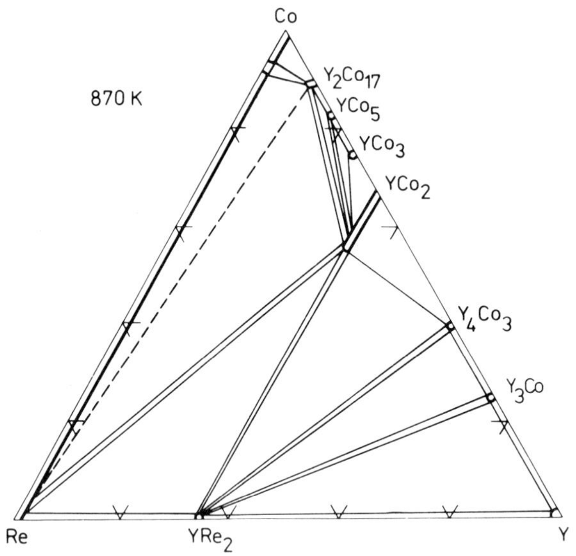

Fig. 126. Y–Co–Re, isothermal section at 870 K.

Y–Co–Re

No ternary compounds were found by Samyratov et al. (1982) during the phase equilibria investigation (fig. 126).

Y–Co–Ru

The isothermal section at 870 K, proposed by Tolkunova et al. (1974), is shown in fig. 127. YCo_2 dissolves up to 48 a/c Ru with linearly increasing unit-cell volume. The only ternary compound, ~Y_3CoRu, has an unknown crystal structure.

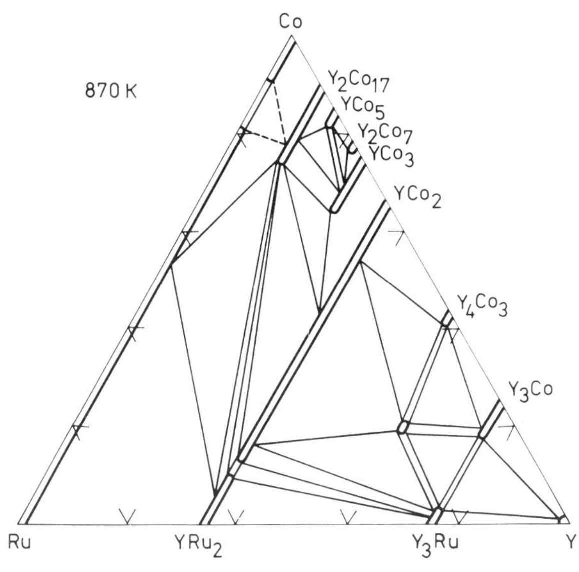

Fig. 127. Y–Co–Ru, isothermal section at 870 K.

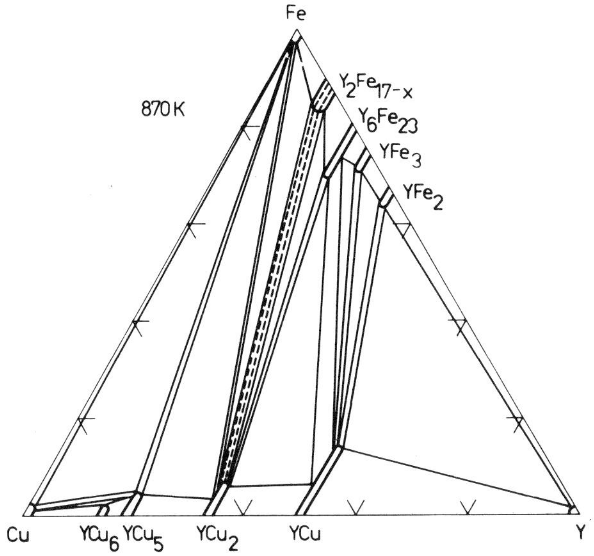

Fig. 128. Y–Cu–Fe, isothermal section at 870 K.

Y–Cu–Fe

The phase diagram is shown in fig. 128 after Kharchenko et al. (1977b). No ternary compounds were found.

Y–Cu–Ni

The isothermal section at 870 K, shown in fig. 129, was derived by Kharchenko et al. (1986). Ternary compounds are as follows: (1) $\sim YCu_{0.6-0.7}Ni_{0.4-0.3}$ with unknown crystal structure, and (2) $\sim YCu_{0.1-0.3}Ni_{0.9-0.7}$ which belong to the FeB type of crystal structure, $a = 7.103$, $b = 4.188$, $c = 5.491$.

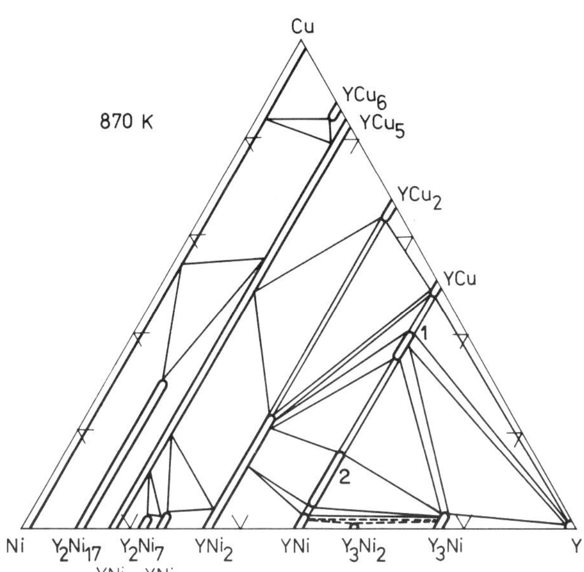

Fig. 129. Y–Cu–Ni, isothermal section at 870 K.

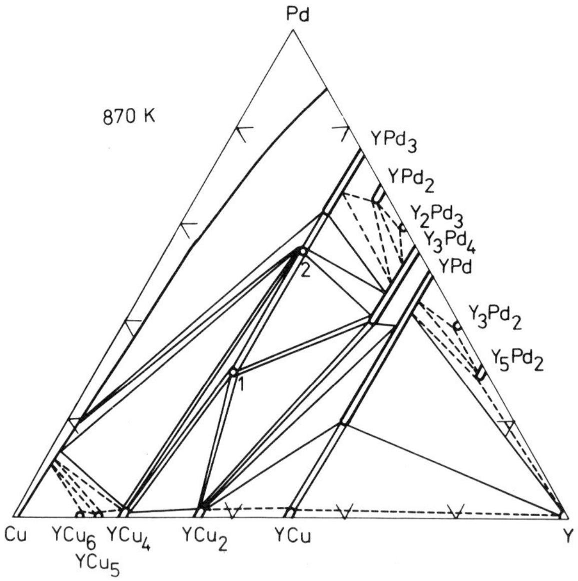

Fig. 130. Y–Cu–Pd, isothermal section at 870 K.

Y–Cu–Pd

The phase diagram is shown in fig. 130, after Mushagi et al. (1983). The existence of two ternary compounds, both with unknown crystal structures, was established: (1) ~ YCu_2Pd, and (2) ~ $YCuPd_2$.

Y–Fe–Hf

Bodak and Berezyuk (1982b) established the phase diagram (fig. 131) and found that there are no ternary compounds or solid solubility in the system Y–Fe–Hf.

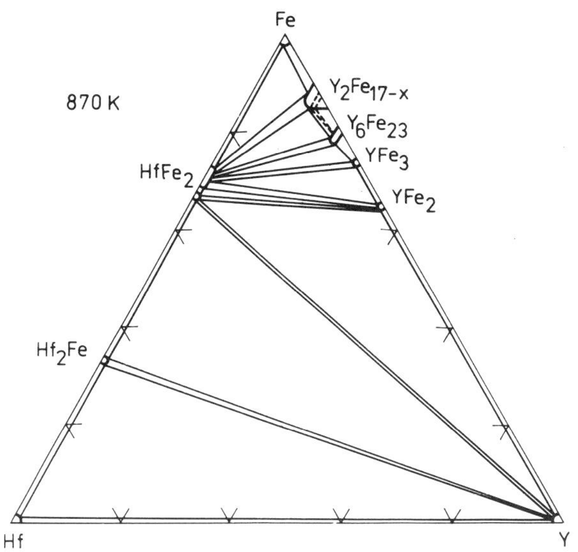

Fig. 131. Y–Fe–Hf, isothermal section at 870 K.

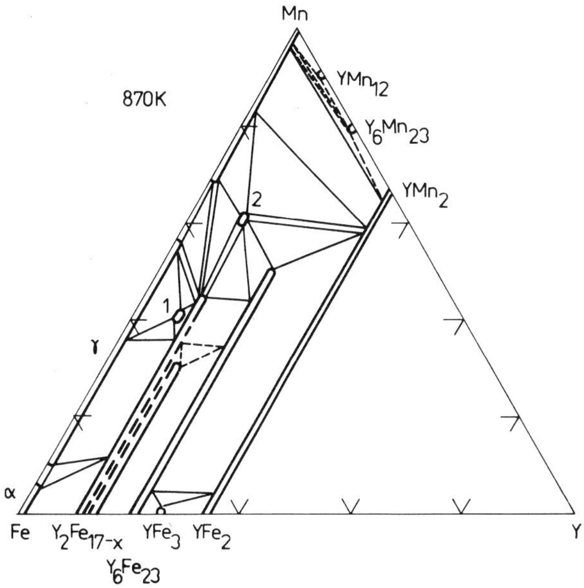

Fig. 132. Y–Fe–Mn, isothermal section at 870 K.

Y–Fe–Mn

Bodak and Kyrchiv (1983) carried out an investigation of the Y–Fe–Mn system and derived the phase diagram (fig. 132). Both of the two existing ternary compounds belong to the $ThMn_{12}$ type of crystal structure: (1) ~ $YFe_{6.9}Mn_{5.1}$ ($a = 8.528$, $c = 4.756$) and (2) ~ $Y(Y_{0.025}Fe_{0.314}Mn_{0.661})_{12}$ ($a = 8.553$, $c = 4.769$).

Y–Fe–Mo

The phase diagram was derived by Bodak and Berezyuk (1981) and is shown in fig. 133. The only ternary compound has variable composition that one can

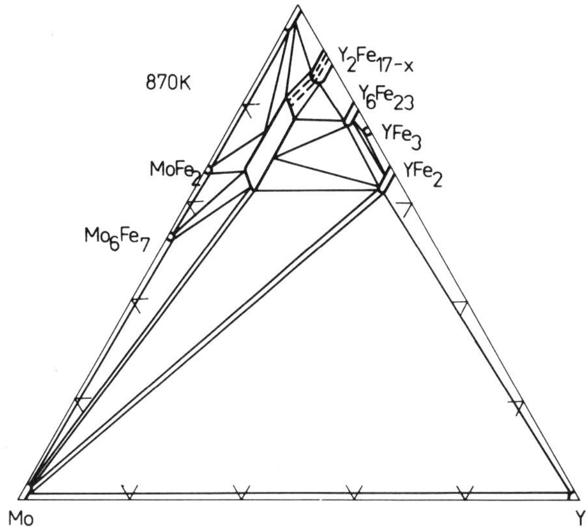

Fig. 133. Y–Fe–Mo, isothermal section at 870 K.

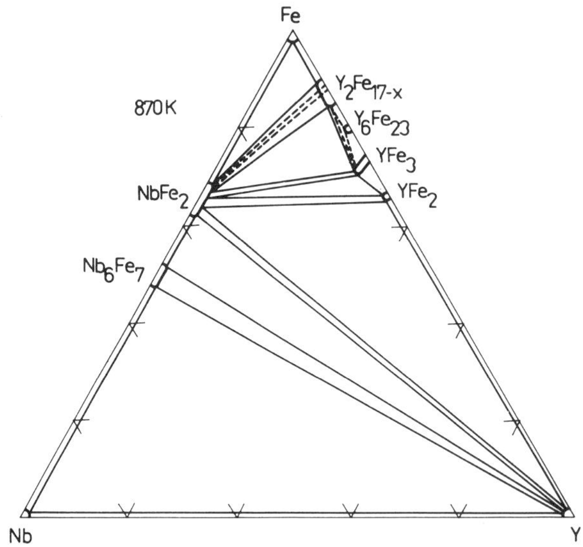

Fig. 134. Y–Fe–Nb, isothermal section at 870 K.

describe as $Y(Y, Mo, Fe)_{12}$, and it belongs to the $ThMn_{12}$ ST, $a = 8.514$–8.588, $c = 4.775$–4.826.

Y–Fe–Nb

The isothermal section at 870 K (fig. 134) was derived by Bodak and Berezyuk (1982a). No ternary compounds or solid solubility in the binary ones were found.

Y–Fe–Ni

The isothermal section at 870 K, shown in fig. 135, was derived by Kharchenko et al. (1977a). No ternary compounds are formed.

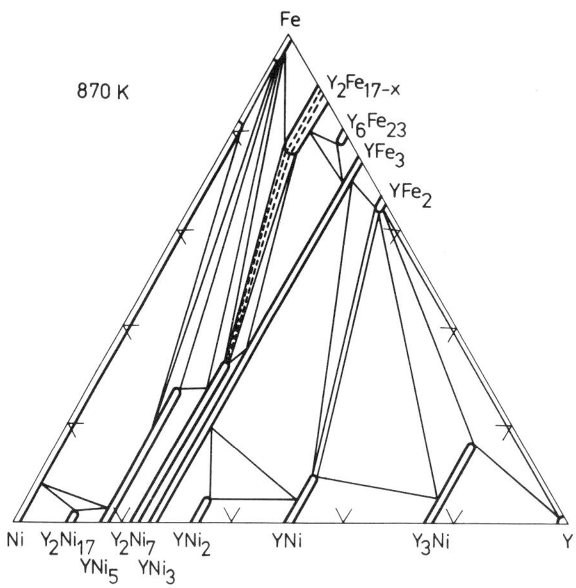

Fig. 135. Y–Fe–Ni, isothermal section at 870 K.

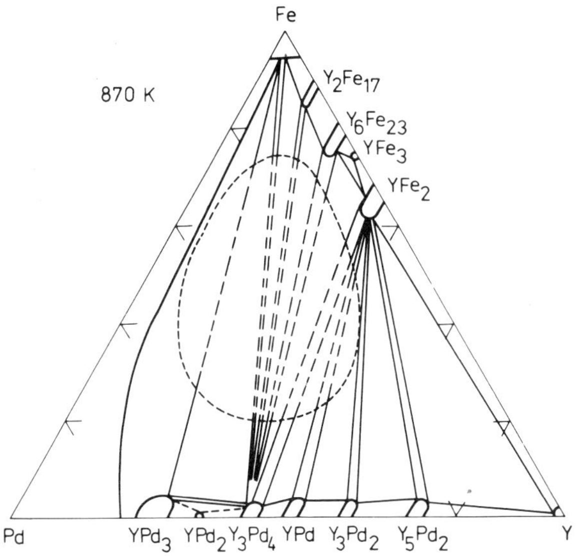

Fig. 136. Y–Fe–Pd, isothermal section at 870 K.

Y–Fe–Pd

The phase diagram, described by Sokolovskaya et al. (1985b), is shown in fig. 136. The liquid immiscibility range is shown by the dotted line.

Y–Fe–Ru

Tolkunova et al. (1974) reported the phase diagram (fig. 137). Several extensive solid solutions in the binary compounds were found.

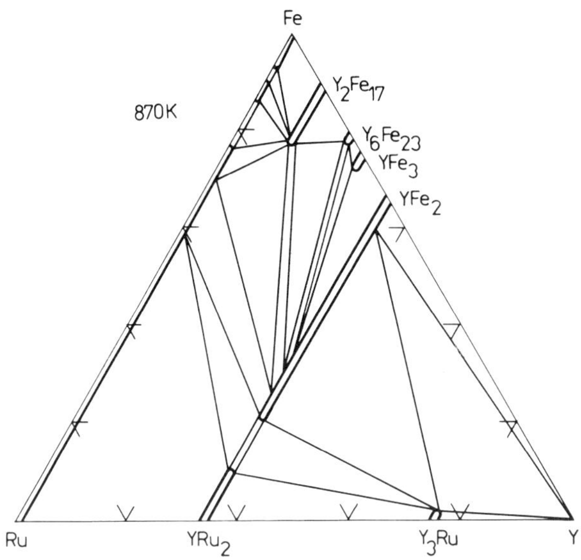

Fig. 137. Y–Fe–Ru, isothermal section at 870 K.

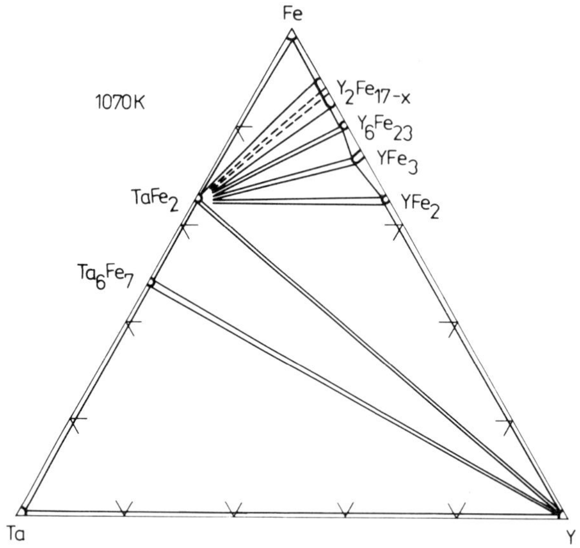

Fig. 138. Y–Fe–Ta, isothermal section at 1070 K.

Y–Fe–Ta

No solid solubility or ternary compounds were found by Bodak and Berezyuk (1982b) in the system Y–Fe–Ta (fig. 138).

Y–Fe–W

The phase diagram (fig. 139) was derived by Bodak and Berezyuk (1981). The compound $YFe_{10.4}W_{1.6}$ belongs to the $ThMn_{12}$ ST, $a = 8.513$, $c = 4.760$.

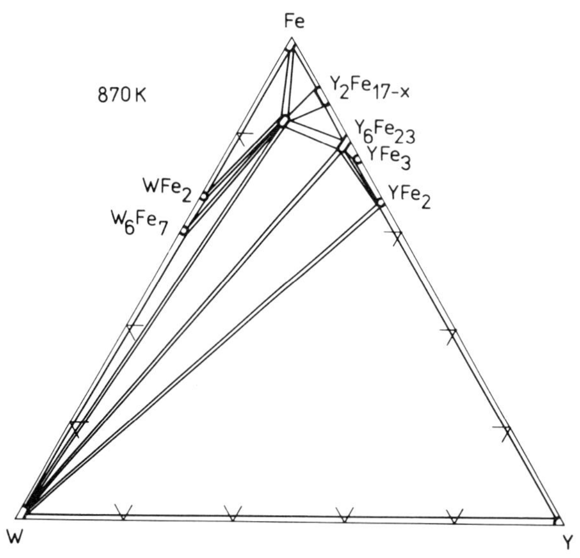

Fig. 139. Y–Fe–W, isothermal section at 870 K.

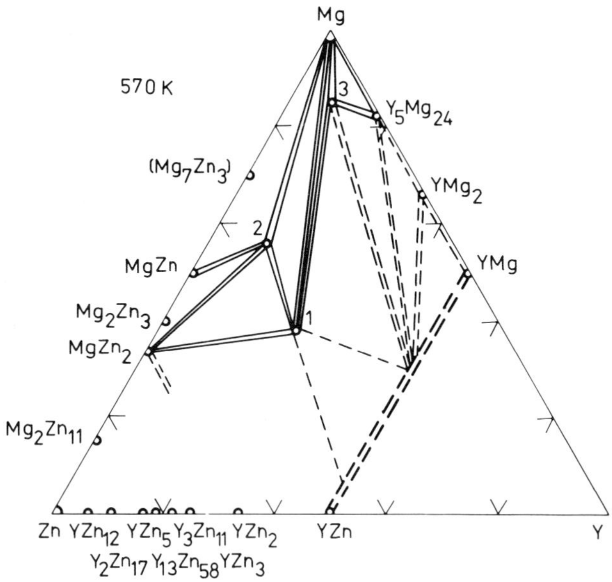

Fig. 140. Y–Mg–Zn, isothermal section at 570 K.

Y–Mg–Zn

The Y–Mg–Zn system was partially investigated by Padezhnova et al. (1982). The isothermal section at 570 K is shown in fig. 140. The ternary compounds are as follows: (1) ~ $Y_2Mg_3Zn_3$ with the $MnCu_2Al$ type structure, $a = 6.848$; (2) ~ YMg_3Zn_6 and (3) ~ $YMg_{12}Zn$, both with unknown crystal structures.

Y–Ni–Pd

Sokolovskaya et al. (1985b) derived the phase diagram, shown in fig. 141. No ternary compounds were found.

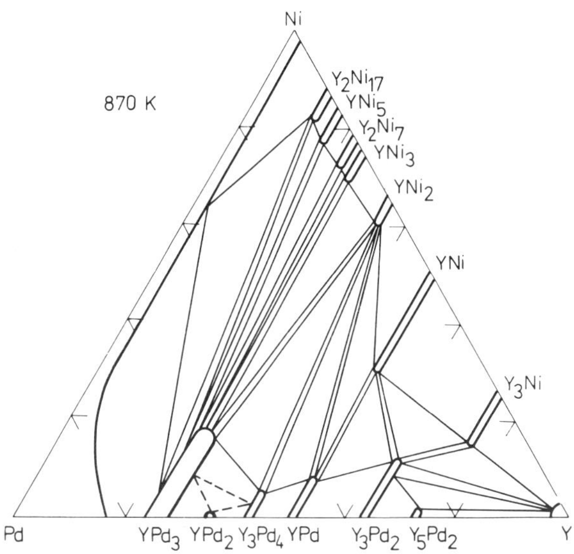

Fig. 141. Y–Ni–Pd, isothermal section at 870 K.

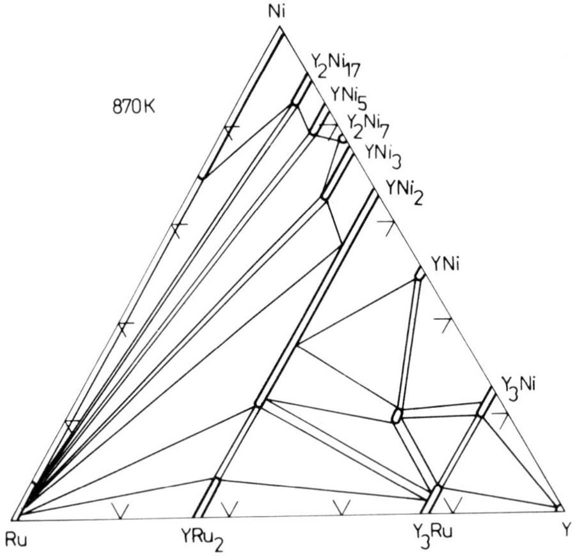

Fig. 142. Y–Ni–Ru, isothermal section at 870 K.

Y–Ni–Ru

An investigation of the system Y–Ni–Ru (fig. 142) was carried out by Tolkunova et al. (1974). Crystal structure of $\sim Y_3NiRu$ is unknown.

La–Co–Fe

The phase diagram (fig. 143) was described by Kharchenko et al. (1975). The only ternary compound $La_2(Co, Fe)_{17-x}$ belongs to the Th_2Zn_{17} ST, $a =$

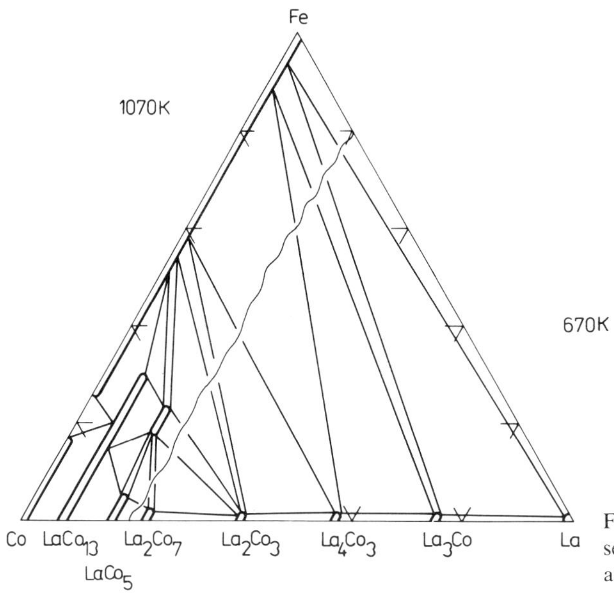

Fig. 143. La–Co–Fe, isothermal sections at 1070 K (0–20 a/c La) and 670 K (20–100 a/c La).

8.524–8.546, $c = 12.370$–12.400. The solubility of Fe (up to 30 a/c) in $LaCo_{13}$ linearly increased the unit cell.

La–Co–Mn

The phase diagram, shown in fig. 144, was derived by Kalychak et al. (1980). Three ternary compounds were found: (1) $\sim La_2(Co_{0.73-0.47}Mn_{0.27-0.53})_{17}$ with Th_2Zn_{17} ST, $a = 8.792$–8.578, $c = 12.70$–12.44; (2) and (3) both with unknown structures, at the approximate compositions are $LaCo_2Mn_{3.3}$ and $LaCoMn_{4.5}$, respectively.

La–Co–Ni

The phase diagram has been reported by Kharchenko (1977) and is shown in fig. 145. The ternary compound $La_{24}(Co, Ni)_{11}$ crystallized in the $Ce_{24}Co_{11}$ structure type, $a = 9.605$, $c = 21.85$.

La–Cu–Mn

Phase equilibria were established by Kalychak et al. (1979b), partially at 870 K and partially at 60 K (fig. 146). The only ternary compound $La(Cu_{0.94-0.78}Mn_{0.06-0.22})_{11}$ has variable composition and belongs to the $Ce(Mn_{0.55}Ni_{0.45})_{11}$ ST, $a = 8.393$–8.436, $c = 4.947$–4.977.

La–Fe–Ni

Kharchenko et al. (1976) reported the absence of ternary compounds. The phase diagram is shown in fig. 147.

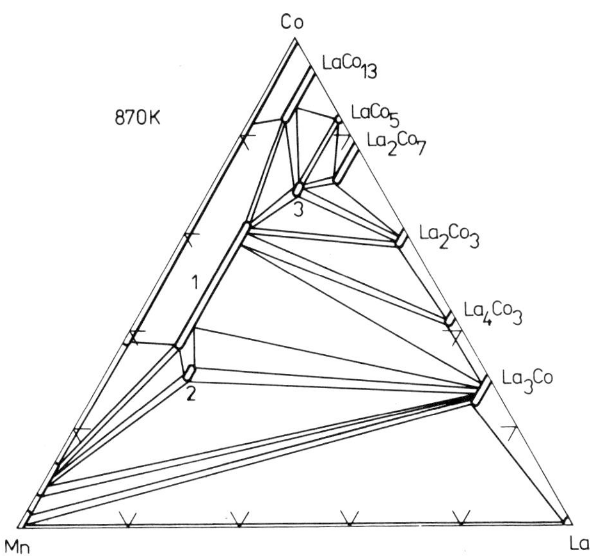

Fig. 144. La–Co–Mn, isothermal section at 870 K.

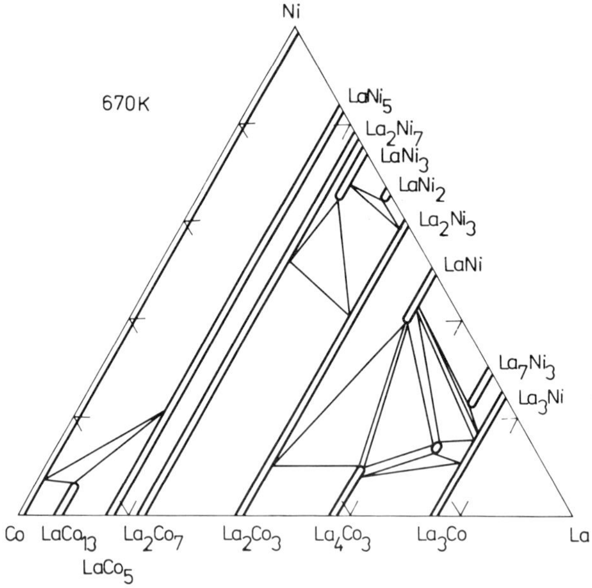

Fig. 145. La–Co–Ni, isothermal section at 670 K.

La–Mg–Zn

The phase diagram investigation was carried out by Dobatkina et al. (1985). The only ternary compound has the composition $\sim La_5Mg_{42}Zn_{53}$ and an unknown crystal structure. Another ternary compound, $La_2Mg_3Zn_3$, with a cubic unit cell, $a = 7.145$, was reported earlier by Melnyk et al. (1978b), but was not confirmed by the former authors (fig. 148).

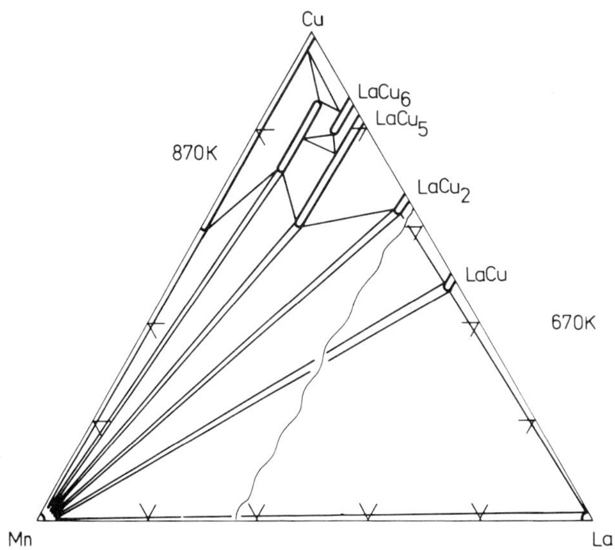

Fig. 146. La–Cu–Mn, isothermal sections at 870 K (0–33 a/c La) and at 670 K (33–100 a/c La).

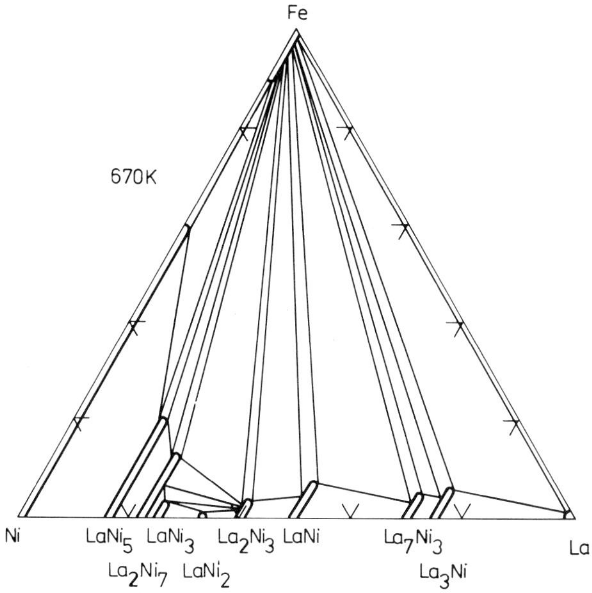

Fig. 147. La–Fe–Ni, isothermal section at 670 K.

La–Mn–Ni

The two parts of the system La–Mn–Ni at two different temperatures were investigated by Kalychak et al. (1980). Ternary compounds are as follows: (1) $\sim LaMn_{4.8-6.6}Ni_{6.2-4.6}$ ($(Ce(Mn_{0.55}Ni_{0.45})_{11}$ ST, $a = 8.356–8.432$, $c = 4.918–4.962$), and (2) $\sim LaMn_{2.8-3.2}Ni_{2.2-1.8}$. Crystal structure for the latter is unknown. The phase diagram is shown in fig. 149.

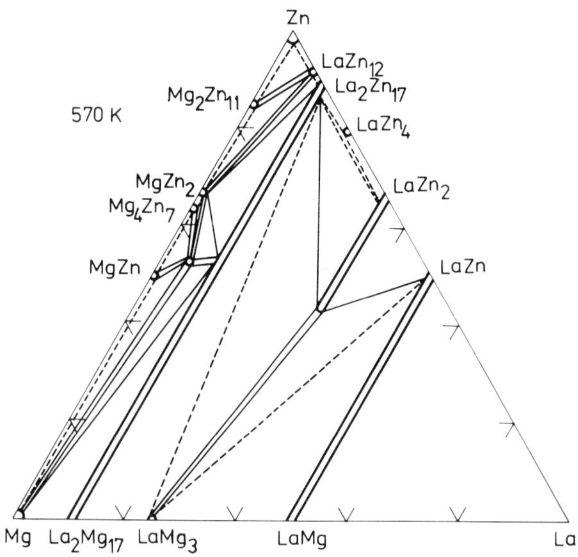

Fig. 148. La–Mg–Zn, isothermal section at 570 K.

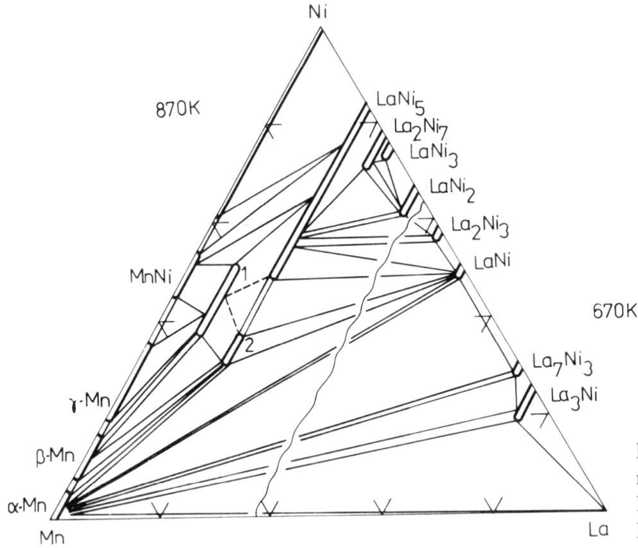

Fig. 149. La–Mn–Ni, isothermal sections at 870 K (0–33 a/c La), and at 670 K (33–100 a/c La).

Ce–Co–Cu

The two ternary compounds: (1) ~ $Ce_{24}Co_{3.5}Cu_{7.5}$ and (2) ~ $Ce_7Co_{2.7-2.5}Cu_{0.3-0.5}$, both with unknown crystal structures, were found by Kharchenko (1981). Substitution of Co atoms by the Cu in the continuous solid solution $Ce(Co, Cu)_5$ increased the unit-cell constant. The phase diagram is shown in fig. 150.

Ce–Co–Mn

The phase diagram (fig. 151) was described by Bodak et al. (1981a). Two

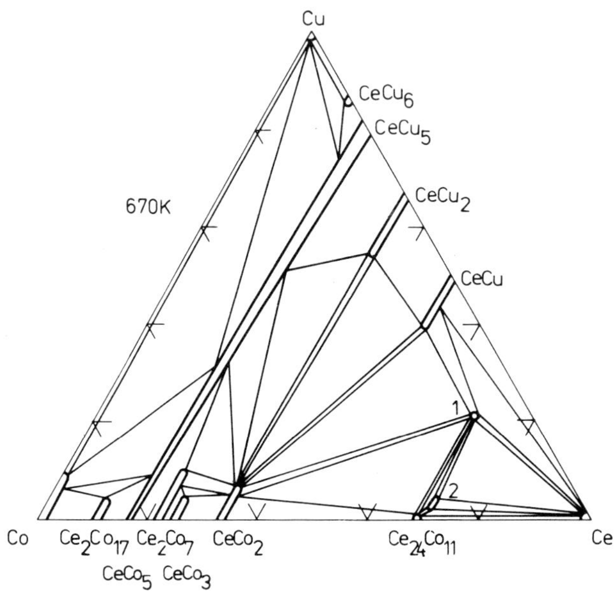

Fig. 150. Ce–Co–Cu, isothermal section at 670 K.

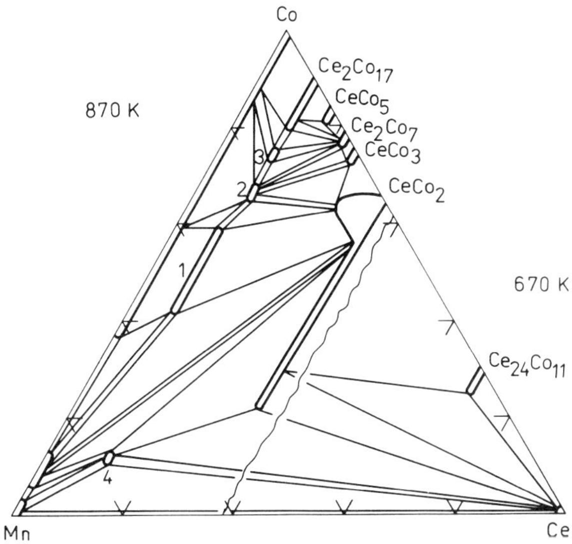

Fig. 151. Ce–Co–Mn, isothermal sections at 870 K (0–33 a/c Ce) and at 670 K (33–100 a/c Ce).

different ternary compounds, both with the same type of crystal structure, $ThMn_{12}$, were found: (1) ~ $CeCo_{7.84-5.76}Mn_{4.16-6.24}$, $a = 8.470$–8.531, $c = 4.730$–4.773, and (4) ~ $Ce(Ce_{0.05}Co_{0.13}Mn_{0.82})_{12-x}$, $x = 0.84$, $a = 8.542$, $c = 4.773$. A third one belongs to the Th_2Ni_{17} ST: (3) ~ $Ce_2(Co_{0.83}Mn_{0.17})_{17}$, $a = 8.392$, $c = 8.230$. The crystal structure of (2) ~ $Ce_2(Co, Mn)_{17}$ was not reported.

Ce–Co–Pu

The phase diagram is shown in fig. 152 after Ellinger et al. (1966). The crystal structure of (1) ~ $Ce_{0.3-1.9}Co_3Pu_{4.7-3.1}$ was established by Larson et al.

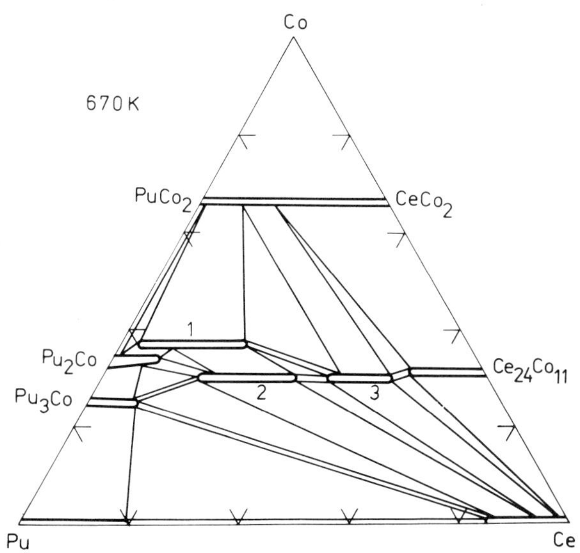

Fig. 152. Ce–Co–Pu, isothermal section at 670 K.

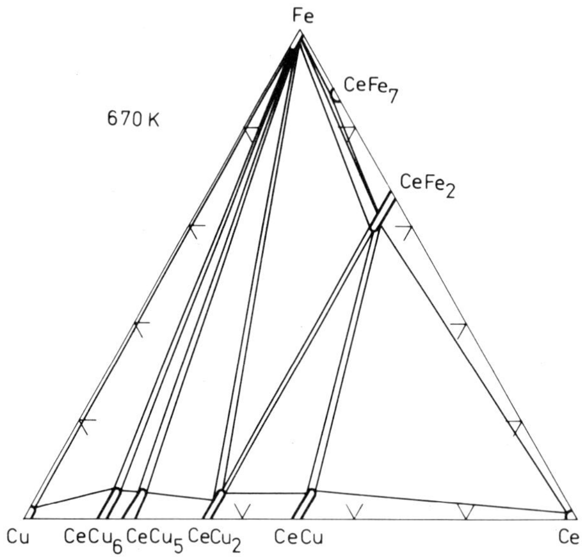

Fig. 153. Ce–Cu–Fe, isothermal section at 670 K.

(1964) and belongs to the W_5Si_3 ST, $a = 10.72–10.73$, $c = 5.346–5.359$; (2) ~ $Ce_{1.8–3.5}Co_3Pu_{5.2–3.5}$ belongs to the Th_7Fe_3 ST, $a = 9.298–9.359$, $c = 5.887–5.986$; and (3) ~ $Ce_{4.2–5.3}Co_3Pu_{2.8–1.7}$ has a cubic unit cell, $a = 13.46–13.49$.

Ce–Cu–Fe

Kharchenko et al. (1977b) described the isothermal section at 670 K of the system Ce–Cu–Fe, shown in fig. 153. The third component solubility was found to be less than 5 a/c.

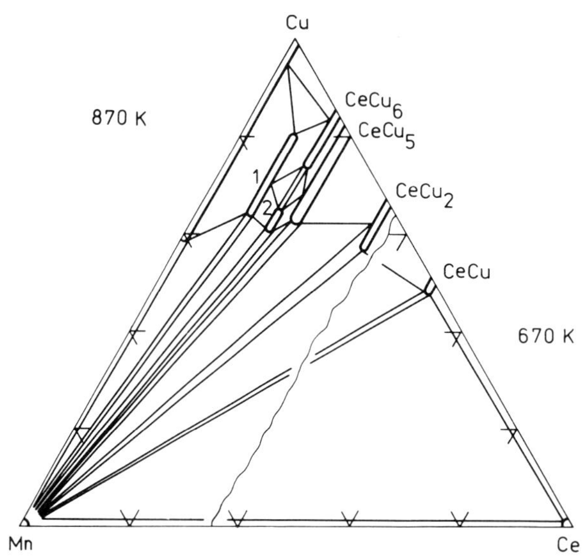

Fig. 154. Ce–Cu–Mn, isothermal sections at 870 K (0–33 a/c Ce) and at 670 K (33–100 a/c Ce).

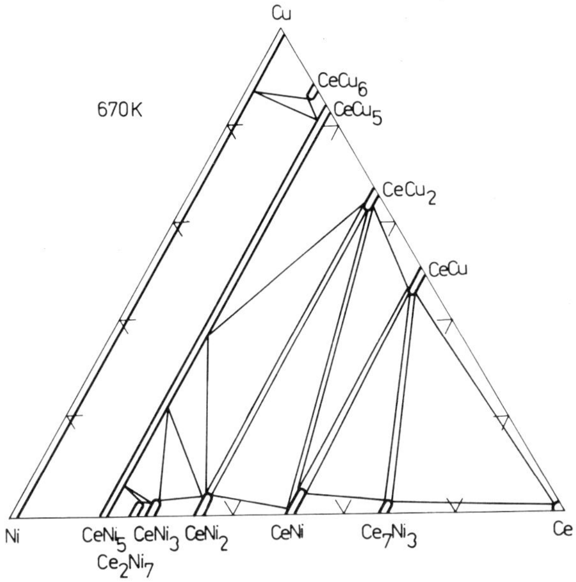

Fig. 155. Ce–Cu–Ni, isothermal section at 670 K.

Ce–Cu–Mn

Kalychak et al. (1979a) reported the phase diagram shown in fig. 154. The two ternary compounds found are: (1) ~ $CeCu_{9.92-8}Mn_{1.08-3}$ ($(Ce(Mn_{0.55}Ni_{0.45})_{11}$ ST, $a = 8.353–8.432$, $c = 4.945–4.987$), and (2) ~ $CeCu_{4.7-4.32}Mn_{1.3-1.68}$ (unknown crystal structure).

Ce–Cu–Ni

Kharchenko et al. (1986) did not find any ternary compounds during the phase equilibria study (see fig. 155).

Ce–Cu–Pu

Wittenberg et al. (1965) established that there are two continuous solid solutions in the Ce–Cu–Pu system: $CeCu_2$–$PuCu_2$ and $CeCu_6$–Pu_2Cu_{11}.

Ce–Cr–Ni

Kalychak et al. (1976a) found that the region of liquid immiscibility reach up to 35 a/c Ni in the ternary system Ce–Cr–Ni at 870 K (fig. 156).

Ce–Fe–Hf

The phase diagram was derived by Bodak and Berezyuk (1982b), and is shown in fig. 157. No ternary compounds or solid solutions were observed.

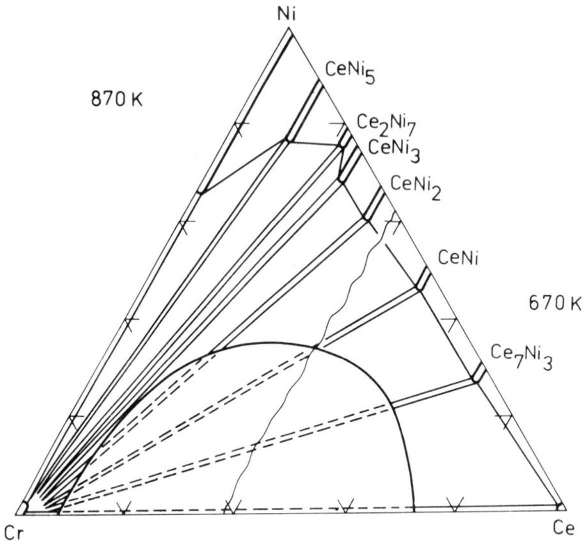

Fig. 156. Ce–Cr–Ni, isothermal sections at 870 K (0–33 a/c Ce) and 670 K (33–100 a/c Ce).

Ce–Fe–Mn

Kalychak et al. (1974) found that two ternary compounds exist in the Ce–Fe–Mn system (fig. 158). They are: (1) ~ $CeFe_{7.04-2.35}Mn_{4.96-9.65}$ with $ThMn_{12}$ type structure, $a = 8.506$–8.573, $c = 4.747$–4.777, and (2) ~ $Ce_2Fe_{12.65-9.40}Mn_{4.35-7.60}$ with the Th_2Ni_{17} ST, $a = 8.473$–8.502, $c = 8.356$–8.381.

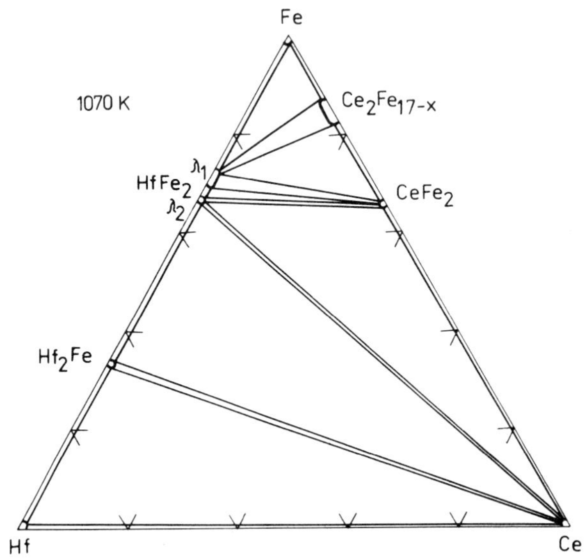

Fig. 157. Ce–Fe–Hf, isothermal section at 1070 K.

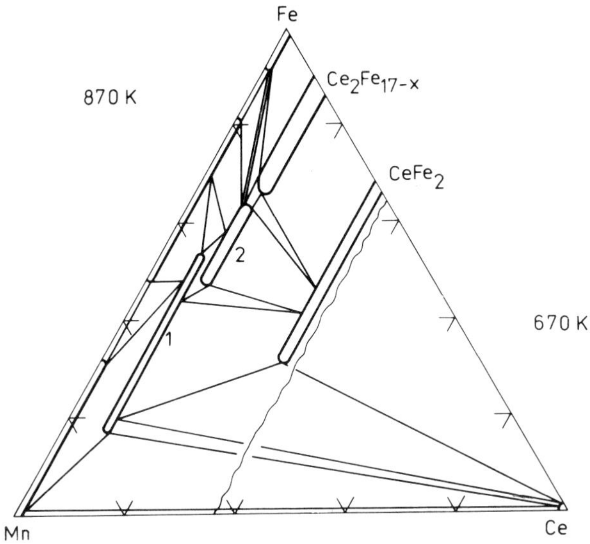

Fig. 158. Ce–Fe–Mn, isothermal sections at 870 K (0–33 a/c Ce) and 670 K (33–100 a/c Ce).

Ce–Fe–Mo

The phase diagram is shown in fig. 159, after Bodak and Berezyuk (1985). The only ternary compound has variable composition, which one can describe as $Ce(Ce_{0.025}Fe_{0.845-0.715}Mo_{0.130-0.260})_{12}$. The crystal structure of it has been determined: $ThMn_{12}$ ST, $a = 8.560-8.507$, $c = 4.776-4.829$. A small fraction of the large cerium atoms are randomly mixed with the smaller ones – Fe and Mo, on the Mn sites of the $ThMn_{12}$ ST.

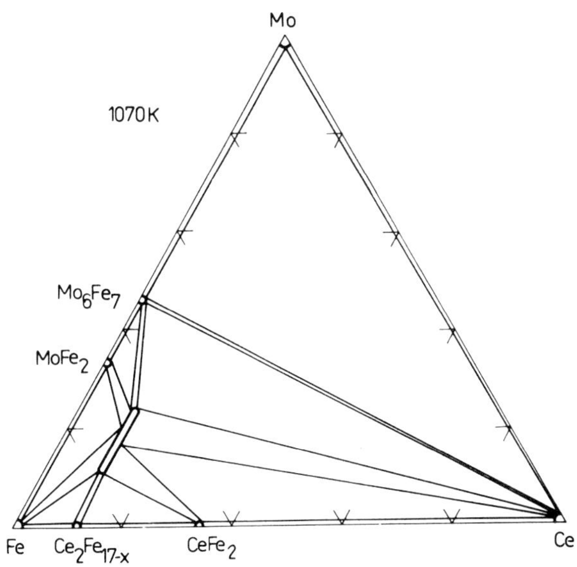

Fig. 159. Ce–Fe–Mo, isothermal section at 1070 K.

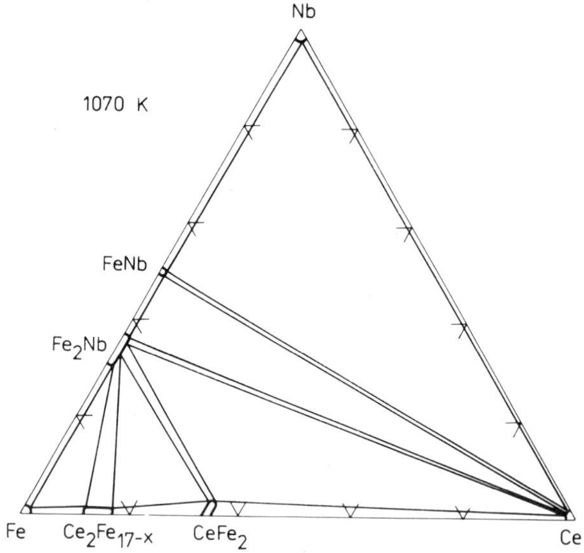

Fig. 160. Ce–Fe–Nb, isothermal section at 1070 K.

Ce–Fe–Nb

Bodak and Berezyuk (1982a) derived the phase diagram shown in fig. 160. No solid solubility or ternary compounds were found.

Ce–Fe–Pu

A continuous solid solution $Ce(FePu)_2$ was found by Tucker et al. (1967) in the Ce–Fe–Pu system.

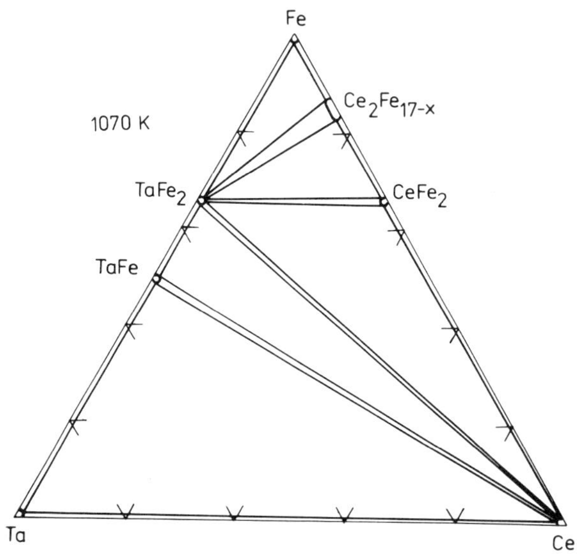

Fig. 161. Ce–Fe–Ta, isothermal section at 1070 K.

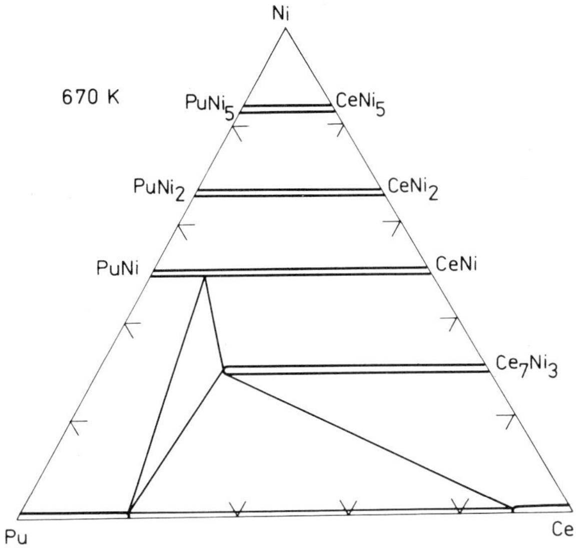

Fig. 162. Ce–Ni–Pu, isothermal section at 670 K.

Ce–Fe–Ta

Phase equilibria (fig. 161) were studied by Bodak and Berezyuk (1982b). There are no ternary compounds in this system.

Ce–Ni–Pu

An investigation of the system Ce–Ni–Pu (fig. 162) was carried out by Menshykova et al. (1967). The unit-cell volume decreased linearly with the increasing of Pu content in the solid solution regions.

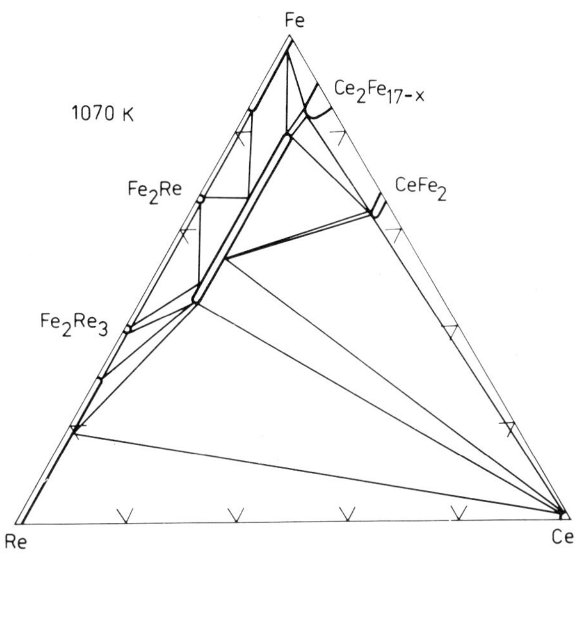

Fig. 163. Ce–Fe–Re, isothermal section at 1070 K.

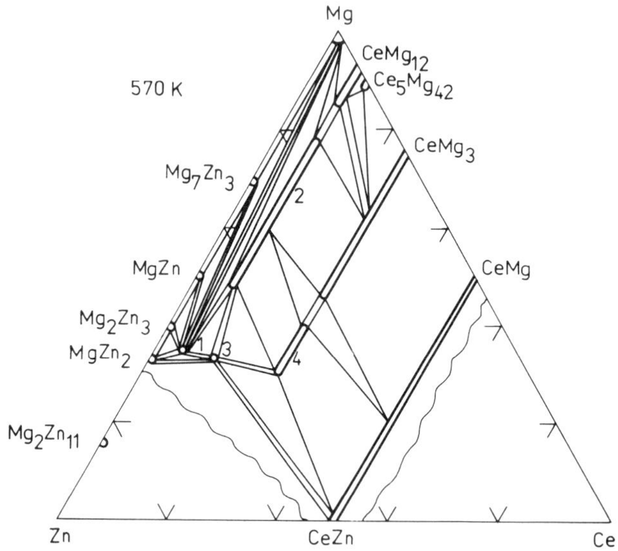

Fig. 164. Ce–Mg–Zn, partial isothermal section at 570 K (80–0 a/c Zn). Ternary compounds with unknown crystal structures: (1) ~ $CeMg_7Zn_{12}$; (2) ~ $Ce(Mg_{0.5-0.85}Zr_{0.5-0.15})_9$; (3) ~ $CeMg_3Zn_5$.

Ce–Fe–Re

The phase diagram at 1070 K (fig. 163) was described by Bodak et al. (1981b). The only ternary compound $CeFe_{10.64-5.18}Re_{1.36-6.82}$ belongs to the $ThMn_{12}$ ST, $a = 8.547–8.745$, $c = 4.746–4.782$.

Ce–Mg–Zn

A total of 4 ternary compounds were found by Melnyk et al. (1978a) during partial (more than 30 a/c Zn) phase-diagram investigations (fig. 164). The crystal

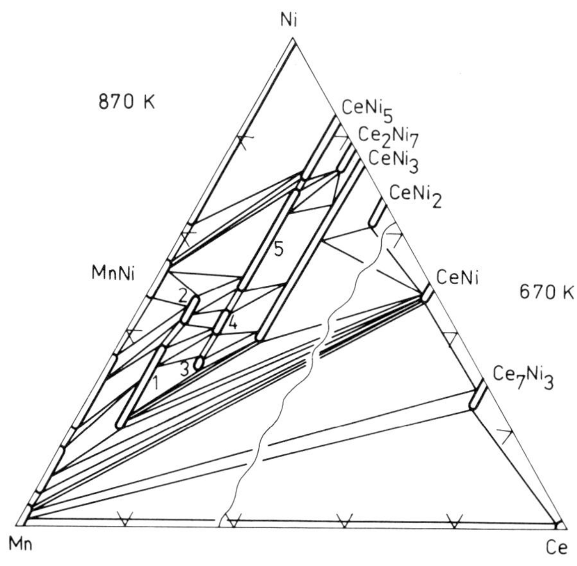

Fig. 165. Ce–Mn–Ni, isothermal sections at 870 K (0–33 a/c Ce) and at 670 K (33–100 a/c Ce).

structure of (4) ~ $Ce_2Mg_3Zn_3$ is unknown, but it was reported that it has a cubic unit cell, $a = 7.064$.

Ce–Mn–Ni

Phase equilibria after Kalychak et al. (1976a) are shown in fig. 165. Five ternary compounds are formed here: (1) ~ $CeMn_{7.16-9.49}Ni_{4.84-2.51}$ ($ThMn_{12}$ ST, $a = 8.684$, $c = 4.863$ for an Mn content 8.45); (2) ~ $CeMn_{5.52-6.48}Ni_{5.48-4.52}$ ($Ce(Mn_{0.55}Ni_{0.45})_{11}$ a new structure type, solved by Kalychak et al. (1975): P4/mbm, $a = 8.330–8.367$, $c = 4.912–4.926$); (3) ~ $CeMn_3Ni_2$ (unknown crystal structure); (4) ~ $CeMn_{2.4-2.7}Ni_{2.6-2.3}$ ($CaCu_5$ ST, $a = 5.098$, $c = 4.171$ for an Mn content 2.58); (5) ~ $CeMn_{0.9-2.1}Ni_{4.1-2.9}$ ($AuBe_5$ ST, $a = 6.957–7.070$).

Pr–Co–Mn

The phase diagram is shown in fig. 166, after Kalychak et al. (1983c). A total of 5 ternary compounds were described: (1) ~ $Pr(Co_{0.57-0.46}Mn_{0.43-0.54})_{12}$ with the $ThMn_{12}$ ST, $a = 8.520–8.583$, $c = 4.768–4.804$; (2) ~ $Pr_2(Co_{0.31-0.08}Mn_{0.69-0.92})_{17}$ probably with the Th_2Ni_{17} ST, $a = 8.704$, $c = 8.536$ at 70 a/c Mn; (3) ~ $Pr_2(Co_{0.44-0.33}Mn_{0.56-0.67})_{17}$ probably with the Th_2Zn_{17} ST, $a = 8.684$, $c = 12.63$ at 55 a/c Mn; (4) ~ $Pr_2(Co_{0.69-0.53}Mn_{0.31-0.47})_{17}$ with an unknown crystal structure, and the last (5) ~ $Pr_6(Co_{0.35-0.1}Mn_{0.65-0.9})_{23}$ with the Th_6Mn_{23} ST, $a = 12.612–12.632$. Earlier it was reported by Kalychak and Oniskovetz (1978) as $Pr_6Co_6Mn_{17}$ with the same crystal structure.

Pr–Cu–Mn

The system Pr–Cu–Mn was investigated by Kalychak et al. (1982), the phase diagram is shown in fig. 167. Ternary compounds are: (1) ~

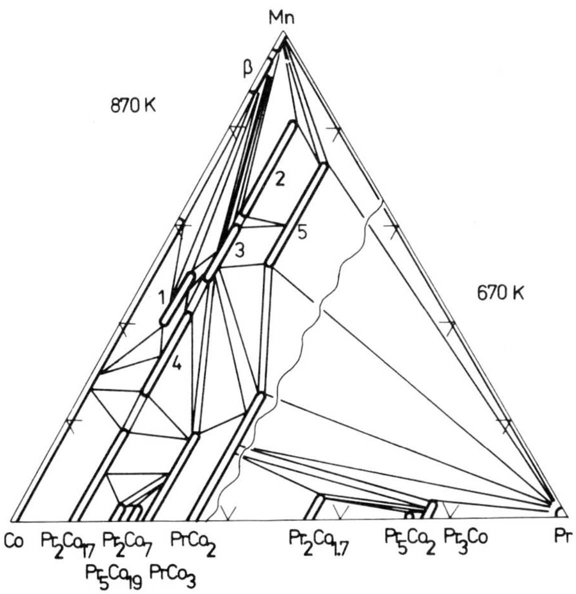

Fig. 166. Pr–Co–Mn, isothermal sections at 870 K (0–33 a/c Pr) and at 670 K (33–100 a/c Pr).

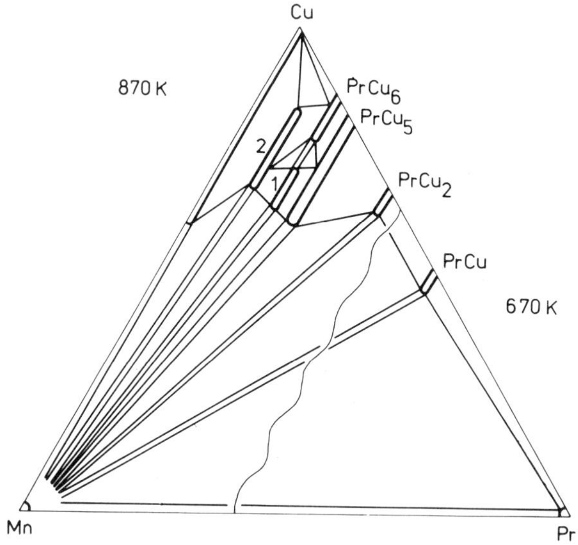

Fig. 167. Pr–Cu–Mn, isothermal sections at 870 K (0–33 a/c Pr) and at 670 K (33–100 a/c Pr).

$PrCu_{10.4-8.12}Mn_{0.96-2.88}$ belonging to the $Ce(Mn_{0.55}Ni_{0.45})_{11}$ ST, $a = 8.330–8.412$, $c = 4.914–4.965$, and (2) ~ $Pr(Cu, Mn)_6$ with an unknown crystal structure. The homogeneity range for the last compound reaches from 14 to 22 a/c Mn.

Pr–Mg–Zn

Kynghybalo et al. (1985) proposed the phase diagram shown in fig. 168 and established the existence of 4 ternary compounds: (1) ~ $PrMg_6Zn$, (2) ~

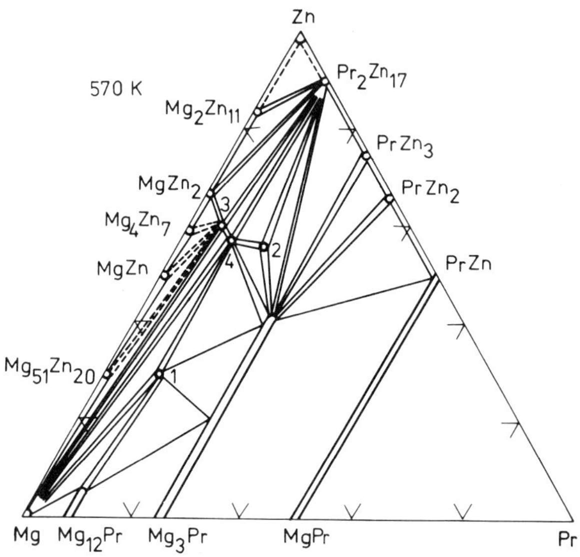

Fig. 168. Pr–Mg–Zn, isothermal section at 570 K.

$Pr_2Mg_7Zn_{11}$, (3) ~ $PrMg_7Zn_{12}$, and (4) ~ $Pr_3Mg_8Zn_{11}$. It was assumed that (3) has a hexagonal unit cell with $a = 14.63$, $c = 8.78$. The existence of $Pr_2Mg_3Zn_3$ with the cubic unit cell, $a = 7.041$, which was found by Melnyk et al. (1978b), was not confirmed.

Pr–Mn–Ni

Kalychak and Oniskovetz (1981) determined the phase diagram (fig. 169). A total of 5 ternary compounds were found, all of them with less than 20 a/c Pr: (1) ~ $PrMn_{6.9-8.84}Ni_{5.1-3.16}$ ($ThMn_{12}$ ST, $a = 8.677$–8.774, $c = 4.848$–4.898); (2) ~ $PrMn_{4.8-6}Ni_{6.2-5}$ ($Ce(Mn_{0.55}Ni_{0.45})_{11}$ ST, $a = 8.358$–8.363, $c = 4.926$–4.932); (3) ~ $Pr_2Mn_{13.3}Ni_{3.7}$ (Th_2Ni_{17} ST, $a = 8.795$, $c = 8.593$); (4) ~ $PrMn_{2.4-3.2}Ni_{2.6-1.8}$ (unknown structure), and (5) ~ $Pr_2(Mn_{0.53}Ni_{0.21})_{13.5}$ (Th_2Zn_{17} ST, $a = 8.836$, $c = 12.84$).

Nd–Co–Mn

The investigation of phase equilibria has been carried out by Kalychak et al. (1983a). The phase diagram is shown in fig. 170. The ternary compounds are as follows: (1) ~ $Nd(Co_{0.65-0.3}Mn_{0.35-0.7})_{12}$ with the $ThMn_{12}$ ST, $a = 8.508$–8.655, $c = 4.758$–8.840. (The existence of this compound was also reported by Kalychak et al. (1976c)); (2) ~ $Nd_2(Co_{0.72}Mn_{0.28})_{17}$ belongs to the Th_2Ni_{17} ST, $a = 8.473$, $c = 8.314$, and (3) ~ $Nd_2(Co_{0.36-0.4}Mn_{0.64-0.6})_{17}$, the crystal structure for this compound was not determined.

Nd–Co–Ru

The system was partially investigated by Efremenko (1979). The phase diagram is shown in fig. 171.

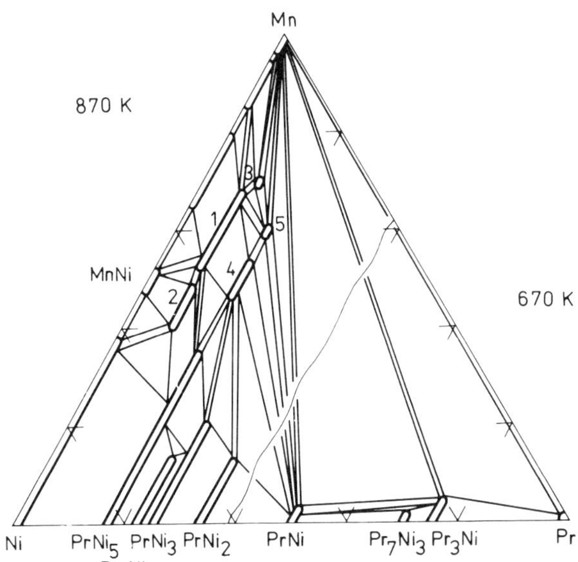

Fig. 169. Pr–Mn–Ni, isothermal sections at 870 K (0–33 a/c Pr) and at 670 K (33–100 a/c Pr).

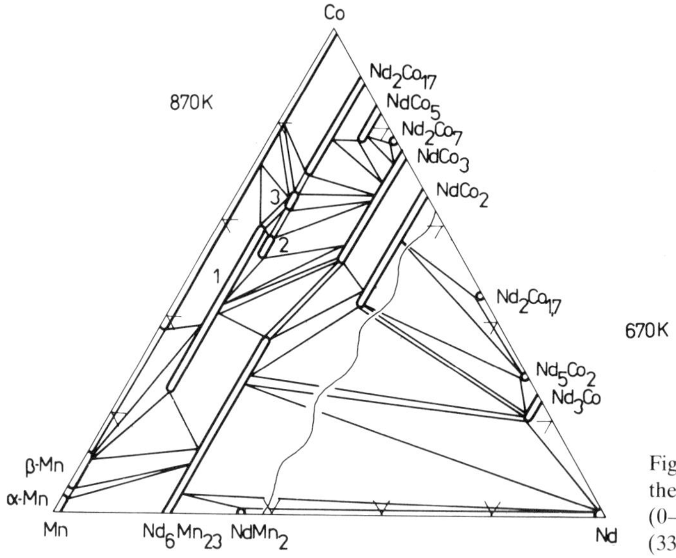

Fig. 170. Nd–Co–Mn, isothermal sections at 870 K (0–33 a/c Nd) and at 670 K (33–100 a/c Nd).

Nd–Cu–Mn

The phase equilibria are shown in fig. 172 after Kalychak et al. (1982). Compound (1) ~ $NdCu_{9.56-7.64}Mn_{1.44-3.36}$ crystallized in the $Ce(Mn_{0.55}Ni_{0.45})_{11}$ ST, $a = 8.357–8.413$, $c = 4.947–4.978$. The crystal structure for (2) ~ $Nd(Mn, Cu)_6$ is unknown, but its homogeneity range extends from 12 to 28 a/c Mn.

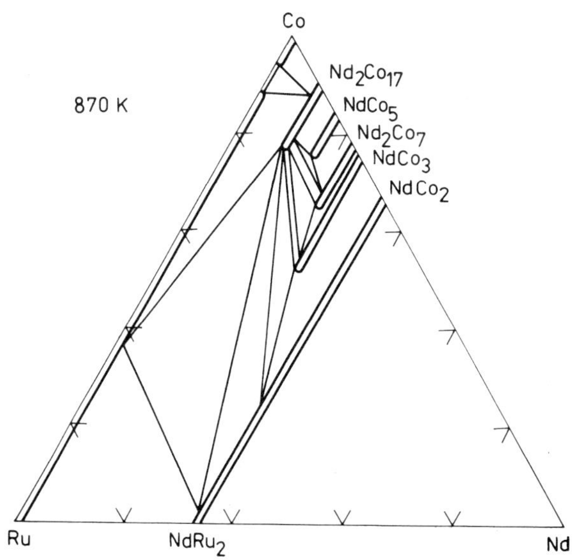

Fig. 171. Nd–Co–Ru, partial isothermal section at 870 K (0–33 a/c Nd).

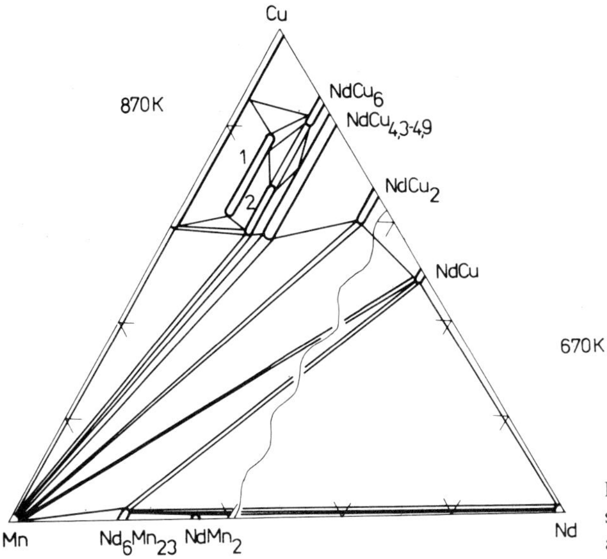

Fig. 172. Nd–Cu–Mn, isothermal sections at 870 K (0–33 a/c Nd) and at 670 K (33–100 a/c Nd).

Nd–Fe–Ru

Efremenko (1979) did not find any ternay compounds in the 0–33 a/c Nd region (fig. 173).

Nd–Mg–Zn

The phase diagram (fig. 174) was derived by Kynghybalo et al. (1985). Four ternary compounds were observed at approximate compositions: (1) ~ $NdMg_6Zn$,

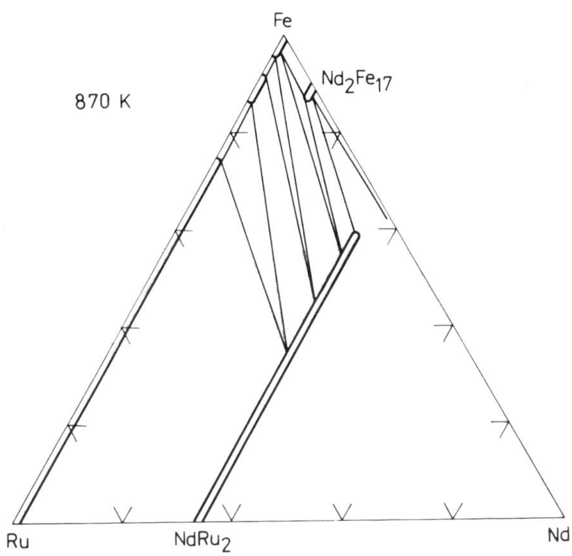

Fig. 173. Nd–Fe–Ru, partial isothermal section at 870 K (0–33 a/c Nd).

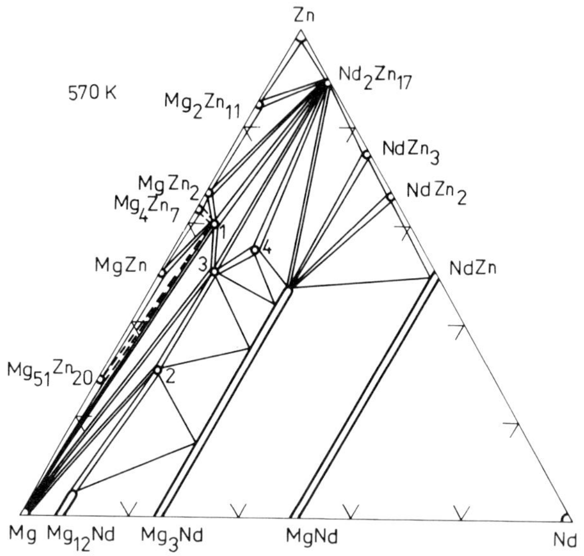

Fig. 174. Nd–Mg–Zn, isothermal section at 570 K.

(2) ~ $Nd_2Mg_7Zn_{11}$, (3) ~ $NdMg_7Zn_{12}$, and (4) ~ $Nd_3Mg_8Zn_{11}$. Lattice parameters were reported only for (3): it was stated that (3) possibly had a hexagonal lattice: $a = 14.70$, $c = 8.78$. The existence of: ~$Nd_4Mg_9Zn_5$, ~$Nd_2Mg_2Zn_9$ and ~$Nd_2Mg_6Zn_7$ earlier reported by Dritz et al. (1974), and ~$Nd_2Mg_3Zn_3$ with cubic unit cell, $a = 6.967$, earlier found by Melnyk et al. (1978b), were not confirmed.

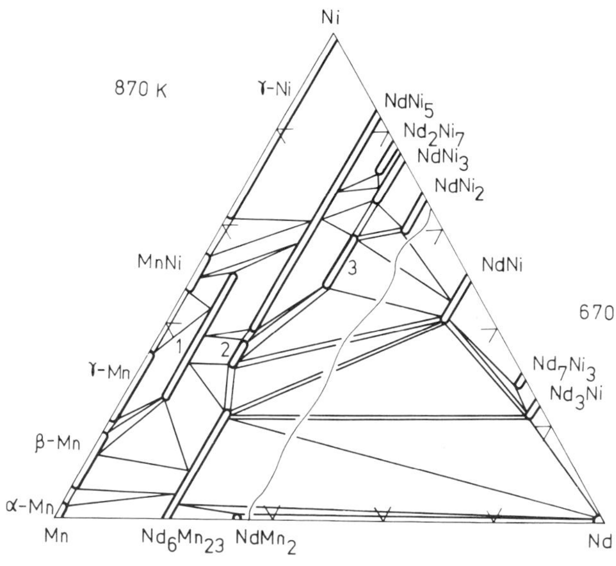

Fig. 175. Nd–Mn–Ni, isothermal sections at 870 K (0–33 a/c Nd) and at 670 K (33–100 a/c Nd).

Nd–Mn–Ni

The phase diagram was described by Kalychak et al. (1983b) and is shown in fig. 175. The crystal structure of (1) ~ $Nd(Mn_{0.45-0.70}Ni_{0.55-0.30})_{12}$ belongs to the $ThMn_{12}$ ST, $a = 8.593–8.719$, $c = 4.818–4.887$. Compound (3) ~ $Nd(Mn, Ni)_3$, where the Mn content changes in the range from 17 to 27 a/c, was reported to have a hexagonal lattice with $a = 5.187$, $c = 16.62$ (at 25 a/c Mn). The authors also made the assumption that (3) has the $CeNi_3$ type of crystal structure. No structure details were given for compound (2) ~ $Ce(Mn, Ni)_5$, with a variable composition from 47 to 52 a/c Mn.

Nd–Ni–Ru

It was established by Efremenko (1979), that the unit cell volume changes linearly in the continuous solid-solution region $Nd(Ni, Ru)_2$. The phase diagram is shown in fig. 176.

Sm–Co–Cu

Partial isothermal sections at 1070 K (after Perry 1977) and at 1120 K (after Giardon and Kurz 1979) are shown in fig. 177. These two sections are very similar. Shkatova et al. (1979) derived the partial one at 1170 K, which is also shown in fig. 177. The third is quite different from the first two. According to Shkatova et al. (1979) the ternary compound $SmCo_{3.3}Cu_{1.7}$ belongs to the $CaCu_5$ type of crystal structure. The unit cell was not determined.

Sm–Co–Mn

The system Sm–Co–Mn was investigated by Bodak et al. (1981a). The ternary compounds are as follows: (1) ~ $SmCo_{8.23-3.29}Mn_{3.77-8.71}$ ($ThMn_{12}$ ST, $a =$

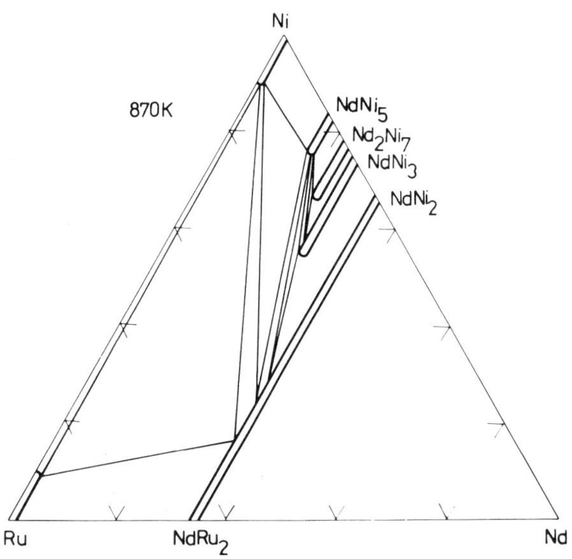

Fig. 176. Nd–Ni–Ru, partial isothermal section at 870 K (0–33 a/c Nd).

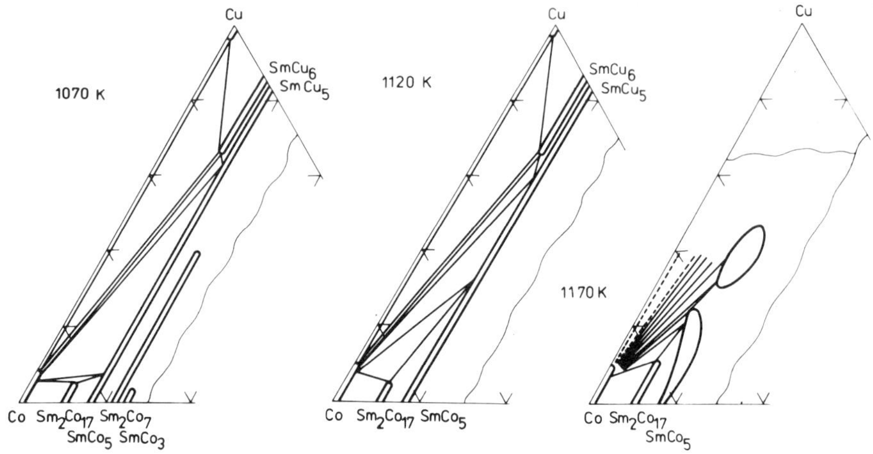

Fig. 177. Sm–Co–Cu, partial isothermal sections at 1070, 1120 and 1170 K.

8.473–8.611, $c = 4.746$–4.850), and (2) ~ $Sm_2(Co,Mn)_{17}$ (Th_2Zn_{17} ST, $a =$ 8.421–8.450, $c = 12.61$–12.38). See fig. 178.

Sm–Co–Ru

Sokolovskaya et al. (1978) described the partial isothermal section at 870 K (fig. 179). Ternary compound (1) is the high-temperature modification of Sm_2Co_{17} with Th_2Zn_{17} ST, stabilized by the Ru, and (2) ~ $SmCo_{1.4}Ru_6$ belongs to the $CeNi_3$ ST, $a = 5.18$, $c = 17.00$.

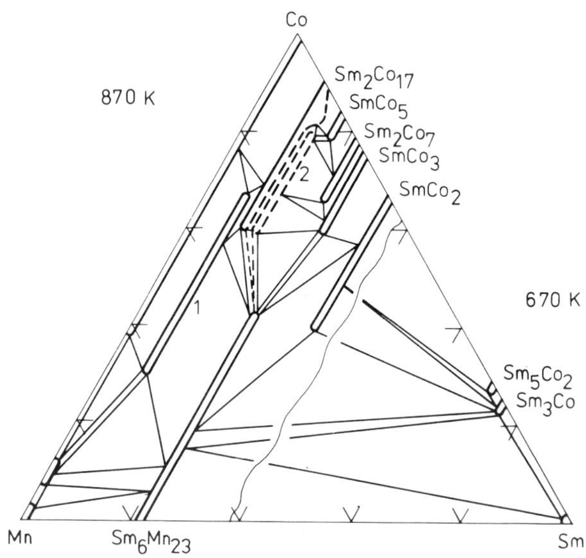

Fig. 178. Sm–Co–Mn, isothermal section at 870 K (0–33 a/c Sm) and at 670 K (33–100 a/c Sm).

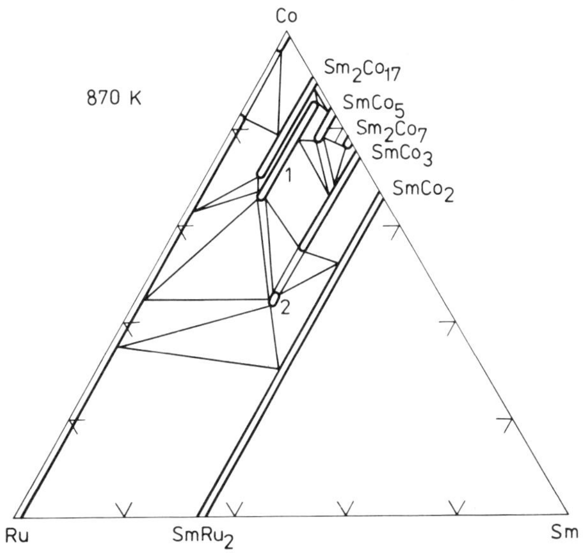

Fig. 179. Sm–Co–Ru, partial isothermal section at 870 K.

Sm–Fe–Ru

The system Sm–Fe–Ru (fig. 180) was partially investigated by Sokolovskaya et al. (1978). The solubility of Ru in the Sm_2Fe_{17} is less than 5 a/c.

Sm–Mn–Ni

Bodak et al. (1981b) described the phase diagram shown in fig. 181, and reported the existence of two ternary compounds. (1) ~ $SmMn_{9.9-8.06}Ni_{8.1-3.94}$

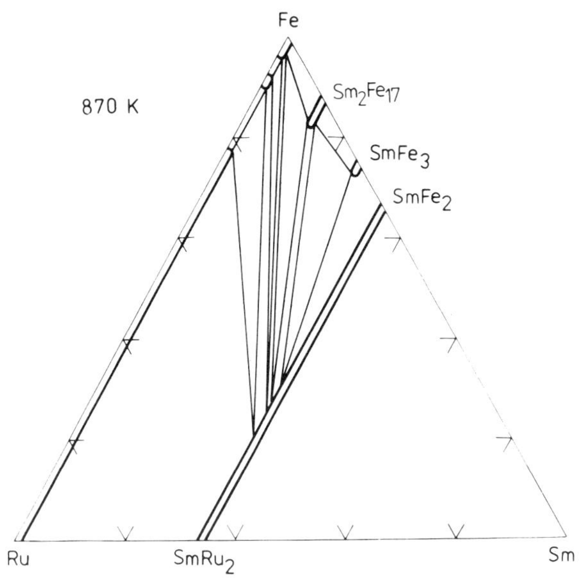

Fig. 180. Sm–Fe–Ru, partial isothermal section at 870 K.

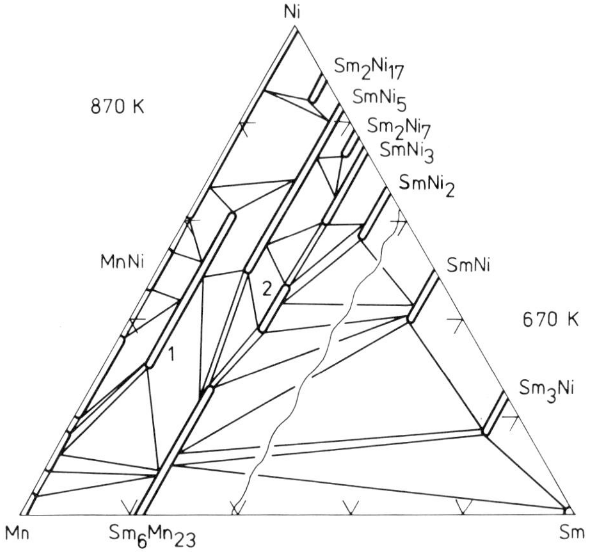

Fig. 181. Sm–Mn–Ni, isothermal sections at 870 K (0–33 a/c Sm) and at 670 K (33–100 a/c Sm).

($ThMn_{12}$ ST, $a = 8.465$–8.693, $c = 4.766$–4.864), and (2) ~ $Sm(Mn, Ni)_3$ (hexagonal unit cell with $a = 5.174$, $c = 33.62$).

Sm–Mn–Ru

The system was partially described by Raevskaya and Sokolovskaya (1979). The phase diagram is shown in fig. 182. No ternary compounds were found.

Sm–Ni–Ru

The system has been studied by Sokolovskaya et al. (1978). The isothermal section at 870 K is shown in fig. 183. The only ternary compound,

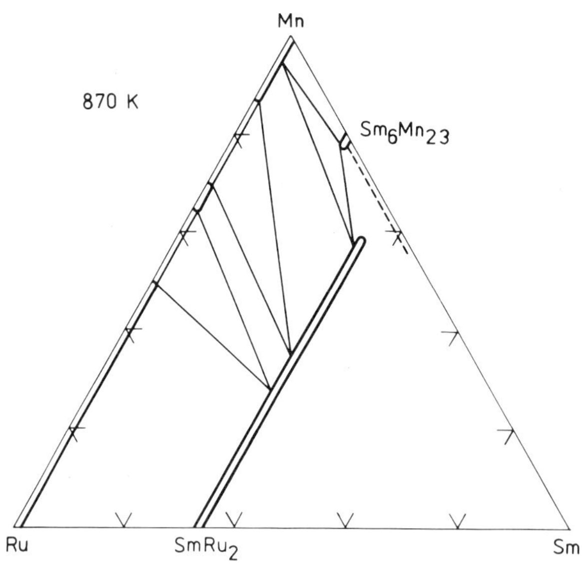

Fig. 182. Sm–Mn–Ru, partial isothermal section at 870 K.

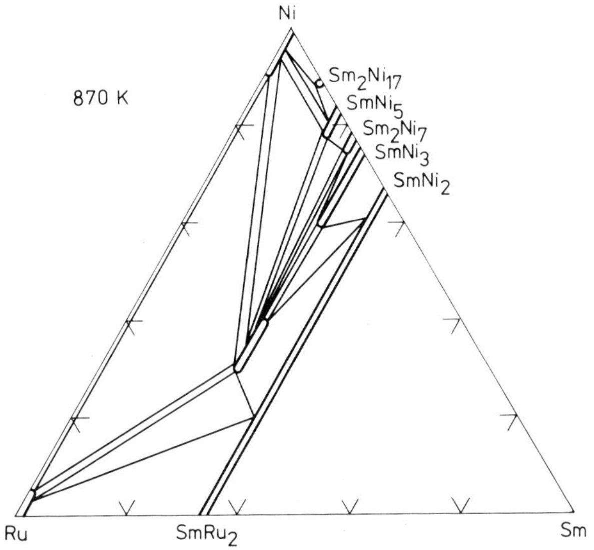

Fig. 183. Sm–Ni–Ru, partial isothermal section at 870 K.

$SmNi_{1.6-1.2}Ru_{1.4-1.8}$, belongs to the $CeNi_3$ ST, $a = 5.13$, $c = 16.68$ with an unknown composition.

Dy–Ag–Pd

Slavev et al. (1977b) investigated the system Dy–Ag–Pd. The derived isothermal section at 870 K is shown in fig. 184. The structures of both of the existing ternary compounds are unknown: (1) ~ $DyAg_{1.3}Pd_{1.7}$ and (2) ~ $DyAg_2Pd$.

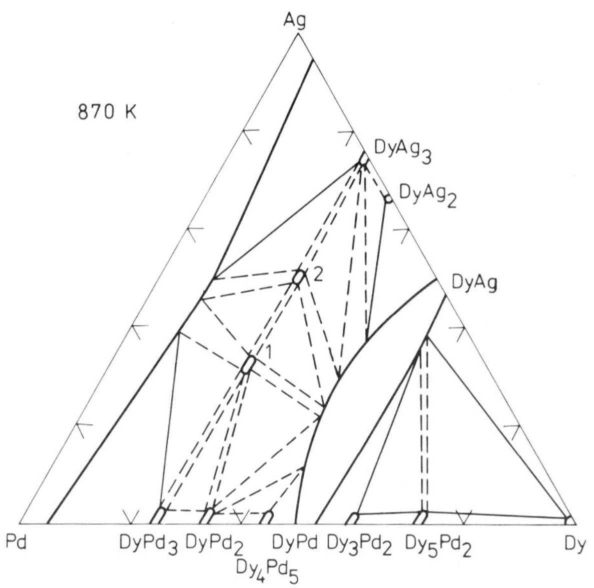

Fig. 184. Dy–Ag–Pd, isothermal section at 870 K.

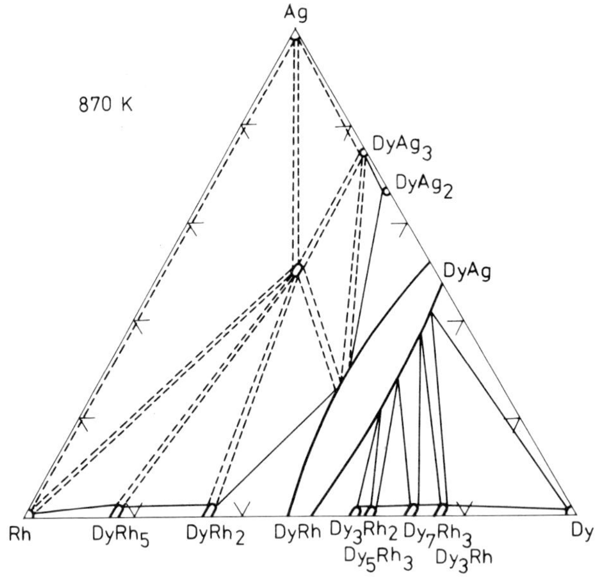

Fig. 185. Dy–Ag–Rh, isothermal section at 870 K.

Dy–Ag–Rh

Only one ternary compound, ~$DyAg_2Rh$, with an unknown crystal structure was found in the phase diagram (fig. 185), derived by Slavev et al. (1977a).

Dy–Co–Re

The two ternary compounds were described by Sokolovskaya et al. (1985a) in the system Dy–Co–Re. The phase diagram is shown in fig. 186. The first compound, ~$Dy_2(Co_{0.94-0.92}Re_{0.06-0.08})_{17}$ represents a solid solution of rhenium in the high-temperature modification of Dy_2Co_{17}, and the second has variable

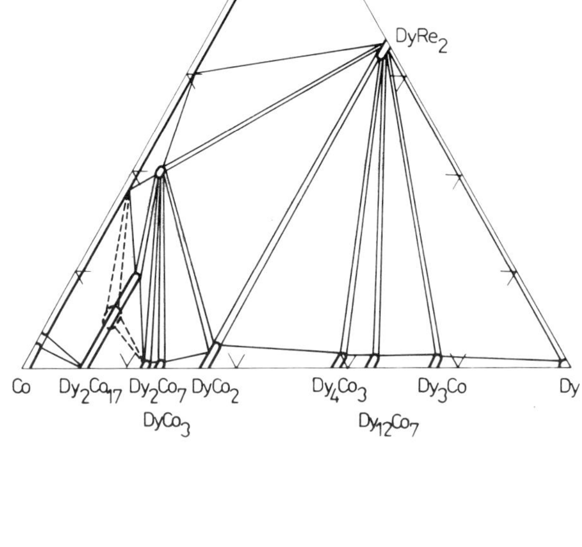

Fig. 186. Dy–Co–Re, isothermal section at 870 K.

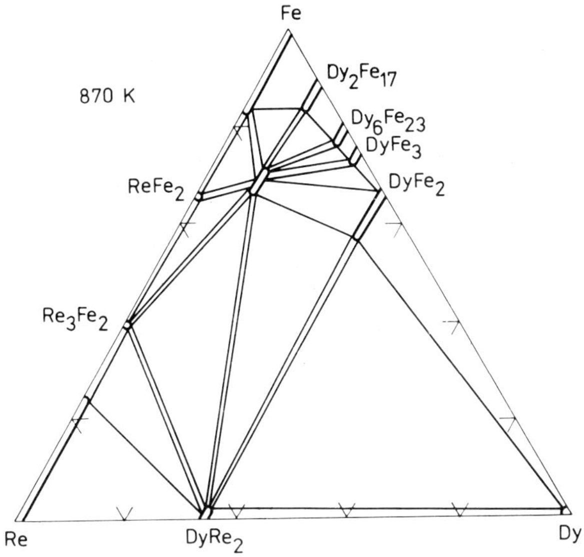

Fig. 187. Dy–Fe–Re, isothermal section at 870 K.

composition $Dy(Co_{0.38-0.44}Re_{0.62-0.56})_{12}$ and the $ThMn_{12}$ ST. Unit cell dimensions were not given.

Dy–Fe–Re

The existence of ternary compound $Dy(Dy_{0.03}Fe_{0.2-0.3}Re_{0.8-0.7})_{12}$ with the $ThMn_{12}$ ST and unknown unit cell dimensions in the system Dy–Fe–Re (fig. 187) was reported by Sokolovskaya et al. (1985a).

Dy–Ni–Re

The phase diagram was derived by Sokolovskaya et al. (1985a) and is shown in fig. 188. No ternary compounds were found.

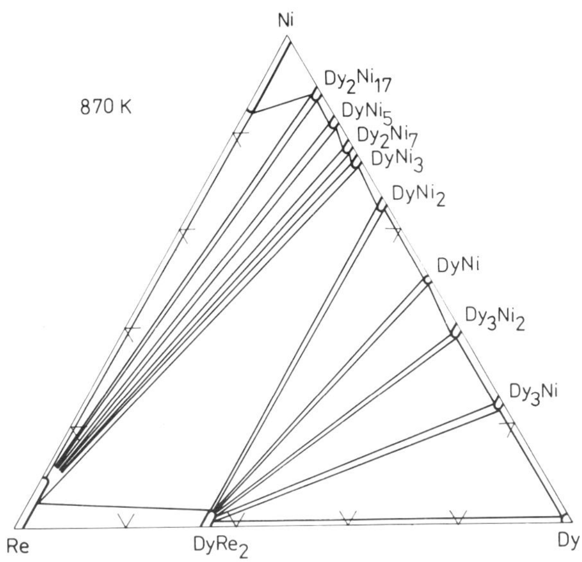

Fig. 188. Dy–Ni–Re, isothermal section at 870 K.

TABLE 6
Ternary compounds found in the partially investigated systems R–M–M′.

Compound	Structure type	Lattice parameters *a*	*b*	*c*	Refs.
$ScAuNi_4$	$MgCu_4Sn$	7.128	–	–	Dw, 75
$YAuNi_4$	$MgCu_4Sn$	7.059	–	–	Dw, 75
$Y_3Cd_2Zn_3$	Hexagonal	15.103	–	14.985	HaR, 69
$Y(Co_{0.84}Mo_{0.16})_{12}$	$ThMn_{12}$	8.421	–	4.720	Be, 83
$Y(Fe_xMo_{1-x})_{12}$	$ThMn_{12}$	8.514–	–	4.775–	Be, 83
$(x = 0.89 - 0.73)$		8.569		4.813	
$YMn_{1.5}Ni_{3.5}$	$AuBe_5$	7.025	–	–	Ka, 75
$Ce(Ce_{0.03}Fe_{0.87}W_{0.10})_{12}$	$ThMn_{12}$	8.567	–	4.815	BoB, 83
$PrFe_{6.4}Mn_{5.6}$	$ThMn_{12}$	8.552	–	4.789	FrK, 78
$Pr(Pr_{0.03}Fe_xMo_{0.97-x})_{12}$	$ThMn_{12}$	8.593–	–	4.777–	BoB, 85
$(x = 0.87–0.72)$		8.620		4.805	
$Pr(Pr_{0.03}Fe_{0.87}W_{0.10})_{12}$	$ThMn_{12}$	8.617	–	4.800	BoB, 83
$Nd(Nd_{0.03}Fe_xMo_{0.97-x})_{12}$	$TbMn_{12}$	8.591–	–	4.791–	BoB, 85
$(x = 0.81–0.70)$		8.632		4.822	
$Nd(Nd_{0.03}Fe_{0.87}W_{0.10})_{12}$	$ThMn_{12}$	8.597	–	4.812	BoB, 83
$SmCu_{5.3}Mn_{0.7}$	...	...	...	...	Ka, 77
$SmCu_{9.2}Mn_{1.8}$	$Ce(Mn, Ni)_{11}$	8.357	–	4.939	Ka, 77
$Sm(Sm_{0.03}Fe_xMo_{0.97-x})$	$ThMn_{12}$	8.563–	–	4.771–	BoB, 85
$(x = 0.83–0.76)$		8.573		4.780	
$Sm(Sm_{0.03}Fe_{0.87}W_{0.10})_{12}$	$ThMn_{12}$	8.567	–	4.821	BoB, 83
$Sm_2Mg_3Zn_3$	Cubic	6.907	–	–	MeKPD, 78b
$SmMg_{9.54}Zn_{5.21}$	...	...	...	...	DrRAKT, 85
$SmMg_3Zn_6$	Hexagonal	14.62	–	8.78	DrRAKT, 85
$GdAuNi_4$	$MgCu_4Sn$	7.128	–	–	Dw, 75
$GdCo_{0.5}Cu_{0.5}$	YNi	4.202	7.390	5.551	Kh, 77
		$(\gamma = 90.0°)$			
$GdCo_{0.7}Cu_{0.3}$	MoB	3.905	–	20.63	Kh, 77
$Gd_2(Co_{0.5}Cu_{0.5})_7$	...	...	...	...	TrBCW, 72
$Gd(Co_{0.84}Mo_{0.16})_{12}$	$ThMn_{12}$	8.445	–	4.740	Be, 83
$GdCo_{0.6}Ni_{0.4}$	MoB	3.939	–	21.12	Kh, 77
$GdCu_{5.3}Mn_{1.7}$	...	...	...	...	Ka, 77
$GdCu_{9.2}Mn_{1.8}$	$Ce(Mn, Ni)_{11}$	8.314	–	4.916	Ka, 77
$GdCu_xNi_{1-x}$	FeB	7.033–	4.230–	5.410–	Ra, 68
$(x = 0.5–0.75)$		7.030	4.212	5.490	
$Gd(Fe_xMo_{1-x})_{12}$	$ThMn_{12}$	8.540–	–	4.772–	Be, 83
$(x = 0.91–0.62)$		8.600		4.812	
$Gd_2Mg_3Zn_3$	Cubic	6.857	–	–	MeKPD, 78b
$GdMn_3Ni_2$	Hexagonal	5.252	–	17.17	Ka, 77
$TbAuNi_4$	$MgCu_4Sn$	7.074	–	–	Dw, 75
$TbCo_{0.5}Cu_{0.5}$	YNi	4.196	7.350	5.511	Kh, 77
		$(\gamma = 90.0°)$			
$TbCo_{0.7}Cu_{0.3}$	MoB	3.885	–	20.44	Kh, 77
$Tb(Co_{0.84}Mo_{0.16})_{12}$	$ThMn_{12}$	8.415	–	4.711	Be, 83
$TbCo_{0.6}Ni_{0.4}$	MoB	3.920	–	20.96	TrBCW, 72
$Tb(Fe_xMo_{1-x})_{12}$	$ThMn_{12}$	8.500–	–	4.773–	Be, 83
$(x = 0.91–0.68)$		8.586	–	4.821	
$Tb_2Mg_3Zn_3$	Cubic	6.837	–	–	MeKPD, 78b
$TbMn_{1.5}Ni_{3.5}$	$AuBe_5$	7.020	–	–	Ka, 77
$DyAuNi_4$	$MgCu_4Sn$	7.036	–	–	Dw, 75

TABLE 6 (cont'd)

Compound	Structure type	Lattice parameters *a*	*b*	*c*	Refs.
$Dy_2(Co_{0.5}Cu_{0.5})_7$	...	...	...	...	TrBCW, 72
$Dy(Co_{0.84}Mo_{0.16})_{12}$	$ThMn_{12}$	8.415	–	4.707	Be, 83
$DyCo_{0.6}Ni_{0.4}$	MoB	3.905	–	20.84	KhBG, 77a
$DyFe_3Mn$	...	...	...	...	WiSWN, 64
$Dy(Fe_xMo_{1-x})_{12}$ ($x = 0.89$–0.71)	$ThMn_{12}$	8.513–8.574	–	4.778–4.800	Be, 83
$DyMn_{1.5}Ni_{3.5}$	$AuBe_5$	6.984	–	–	Ka, 75
$HoAuNi_4$	$MgCu_4Sn$	6.999	–	–	Dw, 75
$Ho(Co_{0.84}Mo_{0.16})_{12}$	$ThMn_{12}$	8.418	–	4.726	Be, 83
$HoCo_{0.6}Ni_{0.4}$	MoB	3.897	–	20.70	KhBG, 77a
$HoFe_3Mn$	...	...	...	...	WiSWN, 64
$Ho(Fe_xMo_{1-x})_{12}$ ($x = 0.89$–0.71)	$ThMn_{12}$	8.504–8.572	–	4.774–4.812	Be, 83
$HoMn_{1.5}Ni_{3.5}$	$AuBe_5$	6.975	–	–	Ka, 75
$ErAuNi_4$	$MgCu_4Sn$	6.993	–	–	Dw, 75
$Er_2(Co_{0.5}Cu_{0.5})_7$	...	...	...	...	TrBCW, 72
$Er(Co_{0.84}Mn_{0.16})_{12}$	$ThMn_{12}$	8.401	–	4.709	Be, 83
$ErCo_{0.6}Ni_{0.4}$	MoB	3.886	–	20.60	KhBG, 77a
$ErCu_xNi_{1-x}$ ($x = 0.5$–0.75)	CrB	4.225–4.193	10.57–10.35	3.851–3.787	Ra, 68
$ErFe_3Mn$	...	...	...	...	WiSWN, 64
$Er(Fe_xMo_{1-x})_{12}$ ($x = 0.89$–0.73)	$ThMn_{12}$	8.500–8.560	–	4.772–4.806	Be, 83
$Er_2Mg_3Zn_3$	cubic	6.769	...	...	MeKPD, 78b
$ErMn_{1.5}Ni_{3.5}$	$AuBe_5$	6.972	–	–	Ka, 75
$TmAuNi_4$	$MgCu_4Sn$	6.967	–	–	Dw, 75
$Tm(Co_{0.84}Mo_{0.16})_{12}$	$ThMn_{12}$	8.404	–	4.713	Be, 73
$TmCo_{0.6}Ni_{0.4}$	MoB	3.880	–	20.59	KhBG, 77a
$TmFe_3Mn$	...	...	...	...	WiSWN, 64
$Tm(Fe_xMo_{1-x})_{12}$ ($x = 0.78$–0.71)	$ThMn_{12}$	8.527–8.534	–	4.790–4.800	Be, 83
$TmMn_{1.5}Ni_{3.5}$	$AuBe_5$	6.962	–	–	Ka, 75
$YbAuNi_4$	$MgCu_4Sn$	6.969	–	–	Dw, 75
$LuAuNi_4$	$MgCu_4Sn$	6.948	–	–	Dw, 75
$Lu(Co_{0.84}Mo_{0.16})_{12}$	$ThMn_{12}$	8.367	–	4.684	Be, 83
$Lu(Fe_xMo_{1-x})_{12}$ ($x = 0.89$–0.75)	$ThMn_{12}$	8.480–8.544	–	4.763–4.802	Be, 83

References

Berezyuk, D.A., 1983, New representatives of the structure type $ThMn_{12}$ in the systems R–Fe,Co–Mo, in: Chetvertaya Vsesoyuzn. Konf. Kristallokhim. Intermetal. Soedin., Tezisy Dokl., 1983, Lvov (Vyscha Shkola, Lvov) p. 43.

Bodak, O.I., and D.A. Berezyuk, 1981, Dopov. Akad. Nauk Ukr. RSR, Ser. A (3), 82.

Bodak, O.I., and D.A. Berezyuk, 1982a, Dopov. Akad. Nauk Ukr. RSR, Ser. A (3), 70.

Bodak, O.I., and D.A. Berezyuk, 1982b, Dopov. Akad. Nauk Ukr. RSR, Ser. A (5), 44.

Bodak, O.I., and D.A. Berezyuk, 1983, The systems Mo,W–Fe–Ce, in: Y Vsesoyuzn. Sov. Khim. Tekhnolog. Mo, W., Tezisy Dokl., part. 1, Ulan Ude, p. 78.

Bodak, O.I., and D.A. Berezyuk, 1985, Dopov. Akad. Nauk Ukr. RSR, Ser. A (5), 83.
Bodak, O.I., and G.I. Kyrchiv, 1983, Izv. Akad. Nauk SSSR, Met. (2), 220.
Bodak, O.I., Ya.M. Kalychak and V.P. Skvorchuk, 1981a, Dopov. Akad. Nauk Ukr. RSR, Ser. A (10), 83.
Bodak, O.I., V.K. Pecharsky, Ya.M. Kalychak, O.I. Kharchenko, I.R. Mokraya, L.A. Muratova, D.A. Berezyuk and M.M. Shevchuk, 1981b, Some ternary systems with rare earth metals, in: Phase equilibria in metallic alloys, 1981, Moscow (Nauka, Moscow) pp. 57–63.
Dobatkina, T.V., E.V. Melnyk, A.T. Tyvanchuk and E.V. Muratova, 1985, Phase equilibria in the system Mg–Zn–La, in: Stable and Metastable phase equilibria in metallic alloys, 1985, Moscow (Nauka, Moscow) pp. 75–79.
Dritz, M.E., E.M. Padezhnova and N.F. Miklina, 1974, Izv. Akad. Nauk SSSR, Met. (3), 225.
Dritz, M.E., L.L. Rokhlina, N.P. Abrukina, V.V. Kynghybalo and A.T. Tyvanchuk, 1985, Izv. Akad. Nauk SSSR, Met. (6), 134.
Dwight, A.E., 1975, J. Less-Common Met. **43**, 117.
Efremenko, N.E., 1979, Physical chemistry investigation of interaction of Nd with some elements of the VIIIth group, in: Mater. Konf. Molod. Uchen. Posv. 60 let. VLKSM (Moscow University Press, Moscow) pp. 191–195.
Ellinger, F.H., C.C. Land, K.A. Johnson and V.O. Strueberg, 1966, Trans. Met. Soc. AIME **236**, 1577.
Frankevich, D.P., and Ya.M. Kalychak, 1978, Intermetallic phases in the system Pr–Mn–Fe, in: Tretya Vsesoyuzn. Konf. Kristallokhim. Intermetal. Soedin., Tezisy Dokl., 1978, Lvov (Vyscha Shkola, Lvov) pp. 67,68.
Giardon, R., and W. Kurz, 1979, Z. Metallkd. **70**, 386.
Hahn, B.H., and E. Ryba, 1969, J. Less-Common Met. **18**, 319.
Kalychak, Ya.M., 1975, Vestn. Lvovsk. Univ., Ser. Khim. **17**, 20.
Kalychak, Ya.M., 1977, The investigation of ternary systems Ce–Mn–Fe,Co,Ni,Cu and other related systems (phase equilibria and crystal structures), Ph.D. Chemistry thesis, Lvov, 1977 (State University, Lvov) pp. 1–19.
Kalychak, Ya.M., and B.D. Oniskovetz, 1978, Crystal structure of compound $Pr_6(Co,Mn)_{23}$, in: Tretya Vsesoyuzn. Konf. Kristalokhim. Intermetal. Soedin., Tezisy Dokl., Lvov, 1978 (Vyscha Shkola, Lvov) pp. 73, 74.
Kalychak, Ya.M., and B.D. Oniskovetz, 1981, Vestn. Lvovsk. Univ., Ser. Khim. **23**, 54.
Kalychak, Ya.M., O.I. Bodak and E.I. Gladyshevsky, 1974, Vestn. Lvovsk. Univ., Ser. Khim. **16**, 11.
Kalychak, Ya.M., L.G. Akselrud, Ya.P. Yarmolyuk, O.I. Bodak and E.I. Gladyshevsky, 1975, Kristallografiya **20**, 1045.
Kalychak, Ya.M., O.I. Bodak and E.I. Gladyshevsky, 1976a, Izv. Akad. Nauk SSSR, Neorg. Mater. **12**, 1149.
Kalychak, Ya.M., O.I. Bodak and L.M. Pastelyak, 1976b, Vestn. Lvovsk. Univ., Ser. Khim. **18**, 9.
Kalychak, Ya.M., O.I. Bodak and E.I. Gladyshevsky, 1976c, Kristallografiya **21**, 507.
Kalychak, Ya.M., O.I. Bodak and G.V. Krasovskaya, 1979a, Dopov. Akad. Nauk Ukr. RSR, Ser. A, p. 867.
Kalychak, Ya.M., O.I. Bodak, B.D. Oniskovetz and E.I. Gladyshevsky, 1979b, Vestn. Lvovsk. Univ., Ser. Khim. **21**, 38.
Kalychak, Ya.M., O.I. Bodak and E.I. Gladyshevsky, 1980, Izv. Akad. Nauk SSSR, Met. (5), 245.
Kalychak, Ya.M., O.I. Bodak and I.V. Suslova, 1982, Dopov. Akad. Nauk Ukr. RSR, Ser. A (3), 74.
Kalychak, Ya.M., B.D. Oniskovetz and O.I. Bodak, 1983a, Dopov. Akad. Nauk Ukr. RSR, Ser. B (2), 32.
Kalychak, Ya.M., B.D. Oniskovetz and O.I. Bodak, 1983b, Izv. Vyssh. Uchebn. Zaved. Tsvetn. Metall. (3), 74.
Kalychak, Ya.M., B.D. Oniskovetz, O.I. Bodak and E.I. Gladyshevsky, 1983c, Izv. Akad. Nauk SSSR, Metall. (1), 169.
Kharchenko, O.I., 1977, Investigation of the ternary systems of yttrium, cerium and lanthanum with iron, cobalt, nickel and copper: phase equilibria, crystal structures and some physical properties of the alloys, Ph.D. Chemistry thesis, 1977, Lvov (Lvov State University, Lvov) pp. 1–20.

Kharchenko, O.I., 1981, Vestn. Lvovsk. Univ., Ser. Khim. **23**, 58.

Kharchenko, O.I., O.I. Bodak, E.I. Gladyshevsky and B.I. Mazurenko, 1975, Vestn. Lvovsk. Univ., Ser. Khim. **17**, 16.

Kharchenko, O.I., O.I. Bodak, E.I. Gladyshevsky and L.V. Bondarenko, 1976, Ann. Univ., Sect. AA, ed. M. Curie-Sklodowska, 1976/1977, Lublin Poland, **31–32**, 131.

Kharchenko, O.I., O.I. Bodak and E.I. Gladyshevsky, 1977a, Izv. Akad. Nauk SSSR, Metall. (1), 200.

Kharchenko, O.I., O.I. Bodak and M.Z. Palasyuk, 1977b, Vestn. Lvovsk. Univ., Ser. Khim. **19**, 71.

Kharchenko, O.I., L.M. Kondratyuk and M.M. Rak, 1986, Vestn. Lvovsk. Univ., Ser. Khim. **27**, 50.

Kynghybalo, V.V., A.T. Tyvanchuk and E.V. Melnyk, 1985, Investigation of the ternary systems Mg–Zn–Pr and Mg–Zn–Nd, in: Stable and metastable phase equilibria in metallic alloys, 1985, Moscow (Nauka, Moscow) pp. 70–74.

Larson, A.C., R.B. Root and D.T. Cromer, 1964, Acta Crystallogr. **17**, 1382.

Maslennikov, S.B., and G.S. Braslavskaya, 1986, Izv. Akad. Nauk SSSR, Metall. (5), 213.

Melnyk, E.V., M.F. Kostina, Ya.P. Yarmolyuk and O.F. Zmij, 1978a, The investigation of ternary systems Mg–Zn–Ce and Mg–Zn–Ca, in: Magnesium alloys, 1978, Moscow (Nauka, Moscow) pp. 95–99.

Melnyk, E.V., V.V. Kynghybalo, E.M. Padezhnova and T.V. Dobatkina, 1978b, The new ternary compounds with cubic face centered lattice in the systems Mg–Zn–R, in: Tretya Vsesoyuzn. Konf. Kristallokhim. Intermetal. Soedin., Tezisy Dokl., Lvov, 1978 (Vyscha Shkola, Lvov) p. 73.

Menshykova, T.S., V.S. Kurylo, N.T. Chebotarev and N.P. Petrov, 1967, Investigation of ternary alloys Pu–Ce–Ni, in: Plutonium reactor Fuel, 1967, Vienna, pp. 151–164.

Mushagi, A., M.V. Raevskaya, T.P. Loboda and O.I. Bodak, 1983, Izv. Akad. Nauk SSSR, Metall. (1), 200.

Padezhnova, E.M., E.V. Melnyk, R.A. Milevsky, T.V. Dobatkina and V.V. Kynghybalo, 1982, Izv. Akad. Nauk SSSR, Metall. (4), 204.

Perry, A.J., 1977, J. Less-Common. Met. **51**, 153.

Raevskaya, M.V., and E.M. Sokolovskaya, 1979, Physical chemistry of the ruthenium and its alloys, 1979 (Moscow University Press, Moscow) p. 227.

Raman, A., 1968, Anorg. Chem. **7**, 973.

Samyratov, S.T., M.V. Raevskaya, I.G. Sokolova, E.M. Sokolovskaya and U.M. Mukhanov, 1982, Vestn. Mosk. Univ. Khim. **23**, 406.

Savitzky, E.M., Yu.V. Efimov and O.I. Zvolinsky, 1975, The system Nb–V–Sc, in: Rare earth metals in the alloys, 1975 (Nauka, Moscow) pp. 26–31.

Shkatova, T.M., V.F. Terekhova and E.M. Savitzky, 1979, Investigation of the Sm–Co–Cu system alloys, in: Alloys of rare and high-melting metals with unique physical behaviours, 1979 (Nauka, Moscow) pp. 179–182.

Slavev, A.G., V.K. Portnoj, M.V. Raevskaya and E.M. Sokolovskaya, 1977a, Isothermal section of the phase diagram Ag–Rh–Dy at 600°C. Manuscript referred to VINITI, 1977, No. 906-77Dep, pp. 1–12.

Slavev, A.G., V.K. Portnoj, M.V. Raevskaya and E.M. Sokolovskaya, 1977b, Vestn. Mosk. Univ. Khim. **18**, 695.

Slavev, A.G., V.K. Portnoj, M.V. Raevskaya and E.M. Sokolovskaya, 1978, Vestn. Mosk. Univ. Khim. **19**, 289.

Sokolovskaya, E.M., M.V. Raevskaya, O.I. Bodak and E.I. Yaropolova, 1978, Vestn. Mosk. Univ. Khim. **19**, 285.

Sokolovskaya, E.M., M.V. Raevskaya, A.I. Ilias, M.A. Pastushkova and O.I. Bodak, 1985a, Izv. Akad. Nauk SSSR, Metall. (5), 197.

Sokolovskaya, E.M., M.V. Raevskaya, E.F. Kazakova, M.A. Pastushkova, O.I. Bodak and M.M. Kandelaky, 1985b, Izv. Akad. Nauk SSSR, Metall. (6), 201.

Tolkunova, E.F., V.V. Burnashova, M.V. Raevskaya and E.M. Sokolovskaya, 1974, Vestn. Mosk. Univ. Khim. **5**, 559.

Troc, R., R.A. Butera, R.S. Craig and W.E. Wallace, 1972, Magnetic characteristics of some rare earth cobalt intermetallics with copper, nickel and uranium, in: AIP Conf. Proc., 1972 (American Institute of Physics, New York) p. 1425.
Tucker, P.A., D.A. Etter and J.M. Gebhardt, 1967, Phase equilibria in the ternary system plutonium–cerium–iron, in: Plutonium, 1965, London, pp. 392–404.
Williams, H.J., R.C. Sherwood, J.H. Wernick and E.A. Nesbitt, 1964, J. Appl. Phys. **11**, 1032.
Wittenberg, L.J., D.A. Etter, J.E. Selle and P.A. Tucker, 1965, Nucl. Sci. Eng. **23**, 1.

3. Some peculiarities on the interactions of rare earth metals with other metals

According to sections 2.1–2.6 we have to take into account a total of 215 ternary rare earth systems with metallic elements for which phase equilibria have been investigated, and partial or complete phase diagrams at one or more temperatures have been presented. Moreover, another 539 ternary systems were investigated only partially, without studying any of the phase-diagram details. As the result of this, a total of 1857 ternary compounds with rare earths were described, of these the crystal structures of 1639 were solved.

The majority of experimental investigations deals with Al, Sn and metals of the fourth period in the Mendeleev's Periodic System. Therefore, our discussion will be based on the interaction peculiarities of rare earths with the above-mentioned elements.

3.1. *Peculiarities of the interactions with aluminium, gallium and indium*

The results of the investigations of ternary R–M–Al systems, where M is a metal from the 2–5th periods are briefly summarized in table 7. It is difficult to make a completely correct comparison and conclusions about total number of ternary compounds formed, because of many partial investigations of the phase diagrams. Nevertheless, one can see, that maximum number of ternary compounds formed occurs with Ni and Cu.

Ternary aluminium systems can be divided into two groups. Systems with transition metals belong to the first, and those with Be, Mg, Ga, Ge, Sr, Ba to the second one.

The first group is characterized by the presence of three components with much different crystallochemical properties, these are:

(i) atomic size;

(ii) electronic structure;

(iii) electronegativity, etc.

This is the main reason for the formation of a large number of ternary compounds.

The greatest similarities are observed in the ternary systems with Cr, Mn and Fe.

(1) The binary R–M compounds dissolve large quantities of Al.

(2) The homogeneity regions of some ternary compounds stretched out, always with constant rare earth contents.

TABLE 7

Ternary R–M–Al systems, where *M* is a metal from the 2–5th periods in the periodic table. The results of experimental studies.*

M \ R	Sc	Y	La	Ce	Pr	Nd	Sm	Eu	Gd	Tb	Dy	Ho	Er	Tm	Yb	Lu
Li																
Be		<u>1</u>	<u>1</u>	<u>1</u>												
Na																
Mg		<u>1</u>		<u>1</u>												
K																
Ca																
Ti		<u>0</u>														
V		<u>2</u>	<u>1</u>	<u>1</u>	1	1					<u>2</u>					
Cr		<u>3</u>	<u>3</u>	<u>3</u>	2	1			2		<u>3</u>	1	2		1	
Mn		<u>2</u>	<u>3</u>	<u>2</u>	1			1	1				1		1	
Fe	<u>2</u>	<u>3</u>	<u>6</u>	<u>4</u>	1	<u>3</u>	<u>3</u>	2	<u>3</u>	<u>8</u>	<u>4</u>	<u>4</u>	<u>4</u>	2	2	2
Co	2	<u>4</u>	6	<u>4</u>	3	1	1		1	1	1	1	1	1	1	1
Ni	<u>4</u>	<u>8</u>	6	<u>7</u>	<u>8</u>	<u>8</u>	3		<u>9</u>	<u>9</u>	<u>9</u>	<u>8</u>	<u>9</u>	<u>9</u>	3	<u>8</u>
Cu	<u>9</u>	<u>3</u>	<u>3</u>	<u>3</u>	<u>6</u>	<u>6</u>	<u>6</u>	3	2	1	1	1	2		2	1
Zn				1												
Ga		<u>3</u>	<u>2</u>	<u>2</u>												
Ge		<u>3</u>	<u>5</u>	<u>4</u>	2	3	2	1	<u>3</u>	2	2	2	2	2	2	1
Rb																
Sr						<u>0</u>										
Zr			<u>0</u>	<u>0</u>	<u>0</u>	<u>0</u>			<u>0</u>	<u>2</u>	<u>1</u>	<u>1</u>				<u>0</u>
Nb			<u>0</u>	<u>0</u>	<u>0</u>	<u>0</u>			<u>0</u>	<u>0</u>	<u>0</u>	<u>0</u>				<u>0</u>
Mo		1							1	1	1	1	1	1	1	1
Tc																
Ru	<u>6</u>	<u>6</u>														
Rh	1															
Pd	<u>4</u>															
Ag																
Cd																
In																
Sn			1	1												

*An open square means that the system was not investigated; the number in the square shows the number of ternary compounds formed; an underlined number means that the phase diagram (partial or complete) was derived.

(3) Crystal structures of ternary compounds often belong to the structure types of binary R–M compounds. The crystallochemical behaviour of aluminum is the same with the metals Cr, Mn and Fe.

A different behavior is noted with the compositions and crystal structures of ternary compounds in the systems with Co and Ni.

The most extensive types of crystal structures observed in the systems of the first group are: $MgZn_2$, $ThMn_{12}$ (or $CeMn_4Al_8$ superstructure), Th_2Zn_{17}, Fe_2P (or ZrNiAl superstructure), $ZrZn_{22}$ (or $CeCr_2Al_{20}$ superstructure), etc.

The forming of compounds with the $R(M, Al)_2$ composition is one of the more common features for the systems with the transition metals. Iron and cobalt always show the $MgZn_2$ type of crystal structure, which is typically an intermetallic structure. Nickel and copper show the ZrNiAl (or Fe_2P) type, which is more typical for the compounds with the semimetallic elements, because of trigonal-prismatic coordination. It is possible that in the systems R–{Ni, Cu}–Al, the crystallochemical differences of components are distinctly displayed, and that this leads to the forming of a large number of ternary compounds.

The second group is characterized by a smaller number of ternary compounds, which may be explained by the similarity of the crystallochemical behaviours of the two components: Be, Mg, Ga or Ge, and Al. As one can see, scandium generally forms a small number of compounds in the investigated systems (Sc is rather similar with a 3d transition metal, than with the other rare earths). All of the ternary compounds with solved crystal structures are characterized by structure types of the binary compounds or closely related ones. Systems with Ge and Al show the La_2O_2S type of crystal structure. This may be explained by the partially semimetallic properties of Al in these systems.

Variable compositions of the compounds $R(M, Al)_4$ with the $BaAl_4$ ($CeAl_2Ga_2$) type of crystal structure may be due to the influence of electron concentration in forming these compounds. Thus, in R–Be–Al systems the compositions are: $RBe_{1.75-1}Al_{2.25-3}$; in R–Ga–Al: $RGa_{2.5-2.0}Al_{1.5-2.0}$ and in R–Cu–Al: $RCu_{1-0.5}Al_{3-3.5}$, and the conclusion is that the larger the number of valence electrons in M atom the smaller the amount of Al necessary to form the phase. Differences in $LaBe_{1.75-1}Al_{2.25-3}$ and $CeBe_{1.4-1.2}Al_{2.6-2.8}$ is possibly caused by interconfiguration fluctuations of Ce ($Ce^{+3} \leftrightharpoons Ce^{+4}$).

The ternary systems R–M–Ga are characterized by the various stoichiometries and crystal structures of the existing ternary compounds (see sect. 2.2 and table 2). The systems Sc–M–Ga and R–{Fe, Co, Ni}–Ga have been studied more or less systematically (table 8). The number of compounds formed goes through a maximum at Co or Ni in the series from Ti to Cu.

Phase equilibria in the systems Sc–{Co, Ni}–Ga (figs. 71, 72), {Ce, Pr, Ho}–Ni–Ga (figs. 83, 85, 91) and Ho–Co–Ga (fig. 90) are very intricate. The homogeneity ranges of the ternary compounds and solid solutions are, as a rule, streteched out with constant or approximately constant R contents.

Crystal structures of the ternary compounds are characterized as being very different: from such simple structures as $MgZn_2$, $CaIn_2$, Fe_2P, $HoCoGa_5$, etc., to such complex structures as Y_3Rh_2, $Sm_{26}(Co, Ga)_{17}$, $Ho_6Co_7Ga_{21}$, $Ho_4Ni_{10}Ga_{21}$, etc.

TABLE 8

Ternary R–M–Ga systems, where M is a metal from the 2–5th periods in the periodic table. The results of experimental studies.*

M \ R	Sc	Y	La	Ce	Pr	Nd	Sm	Eu	Gd	Tb	Dy	Ho	Er	Tm	Yb	Lu
Li																
Be																
Na																
Mg			1	1												
K																
Ca																
Ti	3									1	1	1	2	1		1
V	1	1											1			
Cr	1															
Mn	3	6											8			
Fe	7	1	1	2	3	3	3		1	6	2	8	7	2	1	1
Co	15	13	4	5	8	8	12	2	12	11	12	18	10	5	4	8
Ni	14	6	9	22	25	10	10	1	10	10	11	24	10	11	5	8
Cu	7	7					11									
Zn																
Ge																
Rb																
Sr																
Zr	1															
Nb	2	0														
Mo																
Tc																
Ru																
Rh																
Pd		1		1						1		1	1	1	1	1
Ag																
Cd																
In																
Sn																

*See footnote to table 7.

A large number of the types of crystal structures found for ternary gallides are connected to one another by various relations. Thus, at a R content equal to 33.3 a/c, the compounds with general composition $R(M_xGa_{1-x})_2$ exist (x varies over a wide range). These compounds are characterized by being closely related types of crystal structures: AlB_2 (for $x=0$) $\rightarrow CaIn_2 \rightarrow KHg_2 \rightarrow$ TiNiSi. These structures may be derived from one another by various deformations. There are many examples of closely related compounds, such as $Sm_{26}(CoGa)_{17} \rightarrow Y_3Rh_2$, $Y_2Co_2Ga \rightarrow (Mo_2NiB_2) \rightarrow Y_3Co_3Ga$ $(W_3CoB_3) \rightarrow Y_4Co_4Ga \rightarrow Y_5Co_5Ga$, etc. Very often these compounds with closely related crystal structures exist in equilibrium with one another.

In comparison with Al, the systems R–M–Ga are characterized by the greater number of ternary compounds and structure types. Both Al and Ga often show the $ThMn_{12}$, $BaAl_4$, Fe_2P, Mo_2NiB_2, etc., structure types. Both elements show prevailing metallic behaviours; statistical mixtures of Ga or Al with another metal often occur.

Phase equilibria in the systems with indium have not been investigated. Ternary compounds crystallized in both simple structure types as Fe_2P, CsCl, $CaIn_2$, $MnCu_2Al$, and complex ones such as $Pr_5Ni_6In_{11}$, $Ho_{10}Ni_9In_{20}$, etc.

3.2. *Peculiarities of the interactions with germanium and tin*

Ternary systems with germanium more or less were studied with M = Li, Mn, Fe, Co, Ni, Cu, Ru, Rh, Pd, Os, Ir, Pt (see tables 9 and 4). A maximum number of ternary compounds are observed for cobalt and nickel.

Practically, for all of the partially or completely investigated ternary systems with a transition metal, the compounds RM_2Ge_2 with $CeAl_2Ga_2$ ST, and RMGe with TiNiSi or PbFCl ST are formed. The clearest differences between Al (or Ga) and Ge in the ternary R–M–X systems is observed just for these simple compounds. Germanides RM_2Ge_2 with a crystal structure of $CeAl_2Ga_2$ type are formed as a rule with the ideal stoichiometry, and an ordered M and Ge atomic distribution in the unit cell. Aluminides and gallides, on the contrary exhibit a random M and Al{Ga} distribution and homogeneity regions which rarely includes the $RM_2(Al\{Ga\})_2$ composition. Aluminides and gallides often include the RMX_3 composition with $BaAl_4$ type of crystal structure; germanides with the appropriate compositions are characterized by the $BaNiSn_3$ or $SmNiGe_3$ types of crystal structure. It is possible that the smaller amounts of germanium, which are needed to form these (RM_2Ge_2) compounds are caused by the greater number of valence electrons in Ge, in comparison with Al and Ga.

The compounds RMGe and RMAl{Ga} also have differences in crystal structures: in the series Al $\rightarrow$ Ga $\rightarrow$ Ge one can observe a transition from a prevailing $MgZn_2$ and Fe_2P (ZrNiAl) structure types through $CaIn_2$ and KHg_2 to the TiNiSi structure type. The AlB_2 type of crystal structure becomes readily observed in the ternary germanides at the compositions $R(M_xGe_{1-x})_2$. Only a few of the aluminides and one gallide have this same structure.

The formation of various defective compounds is one more significant peculiarity that is observed in the R–M–Ge systems. The examples are: $RCo_{1-x}Ge_2$ with

TABLE 9
Ternary R–M–Ge systems, where M = metal from the 2–5th periods in the periodic table. The results of experimental studies.*

M \ R	Sc	Y	La	Ce	Pr	Nd	Sm	Eu	Gd	Tb	Dy	Ho	Er	Tm	Yb	Lu
Li			1	4	2	1	1	1							1	
Be																
Na																
Mg								1								
K																
Ca																
Ti																
V	$\underline{1}$															
Cr	2															
Mn	3	1	1	1	1	1	1	1	1	1	1	1	1		1	
Fe	5	1	2	1	$\underline{9}$	2	3	$\underline{0}$	2	$\underline{6}$	3	2	3	$\underline{10}$	2	3
Co	6	6	2	4	$\underline{13}$	6	7	$\underline{3}$	8	$\underline{10}$	7	8	9	$\underline{15}$	6	6
Ni	6	4	8	5	$\underline{11}$	4	4	$\underline{6}$	6	$\underline{10}$	7	7	7	$\underline{12}$	5	4
Cu	$\underline{2}$	3	2	2	2	3	3	$\underline{4}$	4	$\underline{4}$	2	3	3	3	2	2
Zn																
Rb																
Sr																
Zr																
Nb	$\underline{1}$															
Mo																
Tc																
Ru	1	2	2	2	2	2	2	1	2	2	2	2	2	2	1	2
Rh	1	3	2	2	2	2	2	1	3	3	4	4	3	4	3	2
Pd		1	2	2	1	1	1	2	2	1	2	1	1	1	1	1
Ag				1												
Cd																
In																
Sn																

*See footnote to table 7.

TABLE 10
Ternary R–M–Sn systems, where M = metal from 2–5th periods in the periodic table. The results of experimental studies.*

M \ R	Sc	Y	La	Ce	Pr	Nd	Sm	Eu	Gd	Tb	Dy	Ho	Er	Tm	Yb	Lu
Li																
Be																
Na																
Mg																
K																
Ca																
Ti																
V																
Cr																
Mn																
Fe		1	1						1	1	1	1	1	1		1
Co		2	3	2	1	1	1		2	2	1	2	2	1	1	1
Ni		2	6	10	4	4	3		15	3	2	2	2	2	2	11
Cu		1	3	3	11	2	2		6	1	1	1	1	1	1	3
Zn																
Rb																
Sr																
Zr																
Nb																
Mo																
Tc																
Ru		1	2	1	1	1							1			1
Rh	1	2	1	1	1	1	2	1	2	2	2	2	1	2	3	1
Pd	1	5		1	1	1	1		1	1	1	2	1	1		1
Ag				1												
Cd																

*See footnote to table 7.

the $CeNiSi_2$ type of crystal structure, which is formed with various cobalt deficiencies; other compounds are $RFe_{1-x}Ge_2$ and $RNi_{1-x}Ge_2$ with the same structure; $Pr_{0.7}(Co, Ge)_{13}$ with the $NaZn_{13}$ type of crystal structure, and Pr deficiencies, etc.

The various structure types which are observed in the ternary germanium systems are connected with one another by different relations, similar to those observed in the gallium systems. The set of structure types range from the simple ones, such as AlB_2, $CeAl_2Ga_2$, TiNiSi, $CeNiSi_2$, to the complex ones, such as La_8NiGe_5, $La_{11}Ni_4Ge_6$, $Sc_6Ni_8Si_{11}$, etc. A unique crystal structure is found in the Tb–Fe–Ge system, i.e., $Tb_{117}Fe_{52}Ge_{112}$, which has a gigantic unit cell (the unit cell volume is near 23 000 $Å^3$, and the total unit cell contains 1124 atoms).

Ternary systems with tin have been less investigated than those containing germanium (see table 10). The maximum number of ternary compounds also are formed with Ni and Co. The variety of crystal structure types found is less than that with germanium and gallium and is comparable with aluminium. The structure types Fe_2P, TiNiSi, $Pr_3Rh_4Sn_{13}$, $CeAl_2Ga_2$, are widely represented in the R–M–Sn systems.

In comparison with aluminium and gallium, the germanium ternary compounds exhibit semimetallic behaviours, and the tin phases are rather metallic ones.

3.3. *Peculiarities of the interactions in the systems R–M–M′*

The systems R–M–M′ are always characterized by the presence of two components with similar crystallochemical behaviours, and that is the main reason why limited or continuous solid solutions are formed between the isotypical binary compounds, and why there are small numbers of ternary compounds. These solid solutions often show various deviations from linearity in the unit cell volume versus concentration plots.

Ternary compounds were found only in some systems, and as a rule they have the crystal structures of binary R–M compounds: $ThMn_{12}$, Th_2Zn_{17}, Th_2Ni_{17}, $AuBe_5$, MoB, etc. The only exception is represented by the $Ce(Mn_{0.55}Ni_{0.45})_{11}$ type of crystal structure. The maximum number of compounds are observed in the R–Mn–Ni systems.

The crystal structures of compounds in the R–Mg–Zn systems practically have not been investigated, but it seems that they must be different from the ones in systems with two transition metals.

3.4. *General conclusions*

Independent of the nature of the R and X components, the three iron-group metals (Fe, Co and Ni) give the maximum numbers of ternary compounds, this is shown in fig. 189.

Unfortunately, systems with the other two triads; Ru, Rh, Pd and Os, Ir, Pt, have not been studied sufficiently. Only the ternary germanides have been more or less investigated. We think, nevertheless, that these metals would show the similar complex interactions in the R–M–X ternary systems as found in the Fe, Co and Ni systems. As it was said above, the main reason for these complex interactions is the appreciable crystallochemical differences between the R, M and X components.

The previous discussion was centered on the differences between the X components. And what about the influence and the role of the R component? As

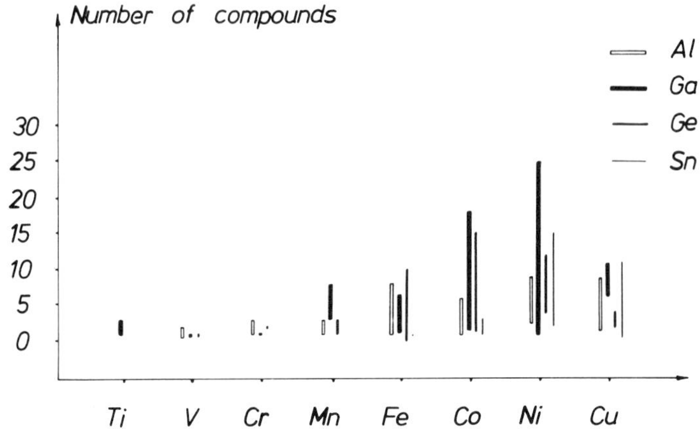

Fig. 189. The numbers of the existing compounds in ternary systems R–M–X, where M is a metal from the 4th period.

one can see, yttrium and heavy lanthanides from Gd through Lu show many similarities in their interactions. Scandium never shows a close similarity to the other rare earths and in only a few cases forms isotypical ternary compounds. Individual behaviours of rare earth metals are more appreciable for the light lanthanides, especially for lanthanum.

4. Crystal chemistry of ternary rare earth compounds with metallic elements

As one can see, the ternary rare earth compounds listed above show a variety of types of crystal structures. There are more than 130 different crystal structures. Some of these types have already been described and discussed by Gladyshevsky and Bodak (1982) and/or Parthé and Chabot (1984). Therefore, our discussion will not be concerned with those types that have been considered in the above-mentioned two surveys.

A list of atomic coordinates for 46 new types of crystal structures, which were solved after the above-mentioned papers were published, are given in the appendix.

4.1. *Ternary rare earth structure type relationships*

The study and derivation of structure types relationships is one of the main tasks for crystal chemistry. Krypyakevich (1977) offered a classification based on the various relations that one can observe between different types of crystal structure. This approach is very useful and allows one to systematize the majority of known crystal structures. This approach was used by Gladyshevsky and Bodak (1982) and Parthé and Chabot (1984).

TABLE 11
Main relationships between different structure types of rare earth intermetallides.

Common features of structure types	Way of transformation (always may be a direct or a reverse way)	Examples
(1) Equal positions for all of the atoms	Ordered substitution (superstructure formation)	$Sm_5Ge_5 \rightarrow Ce_2Sc_3Si_4$; $Ca_3Ag_8 \rightarrow Ce_3Ni_6Si_2$; $CeNi_3 \rightarrow Ce_3Co_8Si$, etc.
	Redistribution of different atom types	$MgCu_2 \rightarrow AuBe_5$; $MgCu_2 \rightarrow MgCu_4Sn$, etc.
(2) Approximately equal positions for all of the atoms	Internal and/or external deformation	$AlB_2 \rightarrow EuGe_2 \rightarrow CaIn_2 \rightarrow KHg_2 \rightarrow TiNiSi \rightarrow$, etc.
(3) Equal positions only for the part of the atoms	Multiple substitution	$CaCu_5 \rightarrow ThMn_{12}$; $Tm_9Fe_{10}Ge_{10} \rightarrow Zr_4Co_4Ge_7$
	Atomic or substituting atomic group redistribution	$Th_2Zn_{17} \rightarrow Th_2Ni_{17}$
	Inclusion or elimination	$AuZn_5 \rightarrow Pr_3Rh_4Sn_{13}$
(4) Equal structure details	Homeotectics (closely packed network stacking)	$MgZn_2 \rightarrow MgCu_2$; $HoAl_3 \rightarrow AuCu_3$, etc.
	Certain mode stacking	$AlB_2 \rightarrow \alpha$-$ThSi_2$ $CeMg_2Si_2 \rightarrow CeAl_2Ga_2$
(5) Different structure details	One-dimensional, or two-dimensional, or three-dimensional interchanging	$CeNiSi_2$, $SmNiGe_3$, etc. $YIrGe_2$, $Sc_5Co_4Si_{10}$, etc. $Tb_{117}Fe_{52}Ge_{112}$

The main relations that can be observed in the different rare earth structures, and how they are connected with one another are shown in table 11.

Many different types of crystal structure are simultaneously related with one another by several relationships. In these cases we would describe the way of transformation which explains the most significant differences or similarities between the related types. Some types of crystal structures can often be described by various transformations of a few initial structures. We would like to extract the closest relation(s) which save(s) the most common features.

So, this point of view will be the main one used in our considerations. Also, we will not discuss all of the new structures types, only the most interesting, and distinguishing relationships will be shown below.

References are not given here, because they have been cited and listed in section 2.

4.1.1. *Structure types with equal or partially equal (all) atomic positions*

The very closely related structure types (as a rule binary with ternary) are referred to this group.

This ordinary relation is represented in the crystal structures of the compounds La_2NiGe, $Tm_3Ni_{11}Ge_4$, $EuCo_2Ga_3$, Ce_5NiGe_2 and YCo_3Sn. All of them are superstructures.

The first one, La_2NiGe, can be derived from the initial type Cr_5B_3 (or hypothetical "La_5Ge_3") with the atomic distribution "$La1^{16(l)}La2^{4(c)}Ge1^{8(h)}Ge2^{4(a)}$". Ordered substitution gives the full formula $La^{16(l)}Ge1^{4(c)}Ni^{8(h)}Ge2^{4(a)}$ and also leads to the stoichiometry changing from 62.5 to 50 a/c of R. This compound shows a relatively rare type of ordered substitution, when the large R atoms (La) are replaced by the small ones (Ge). Ternary rare earth superstructures to the Cr_5B_3 type were unknown earlier.

The next three compounds: $Tm_3Ni_{11}Ge_4$, $EuCo_2Ga_3$ and Ce_5NiGe_2 are formed as a result of an ordered substitution of the smallest atoms in the initial binary structure types by the smallest ones in the resulting superstructures. The initial binary types are: $EuMg_5$, $BaZn_5$ and Nd_5Ir_3, respectively. This type of substitution (small atoms by small ones) is well-known for different kinds of intermetallics.

The last one, YCo_3Sn is related to the $BaLi_4$ binary type. The special feature of this case consists of the partially ordered substitution of the Li atomic positions by Co and Sn. One of the atomic positions [12(k), see the appendix] is statistically filled by Co and Sn.

One more structure type relationship, that of YNi_9In_2 to $Ce(Mn, Ni)_{11}$. The first is formed as the result of ordering in the second. As one can see (sect. 2.6 and table 6), the structure type $Ce(Mn, Ni)_{11}$ is typical for ternary compounds of rare earths with the two metallic elements. The appropriate binary compound (which is a substructure), is unknown. On the other hand, the forming of the YNi_9In_2 occurs through the redistribution of the atoms with different atomic sizes (Ni and In).

The structure type $LaRuSn_3$ can be derived from $Pr_3Rh_4Sn_{13}$ (it was also listed as $PrRh_{1.33}Sn_{4.33}$) when the 2(a) position of Sn in the second is occupied by the largest atoms (La) in the first one. This is also an example of atomic redistribution.

4.1.2. *Structure types with approximately equal positions for all of the atoms*

There are many different types of crystal structure, connected with one another by various deformations, and therefore they show the approximately equal positions for all of the atoms.

The $CePt_{0.95}Ga_{3.05}$ structure together with initial structure type $BaAl_4$ are shown on the left-hand side of fig. 190. The ordering, which is observed partially in two atomic positions 4(a), also causes a slight internal deformation, and this in turn causes the external unit cell deformation. Thus, the unit cell becomes orthorhombic, which is shown on the right-hand side of the figure. The initial space group I4/mmm is transformed to its subgroup Fmm2 through I4mm. The symmetry, therefore, become much lower.

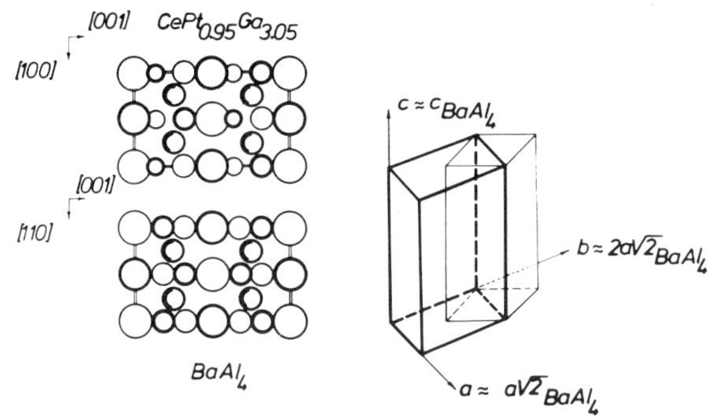

Fig. 190. Structure type $CePt_{0.95}Ga_{3.05}$ as the result of deformation with partial atomic ordering in the $BaAl_4$ structure type. The unit-cell connection is shown on the right-hand side (the $BaAl_4$ unit cell is shown by the thin lines, and the $CePt_{0.95}Ga_{3.05}$ by the heavy ones). Note that on this, and the other figures the R atoms are shown as the largest circles, M as the smallest, and X as the medium ones.

Another kind of deformation is found for the $MgCuAl_2$ structure type, which is widely represented among the R–M–{Al, Ga, In} ternary compounds, and the $CaLiSi_2$ structure, which is formed by some $RLiGe_2$ compounds (fig. 191). Both of them ($MgCuAl_2$ and $GeLiGe_2$) can be described as a planar interchanging of the infinite trigonal prism columns, shifted relative to one another on one-half of the shortest lattice dimension. Columns are stretched normal to the plane of projection on the figure. One can see that the prisms are different and the $CeLiGe_2$ trigonal prisms can be derived from the $MgCuAl_2$ ones by atomic redistribution: the X atoms from a prism corner are interchanged with the inside M atoms. Thus, all trigonal prisms in the $CeLiGe_2$ structure are deformed. However, the deformation does not appear only in the trigonal prisms, but also is concerned with the arrangement of the trigonal prism columns. The arrows on the lower side of fig. 191 shows the scheme of the relative deformation of the $CeLiGe_2$ trigonal prism columns arrangement in the comparison with $MgCuAl_2$. The choices of the unit cells are shown in dotted lines. Symmetry is changed from the Cmcm to Pnma, which is a subgroup of Cmcm. The relative lengths of the

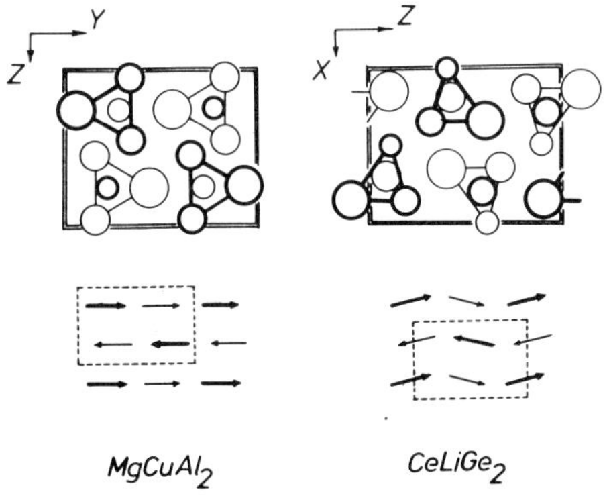

Fig. 191. Deformational relations between the $MgCuAl_2$ and $CeLiGe_2$ structures.

unit cells are left the same. Therefore, the kind of deformation discussed above, cannot be named as only a trivial one (concerned with relative atomic coordinates) but also includes the deformation of atomic group (trigonal prism columns) arrangements.

The AlB_2 type of crystal structure gives the rare earth compounds an excellent building material – trigonal prisms, which are formed by the six large (R) atoms at the corners, and the small seventh one (M or X) inside it. New types, $Tb_3Co_2Ge_4$ and Lu_2CoGa_3, can be derived as different deformations of the AlB_2 father type. Figure 192 shows some of the other known structure types, which gives the whole picture of gradual deformation of AlB_2 type. The AlB_2 type is characterized by ideal trigonal prisms with an ideal inside-atom distribution. The next ones: $EuGe_2$ and $CaIn_2$ saves the ideal prisms and the deformation only affects the small inside atoms. The $CaIn_2$ and $EuGe_2$ are distinguished by different kinds of deformation (see fig. 192). Then, the KHg_2 type shows the beginning of the deformation through the large atoms at the corners of the trigonal prisms. The small atoms return to be at approximately the ideal positions. The next type, TiNiSi, already can be described by distorted KHg_2. The distortion is concerned additionally with the small inside atoms. The two last types (shown in fig. 192): $La_3Ni_4Ga_2$ and $Tb_3Co_2Ge_4$ both represent deep deformations of the TiNiSi (or AlB_2, as a father structure type). The Lu_2CoGa_3

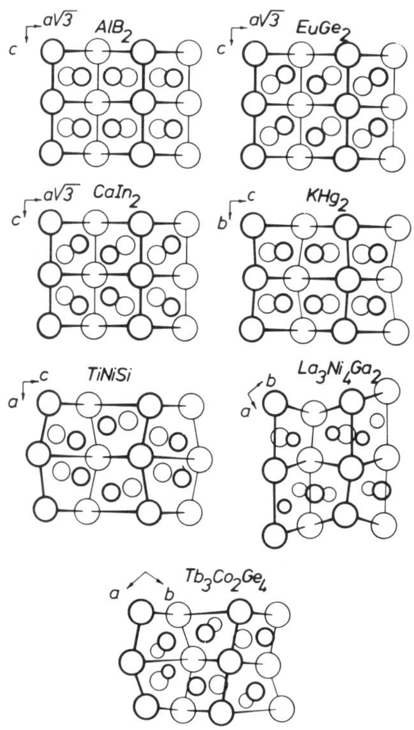

Fig. 192. A group of structure types that are related with the father structure type AlB_2 by various deformations.

also can be described as a distorted AlB_2 type, and it contains several of the above described types of deformations.

A slight external deformation may be observed as a relationship which connects two other structures: the $Lu_2Co_3Si_5$ type and the $U_2Co_3Si_5$ one. The first type is a monoclinically distorted derivative of the second one.

4.1.3. *Equal positions for only a part of the atoms*

The structure type $Tm_9Fe_{10}Ge_{10}$ is closely related with the $Zr_4Co_4Ge_7$ (fig. 193). The first is orthorhombic and the second is tetragonal. Most of the atoms are equally placed, and the only difference is encountered at the closely occupied networks (see the middle part of fig. 193): some of the large (R) atoms in $Tm_9Fe_{10}Ge_{10}$ are substituted for a pair of small (M) atoms in $Zr_4Co_4Ge_7$. The rarely occupied networks (see the right-hand side of the figure) are completely equal. Thus, this kind of relationship represents the multiple substitution of one atom by a pair of atoms.

There are a number of derivative structure types, which can be obtained from simpler ones as a result of inclusion (and, conversely, elimination). The included atoms are, as a rule, the smallest ones. An example of this kind of relationship (inclusion–elimination) is represented by the $U_4Re_7Si_6$ structure type, which can be described as the result of an inclusion of M atoms into the octahedral holes in the $AuCu_3$ structure type. The inclusion leads to eight-fold increase of the initial unit cell.

The $Pr_3Rh_4Sn_{13}$ and $LaRuSn_3$ types are included derivatives of the binary $AuZn_3$ structure type. The third ternary structure type, $Y_3Co_4Ge_{13}$, is dis-

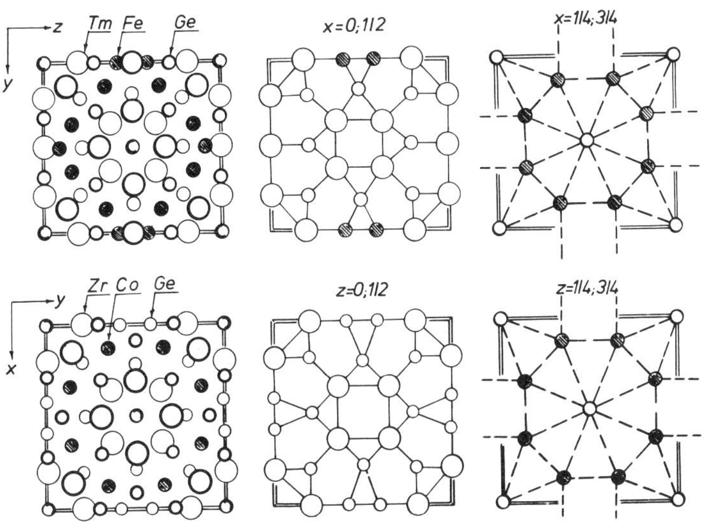

Fig. 193. The related structure types: $Tm_9Fe_{10}Ge_{10}$ (the upper part) and $Zr_4Co_4Ge_7$ (the lower one). The flat atomic networks are shown at 0, $\frac{1}{2}$ and $\frac{1}{4}$ ($\equiv\frac{3}{4}$) parts of the shortest period along the axis of projection.

tinguished from the $Pr_3Rh_4Sn_{13}$ by the statistical placing of the 24(k) germanium atoms, and, therefore, the first is related to the second one, as two structures with partially equal atomic arrangements.

4.1.4. *Equal structure details*

This kind of relationship can be described by the several structure types: $Y_2Co_3Ga_9$, Pr_8CoGa_3, and La_8NiGe_5.

The $Y_2Co_3Ga_9$ is built from close-packed networks, which are stacked along *z*-axis. The order of stacking can be described by the formula . . . ABCBACBC . . . , or . . . cchccchc . . . = $(hccc)_2$. The part of large R atoms in the networks are replaced by three smallest ones.

The Pr_8CoGa_3 also is built from the slightly distorted close-packed networks. The order of stacking along the *z* axis can be described by the formula . . . BACA . . . , or . . . chcc . . . ; some of the networks include the smallest cobalt and gallium atoms.

Therefore, both of above-described ternary compounds are related to one another, and also with other close-packed structures. The first one ($Y_2Co_3Ga_9$) also contains the elements of multiple substitution (one R by three X), and the second structure – the elements of inclusion.

The last ternary compound, La_8NiGe_5, is closely connected with the binary Sm_5Ge_4 structure type. Both of them are shown in fig. 194. The right-hand side of the figure shows the unit cell of Sm_5Ge_4 and a cross-section to eliminate the structure slabs. These slabs become the building material for the La_8NiGe_5 type structure. Only one fragment with the mode of stacking, represented as a reflection in the mirror plane, gives the La_8NiGe_5. Thus, the structure of the ternary compound can be described as "twinned through a mirror" slabs from the Sm_5Ge_4 structure type.

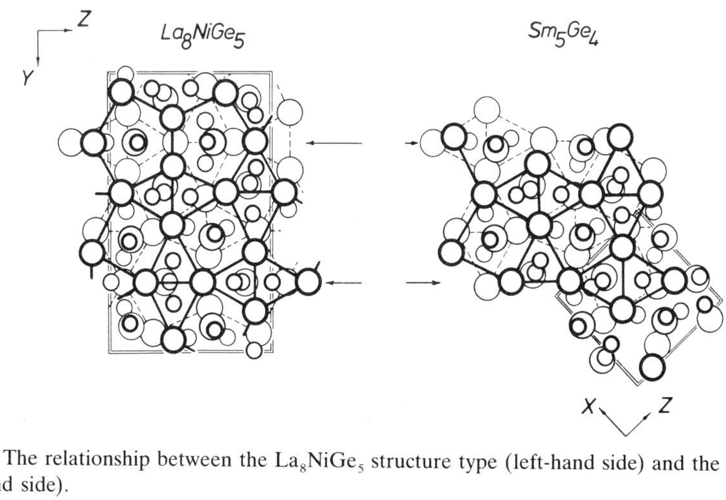

Fig. 194. The relationship between the La_8NiGe_5 structure type (left-hand side) and the Sm_5Ge_4 one (right-hand side).

4.1.5. *Different structure details*

The relationships that one can explain by different structure details are widely represented through the ternary rare earth compounds. The simplest one usually is described as a one-dimensional stacking of different slabs, taken from the various initial (also may be named the father) structures.

The first seven new structure types of ternary rare earth intermetallics, which are related to one another in this way, are shown in fig. 195. All of them can be named as long-period one-dimensional nonhomogeneous homologous structures. Each of them is described by the various stacking of different structure details (usually the slabs are eliminated from more simple related structure types).

$Ce_4Ni_2Ga_{17}$ (see fig. 195) consists of slabs derived from the two father types: $BaAl_4$ and CaF_2; the $Ce_3Ni_2Ga_{15}$ is also built from the two mentioned ones and from slabs taken from the Cu type. One more difference that can be seen, is concerned with the relative length of the $BaAl_4$ slabs: it is $\frac{7}{5}$ times longer in the first compound.

The next compound, $Sm_4Co_{1-x}Ge_7$ consists of trigonal prisms and of filled or unfilled Archimedean antiprism slabs. They are fragments taken from the α-$GdSi_2$ (it is better to say α-$GdSi_2$ than AlB_2 in this case, because the mode of trigonal prisms packing in the slab is the same as in the first), $BaAl_4$ and $ZrSi_2$ type structures, respectively.

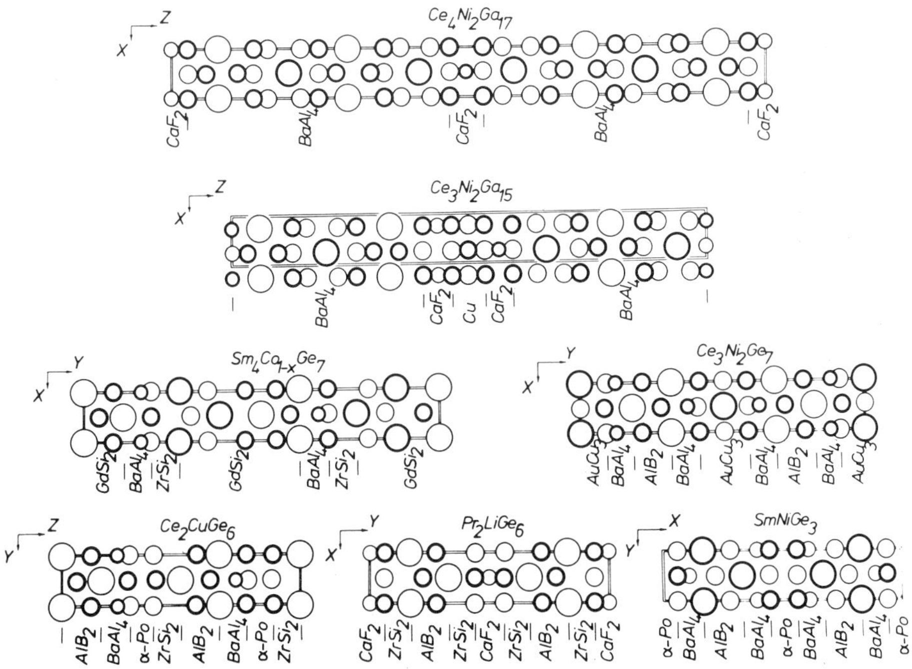

Fig. 195. The stacking variants of different slabs in the structure types: $Ce_4Ni_2Ga_{17}$, $Ce_3Ni_2Ga_{15}$, $Sm_4Co_{1-x}Ge_7$, $Ce_3Ni_2Ge_7$ ($U_3Fe_2Si_7$ ST), Ce_2CuGe_6, Pr_2LiGe_6 and $SmNiGe_3$.

The $Ce_3Ni_2Ge_7$ compound ($U_3Fe_2Si_7$ structure type) can be described as the stacking of AlB_2, $BaAl_4$ and $AuCu_3$ slabs. The Ce_2CuGe_6, Pr_2CuGe_6 and $SmNiGe_3$ are distinguished one from the other by the filling of the cubes (the unfilled cubes are the slabs taken from the α-Po type and the half-filled ones come from the CaF_2 type), and the Archimedean antiprisms (the unfilled ones are slabs from the $ZrSi_2$ type, the filled ones are from the $BaAl_4$ type).

The crystal structure of the compound $Ce_2Li_2Ge_3$ [Czybulka and Schuster (1979), which is the result of an erroneous crystal structure determination which gave the composition as $Ce_5Li_3Ge_4$; for other details see Pavlyuk et al. (1988) the reference is from section 2.4] is closely related to the $Gd_6Cu_8Ge_8$, $MgCuAl_2$ and AlB_2 types, as is shown in fig. 196. The $Gd_6Cu_8Ge_8$ and $Ce_2Li_2Ge_3$ structures are distinguished from one another by the number of AlB_2 slabs. The trigonal prisms, taken from the $MgCuAl_2$ father structure type, in both $Gd_6Cu_8Ge_8$ and $Ce_2Li_2Ge_3$ are characterized by the M–X redistribution. Thus, the composition of the $MgCuAl_2$ fragments become RM_2X.

The $Gd_6Cu_8Ge_8$ and $Ce_2Li_2Ge_3$ structures are two representatives of a new structural series with the general composition formula $R_{2m+n}M_{4m}X_{2m+2n}$, where m and n are the numbers of nonequivalent $MgCuAl_2$ and AlB_2 fragments in the unit cell, respectively. The father structure types are: $MgCuAl_2$, characterized by $m = 1$, and $n = 0$; AlB_2 by $m = 0$, and $n = 1$. The linear homologues are: $Gd_6Cu_8Ge_8$, characterized by $m = 1$, and $n = 1$, giving the resulting composition

$$R_{2+1}M_4X_{2+2} = R_3M_4X_4\ ;$$

and $Ce_2Li_2Ge_3$, characterized by $m = 1$, and $n = 2$

$$R_{2+2}M_4X_{2+4} = 2R_2M_2X_3\ .$$

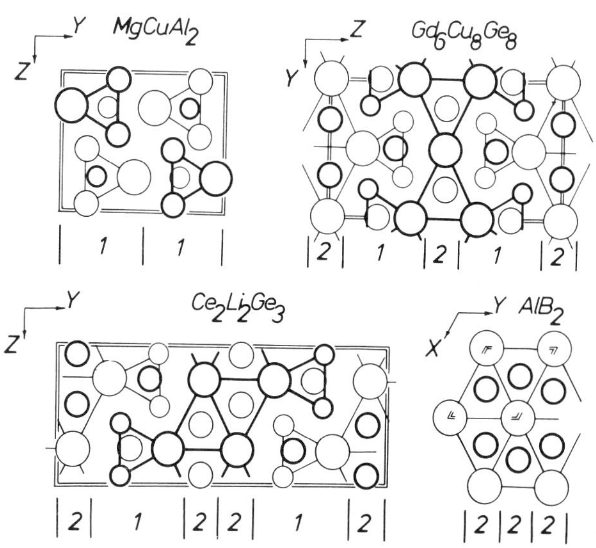

Fig. 196. The structure types $Gd_6Cu_8Ge_8$ and $Ce_2Li_2Ge_3$ as linear nonhomogeneous homologues of the $MgCuAl_2$ and AlB_2 fragments.

It is obvious that the total number of possible different stacked variants is infinite, and that new homologues with different m and n may be found in other ternary systems.

The two new types of crystal structures, Y_4Co_4Ga and Y_5Co_5Ga, together with Y_2Co_2Ga (Mo_2NiB_2 structure type) and Y_3Co_3Ga (W_3CoB_3 structure type) belong to another series of linear nonhomogeneous homologous structures. This series is described by the general compositional formula $Y_{2m+2n}Co_{2m+2n}Ga_{2n}$, where m is the number of nonequivalent α-ITl slabs, and n the number of UPt_2 slabs in a unit cell of the hybride structure. The above-mentioned structures and the father ones are shown in fig. 197. For the YCo (α-ITl structure type) $m = 1$, $n = 0$, and for the hypothetically ordered UPt_2 ("YCoGa") type $m = 0$, $n = 1$. All of the Y_kCo_kGa phases (k varies from 5 to 2) have $n = 1$, and m from 4 to 1, respectively.

The trigonal prism columns are also found in the next structure type: $LaNi_3In_6$ (fig. 198). This one is described as a stacking of $MgCuAl_2$ and hypothetical M_2X_4 slabs. The second consists of infinite unfilled octahedra columns, whereby each octahedron is connected with two others in a column through the triangle faces. The columns are connected through the octahedra cornerns, and therefore a wall is formed. Holes between the trigonal prism columns and octahedra walls are filled with the isolated indium atoms.

The two-dimensional packing of different structure fragments also is widely represented by ternary rare earth intermellides. One of the new types, $La_{11}Ni_4Ge_6$, is a member of the two-dimensional series which is based on the primitive father structures AlB_2 and α-Fe. Some of these structures are shown in fig. 199. As one can see, they differ by the number of trigonal prisms and

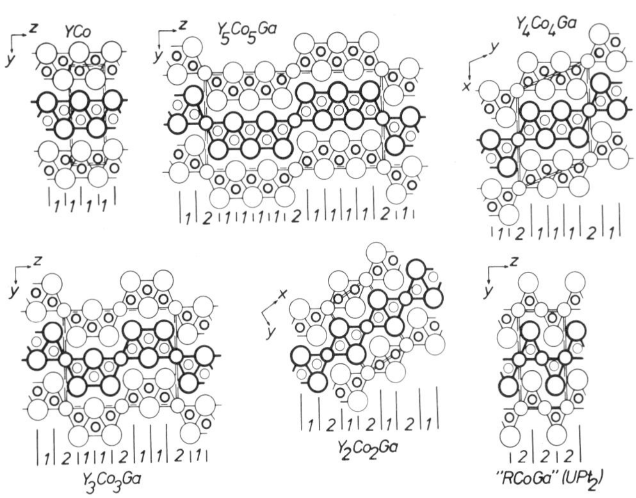

Fig. 197. The α-ITl and UPt_2 structure slabs stacking in the unit cells of Y_5Co_5Ga, Y_4Co_4Ga, Y_3Co_3Ga and Y_2Co_2Ga.

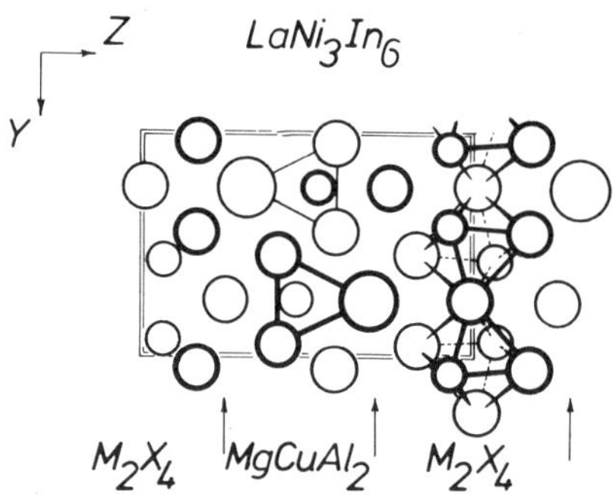

Fig. 198. The $LaNi_3In_6$ structure type as a linear homologue which is built from $MgCuAl_2$ and hypothetical M_2X_4 slabs.

octahedra (or half-octahedra) in the unit cells and by the modes of their arrangements.

The crystal structure of the $YIrGe_2$ is related to the $CeNiSi_2$ type in a very interesting way, which is shown in fig. 200. The $CeNiSi_2$ slabs, which are obtained by sections through the (022) plane in the lower part of fig. 200, are built into the penetrating walls (or as to say the crossed walls), like a twinned crystal, of the

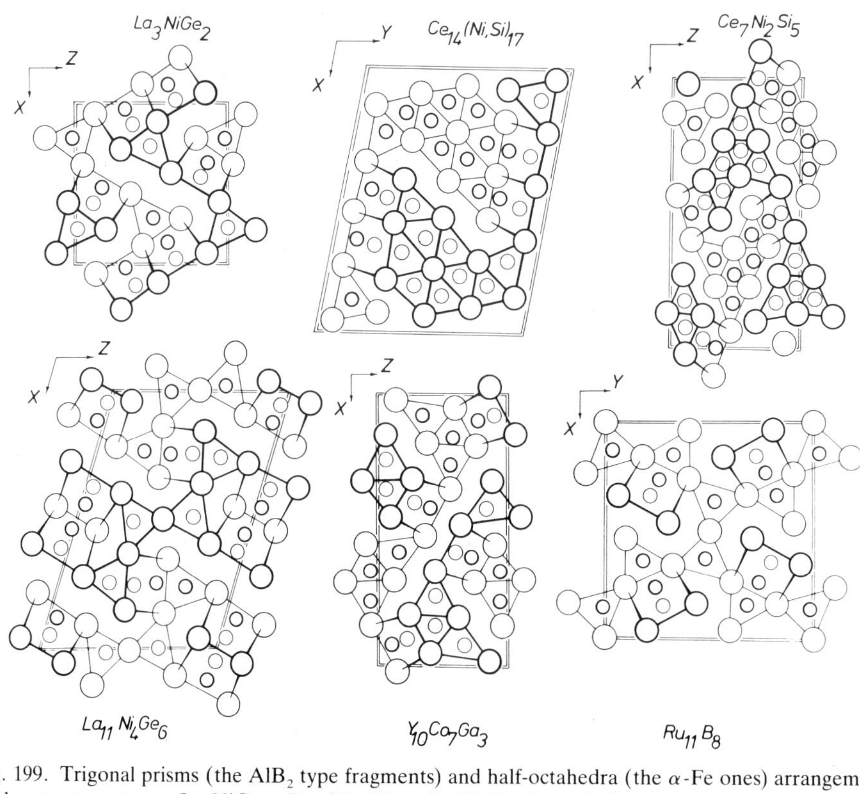

Fig. 199. Trigonal prisms (the AlB_2 type fragments) and half-octahedra (the α-Fe ones) arrangements in the structure types: La_3NiGe_2, $Ce_{14}(Ni, Si)_{17}$, $Ce_7Ni_2Si_5$, $La_{11}Ni_4Ge_6$, $Y_{10}Co_7Ga_3$ and $Ru_{11}B_8$.

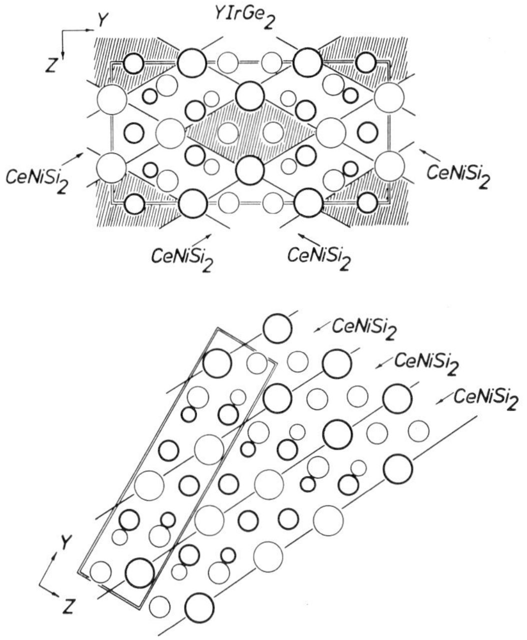

Fig. 200. The related structures $YIrGe_2$ and $CeNiSi_2$.

$YIrGe_2$ type (the upper part of figure). The resulting columns (shaded in the figure) are the parts with $MoSi_2$ type structure.

The $Nd_6Co_5Ge_{2.2}$ structures (see fig. 201) can be described as the result of the two-dimensional interchanging of the trigonal prisms, icosahedra and half-octahedra columns, linked with one another in the way shown. The trigonal prisms are observed in two different orientations. Thus, this type of crystal structure is related to AlB_2, $CaCu_5$ and α-Fe father types.

The three-dimensional packing of various structure fragments for the ternary rare earth compounds also are known. We show only two of them: a very primitive one and a complex one.

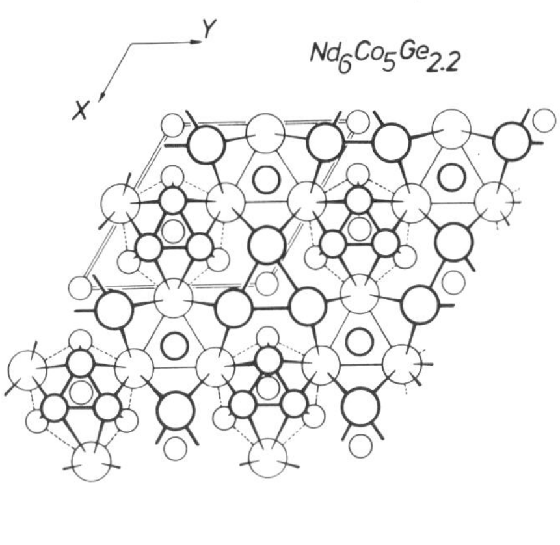

Fig. 201. The $Nd_6Co_5Ge_{2.2}$ structure type as two-dimensional packing of trigonal prisms, icosahedra and halves of the unfilled octahedra.

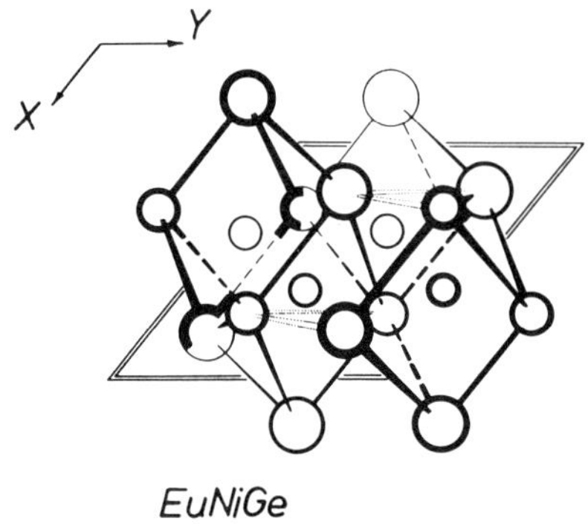

Fig. 202. The three-dimensional trigonal prismatic arrangement in the EuNiGe structure type. The unfilled trigonal prism is indicated by the opened dotted lines.

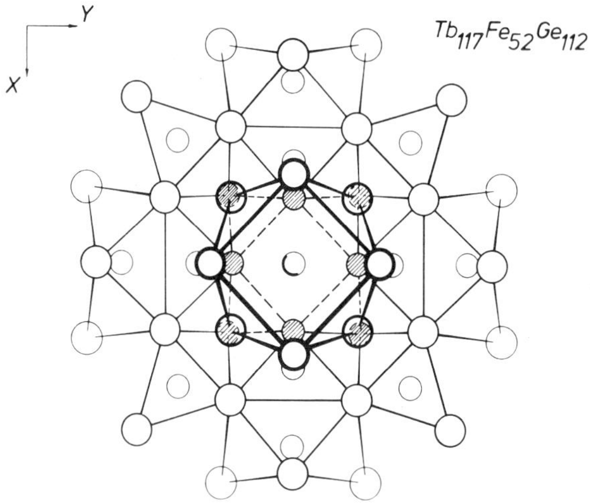

Fig. 203. The three-dimensional polyhedra packing fragment in the structure type $Tb_{117}Fe_{52}Ge_{112}$. The Fe atoms are indicated by shaded circles.

The first, primitive, example is represented in the EuNiGe type (the total unit cell contains 12 atoms). Figure 202 shows the three-dimensional arrangement of the filled and unfilled, half-substituted upon corners, trigonal prisms [$NiEu_3Ge_3$] and [hole Eu_3Ge_3].

The second example is the $Tb_{117}Fe_{52}Ge_{112}$ type structure (the total unit cell content is 1124 atoms), and is shown in fig. 203. Here one can see the fragment of three-dimensional packing formed by one cube [$TbFe_8$], six Archimedean antiprisms [$GeTb_4Fe_4$] and six of another type antiprism [$GeTb_8$], and four trigonal prisms [$GeTb_6$].

4.2. *Valences and chemical bonding in rare earth intermetallics*

It is well-known that the trivalent state is the usual one for the rare earth metals. Cerium sometimes shows a greater valence, equal to +4; europium and ytterbium show smaller ones, equal to +2. Rare earth intermetallics give many examples of the so-called "valence" (or interconfiguration) fluctuation state between the normal +3 valence and +2 (europium and ytterbium) or +4 (cerium) ones. It is obvious that "pure" crystal chemistry (i.e., solving and analyzing crystal structures) cannot give the true answer to the question: what is the valent state of a rare earth metal in the ternary (or binary) compound? However, the effective experimentally determined valence and the calculated valence for different binary and ternary rare earth intermetallic compounds are closely correlated with the mean atomic volume values. As was shown by Gladyshevsky and Bodak (1984), an increase of the cerium effective valence is always accompanied by the display of an appropriate minimum for the mean atomic volume in comparison with the neighbouring rare earth metals, and a decrease of the europium and ytterbium valence states with corresponding maxima.

Therefore, the consideration of the variation of the lattice constants in a series of isotypic compounds may be first (unfortunately only qualitative and not quantitative) property that shows the possibility of a rare earth interconfiguration fluctuation state.

Only a few complete series (from La through Lu, Pm is not considered) of isotypic compounds can be found in the above-described ternary rare earth systems. Some of them are shown in fig. 204. As one can see, the cerium compounds do not display any deviations from the decreasing trivalent lanthanide mean atomic volume. Europium, however, shows large positive deviations that are caused by the europium valence fluctuation state. The effective valence of europium is +2.28 and +2.0 for $EuNi_2Ge_2$ and $EuCu_2Ge_2$, respectively, according to Felner and Nowik (1978). Positive deviations are also observed for YbPdIn, $YbGe_2Al_2$ and $YbCu_2Ge_2$.

An experimental investigation of the deformed electron density distribution (DEDD) was performed by Belsky et al. (1985), and Belsky et al. (1986) for only one ternary germanide, YNi_2Ge_2, and one binary compound, YGe.

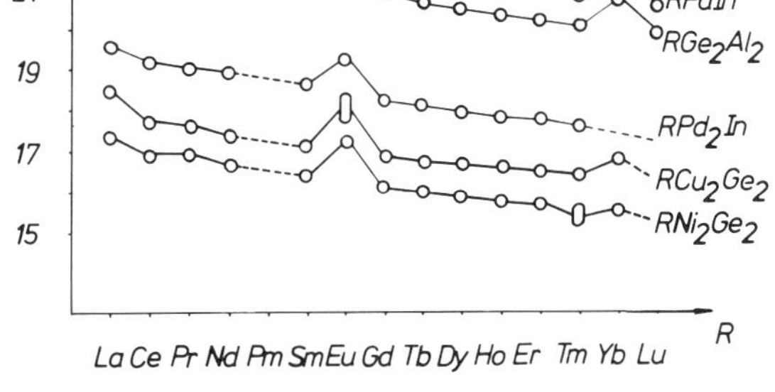

Fig. 204. Mean atomic volume in some series of isotypic ternary compounds.

The DEDD in $[Ni_2Ge_4]$ hexagon for YNi_2Ge_2 ternary compound is shown in fig. 205. The maximum, which is located between two germanium atoms, has a height of $0.7\,e \times Å^{-3}$ and is typical for a covalent-type bond of Ge–Ge atom pairs. The corresponding interatomic distance in YNi_2Ge_2 has a value of 2.502 Å, which is 0.052 Å (or 2.1%) greater than the appropriate one in pure germanium with diamond crystal structure. The strong asphericity of deformation electron density is observed near all of the atoms.

The Ni–Ge bonds do not show any deformation density maximum, although the interatomic distance is 2.347 Å, which is nearly 12% less than the corresponding atomic radii sum.

YNi_2Ge_2 ($CeAl_2Ga_2$ type structure) also can be described as a monatomic flat squared network stacking along the largest (c) axis. The DEDD map for one of them (consisting of Ni atoms) is shown in fig. 206. As one can see, the negative DED values are absent here. The zero level goes through the centers of the Ni atoms. Maxima on the Ni–Ni bonds are not observed. Also there are no maxima for the Y–Ge and Y–Ni bonds.

Figures 205 and 206 show that the deformation electron density is concentrated inside of the flat squared monatomic networks.

Although the binary rare earth compounds are not the subject of our survey, we have considered a unique one – YGe (CrB structure type). The DEDD for germanium chains is shown in fig. 207. Any maximum between the germanium pairs is not observed. The conclusion (about the absence of the covalent Ge–Ge interaction for the binary YGe compound contrary to that found in the ternary YNi_2Ge_2 phase) is reasonable because of:

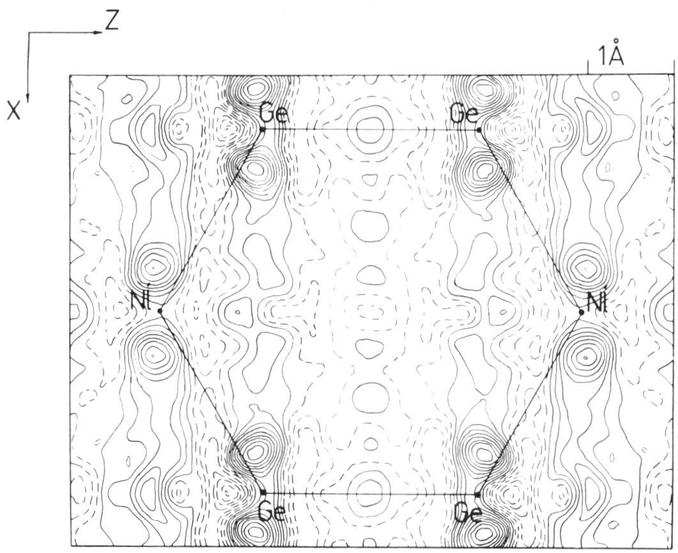

Fig. 205. The deformation electron density (DED) map in the $[Ni_2Ge_4]$ hexagon of the YNi_2Ge_2 compound. Atoms are indicated by points. Contours with negative DED are shown with dotted lines. The steps in the contours are equal to $0.5\,e \times Å^{-3}$.

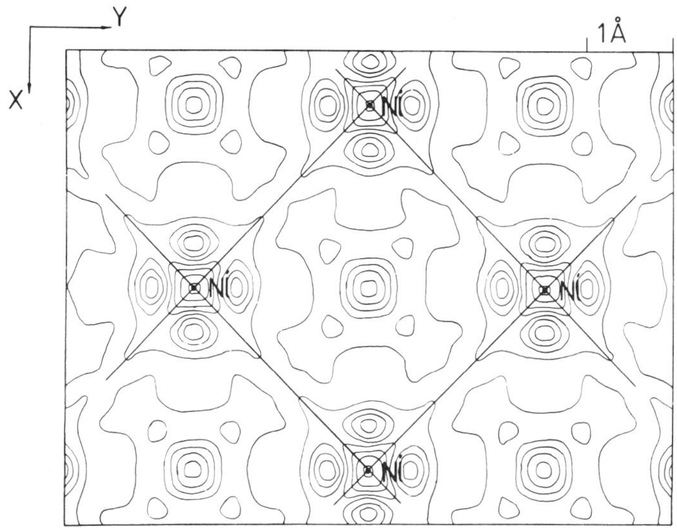

Fig. 206. The deformation electron density map in the flat squared nickel atoms network. Designations are as in fig. 205.

(i) the appropriate deformation electron density distribution maps (figs. 205 and 207); and

(ii) the interatomic Ge–Ge distances, which are 2.502 Å for YNi_2Ge_2 and 2.673 Å for YGe.

The doubled maximum, which is separated by a minimum and is located on the Y–Ge (2.958 Å distance) bond (see fig. 207), cannot be explained as a covalent bond.

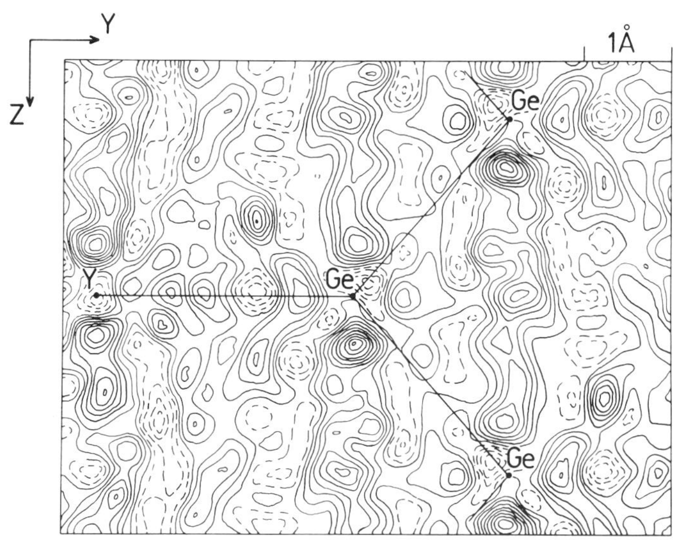

Fig. 207. The deformation electron density map in the germanium atoms chain for YGe. Designations are as in fig. 205.

Both YNi_2Ge_2 and YGe show complex deformation electron density distributions as seen on the present maps. This may be evidence of a complex and mixed types of interatomic interactions and bonding in both germanides. Nevertheless, the following conclusions seem to be possible:

(i) Covalent Ge–Ge bonding is observed in YNi_2Ge_2. The other binary metal–metal interactions in YNi_2Ge_2 and all of them in YGe are free from covalent bonding;

(ii) The strong DED maxima, which are displayed inside of the large atomic groups, may be evidence of the forming of special multicenter covalent bonds. Some of them are located inside of trigonal prisms (YGe) or Archimedean antiprisms (YNi_2Ge_2), and this may explain the excellent stability of these polyhedra and account for these polyhedra as "building" materials for the majority of rare earth ternary and binary intermetallic compounds;

(iii) The reduction of some interatomic distances, widely observed in different rare earth intermetallics, is a necessary but not a sufficient condition for the formation of binary covalent interactions. So, the Ge–Ge and Ni–Ge interatomic distances for YNi_2Ge_2 are reduced approximately equally (nearly 10% for the both in comparison with the sum of the metallic atomic radii), but the first shows, and the second never shows, the appropriate maximum in the DEDD maps. The Ge–Ge interatomic distance becomes approximately normal for YGe, and as a result of this we cannot find covalent bonds here.

Therefore, the above discussed peculiarities of DEDD for YNi_2Ge_2 concerning the X–X interaction, are very interesting. As one can see, the $CeAl_2Ga_2$ type of crystal structure is a most representative one, for the above described ternary rare earth intermetallics. Unfortunately, the number of compounds with the refined Ge atomic parameters is small (most of the ternary germanides were studied without any structure refinement). The related ternary silicides (RM_2Si_2) crystal structures with the $CeAl_2Ga_2$ structure type were better refined. It is easy to see that the interatomic Si–Si (or Ge–Ge) distances are approximately constant and show slight oscillations near the value that is typical for pure silicon (and germanium) with the diamond crystal structure (fig. 208). We think that this is significant evidence for X–X covalent bonding, where X is Si or Ge.

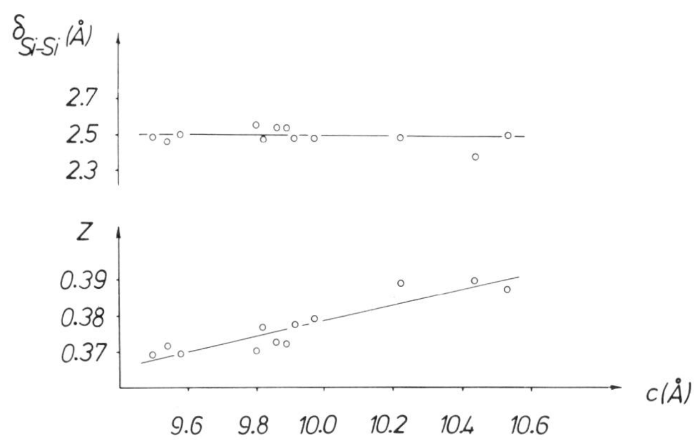

Fig. 208. The relationship of the c unit cell dimension with the silicon atomic parameter (z) and the values of interatomic Si–Si distance(s) (δ) for the RM_2Si_2 compounds.

References

Belsky, V.K., O.I. Bodak, E.I. Gladyshevsky, V.E. Zavodnik and V.K. Pecharsky, 1985, Deformation electron density in the YNi_2Ge_2 and YGe, in: 14 Vsesoyuzn. Sovesch. Primeneniyu Rentgen. Luchey Issledov. Mater., Tezisy Dokl., 1985, Kishinev (Shtiinitza, Kishinev) pp. 50,51.
Belsky, V.K., O.I. Bodak, E.I. Gladyshevsky, V.E. Zavodnik and V.K. Pecharsky, 1986, Dokl. Akad. Nauk SSSR **291**, 1362.
Czybulka, A., and H.-U. Schuster, 1979, Z. Naturforsch. B **34**, 1234.
Felner, I., and I. Nowik, 1978, J. Phys. Chem. Solids **39**, 763.
Gladyshevsky, E.I., and O.I. Bodak, 1982, Crystal chemistry of rare earth intermetallic compounds (Vyscha Shkola, Lvov) pp. 1–253.
Gladyshevsky, E.I., and O.I. Bodak, 1984, Crystal chemistry of ternary rare earth intermetallic compounds, in: Crystal chemistry problems (Nauka, Moscow) pp. 72–101.
Krypyakevich, P.I., 1977, Structure types of intermetallic compounds (Nauka, Moscow) pp. 1–288.
Parthe, E., and B. Chabot, 1984, Crystal structures and crystal chemistry of ternary rare earth–transition metal borides, silicides and homologues, in: Handbook on the physics and chemistry of rare earths, Vol. 6, eds K.A. Gschneidner Jr and L. Eyring, 1984 (North-Holland Physics Publishing, Amsterdam) pp. 113–334.

Appendix. Atom coordinates for the structure types found in ternary rare earth systems with metallic elements

The atom coordinates for the new structure types found in the above-described ternary systems are listed below. The types are arranged according to the descending contents of the R, then M, and then X components.

To find the literature reference for a given structure see sect. 2 of this chapter. Atom coordinates and unit cell choices correspond to those given in the original papers.

Every structure type is described as follows:
The first line: structure type name.
The second line: space group symbol, followed by a, b, c, β or γ.
The third and following lines: number of the atoms, position in Wyckoff notation, and the fractional coordinates in the conventional cell: x, y, z. Fixed parameters are given without a decimal point. Related parameters for atoms in special positions are given through the free ones.

Number and type of atoms	Wyckoff notation	Fractional coordinates x	y	z
Pr_8CoGa_3				
$P6_3mc$, $a = 10.489$,	$c = 6.910$			
6 Pr1	6(c)	0.1706	$-x$	0
6 Pr2	6(c)	0.4650	$-x$	0.287
2 Pr3	2(b)	$\frac{1}{3}$	$\frac{2}{3}$	0.664
2 Pr4	2(a)	0	0	0.285
2 Co	2(b)	$\frac{1}{3}$	$\frac{2}{3}$	0.063
6 Ga	6(c)	0.1639	$-x$	0.530

Number and type of atoms	Wyckoff notation	Fractional coordinates		
		x	y	z
Ce_5NiGe_2				
P4/ncc, $a = 11.760$, $c = 6.429$				
16 Ce1	16(g)	0.4642	0.1636	0.3372
4 Ce2	4(b)	$\frac{3}{4}$	$\frac{1}{4}$	0
4 Ni	4(c)	$\frac{1}{4}$	$\frac{1}{4}$	0.582
8 Ge	8(f)	0.5905	$-x$	$\frac{1}{4}$
La_8NiGe_5				
Pmmn, $a = 15.586$, $b = 18.384$, $c = 11.351$				
2 La1	2(b)	$\frac{1}{4}$	$\frac{3}{4}$	0.8402
2 La2	2(b)	$\frac{1}{4}$	$\frac{3}{4}$	0.3581
4 La3	4(e)	$\frac{1}{4}$	0.4072	0.8606
4 La4	4(e)	$\frac{1}{4}$	0.4117	0.3575
4 La5	4(f)	0.1039	$\frac{1}{4}$	0.4201
4 La6	4(f)	0.1155	$\frac{1}{4}$	0.7727
4 La7	4(f)	0.0804	$\frac{3}{4}$	0.0926
8 La8	8(g)	0.1111	0.3561	0.1158
8 La9	8(g)	0.1003	0.4490	0.6028
8 La10	8(g)	0.1042	0.5641	0.9219
8 La11	8(g)	0.1198	0.5753	0.2726
8 La12	8(g)	0.1186	0.6570	0.6031
4 Ni1	4(e)	$\frac{1}{4}$	0.5585	0.4895
4 Ni2	4(f)	0.0515	$\frac{3}{4}$	0.3855
2 Ge1	2(a)	$\frac{1}{4}$	$\frac{1}{4}$	0.2322
2 Ge2	2(a)	$\frac{1}{4}$	$\frac{1}{4}$	0.9876
4 Ge3	4(e)	$\frac{1}{4}$	0.6537	0.0909
4 Ge4	4(e)	$\frac{1}{4}$	0.5543	0.7286
4 Ge5	4(e)	$\frac{1}{4}$	0.4921	0.1068
4 Ge6	4(e)	$\frac{1}{4}$	0.3264	0.5989
4 Ge7	4(f)	0.0494	$\frac{3}{4}$	0.8207
8 Ge8	8(g)	0.0435	0.4258	0.3453
8 Ge9	8(g)	0.0445	0.3993	0.8767
$La_{11}Ni_4Ge_6$				
C2/m, $a = 18.637$, $b = 4.384$, $c = 14.91$, $\beta = 106.13°$				
2 La1	2(c)	0	0	$\frac{1}{2}$
4 La2	4(i)	0.1422	0	0.3494
4 La3	4(i)	0.0957	0	0.7840
4 La4	4(i)	0.2160	0	0.0999
4 La5	4(i)	0.4515	0	0.8738
4 La6	4(i)	0.3399	0	0.4065
4 Ni1	4(i)	0.0594	0	0.9752
4 Ni2	4(i)	0.2477	0	0.5503
4 Ge1	4(i)	0.4864	0	0.6512
4 Ge2	4(i)	0.3713	0	0.0490
4 Ge3	4(i)	0.2614	0	0.7345

Number and type of atoms	Wyckoff notation	Fractional coordinates x	y	z
La_2NiGe (superstructure to the Cr_5B_3 type)				
I4/mcm, $a = 7.947$, $c = 14.262$				
16 La	16(l)	0.1678	$\frac{1}{2} + x$	0.1457
8 Ni	8(h)	0.3849	$\frac{1}{2} + x$	0
4 Ge1	4(c)	0	0	0
4 Ge2	4(a)	0	0	$\frac{1}{4}$
$Ce_{26}Li_5Ge_{22+x}$				
P4/nmm, $a = 15.402$, $c = 10.900$				
8 Ce1	8(i)	$\frac{1}{4}$	0.4349	0.0529
8 Ce2	8(j)	0.3885	x	0.3409
4 Ce3	4(f)	$\frac{3}{4}$	$\frac{1}{4}$	0.2652
16 Ce4	16(k)	0.4214	0.6170	0.1796
8 Ce5	8(i)	0.5634	$\frac{3}{4}$	0.3679
8 Ce6	8(h)	0.3674	$-x$	$\frac{1}{2}$
8 Li1	8(i)	$\frac{3}{4}$	0.381	0.351
2 Li2	2(b)	$\frac{1}{4}$	$\frac{3}{4}$	$\frac{1}{2}$
2 Ge1	2(c)	$\frac{1}{4}$	$\frac{1}{4}$	0.470
8 Ge2	8(i)	$\frac{1}{4}$	0.5322	0.3170
8 Ge3	8(j)	0.5382	x	0.4002
8 Ge4	8(j)	0.6239	x	0.1544
8 Ge5	8(j)	0.4450	x	0.0353
2 Ge6	2(c)	$\frac{1}{4}$	$\frac{1}{4}$	0.204
8 Ge7	8(i)	0.6325	$\frac{1}{4}$	0.0149
0.74 Ge8	2(c)	$\frac{3}{4}$	$\frac{3}{4}$	0.147
Y_5Co_5Ga				
Cmcm, $a = 4.123$, $b = 10.289$, $c = 20.539$				
8 Y1	8(f)	0	0.009	0.1572
8 Y2	8(f)	0	0.284	0.0656
4 Y3	4(c)	0	0.279	$\frac{1}{4}$
8 Co1	8(f)	0	0.694	0.151
8 Co2	8(f)	0	0.569	0.053
4 Co3	4(c)	0	0.587	$\frac{1}{4}$
4 Ga	4(a)	0	0	0
$Nd_6Co_5Ge_{2.2}$				
$P\bar{6}m2$, $a = 9.272$, $c = 4.188$				
3 Nd1	3(k)	0.1304	$-x$	$\frac{1}{2}$
3 Nd2	3(j)	0.4821	$-x$	0
3 Co1	3(j)	0.8481	$-x$	0
1 Co2	1(e)	$\frac{2}{3}$	$\frac{1}{3}$	0
1 Co3	1(a)	0	0	0
1 Ge1	1(d)	$\frac{1}{3}$	$\frac{2}{3}$	$\frac{1}{2}$
1.2 Ge2	3(k)	0.7641	$-x$	$\frac{1}{2}$

Number and type of atoms	Wyckoff notation	Fractional coordinates		
		x	*y*	*z*
Y_4Co_4Ga				
B2/m, *a* = 10.233, *b* = 8.811, *c* = 4.115, γ = 109.15°				
4 Y1	4(i)	0.3884	0.1167	0
4 Y2	4(i)	0.7401	0.3388	0
4 Co1	4(i)	0.0343	0.3621	0
4 Co2	4(i)	0.0881	0.1219	0
2 Ga	2(d)	0	$\frac{1}{2}$	$\frac{1}{2}$
$La_3Ni_2Ga_2$				
Pbcm, *a* = 5.746, *b* = 8.210, *c* = 13.478				
8 La1	8(e)	0.1500	0.1043	0.0982
4 La2	4(d)	0.6546	0.2515	$\frac{1}{4}$
8 Ni1	8(e)	0.382	0.4655	0.0876
4 Ga1	4(c)	0.642	$\frac{1}{4}$	0
4 Ga2	4(d)	0.171	0.401	$\frac{1}{4}$
$Tb_{117}Fe_{52}Ge_{112}$				
Fm3m, *a* = 28.580				
96 Tb1	96(k)	0.1793	*x*	0.4084
96 Tb2	96(k)	0.1995	*x*	0.0663
96 Tb3	96(k)	0.0681	*x*	0.1557
96 Tb4	96(j)	0.2550	0.1036	0
48 Tb5	48(i)	0.1158	*x*	$\frac{1}{2}$
24 Tb6	24(e)	0.3384	0	0
8 Tb7	8(c)	$\frac{1}{4}$	$\frac{1}{4}$	$\frac{1}{4}$
4 Tb8	4(a)	0	0	0
96 Fe1	96(k)	0.1676	*x*	0.2302
32 Fe2	32(f)	0.3983	*x*	*x*
48 Fe3	48(h)	0.0723	*x*	0
32 Fe4	32(f)	0.4503	*x*	*x*
48 Ge1	48(h)	0.1447	*x*	0
48 Ge2	48(i)	0.2906	*x*	$\frac{1}{2}$
24 Ge3	24(e)	0.2150	0	0
32 Ge4	32(f)	0.1464	*x*	*x*
32 Ge5	32(f)	0.3088	*x*	*x*
96 Ge6	96(k)	0.1071	*x*	0.2423
24 Ge7	24(e)	0.1152	0	0
48 Ge8	48(g)	$\frac{1}{4}$	$\frac{1}{4}$	0.1395
96 Ge9	96(k)	0.0738	*x*	0.3228
$Gd_2Co_{2.9}Ga_{0.1}$ (structure type La_2Ni_3)				
Cmca, *a* = 5.315, *b* = 9.613, *c* = 7.169				
8 Gd	8(f)	$\frac{1}{4}$	0.411	0.089
6 Co	4(a)	$\frac{1}{4}$	$\frac{1}{4}$	0
8 X	8(e)	0	0.174	$\frac{1}{4}$
X = 0.97Co + 0.03Ga				

Number and type of atoms	Wyckoff notation	Fractional coordinates x	y	z
Yb_8LiGe_{13}				
$P2_1/b$, $a = 11.317$,	$b = 15.577$,	$c = 10.817$,	$\gamma = 106.24°$	
4 Yb1	4(e)	0.3801	0.6974	0.6459
4 Yb2	4(e)	0.3715	0.0666	0.5178
4 Yb3	4(e)	0.1186	0.8188	0.6878
4 Yb4	4(e)	0.4054	0.4316	0.2707
4 Yb5	4(e)	0.3579	0.3268	0.5634
4 Yb6	4(e)	0.1452	0.4581	0.4876
4 Yb7	4(e)	0.1667	0.6958	0.3906
4 Yb8	4(e)	0.0947	0.8996	0.3569
4 Li	4(e)	0.116	0.129	0.425
4 Ge1	4(e)	0.4296	0.8971	0.6856
4 Ge2	4(e)	0.0108	0.5386	0.2883
4 Ge3	4(e)	0.3227	0.8636	0.4793
4 Ge4	4(e)	0.1885	0.1619	0.6519
4 Ge5	4(e)	0.4137	0.5385	0.5072
4 Ge6	4(e)	0.4493	0.7697	0.3771
4 Ge7	4(e)	0.3054	0.4940	0.7165
4 Ge8	4(e)	0.0982	0.6199	0.6309
4 Ge9	4(e)	0.3460	0.7214	0.1709
4 Ge10	4(e)	0.1685	0.2992	0.3362
4 Ge11	4(e)	0.0639	0.2907	0.5435
4 Ge12	4(e)	0.2530	0.5586	0.2526
4 Ge13	4(e)	0.0936	−0.0007	0.5779
CeCoGa (structure type CeCoAl)				
C2/m, $a = 11.098$,	$b = 4.410$,	$c = 4.807$,	$\beta = 102.99°$	
4 Ce	4(c)	0.1185	0	0.316
4 Co	4(c)	0.172	0	0.852
4 Ga	4(c)	0.412	0	0.166
EuNiGe				
$P2_1/b$, $a = 6.996$,	$b = 7.581$,	$c = 6.187$,	$\gamma = 130.22°$	
4 Eu	4(e)	0.8014	0.1943	0.8606
4 Ni	4(e)	0.3885	0.0640	0.5957
4 Ge	4(e)	0.2715	0.1983	0.8564
$Tb_3Co_2Ge_4$				
B2/m, $a = 10.692$,	$b = 8.067$,	$c = 4.164$,	$\gamma = 107.22°$	
4 Tb1	4(i)	0.3771	0.3076	0
2 Tb2	2(a)	0	0	0
4 Co	4(i)	0.6904	0.3688	0
4 Ge1	4(i)	0.0964	0.3998	0
4 Ge2	4(i)	0.7105	0.0796	0

Number and type of atoms	Wyckoff notation	Fractional coordinates x	y	z
Lu_2CoGa_3				
$P6_3/mmc$, $a = 8.659$, $b = 6.823$				
6 Lu1	6(h)	0.5231	2x	$\frac{1}{4}$
2 Lu2	2(b)	0	0	$\frac{1}{4}$
4 Co	4(f)	$\frac{1}{3}$	$\frac{2}{3}$	0.450
12 Ga	12(k)	0.1692	2x	0.0432
$Sm_4Co_{1-x}Ge_7$				
2 Sm1	2(a)	0	0	0
2 Sm2	2(a)	0	0.6119	0
2 Sm3	2(b)	0	0.2667	$\frac{1}{2}$
2 Sm4	2(a)	0	0.8801	0
1.28 Co	2(b)	0	0.158	$\frac{1}{2}$
2 Ge1	2(a)	0	0.7965	0
2 Ge2	2(b)	0	0.5356	$\frac{1}{2}$
2 Ge3	2(b)	0	0.0835	$\frac{1}{2}$
2 Ge4	2(a)	0	0.1883	0
2 Ge5	2(a)	0	0.3500	0
2 Ge6	2(b)	0	0.4515	$\frac{1}{2}$
2 Ge7	2(b)	0	0.6873	$\frac{1}{2}$
$Tm_9Fe_{10}Ge_{10}$				
Immm, $a = 5.386$, $b = 13.306$, $c = 13.916$				
8 Tm1	8(l)	$\frac{1}{2}$	0.1462	0.1356
4 Tm2	4(i)	0	0	0.1948
4 Tm3	4(g)	0	0.2164	0
2 Tm4	2(d)	$\frac{1}{2}$	0	$\frac{1}{2}$
16 Fe1	16(o)	0.2478	0.1438	0.3459
4 Fe2	4(i)	0	0	0.4096
4 Ge1	4(e)	0.249	0	0
8 Ge2	8(l)	0	0.2095	0.2049
4 Ge3	4(j)	$\frac{1}{2}$	0	0.2901
4 Ge4	4(h)	$\frac{1}{2}$	0.3252	0
$La_6Co_{11}Ga_3$				
I4/mcm, $a = 8.166$, $c = 23.132$				
8 La1	8(h)	0.1691	$\frac{1}{2}+x$	0.1860
8 La2	8(f)	$\frac{1}{2}$	$\frac{1}{2}$	0.3978
8 Co1	8(h)	0.1767	$\frac{1}{2}+x$	0.0557
4 Co2	4(d)	0	$\frac{1}{2}$	0
16 Co3	16(k)	0.0652	0.2103	0
4 Ga	4(a)	0	0	$\frac{1}{4}$
8 X	8(h)	0.3824	$\frac{1}{2}+x$	0.0917
$X = \frac{1}{2}Co + \frac{1}{2}Ga$				

Number and type of atoms	Wyckoff notation	Fractional coordinates		
		x	y	z
$Ce_2Li_2Ge_3$				
Cmcm, $a = 4.4795$,	$b = 18.846$,	$c = 6.9570$		
4 Ce1	4(c)	$\frac{1}{2}$	0.4484	$\frac{1}{4}$
4 Ce2	4(c)	0	0.1583	$\frac{1}{4}$
8 Li	8(f)	0	0.309	0.441
8 Ge1	8(f)	0	0.4389	0.5688
4 Ge2	4(c)	0	0.2199	$\frac{3}{4}$
$Ho_{10}Ni_9In_{20}$				
P4/nmm, $a = 13.286$,	$c = 9.083$			
2 Ho1	2(c)	$\frac{1}{4}$	$\frac{1}{4}$	0.6382
2 Ho2	2(c)	$\frac{1}{4}$	$\frac{1}{4}$	0.1696
8 Ho3	8(i)	$\frac{1}{4}$	0.5277	0.2324
8 Ho4	8(j)	0.4538	x	0.7334
2 Ni1	2(a)	$\frac{1}{4}$	$\frac{3}{4}$	0
8 Ni2	8(i)	$\frac{1}{4}$	0.0272	0.5899
8 Ni3	8(j)	0.5948	x	0.8997
8 In1	8(g)	0.0964	$-x$	0
8 In2	8(h)	0.6207	$-x$	$\frac{1}{2}$
8 In3	8(i)	$\frac{1}{4}$	0.0824	0.9060
8 In4	8(i)	$\frac{1}{4}$	0.8654	0.7659
8 In5	8(j)	0.3820	x	0.4101
$YIrGe_2$				
Immm, $a = 4.2695$,	$b = 15.979$,	$c = 8.813$		
4 Y1	4(i)	0	0	0.2607
4 Y2	4(h)	0	0.2057	$\frac{1}{2}$
8 Ir	8(l)	$\frac{1}{2}$	0.1459	0.2494
4 Ge1	4(h)	$\frac{1}{2}$	0.785	0
4 Ge2	4(g)	$\frac{1}{2}$	0.0770	$\frac{1}{2}$
8 Ge3	8(l)	$\frac{1}{2}$	0.2998	0.3474
$CeLiGe_2$ (structure type $CaLiSi_2$)				
Pnma, $a = 7.815$,	$b = 3.965$,	$c = 10.748$		
4 Ce	4(c)	0.142	$\frac{1}{4}$	0.136
4 Li	4(c)	0.998	$\frac{1}{4}$	0.613
4 Ge1	4(c)	0.217	$\frac{1}{4}$	0.428
4 Ge2	4(c)	0.660	$\frac{1}{4}$	0.683
$LuNiSn_2$				
Pnma, $a = 15.944$,	$b = 4.361$,	$c = 14.345$		
4 Lu1	4(c)	0.8483	$\frac{1}{4}$	0.5267
4 Lu2	4(c)	0.3743	$\frac{1}{4}$	0.2263
4 Lu3	4(c)	0.1457	$\frac{1}{4}$	0.1024
4 Ni1	4(c)	0.554	$\frac{1}{4}$	0.892

Number and type of atoms	Wyckoff notation	Fractional coordinates		
		x	y	z
4 Ni2	4(c)	0.807	$\frac{1}{4}$	0.755
4 Ni3	4(c)	0.301	$\frac{1}{4}$	0.456
4 Sn1	4(c)	0.1833	$\frac{1}{4}$	0.3190
4 Sn2	4(c)	0.4570	$\frac{1}{4}$	0.4473
4 Sn3	4(c)	0.0223	$\frac{1}{4}$	0.4215
4 Sn4	4(c)	0.7153	$\frac{1}{4}$	0.8931
4 Sn5	4(c)	0.9693	$\frac{1}{4}$	0.7660
4 Sn6	4(c)	0.6751	$\frac{1}{4}$	0.6325

$Ce_3Ni_2Ge_7$ (structure type $U_3Fe_2Si_7$)

Cmmm, $a = 4.243$, $b = 25.758$, $c = 4.289$

4 Ce1	4(i)	0	0.3151	0
2 Ce2	2(d)	0	0	$\frac{1}{2}$
4 Ni	4(j)	0	0.1272	$\frac{1}{2}$
2 Ge1	2(b)	$\frac{1}{2}$	0	0
4 Ge2	4(j)	0	0.4137	$\frac{1}{2}$
4 Ge3	4(i)	0	0.090	0
4 Ge4	4(j)	0	0.222	$\frac{1}{2}$

$Y_4Ir_7Ge_6$ ($U_4Re_7Si_6$ structure type)

Im3m, $a = 8.324$

8 Y	8(c)	$\frac{1}{4}$	$\frac{1}{4}$	$\frac{1}{4}$
2 Ir1	2(a)	0	0	0
12 Ir2	12(d)	$\frac{1}{4}$	0	$\frac{1}{2}$
12 Ge	12(e)	0.290	0	0

$Pr_5Ni_6In_{11}$

Cmmm, $a = 14.579$, $b = 14.579$, $c = 4.400$

8 Pr1	8(p)	0.3630	0.3268	0
2 Pr2	2(c)	$\frac{1}{2}$	0	$\frac{1}{2}$
8 Ni1	8(p)	0.4162	0.1374	0
4 Ni2	4(j)	0	0.4066	$\frac{1}{2}$
8 In1	8(q)	0.3074	0.1559	$\frac{1}{2}$
4 In2	4(g)	0.1954	0	0
4 In3	4(h)	0.1608	0	$\frac{1}{2}$
4 In4	4(j)	0	0.2205	$\frac{1}{2}$
2 In5	2(a)	0	0	0

Pr_2LiGe_6

Cmmm, $a = 4.1268$, $b = 21.0520$, $c = 4.3455$

4 Pr	4(i)	0	0.1657	0
2 Li	2(a)	0	0	0
4 Ge1	4(j)	0	0.2861	$\frac{1}{2}$
4 Ge2	4(j)	0	0.0579	$\frac{1}{2}$
4 Ge3	4(i)	0	0.4431	0

Number and type of atoms	Wyckoff notation	Fractional coordinates		
		x	y	z
Pr_2NiGe_6 (structure type Ce_2CuGe_6)				
Amm2, $a = 4.0343$, $b = 4.1124$, $c = 21.722$				
2 Pr1	2(a)	0	0	0
2 Pr2	2(a)	0	0	0.665
2 Ni	2(b)	$\frac{1}{2}$	0	0.230
2 Ge1	2(a)	0	0	0.267
2 Ge2	2(a)	0	0	0.382
2 Ge3	2(b)	$\frac{1}{2}$	0	0.558
2 Ge4	2(b)	$\frac{1}{2}$	0	0.123
2 Ge5	2(b)	$\frac{1}{2}$	0	0.889
2 Ge6	2(b)	$\frac{1}{2}$	0	0.775
YCo_3Sn (partially ordered $BaLi_4$ structure type)				
$P6_3/mmc$, $a = 11.026$, $c = 8.891$				
6 Y	6(h)	0.476	$2x$	$\frac{1}{4}$
6 Co1	6(h)	0.097	$2x$	$\frac{1}{4}$
4 Co2	4(f)	$\frac{1}{3}$	$\frac{2}{3}$	0.539
2 Co3	2(a)	0	0	0
12 X	12(k)	0.167	$2x$	0.573
$X = \frac{1}{2}Co + \frac{1}{2}Sn$				
$Ho_2Rh_3Ge_5$ (structure type $Lu_2Co_3Si_5$)				
C2/m, $a = 9.930$, $b = 11.976$, $c = 5.813$, $\beta = 91.73°$				
8 Ho	8(f)	0.2695	0.3630	0.2664
8 Rh1	8(f)	0.1001	0.1424	0.1172
4 Rh_2	4(e)	0	0.4987	$\frac{1}{4}$
8 Ge1	8(f)	0.3313	0.0959	0.3319
4 Ge2	4(e)	0	0.7204	$\frac{1}{4}$
4 Ge3	4(e)	0	0.2886	$\frac{1}{4}$
4 Ge4	4(e)	0	0.9929	$\frac{1}{4}$
$LaRuSn_3$				
Pm3m, $a = 9.773$				
6 La1	6(d)	$\frac{1}{4}$	$\frac{1}{2}$	0
2 La2	2(a)	0	0	0
8 Ru	8(e)	$\frac{1}{4}$	$\frac{1}{4}$	$\frac{1}{4}$
24 Sn	24(k)	0.3075	0.1560	0
$SmNiGe_3$				
Cmmm, $a = 21.657$, $b = 4.102$, $c = 4.092$				
4 Sm	4(j)	0.1682	0	$\frac{1}{2}$
4 Ni	4(i)	0.3908	0	0
4 Ge1	4(i)	0.2836	0	0
4 Ge2	4(j)	0.4432	0	$\frac{1}{2}$
4 Ge3	4(i)	0.0567	0	0

Number and type of atoms	Wyckoff notation	Fractional coordinates x	y	z
$CePt_{0.95}Ga_{3.05}$				
Fmm2, $a = 6.100$, $b = 6.113$, $c = 10.512$				
4 Ce	4(a)	0	0	0
8 Ga	8(b)	$\frac{1}{4}$	$\frac{1}{4}$	0.258
4 X	4(a)	0	0	0.378
4 M	4(a)	0	0	0.637
X = 3.26Ga + 0.74Pt and M = 3.10Pt + 0.90Ga				
$Ho_7Co_6Sn_{23}$				
$P\bar{3}m1$, $a = 9.617$, $c = 9.821$				
1 Ho1	1(a)	0	0	0
6 Ho2	6(i)	0.4790	$-x$	0.3064
6 Co1	6(i)	0.1677	$-x$	0.8015
2 Sn1	2(d)	$\frac{1}{3}$	$\frac{2}{3}$	0.8195
6 Sn2	6(i)	0.2188	$-x$	0.5451
6 Sn3	6(i)	0.1179	$-x$	0.2764
1 Sn4	1(b)	0	0	$\frac{1}{2}$
6 Sn5	6(g)	0.3478	0	0
2 Sn6	2(d)	$\frac{1}{3}$	$\frac{2}{3}$	0.1150
$Ce_4Ni_2Ga_{17}$				
I4/mmm, $a = 4.283$, $c = 47.26$				
4 Ce1	4(e)	0	0	0.0796
4 Ce2	4(e)	0	0	0.3042
2 Ni	2(a)	0	0	0
8 X1	8(g)	0	$\frac{1}{2}$	0.0266
4 X2	4(e)	0	0	0.3915
8 X3	8(g)	0	$\frac{1}{2}$	0.1412
4 X4	4(e)	0	0	0.220
4 X5	4(e)	0	0	0.166
4 X6	4(e)	0	0	0.440
4 X7	4(d)	0	$\frac{1}{2}$	$\frac{1}{4}$
$X = \frac{1}{17}Ni + \frac{16}{17}Ga$				
$Sc_6Ni_{18}Ge_{11}$ ($Sc_6Ni_{18}Si_{11}$ structure type)				
Immm, $a = 18.322$, $b = 12.461$, $c = 8.191$				
4 Sc1	4(f)	0.1280	$\frac{1}{2}$	0
4 Sc2	4(f)	0.3079	$\frac{1}{2}$	0
8 Sc3	8(l)	0	0.1828	0.2036
8 Sc4	8(n)	0.2804	0.1820	0
4 Ni1	4(g)	0	0.3798	0
4 Ni2	4(e)	0.4321	0	0
8 Ni3	8(m)	0.7916	0	0.8372
8 Ni4	8(m)	0.9237	0	0.7325
8 Ni5	8(n)	0.1139	0.0988	0
8 Ni6	8(n)	0.4230	0.3326	0
16 Ni7	16(o)	0.2177	0.3471	0.1664
16 Ni8	16(o)	0.4140	0.1499	0.2169
2 Ge1	2(a)	0	0	0

Number and type of atoms	Wyckoff notation	Fractional coordinates		
		x	y	z
2 Ge2	2(c)	$\frac{1}{2}$	$\frac{1}{2}$	0
4 Ge3	4(j)	$\frac{1}{2}$	0	0.7290
4 Ge4	4(h)	0	0.6772	$\frac{1}{2}$
8 Ge5	8(m)	0.3316	0	0.1810
8 Ge6	8(n)	0.1146	0.2864	0
16 Ge7	16(o)	0.1567	0.1687	0.2647
$Tm_3Ni_{11}Ge_4$ ($Sc_3Ni_{11}Si_4$ structure type)				
$P6_3/mmc$, $a = 8.279$, $c = 8.690$				
6 Tm	6(h)	0.1883	$2x$	$\frac{1}{4}$
12 Ni1	12(k)	0.1719	$2x$	0.594
6 Ni2	6(h)	0.537	$2x$	$\frac{1}{4}$
4 Ni3	4(f)	$\frac{1}{3}$	$\frac{2}{3}$	0.073
6 Ge1	6(g)	$\frac{1}{2}$	0	0
2 Ge2	2(b)	0	0	$\frac{1}{4}$
$EuCo_2Ga_3$ (superstructure to the $BaZn_5$ type)				
Cmcm, $a = 10.126$, $b = 7.691$, $c = 5.191$				
4 Eu	4(c)	0	0.337	$\frac{1}{4}$
8 Co	8(e)	0.209	0	0
8 Ga1	8(g)	0.315	0.236	$\frac{1}{4}$
4 Ga2	4(c)	0	0.935	$\frac{1}{4}$
$Y_3Co_4Ge_{13}$				
Pm3n, $a = 8.769$				
6 Y	6(d)	$\frac{1}{4}$	$\frac{1}{2}$	0
8 Co	8(e)	$\frac{1}{4}$	$\frac{1}{4}$	$\frac{1}{4}$
2 Ge1	2(a)	0	0	0
13.9 Ge2	24(k)	0.327	0.161	0
10.1 Ge3	24(k)	0.273	0.142	0
$Ce_3Ni_2Ga_{15}$				
P4/nmm, $a = 4.278$, $c = 39.08$				
2 Ce1	2(c)	$\frac{1}{4}$	$\frac{1}{4}$	0.0632
2 Ce2	2(c)	$\frac{1}{4}$	$\frac{1}{4}$	0.800
2 Ce3	2(c)	$\frac{1}{4}$	$\frac{1}{4}$	0.3340
2 Ni1	2(c)	$\frac{1}{4}$	$\frac{1}{4}$	0.4390
2 Ni2	2(a)	$\frac{1}{4}$	$\frac{3}{4}$	0
2 Ga1	2(b)	$\frac{1}{4}$	$\frac{1}{4}$	$\frac{1}{2}$
2 Ga2	2(c)	$\frac{1}{4}$	$\frac{1}{4}$	0.968
2 Ga3	2(c)	$\frac{1}{4}$	$\frac{1}{4}$	0.9014
4 Ga4	4(f)	$\frac{1}{4}$	$\frac{3}{4}$	0.133
2 Ga5	2(c)	$\frac{1}{4}$	$\frac{1}{4}$	0.160
2 Ga6	2(c)	$\frac{1}{4}$	$\frac{1}{4}$	0.2270
4 Ga7	4(f)	$\frac{1}{4}$	$\frac{3}{4}$	0.2654
2 Ga8	2(c)	$\frac{1}{4}$	$\frac{1}{4}$	0.703
2 Ga9	2(c)	$\frac{1}{4}$	$\frac{1}{4}$	0.644
4 Ga10	4(f)	$\frac{1}{4}$	$\frac{3}{4}$	0.4041
4 Ga11	4(f)	$\frac{1}{4}$	$\frac{3}{4}$	0.4639

Number and type of atoms	Wyckoff notation	Fractional coordinates		
		x	y	z
$Y_2Co_3Ga_9$				
Cmcm, $a = 12.718$, $b = 7.383$, $c = 9.431$				
8 Y	8(g)	0.1644	0.1676	$\frac{1}{4}$
8 Co1	8(e)	0.3286	0	0
4 Co2	4(a)	0	0	0
16 Ga1	16(h)	0.1665	0.1677	0.5715
8 Ga2	8(g)	0.3934	0.0578	$\frac{1}{4}$
8 Ga3	8(f)	0	0.6655	0.5572
4 Ga4	4(c)	0	0.8784	$\frac{1}{4}$
$GdNi_{2.67}Sn_{5.44}$				
Im3, $a = 11.854$				
8 Gd	8(c)	$\frac{1}{4}$	$\frac{1}{4}$	$\frac{1}{4}$
2 X	2(a)	0	0	0
24 Ni	24(g)	0	0.1562	0.3242
24 Sn1	24(g)	0	0.6287	0.6266
24 Sn2	24(g)	0	0.2524	0.1329
$X = \frac{1}{2}Gd + \frac{1}{2}Sn$				
$LaNi_3In_6$				
Pmmn, $a = 4.388$, $b = 7.574$, $c = 12.110$				
2 La	2(a)	$\frac{1}{4}$	$\frac{1}{4}$	0.3219
2 Ni1	2(b)	$\frac{1}{4}$	$\frac{3}{4}$	0.4658
4 Ni2	4(e)	$\frac{1}{4}$	0.5764	0.0656
2 In1	2(b)	$\frac{1}{4}$	$\frac{3}{4}$	0.2511
2 In2	2(a)	$\frac{1}{4}$	$\frac{1}{4}$	0.0117
4 In3	4(e)	$\frac{1}{4}$	0.4479	0.5836
4 In4	4(e)	$\frac{1}{4}$	0.5430	0.8391
YNi_9In_2 (superstructure to the $Ce(Mn_{0.55}Ni_{0.45})_{11}$ type)				
P4/mbm, $a = 8.216$, $c = 4.827$				
2 Y	2(a)	0	0	0
8 Ni1	8(j)	0.196	0.061	$\frac{1}{2}$
2 Ni2	2(c)	0	$\frac{1}{2}$	$\frac{1}{2}$
8 Ni3	8(k)	0.326	$\frac{1}{2} + x$	0.252
4 In	4(g)	0.111	$\frac{1}{2} + x$	0
$NdCo_3Ga_9$				
Imma, $a = 4.119$, $b = 12.086$, $c = 17.146$				
4 Nd	4(e)	0	$\frac{1}{4}$	0.0294
8 Co1	8(h)	0	0.5537	0.1310
4.0 Co2	8(h)	0	0.5789	0.2697
8 Ga1	8(h)	0	0.1300	0.1956
8 Ga2	8(h)	0	0.0816	0.3964
8 Ga3	8(h)	0	0.6371	0.4239
4 Ga4	4(a)	0	0	0
4 Ga5	4(e)	0	$\frac{1}{4}$	0.3116
4 Ga6	4(e)	0	$\frac{1}{4}$	0.8333

Handbook on the Physics and Chemistry of Rare Earths, Vol. 13
edited by K.A. Gschneidner Jr. and L. Eyring

Chapter 89

PHASE EQUILIBRIUM AND CRYSTAL CHEMISTRY IN RARE EARTH TERNARY SYSTEMS WITH CHALCOGENIDE ELEMENTS

A.A. ELISEEV† and G.M. KUZMICHYEVA
Moscow Instutute of Fine Chemical Technology named after M.V. Lomonosov

Contents

Introduction

Production of compounds with increasingly complex compositions, such as ternary, quaternary, etc., has become a principle direction in modern science of materials. Among multicomponent substances, an important place belongs to the ternary compounds of rare earth elements with various physical and physico-chemical properties. The addition of the third component, as compared to the binary compounds, makes it possible to considerably enlarge the number of compounds and to synthesize substances with preliminary determined properties. In certain cases, the addition of the third component also allows one to overcome a deficit of some rare earth elements and use them economically.

The present chapter deals with a description of the compounds resulting from the reactions between rare earth elements with the chalcogenides and the elements of the main subgroups of the periodic system, from group I to VIII. Uranium- or transuranium elements as well as group-VII elements (halides) are

†Deceased.

not included in this chapter. These elements should be reviewed in a special way due to their great variety. Apart from this, the chapter contains information on more than 2000 compounds, compiled from absolutely different sources. Therefore, this chapter should be considered as an attempt of a systematic approach to these ternary compounds and the chapter is designed to compensate for the lack of information on the subject in both the Soviet and worldwide scientific and technical literature.

The procedures for obtaining ternary compounds are well-known and, therefore, there is no need for a detailed description. In general, for the substances in polycrystalline or monocrystalline states one uses the methods of direct synthesis or various methods of indirect synthesis, such as exchange reactions, from solutions in melts, chemical transport reactions using carbon disulphide, hydrogen sulphide and other media. In order to attain an equilibrium state for the ternary compound specimen it is necessary to subject them to homogenizing and annealing processes, which often takes hundreds or even thousands of hours. This chapter gives the compositions of the compounds and some thermodynamic conditions for their formation as a result of the application of a physicochemical analysis developed by Kurnakov and his school in the Soviet Union. This method has been widely applied in many other countries, such as France, the USA, Switzerland, Italy, for constructing the phase diagrams. The phase diagram of a system makes it possible to establish the conditions of phase crystallization, which is also necessary for obtaining single crystals with required properties.

Only three ternary rare earth systems, namely, Yb–Sb–Te, Gd–Bi–Te, Yb–In–Se, have been completely investigated over the whole range of concentrations, among other ternary systems with the properties indicated above. Information about binary sections of ternary systems is more complete (in many cases these sections are quasi-binary). Some of these sections are given in this chapter as an illustration. The work has been mostly done by chemists from the Academy of Sciences of Azerbaidjan.

The reference data which are included in this chapter are presented in the form of tables and contain information about formula of the compounds, thermodynamic conditions of formation (these are mainly temperatures), the compositions and eutectic temperatures, crystallographic data on the geometric dimensions of unit cells, crystal systems structure types, and space groups.

1. The R–M^I–X systems

The following chemical symbols are used in this chapter: R designates the rare earth elements, including the lanthanides, yttrium and scandium; M^I stands for the elements of group I (lithium, sodium, calcium, rubidium, cesium, copper, silver), X is used to designate chalcogenides (sulphur, selenium, tellurium).

The T–X diagram of the Na_2S–Nd_2S_3 system (fig. 1) has been studied (Shilkina and Vasil'yeva 1986). The section under investigation is found to be quasi-binary. One ternary compound, $NaNdS_2$, has been observed in this system (Shilkina and Vasil'yeva 1984) (table 1). The coordinates of the eutectic observed between

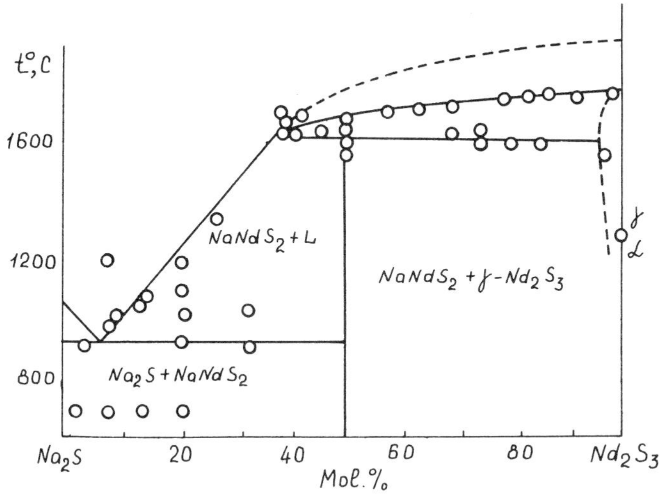

Fig. 1. The phase diagram of the Na_2S–Nd_2S_3 system.

TABLE 1
Crystallographic data for compounds in R–{Li, Na, K, Rb, Cs}–S systems.

Compound	a (Å)	c (Å)	Note	Ref.*
$LiRS_2$ Rhombohedral lattice of the α-$NaFeO_2$ type (the cell parameters are given for hexagonal axes)				
$LiHoS_2$	3.898	18.68		[1]
$LiErS_2$	3.875	18.63		[1]
$LiYbS_2$	3.842	18.54		[1]
MRS_2 Cubic latice of the NaCl type. The Li and R atoms randomly occupy the position of Na in the NaCl-type structure				
$LiPrS_2$	5.686			[1]
$LiNdS_2$	5.628			[1]
$LiSmS_2$	5.588			[1]
$LiEuS_2$	5.606		At 500–700°C, $a = 5.572$ Å, $D_x = 4.35$, $D_m = 4.34$. At 900°C and above, the solid solutions $[Eu^{+2}_{2x}, Eu^{+3}_{1-x}, Li_{1-x}]S_2$, $0 \leqslant x \leqslant 0.07$, are obtained. At 1100°C the composition of solid solutions $[Eu^{+2}_{0.375}Eu^{+3}_{0.717}Li_{0.908}]S_{1.905}$ are: $a = 5.649$ Å, $D_m = 4.30$ (refs. [2, 3])	[1] [2, 3]
$LiGdS_2$	5.530			[1]
$LiTbS_2$	5.505			[1]
$LiDyS_2$	5.474			[1]
$LiYS_2$	5.473			[1]
$NaLaS_2$	5.881			[4]
$NaCeS_2$	5.832			[4]
$NaPrS_2$	5.777			[4]
$NaNdS_2$	5.803			[4]

TABLE 1 (cont'd)

Compound	a (Å)	c (Å)	Note	Ref.*
MRS_2	Rhombohedral lattice of the α-$NaFeO_2$ type (the cell parameters are given for hexagonal axes). The cell parameters (a, c) are related with the cell parameters of NaCl (a_k) through the following ratios: $a = \frac{1}{4}a_0\sqrt{2}$, $c \sim a_0\sqrt{3}$, $a_0 = 2a_k$			
$NaSmS_2$	4.056	19.87		[4]
$NaEuS_2$	4.042	19.92		[4]
$NaGdS_2$	4.009	19.87		[4]
$NaTbS_2$	3.989	19.87		[4]
$NaDyS_2$	3.978	19.92		[4]
$NaHoS_2$	3.949	19.86		[4]
$NaErS_2$	3.939	19.98		[4]
$NaYS_2$	3.968	19.89		[4]
$KLaS_2$	4.464	21.89		[1]
$KCeS_2$	4.223	21.80	The crystal structure has been determined using a single-crystal sample (see ref. [5])	
$KPrS_2$	4.185	21.75		[1]
$KNdS_2$	4.160	21.83		[1]
$KSmS_2$	4.107	21.76		[1]
$KEuS_2$	4.093	21.85		[1]
$KGdS_2$	4.075	21.89		[1]
$KTbS_2$	4.051	21.87		[1]
$KDyS_2$	4.030	21.83		[1]
$KHoS_2$	4.009	21.80		[1]
$KErS_2$	3.993	21.77		[1]
$KYbS_2$	3.964	21.82		[1]
KYS_2	4.022	21.85		[1]
$RbLaS_2$	4.292	22.89		[6]
$RbCeS_2$	4.246	22.80		[6]
$RbPrS_2$	4.222	22.87		[6]
$RbNdS_2$	4.189	22.89		[6]
$RbSmS_2$	4.141	22.86		[6]
$RbEuS_2$	4.119	22.84		[6]
$RbGdS_2$	4.098	22.88		[6]
$RbTbS_2$	4.070	22.80		[6]
$CsLaS_2$	4.306	24.08		[6]
$CsCeS_2$	4.262	23.99		[6]
$NaScS_2$	3.751(1)	19.744(3)		[7]
$LiScS_2$	3.687(1)	18.318(2)		[7]

*References: [1] Ballestracci (1965); [2] Palazzi (1986); [3] Palazzi et al. (1987); [4] Ballestracci and Bertaut (1964); [5] Plug and Verschoor (1976); [6] Bronger et al. (1973); [7] Dijk and Plug (1980).

Na_2S and $NaNdS_2$ are 930 ± 20°C, 6.0 mol.% Nd_2S_3. The solid solution is assumed to be formed on the basis of γ-Nd_2S_3.

The compounds with compositions M^IRS_2 and M^IRSe_2 (where M^I is Li, Na, K, Rb, Cs) crystallize in two structural types: NaCl and α-$NaFeO_2$, depending on the ratio of the ionic radii for R^{+3} and M^{+1}. When $r_{R^{+3}}/r_{M^{+1}} > 0.94$, the phases of the NaCl-type structure are realized, while at $r_{R^{+3}}/r_{M^{+1}} < 0.94$, the phases of the

TABLE 2
Crystallographic data for compounds in R–Na–Se systems.

Compound	a (Å)	c (Å)	Note	Ref.*
$NaRSe_2$	Rhombohedral latice of the α-$NaFeO_2$ type (the cell parameters are given for hexagonal axes)			
$NaLaSe_2$	4.348	20.79	NaCl-type cubic cell, $a = 6.105$ Å, is also given for this composition	[1]
$NaCeSe_2$	4.303	2.076		[1]
$NaPrSe_2$	4.268	20.78		[1]
$NaNdSe_2$	4.250	20.82	Pseudocubic cell, $a = 12.022$ Å	[1]
$NaSmSe_2$	4.202	20.80		[1]
$NaEuSe_2$	4.211	20.88	NaCl-type cubic cell, $a = 6.165$ Å is given for this composition	[1]
$NaGdSe_2$	4.166	20.82		[1]
$NaTbSe_2$	4.142	20.80		[1]
$NaDySe_2$	4.124	20.83		[1]
$NaHoSe_2$	4.107	20.83		[1]
$NaErSe_2$	4.091	20.80		[1]
$NaYSe_2$	4.118	20.82		[1]

*Reference: [1] Ballestracci and Bertaut (1965a).

α-$NaFeO_2$-type structure (Ballestracci 1965, Plug and Verschoor 1976, Plug and Prodan 1978, Brunel et al. 1972) (tables 1 and 2).

The investigation of the section Cu_2S–Yb_2S_3 of the ternary Yb–Cu–S system has shown the formation of the ternary compound $CuYbS_2$ with the ratio range of the components from 1 to 1. The solid solutions based on $CuYbS_2$ are formed in the ratio range of the components from 1:1 to 3:1 (Cu_3YbS_3). Their structure remains unchanged until the composition Cu_5YbS_4 is attained. The $Cu_8YbS_{5.5}$ phase has been established in the system (table 3).

The solid solutions having the Th_3P_4-type structure were found in the Cu_2S–R_2S_3 systems (where R = La to Dy), the compositions ranging from R_2S_3 to CuR_5S_8 (Julien-Pouzol and Guittard 1972). The compounds $CuRS_2$ crystallize either in the monoclinic system (R–La–Tb) or orthorhombic system (R = Dy–Y, Se) (Julien-Pouzol and Guittard 1968b, 1972, Julien-Pouzol et al. 1970). According to Aliyev et al. (1973), the solid solutions with monoclinic structure are formed between the compositions Cu_3RS_3 and $CuRS_2$. In addition to the phases with composition Cu_3RS_3 (R = Tb–Lu, Y, Sc), which crystallize in a trigonal system, the phases with composition $R_xCu_yS_2$ ($3x + y = 4$) were also observed to have the trigonal structure (Ballestracci and Bertaut 1968).

The AgS–R_2S_3 systems (Julien-Pouzol and Guittard 1973) exhibit the compounds AgR_5S_8 (R = La–Dy) with the Th_3P_4 type structure and the compounds on the composition $AgRS_2$ (R = Sm–Yb) form the monoclinic $AgYbS_2$-type structure. For R = Er–Tm, cubic structure is established. The solid solution found in the system are based on the NaCl structure type. Besides, the compound of the composition $Ag_2R_5S_8$ (R = Yb, Lu) with tetragonal symmetry was determined representing a superstructure of the NaCl type (table 3).

The ternary chalcogenides (X = Se, Te) of rare earth elements with copper and

TABLE 3
Crystallographic data for compounds in R–{Cu, Ag}–S systems.

Compound	a (Å)	b (Å)	c (Å)	β	γ	Density (g/cm³) D_x	Density (g/cm³) D_m	Note	Ref.*
$CuRS_2$ Monoclinic lattice									
$CuLaS_2$	6.50	6.91	7.30		98.53°			The crystal structure of $CuLaS_2$ was determined on a single crystal: $a = 6.646(2)$, $b = 6.938(2)$, $c = 7.325(2)$, $\gamma = 98.73(2)°$ $D_x = 5.3$; $D_m = 5.29(8)$ (ref. [2])	[1]
$CuCeS_2$	6.47	6.88	7.25		98.60°				[3, 4, 5]
$CuPrS_2$	6.42	6.86	7.21		98.43°				[3, 4, 5]
$CuNdS_2$	6.43	6.83	7.17		98.88°				[3, 4, 5]
$CuSmS_2$	6.39	6.79	7.10		98.17°			The parameters: $a = 6.496(3)$, $b = 7.133(2)$, $c = 6.799(2)$, $\beta = 98.31(3)°$ were determined using a single crystal (see ref. [6])	[3, 4, 5]
$CuGdS_2$	6.36	6.75	7.05		98.67°				[3, 4, 5]
$CuTbS_2$	6.34	6.72	7.02		98.67°				[3, 4, 5]
$CuRS_2$ Orthorhombic lattice of the $CuErS_2$ type									
$CuDyS_2$	6.33	13.64	3.93						[1, 3, 4]
$CuHoS_2$	6.28	13.61	3.92			–	–		[3]
$CuErS_2$	6.26	13.5	3.92			6.0	5.9		[4]
$CuTmS_2$	6.24	13.53	3.89			–	–		[3]
$CuYbS_2$	6.22	13.50	3.88			–	–		[4]
$CuLuS_2$	6.20	13.43	3.86			–	–		[3]
$CuYS_2$	6.31	13.66	3.94			–	–		[4]
$CuRS_2$ Trigonal lattice									
$CuScS_2$	3.729		6.099					The crystal structure was determined on a single crystal: $a = 3.7333(5)$, $c = 6.098(1)$, space group P3m1 (ref. [7])	[4]
R_2CuS_3 Orthorhombic lattice									
Eu_2CuS_3	10.347(4)	12.863(3)	3.954			5.85	5.7	Crystal structure was determined on a single crystal. Eu in Eu_2CuS_3 was found in two valency states Eu^{+2} and Eu^{+3} (ref. [9])	[8]

CuR_5S_8 Cubic lattice of the Th_3P_4-type. CuR_5S_8 exists for R = La–Dy								
$CuLa_5S_8$	8.689							[4]
$CuCe_5S_8$	8.623							[4]
$CuPr_5S_8$	8.567							[4]
$CuNd_5S_8$	8.520							[4]
$CuDy_5S_8$	8.322							[4]
$Cu_2R_4S_7$ Hexagonal lattice								
$Cu_2La_4S_7$	10.331		5.833					[10]
$Cu_2Ce_4S_7$	10.232		5.677					[10]
$Cu_2Pr_4S_7$	10.184		5.624					[10]
Cu_3RS_3 Monoclinic lattice of the Cu_3SmS_3 type								
Cu_3SmS_3	6.55	7.20	7.00	98°	4.64	4.64		[11, 12]
Cu_3GdS_3	6.39	6.78	7.07	98°	–	5.86		[11, 12]
Cu_3RS_3 Trigonal lattice (space group $P\bar{3}$). The values of cell parameters and of densities are given in ref. [11].								
Cu_3TbS_3	3.888		6.382		5.86	5.90		[13]
Cu_3DyS_3	3.881		6.366		5.99	5.94		[13]
Cu_3HoS_3	3.876		6.349		6.00	6.02		[13]
Cu_3ErS_3	3.874		6.332		6.16	6.20	The crystal structure is determined	[13]
Cu_3TmS_3	3.869		6.303		6.30	6.34		[13]
Cu_3YbS_3	3.871		6.272		6.10	6.40		[13]
Cu_3LuS_3	3.870		6.270		6.40	6.44		[13]
Cu_3YS_3	3.882		6.363		4.90	4.70		[13]
$Cu_2Gd_{2/3}S_2$	3.911		6.414		–	–	Compounds Cu_3RS_3 are considered	[4]
$Cu_2Sc_{2/3}S_2$	3.76		6.171		–	–	in the form of $Cu_2[R_{2/3}\square_{1/3}]S_2$	[4]
$Cu_2Y_{2/3}S_2$	3.89		6.35		–	–		[4]
Cu_5RS_4 Hexagonal lattice of the Cu_5RTe_4 type								
Cu_5GdS_4	11.70		6.50		5.25	5.20	Ref. [14] gives the following cell parameters for this composition: $a = 4.320$, $c = 7.112$	[15]
Cu_5TbS_4	11.68		6.47		5.30	5.39		[15]
Cu_5DyS_4	11.65		6.45		5.32	5.45		[15]
Cu_5HoS_4	11.63		6.46		5.39	5.50		[15]
Cu_5ErS_4	11.63		6.44		5.40	5.62		[15]
Cu_5YbS_4	11.61		6.44		5.47	5.70	For this composition ref. [14] gives the following cell parameters: $a = 3.957$, $c = 6.405$	[15]
Cu_5LuS_4	11.59		6.42		5.60	5.77		[15]

TABLE 3 (cont'd)

Compound	a (Å)	b (Å)	c (Å)	β	γ	Density (g/cm³) D_x	Density (g/cm³) D_m	Note	Ref.*
$AgRS_2$ Monoclinic lattice of the $AgYbS_2$ type									
$AgSmS_2$	7.78	7.78	12.32	88°53′					[16]
$AgGdS_2$	7.73	7.73	12.15	89°13′					[16]
$AgTbS_2$	7.70	7.70	12.12	89°17′					[16]
$AgDyS_2$	7.68	6.68	12.05	89°25′					[16]
$AgHoS_2$	7.65	7.65	11.99	89°32′					[16]
$AgErS_2$	7.63	7.63	11.92	89°35′				For this composition, there exists a phase with the NaCl-type cubic lattice $a = 5.512$ (ref. [16])	[16]
$AgTmS_2$	7.60	7.60	11.86	89°52′				There exists a phase with the NaCl-type cubic lattice; $a = 5.501$	[16]
$AgYbS_2$	7.573	7.573	11.803	90°				Tetragonal lattice, space group 14_1md. For this composition, there exists a phase with the NaCl-type, cubic lattice $a = 5.491$ (ref. [16])	[16]
$AgYS_2$	7.66	7.66	12.02	89°30′					[16]
$AgLuS_2$	5.497	–	–	–				The phase has a cubic symmetry	[16]
AgR_5S_8 Cubic lattice of the Th_3P_4 type. The phases containing AgR_5S_8 are imperfect. The compound formula may be written as follows: $Ag_{1-3\varepsilon}R_{5+\varepsilon}\square_{2\varepsilon}S_8$									
$AgLa_5S_8$	8.699								[17]
$AgCe_5S_8$	8.623								[17]
$AgPr_5S_8$	8.567								[17]
$AgNd_5S_8$	8.520								[17]
$AgSm_5S_8$	8.444								[17]
$AgGd_5S_8$	8.382								[17]
$AgTb_5S_8$	8.350								[17]
$AgDy_5S_8$	8.315								[17]

*References: [1] Ballestracci and Bertaut (1965b); [2] Julien-Pouzol et al. (1981); [3] Julien-Pouzol et al. (1970); [4] Julien-Pouzol and Guittard (1972); [5] Julien-Pouzol and Guittard (1968b); [6] Guseinov et al. (1984); [7] Dismukes and Smith (1971); [8] Guittard et al. (1987); [9] Lemoine et al. (1986); [10] Collin et al. (1968); [11] Aliyev et al. (1973); [12] Aliyev et al. (1972b); [13] Ballestracci and Bertaut (1968); [14] Gamidov and Keiserukhs'kaya (1972); [15] Rustamov et al. (1976); [16] Ballestracci (1966b); [17] Ballestracci (1966a).

silver were studied in detail, and some of the constitutional diagrams of separate sections having been constructed.

The region of solid solutions based on La_2Se_3 was observed in the La_2Se_3–Cu_2Se system (Julien-Pouzol et al. 1972, Julien-Pouzol and Guittard 1972). The compound $LaCuSe_2$ is formed only in case when the components ratio is 1:1 and decomposes according to a peritectic reaction at 1240°C. Another compound, $La_5Cu_{13}Se_{14}$, decomposes at peritectic reaction temperature of 1000°C. The eutectic is formed at 910°C and 90 mol.% Cu_2Se.

The Pr_2Se_3–Cu_2Se system appears to be a quasi-binary section of the ternary Cu–Pr–Se system and is characterized by the formation of one compound $CuPrSe_2$ exhibiting a peritectic reaction at 1140°C,

$$L + Pr_2Se_3 \rightleftarrows CuPrSe_2$$

A eutectic exists between Cu_2Se and Pr_2Se_3 with coordinates: 17 mol.% Pr_2Se_3 and 910°C. The solubility of Pr_2Se_3 in Cu_2Se is about 2 mol.%; the maximum solubility is 8 mol.% at the eutectic temperature (Pribylskii and Gamidor 1984). The compound $CuNdS_2$ decomposes in a peritectic reaction at 1100°C (Julien-Pouzol et al. 1968b).

The R_2Se_3–Cu_2Se systems (R = Sm, Gd, Tb, Yb) exhibit complex chemical interaction resulting in the formation of several intermediate phases or solid solutions based on these phases.

The authors (Pribylskii et al. 1983b) have studied the T–X phase diagram projection of the Tb_2Se_3–Cu_2Se (fig. 2). The Cu_3TbSe_3 phase melts congruently

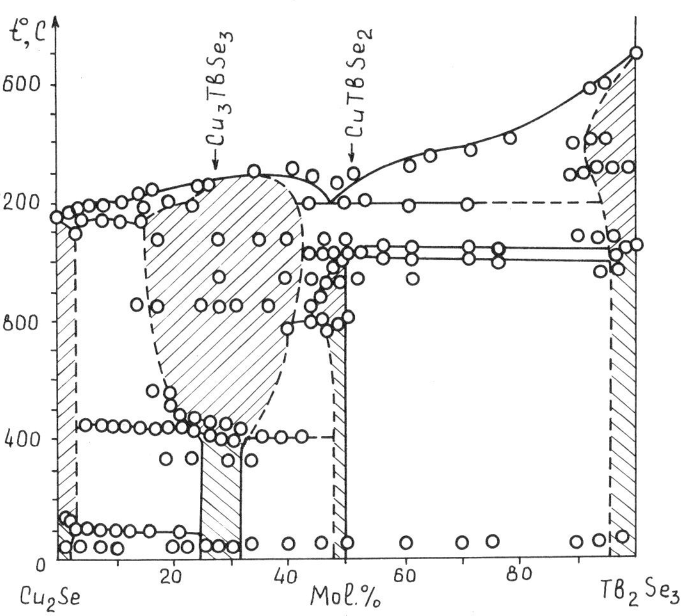

Fig. 2. The phase diagram of the Cu_2Se–Tb_2Se_3 system.

at 1310°C. The maximum melting corresponds to the point with 30 mol.% Tb_2Se_3. The eutectic between Cu_3TbSe_3 and γ-Tb_2Se_3 has coordinates 45.5 mol.% Tb_2Se_3, 1235°C. The phase $CuTbSe_2$ decomposes to solid solutions based on Cu_3TbSe_3 and γ-Tb_2Se_3 at 1050°C. The compounds $CuTbSe_2$ and Cu_3TbSe_3 exhibit polymorphism at 810 and 450–475°C, respectively.

The authors (Pribylskii et al. 1983a) investigated the phase equilibrium in the Cu–Gd–Se system at 600°C and presented the isothermal section of the system. The primary system is divided by the sections Cu_2Se–Gd_2Se_3, Cu–Gd_2Se_3 and Cu_2Gd–GdSe into four secondary systems. The system has the quasi-binary section Cu_2Gd–GdSe, and the partially quasi-binary sections CuGd–GdSe and Cu–$CuGdSe_2$. The phase Cu_3GdSe_3 exhibits a phase transition at 475–495°C and a peritectic reaction at 1205°C. The phase of the constant composition $Cu_6Gd_4Se_9$ decomposes according to a peritectic reaction. The compound $CuGdSe_2$ decomposes according to a peritectic reaction at 920°C.

The Cu_2Se–Sm_2Se_3 system is a quasi-binary section of the ternary system Cu–Sm–Se (Pribylskii and Gamidov 1983). The phase Cu_5SmSe_4 is observed upon heating to 775°C and melts incongruently at 1080°C. The phase is stable in the solid state and decomposes at 1000°C. The maximum solubility of Cu_2Se in Sm_2Se_3 reaches 12 mol.% at 1250°C.

The Yb_2Se_3–Cu_2Se system (Rustamov et al. 1981) exhibits five types of compounds: the cubic phase $CuYb_5Se_8$ with the Th_3P_4-type structure, hexagonal phases $CuYbSe_2$, Cu_3YbSe_3, Cu_5YbSe_4 and the orthorhombic phase $Cu_8YbSe_{5.5}$ with the Cu_3Se_2-type structure (table 4). The authors of an earlier work (Rustamov et al. 1976) isolated a phase with composition Cu_5RSe_4 (R = Sm–Lu) which appears to be a superstructure with respect to Cu_3RSe_3 ($a = a_0\sqrt{3}$, $c = 2c_0$, where a_0 and c_0 are the Cu_3RSe_3 lattice parameters).

The R_2Se–Cu_2Se system (Julien-Pouzol and Guittard 1968a, 1972, Julien-Pouzol et al. 1972) exhibits the following phases CuR_5Se_8 (R = La–Gd) with the Th_3P_4-type structure, $CuRSe_2$(R = La–Gd) with monoclinic symmetry $CuRSe_2$ (R = Tb–Sc) with trigonal symmetry. The structure for the phases R = Dy–Yb is unknown. The phase $CuRSe_2$ has the region of homogeneity between the compositions $CuRSe_2$ and Cu_3RSe_3 ($Cu_2R_{2/3}Se_2$).

The Ag_2Se–R_2Se_3 systems (Julien-Pouzol and Guittard 1969, 1973) exhibit considerable regions of homogeneity between R_2Se_3 and AgR_5Se_8 (R = La–Gd). The phase with composition $AgRSe_2$ crystallizes in two structural types: $AgYbS_2$ (R = Gd–Ho) and $AgErSe_2$ (R = Dy–Lu, Sc). The phase transition temperature is 632°C. The phase with composition $Ag_2R_5Se_8$ having tetragonal symmetry and which is a superstructure with respect to the NaCl type, was observed in the section Ag_2Se–R_5Se_7.

The diagrams of the Cu_2Te–R_2Te_3 sections state (R = La–Sm) were drawn for the systems containing tellurium (Pardo and Flahaut 1971). In all the systems, the phase with the composition $Cu_2R_4Te_7$, which is stable in the temperature range 500–960°C, was observed. This phase breaks up into R_2Te_3 and $CuRTe_2$ at higher temperatures. The compound Cu_xLaTe_2 with a region of homogeneity at x = 0.28–0.6 was found in the Cu–La–Te system (Pardo and Nguyen 1987). The data

TABLE 4

Crystallographic data for compounds in R–{Cu, Ag}–Se systems.

Compound	a (Å)	b (Å)	c (Å)	γ	Density (g/cm^3) D_x	D_m	Note	Ref.*
$CuRSe_2$ Monoclinic lattice (space group $P2_1/b$)								
$CuLaSe_2$	6.73	7.18	7.53	97.68°	6.65	–		[1, 2, 3]
$CuCeSe_2$	6.70	7.17	7.49	97.25°	–	–		[1, 2, 3]
$CuPrSe_2$	6.68	7.12	7.45	97.14°	–	–		[1, 2, 3]
$CuNdSe_2$	6.67	7.11	7.44	97.25°	6.70	6.69		[1, 2, 3]
$CuSmSe_2$	6.59	7.06	7.37	97.23°	–	–		[6]
$CuGdSe_2$	6.62	7.04	7.35	97.33°	–	–	According to ref. [4], $a = 6.60$, $b = 7.04$, $c = 7.33$, $\beta = 97.5°$ (ref. [5])	[1]
$CuRSe_2$ Trigonal lattice (space group $P\bar{3}$). The compound densities are given in ref. [2]								
$CuTbSe_2$	4.083		6.478		6.753			[7]
$CuDySe_2$	4.063		6.466		6.90			[7]
$CuHoSe_2$	4.063		6.466		6.94	–		[7]
$CuErSe_2$	4.043		6.456		7.06	–		[7]
$CuTmSe_2$	4.033		6.440		7.14	7.04		[7]
$CuYbSe_2$	4.021		6.446		7.26	–		[7]
$CuLuSe_2$	4.009		6.434		7.35	–		[7]
$CuScSe_2$	3.889		6.265		5.39	–		[7]
$CuYSe_2$	4.068		6.454		5.57	–		[7]
CuR_5Se_8 Cubic lattice of the Th_3P_4 type								
$CuLa_5Se_8$	9.048							[2, 8]
$CuCe_5Se_8$	8.973							[2, 8]
$CuPr_5Se_8$	8.914							[2, 8]
$CuNd_5Se_8$	8.874							[2, 8]
$CuSm_5Se_8$	8.798							[2, 8]
$CuGd_5Se_8$	8.752							[2, 8]
$Cu_2R_{2/3}Se_2$ Trigonal lattice of the $Cu_2Gd_{2/3}Se_2$ type. Compounds of the Cu_3RSe_3 composition are considered to have the form of $Cu_2(R_{2/3}\square_{1/3})\ Se_2$. The cell parameters for Cu_3RSe_3 are given in ref. [2]								
$Cu_2Gd_{2/3}Se_2$	4.054		6.630		6.66	–		[2, 7]
$Cu_2Tb_{2/3}Se_2$	4.047		6.618		6.92	–		[2, 7]
$Cu_2Dy_{2/3}Se_2$	0.042		6.583		7.01	–		[2, 7]
$Cu_2Ho_{2/3}Se_2$	4.030		6.566		7.10	–		[2, 7]

TABLE 4 (cont'd)

Compound	a (Å)	b (Å)	c (Å)	γ	Density (g/cm³) D_x	D_m	Note	Ref.*
$Cu_2R_{2/3}Se_2$ (cont'd)								
$Cu_2Er_{2/3}Se_2$	4.023		6.561		7.16	–		[2, 7]
$Cu_2Tm_{2/3}Se_2$	4.021		6.551		7.20	7.12		[2, 7]
$Cu_2Yb_{2/3}Se_2$	4.014		6.528		7.30	–		[2, 7]
$Cu_2Lu_{2/3}Se_2$	4.014		6.518		7.33	–		[2, 7]
$Cu_2Sc_{2/3}Se_2$	3.898		6.267		6.34	–		[2, 7]
$Cu_2Y_{2/3}Se_2$	4.038		6.597		6.14	–		[2, 7]
Cu_5RSe_4 Hexagonal lattice of the Cu_5RTe_4 type								
Cu_5GdSe_4	11.92		6.72		6.05	6.12		[9]
Cu_5TbSe_4	11.90		6.72		6.38	6.35		[9]
Cu_5DySe_4	11.88		6.71		6.49	6.47		[9]
Cu_5HoSe_4	11.89		6.65		6.51	6.50		[9]
Cu_5ErSe_4	11.87		6.60		6.60	6.63		[9]
Cu_5YbSe_4	12.12		6.40		6.68	6.70		[9]
Cu_5LuSe_4	12.10		6.32		6.70	6.75		[9]
$Cu_{13}R_5Se_{14}$ Orthorhombic lattice of the $Cu_{13}La_5Se_{14}$ type								
$Cu_{13}La_5Se_{14}$	7.74	24.67	7.01		6.55	6.69		[3]
$AgRSe_2$ Tetragonal lattice of the $AgYbS_2$ type								
$AgGdSe_2$	8.01		12.45		7.03		Transition of the orthorhombic phase to the tetragonal one at 632°C	[10]
$AgTbSe_2$	7.98		12.35		7.17		High-temperature modification	[11]
$AgDySe_2$	7.97		12.28		7.29			[10, 11]
$AgHoSe_2$	7.93		12.26		7.42			[10, 11]
$AgRSe_2$ Orthorhombic lattice of the $AgErSe_2$ type								
$AgDySe_2$	4.20	6.92	13.74		7.12		Low-temperature modification	[10, 11]
$AgHoSe_2$	4.21	6.90	13.73		7.17		There exists solid solution of the $AgHoSe_2$ composition with cubic symmetry and NaCl-type structure $a = 5.762$, $D_x = 7.48$ (ref. [10])	[10, 11]

$AgErSe_2$	4.20	6.84	13.87		7.32	7.25	Ref. [2] gives the following cell parameters for this phase: $a = 6.88$, $b = 13.64$, $c = 4.78$ (ref. [11]). The crystal structure was determined on a single crystal; there exists a solid solution of the $AgErSe_2$ composition with the NaCl-type structure with cubic symmetry $a = 5.744$, $D_x = 7.59$ (ref. [10])	[10, 11]
$AgTmSe_2$	4.19	6.85	13.68		7.35	–		[10, 11]
$AgYbSe_2$	4.16	6.77	13.54		7.64	–		[10, 11]
$AgLuSe_2$	4.17	6.74	13.57		7.64	–		[10, 11]
$AgScSe_2$	4.25	6.73	13.52		5.33	–		[10, 11]
$AgYSe_2$	4.23	6.95	13.84		5.79	–	There exists a solid solution $AgYSe_2$ with the NaCl-type structure with cubic symmetry $a = 5.780$ Å, $D_x = 6.09$ (ref. [10])	[10, 11]
AgR_5Se_8 Cubic lattice of the Th_3P_4 type								
$AgLa_5Se_8$	9.034							[12]
$AgCe_5Se_8$	8.953						The phase contains the region of homogeneity	[12]
$AgPr_5Se_8$	8.894							[12]
$AgNd_5Se_8$	8.858							[12]
$AgSm_5Se_8$	8.788							[12]
$AgGd_5Se_8$	8.740							[12]

*References: [1] Julien-Pouzol and Guittard (1968b); [2] Julien-Pouzol and Guittard (1972); [3] Julien-Pouzol et al. (1972); [4] Pribylskii et al. (1983a); [5] Pribylskii et al. (1983b); [6] Gambarov et al. (1971); [7] Julien-Pouzol et al. (1968); [8] Julien-Pouzol and Guittard (1968a); [9] Rustamov et al. (1976); [10] Julien-Pouzol and Guittard (1969); [11] Julien-Pouzol and Guittard (1973); [12] Ballestracci (1966a).

on the crystal system for $Cu_2R_4Te_7$ are inconsistent: according to Pardo and Nguyen (1987) the symmetry is orthorhombic, whereas according to Collin et al. (1968) it is hexagonal in the series R = La–Pr.

The compounds with the composition $CuRTe_2$ (R = Gd–Tm) crystallize in a trigonal system (Aliyev et al. 1972a, Agayev and Aliyev 1973, Pardo and Flahaut 1971, 1972) and appear to be phases with changeable composition. In addition, phases with the following compositions were observed: Cu_5RTe_4 (R = Gd–Dy) (Rustamov et al. 1976), $Cu_6R_4Te_6$ (R = La, Ce), $Cu_4R_2Te_5$ (R = La) with tetragonal symmetry (Pardo and Flahaut 1971, 1972, Agayev and Aliyev 1973) (table 5).

The compounds $AgRTe_2$ are found to exist in two modifications: tetragonal for R = Gd–Tm, Y and trigonal for R = Dy–Lu, Y of the $CuScS_2$ type (Pardo et al. 1973). The temperature of transition from the low-temperature tetragonal modification to trigonal one is equal to 708, 504, 518, 488°C for R = Gd, Dy, Ho, Er, respectively, and to 520°C for Y. The formation of the solid solution between the compositions $AgGdTe_2$ and $Ag_2Gd_{2/3}Te_2$ was established in the Gd_2Te_3–Ag_2Te system (table 5).

TABLE 5
Crystallographic data for compounds in R–{Cu, Ag}–Te systems.

Compound	a (Å)	b (Å)	c (Å)	Density (g/cm³) D_x	D_m	Note	Ref.*
$Cu_{0.5}RTe_2$ Orthorhombic lattice (space group Pbcm)							
$Cu_{0.5}LaTe_2$	7.69	8.55	6.29	6.84	6.84	Cell parameters of the single crystal of the $Cu_{0.28}LaTe_2$ composition are: $a = 7.72$, $b = 8.60$, $c = 6.25$ (ref. [2])	[1]
$Cu_{0.5}CeTe_2$	7.66	8.50	6.27	–	–		[1]
$Cu_{0.5}PrTe_2$	7.64	8.44	6.24	–	–		[1]
$Cu_{0.5}NdTe_2$	7.61	8.39	6.41	–	–		[1]
$Cu_{0.5}SmTe_2$	7.56	8.31	6.16	–	–		[1]
$CuRTe_2$ Tetragonal lattice							
$CuGdTe_2$	4.18		7.223	6.70	5.93		[3, 4]
$CuTbTe_2$	4.316		6.919	–	–		
$CuDyTe_2$	4.299		6.908	–	–		[4]
$CuHoTe_2$	4.287		6.902	–	–		[4]
$CuErTe_2$	4.274		6.893	–	–		[4]
$CuTmTe_2$	4.266		6.880	–	–		[4]
$CuLuTe_2$	4.256		6.864	–	–		[4]
$CuRTe_2$ Hexagonal lattice of the $CuGdTe_2$ type							
$CuGdTe_2$	4.29		7.223	6.70	6.77		[4]
$CuTbTe_2$	4.271		7.216	6.980	–		[4]
$CuDyTe_2$	4.266		7.200	7.02	–		[4]
$CuHoTe_2$	4.219		7.19	7.12	7.15		[4]
$CuErTe_2$	4.20		7.15	7.30	7.39		[4]
$CuTmTe_2$	4.20		7.071	7.42	7.39		[4]

TABLE 5 (cont'd)

Compound	a (Å)	b (Å)	c (Å)	Density (g/cm^3) D_x	D_m	Note	Ref.*
Cu_3RTe_3 Trigonal lattice (space group $P\bar{3}$)							
Cu_3SmTe_3	4.220		7.300	6.609	6.700	According to ref. [5], $a = 4.340$, $c = 7.255$, $D_x = 6.70$, $D_m = 6.70$	[6]
Cu_3GdTe_3	4.212		7.286	7.220	7.269	$a = 4.32$, $c = 7.248$, $D_x = 6.9$ (ref. [7]); $a = 4.18$, $c = 7.218$, $D_x = 7.1$ (ref. [13]); $a = 4.288$, $c = 6.989$ (ref. [4])	[6]
Cu_3TbTe_3	4.210		7.280	7.252	7.083		[6]
Cu_3DyTe_3	4.200		7.272	7.330	7.200	$a = 4.310$, $c = 7.248$, $D_x = 6.97$ (ref. [7])	[6]
Cu_3HoTe_3	4.190		7.210	7.42	7.37		[6]
Cu_3ErTe_3	4.180		7.160	7.45	7.58		[6]
Cu_3TmTe_3	4.230		7.110	7.46	7.43		[6]
Cu_3YbTe_3	4.226		6.782	7.80	–		[7]
Cu_3YTe_3	4.300		7.240	6.30	6.40		[6]
Cu_5RTe_4 Hexagonal lattice (space group P6/mmm)							
Cu_5GdTe_4	12.63		6.86	6.60	6.65	Parameters were determined on single crystals. For this composition other cell parameters are known: $a = 4.18$, $c = 7.212$ $D_x = 6.72$ (ref. [8]) $a = 4.173$, $c = 8.648$ (ref. [5])	[6]
Cu_5TbTe_4	12.62		6.84	6.88	6.90	Single crystal	[6]
Cu_5DyTe_4	12.62		6.83	7.08	6.99	Single crystal	[6]
$AgRTe_2$ Trigonal lattice of the $CuScS_2$ type							
$AgDyTe_2$	4.319		7.01	7.72		Low-temperature modifications	[4]
$AgHoTe_2$	4.305		7.01	7.80			[4]
$AgErTe_2$	4.299		7.00	7.80		Single crystal	[4]
$AgTmTe_2$	4.288		6.99	7.93			[4]
$AgLuTe_2$	4.279		6.99	8.06			[4]
$AgYTe_2$	4.323		7.03	6.60			[4]
$AgRTe_2$ Tetragonal lattice of the $AgGdTe_2$ type							
$AgGdTe_2$	4.633		7.172	7.02			[4]
$AgDyTe_2$	4.583		7.125	7.50			[4]
$AgHoTe_2$	4.575		7.103	7.60			[4]
$AgErTe_2$	4.553		7.097	7.60		Single crystal	[4]
$AgTmTe_2$	4.532		7.080	7.78			[4]
$AgYTe_2$	4.581		7.126	6.45			[4]

*References: [1] Collin et al. (1968); [2] Pardo and Nguyen (1987); [3] Agayev and Aliyev (1973); [4] Pardo et al. (1973); [5] Gamidov and Keiserukhs'kaya (1972); [6] Rustamov et al. (1981); [7] Aliyev et al. (1972a); [8] Agayev and Aliyev (1973).

References

Agayev, A., and U. Aliyev, 1973, Uch. Zap. Azerb. Univ. **1**, 17.

Aliyev, U., R. Gamidov, G. Guseinov and M. Alidzhanov, 1972a, Izv. Akad. Nauk SSSR, Ser. Neorg. Mater. **8**, 370.

Aliyev, U., R. Gamidov and G. Guseinov, 1972b, Izv. Akad. Nauk SSSR, Ser. Neorg. Mater. **8**, 1855.

Aliyev, U., R. Gamidov, G. Guseinov and M. Alidzhanov, 1973, Izv. Akad. Nauk SSSR, Ser. Neorg. Mater. **9**, 843.

Ballestracci, R., 1965, Bull. Soc. Fr. Mineral. Crystallogr. **88**, 207.

Ballestracci, R., 1966a, C.R. Acad. Sci., Paris, Ser. C. **262**, 1155.

Ballestracci, R., 1966b, C.R. Acad. Sci., Paris, Ser. C. **262**, 1253.

Ballestracci, R., and E. Bertaut, 1964, Bull. Soc. Fr. Mineral. Crystallogr. **84**, 512.

Ballestracci, R., and E. Bertaut, 1965a, Bull. Soc. Fr. Mineral. Crystallogr. **88**, 136.

Ballestracci, R., and E. Bertaut, 1965b, C.R. Acad. Sci., Paris, Ser. C. **261**, 5064.

Ballestracci, R., and E. Bertaut, 1968, Bull. Soc. Fr. Mineral. Crystallogr. **88**, 575.

Bronger, W., R. Elter, E. Maus and T. Schmitt, 1973, Rev. Chim. Miner. **10**, 147.

Brunel, M., F. Bergevin and M. Gondrand, 1972, J. Phys. Chem. Solids **33**, 1927.

Collin, G., F. Rouyer and J. Loriers, 1968, C.R. Acad. Sci., Paris, Ser. C. **266**, 689.

Dijk, M., and C. Plug, 1980, Mater. Res. Bull. **15**, 103.

Dismukes, J., and R. Smith, 1971, J. Phys. Chem. Solids **32**, 913.

Gambarov, D., Z. Karayev, G. Guseinov and M. Alidzhanov, 1971, Uch. Zap. Azerb. Univ. (2), 12.

Gamidov, R., and L. Keiserukhs'kaya, 1972, Izv. Akad. Nauk SSSR, Ser. Neorg. Mater. **8**, 177.

Guittard, M., P. Lemoine, M. Wintenberger and J. Flahaut, 1987, J. Less-Common Met. **127**, 394.

Guseinov, G., U. Amiraslanov, A. Amirov and Kh. Mamedov, 1984, Dokl. Akad. Nauk Azerb. SSR. **40**, 32.

Julien-Pouzol, M., and M. Guittard, 1968a, Bull. Soc. Chim. Fr. (6), 2293.

Julien-Pouzol, M., and M. Guittard, 1968b, C.R. Acad. Sci., Paris, Ser. C **267**, 823.

Julien-Pouzol, M., and M. Guittard, 1969, C.R. Acad. Sci., Paris, Ser. C **269**, 316.

Julien-Pouzol, M., and M. Guittard, 1972, Ann. Chim (Paris) **7**, 253.

Julien-Pouzol, M., and M. Guittard, 1973, Ann. Chim. (Paris) **8**, 139.

Julien-Pouzol, M., M. Guittard and J. Flahaut, 1968, Bull. Soc. Chim. Fr. (2), 533.

Julien-Pouzol, M., M. Guittard and A. Mazurier, 1970, C.R. Acad. Sci., Paris, Ser. C. **271**, 1317.

Julien-Pouzol, M., M. Guittard and M. Rivet, 1972, Bull. Soc. Chim. Fr. (5), 1732.

Julien-Pouzol, M., S. Jaulmes, A. Mazurier and M. Guittard, 1981, Acta Crystallogr. B **37**, 1901.

Lemoine, P., D. Carré and M. Guittard, 1986, Acta Crystallogr. C **42**, 390.

Palazzi, M., 1986, C.R. Acad. Sci., Paris, Ser. 2 **303**, 33.

Pallazzi, M., J. Hodges, G. Jehanno, M. Wintenberger, A. Tomas and J. Flahaut, 1987, J. Less-Common Met. **127**, 393.

Pardo, M.-P., and Nguyen Huy Dung, 1987, C.R. Acad. Sci., Paris, Ser. 2 **304**, 637.

Pardo, M.-P., and J. Flahaut, 1971, Bull. Soc. Chim. Fr. (10), 3411.

Pardo, M.-P., and J. Flahaut, 1972, Bull. Soc. Chim. Fr. (6), 2189.

Pardo, M.-P., M. Julien-Pouzol and J. Flahaut, 1973, C.R. Acad. Sci., Paris, Ser. C **276**, 599.

Plug, C., and A. Prodan, 1978, Acta Crystallogr. A **34**, 250.

Plug, C., and G. Verschoor, 1976, Acta Crystallogr. B **32**, 1856.

Pribil'skii, N., and R. Gamidov, 1983, Zh. Neorg. Khim. **28**, 719.

Pribil'skii, N., and R. Gamidov, 1984, Zh. Neorg. Khim. **29**, 2984.

Pribil'skii, N., R. Gamidov and I. Vasil'yeva, 1983a, Izv. Sib. Otd. Akad. Nauk SSSR, Ser. Khim. Nauk **1/2**, 46.

Pribil'skii, N., R. Gamidov and I. Vasil'yeva, 1983b, Izv. Sib. Otd. Akad. Nauk SSSR, Ser. Khim. Nauk **1/2**, 51.

Rustamov, P., O. Aliyev, G. Guseinov, M. Alidzhanov and M. Agayev, 1976, Izv. Akad. Nauk SSSR, Ser. Neorg. Khim. **12**, 1192.

Rustamov, P., O. Aliyev and T. Kurbanov, 1981, V kn: Troiinyye Khal'kogenidy Redkozemel'nykh Elementov. Izd. "Elm", Baku.
Shilkina, T., and I. Vasil'yeva, 1984, Izv. Sib. Otd. Akad. Nauk SSSR, Ser. Khim. Nauk **8**(3), 35.
Shilkina, T., and I. Vasil'yeva, 1986, Izv. Sib. Otd. Akad. Nauk SSSR, Ser. Khim. Nauk **15**(5), 73.

2. The R–M^{II}–X systems

We will now consider R–M^{II}–X systems, where M^{II} designates magnesium, calcium, strontium, barium, zinc, cadmium, or mercury.

The compounds of the compositions MR_2S_4 and MR_4S_7 were observed in the $M^{II}S$–R_2S_3 systems (where M^{II} = Mg, Ca, Sr, Ba) (Flahaut et al. 1962a, b). There exists a region of solid solutions between MR_2S_4 and R_2S_3, and between MgS and R_2S_3.

The solid solution based on CaS with the NaCl-type structure is formed in the CaS–Y_2S_3 system (Flahaut et al. 1961). Polymorphism of CaY_2S_4 is observed; namely, a low-temperature α-form and a high-temperature β-form, with the transition temperature being 1110°C. The melting point of the β-phase is 1790°C. The solid solution with the Th_3P_4-type structure and based on CaY_2S_4 exists at temperatures above 990°C. The region of its existence broadens at 1120–1370°C. The eutectic coordinates of $CaY_2S_4 + Y_2S_3$ are 32 mol.% and 990°C.

The structures of the polymorphic modifications of the compounds CaR_2S_4 (R = Ho–Lu, Y) are as follows: a low-temperature orthorhombic (α) structure of the Yb_3S_4 type and a high-temperature (β) structure of the MnY_2S_4 type (Flahaut et al. 1961, Patrie and Flahaut 1967). The cubic structure of the Th_3P_4 type was observed for the compounds CaR_2S_4 (R = La–Dy, Ho) (Flahaut et al. 1965, Yim et al. 1973).

The compounds MgR_2S_4 have different structures for different R: the MnY_2S_4 type for R = La–Er, and the spinel type for R = Tm–Lu, Sc (Flahaut et al. 1965, Yim et al. 1973).

The compounds SrR_2S_4 (R = La–Dy) have the Th_3P_4-type structure (Flahaut et al. 1962b, 1965, Sallavuard and René 1971), in the case of R = Tb–Lu, Y, they have the $CaFe_2O_4$-type structure (Patrie et al. 1964b, Vovan et al. 1966).

The compounds of composition BaR_2S_4 (R = Nd–Lu, Y) crystallize in the $CaFe_2O_4$-type structure (Vovan and Khodadad 1969) (table 6).

In the EuS–CaS(SrS, BaS) systems there exists a continuous series of solid solutions with the NaCl-type structure (Vovan and Khodadad 1970). Samarium monosulphid forms a continuous series of solid solutions of the NaCl type with CaS, SrS at 1300°C, whereas with MgS it forms the ternary compound $Mg_xSm_{2-x}S_4$ ($x \approx 0.33$) with the Th_3P_4-type structure. The solid solutions with the NaCl-type structure decompose at 1000°C. At this temperature, the compound $BaSm_2S_4$ with the orthorhombic lattice and an assumed structure of the $CaFe_2O_4$ type is formed in the BaS–SmS system. Such a behaviour of the solid solutions is due to instability of SmS: at a temperature of 1260°C, Sm_3S_4 is one of the decomposition products.

TABLE 6
Crystallographic data for compounds in R–{Mg, Ca, Sr, Ba}–S systems.

Compound	a (Å)	b (Å)	c (Å)	β	Density (g/cm³) D_x	Density (g/cm³) D_m	Note	Ref.*
MgR_2S_4 Orthorhombic lattice of the MnY_2S_4 type								
$MgTb_2S_4$	12.62	12.74	3.78		5.15			[1, 2, 3, 4]
$MgDy_2S_4$	12.59	12.70	3.77		5.26			[1, 2, 3, 4]
$MgHo_2S_4$	12.54	12.66	3.76		5.37			[1, 2, 3, 4]
$MgEr_2S_4$	12.51	12.63	3.74		5.47			[1, 2, 3, 4]
MgY_2S_4	12.60	12.73	3.77		3.63		Low-temperature modification α-MgY_2S_4 crystallizes in an orthorhombic system within the structure type Yb_3S_4, $a = 12.66$, $b = 12.72$, $c = 3.76$ Å, $D_x = 3.64$, $D_m = 3.62$ (ref. [5])	[1, 2, 3, 4]
MgR_2S_4 Cubic lattice								
$MgTm_2S_4$	10.958						$MgAl_2O_4$-type structure	[3]
$MgYb_2S_4$	10.957						$MgAl_2O_4$-type structure	[3]
$MgLu_2S_4$	10.949						$MgAl_2O_4$-type structure	[3]
$MgSc_2S_4$	10.627						$MgAl_2O_4$-type structure	[3]
MgY_2S_4	10.967						$MgAl_2O_4$-type structure	[6]
$MgGd_2S_4$	8.359						Th_3P_4-type structure	[6]
MgR_4S_7 Monoclinic lattice of the Y_5S_7 type								
MgY_4S_7	12.66	3.80	11.45	105.45°				[7]
$MgDy_4S_7$	12.67	3.80	11.45	105.55°				[7]
$MgHo_4S_7$	12.61	3.78	11.42	105.45°				[7]
$MgEr_4S_7$	12.55	3.77	11.38	105.35°				[7]
$MgTm_4S_7$	12.50	3.75	11.34	105.25°				[7]
CaR_2S_4 Cubic lattice of the Th_3P_4 type								
$CaLa_2S_4$	8.687				4.52	4.48	$a = 8.66$ (ref. [9])	[5, 8, 9]
$CaCe_2S_4$	8.615				–	–		[5, 8, 9]
$CaPr_2S_4$	8.578							[5, 8, 9]
$CaNd_2S_4$	8.533							[5, 8, 9]
$CaSm_2S_4$	8.472							[5, 8, 9]
$CaGd_2S_4$	8.423							[5, 8, 9]
$CaTb_2S_4$	8.400							[5, 8, 9]

Compound	a	b	c	D_x	D_m	Remarks	Ref.
$CaDy_2S_4$	8.376						[3]
$CaHo_2S_4$	8.363					The Yb_3S_4 type structure transforms to Th_3P_4 at 900°C	[10]
CaR_2S_4 Orthorhombic lattice of the Yb_3S_4 type							
$CaHo_2S_4$	12.90	13.04	3.86	5.10			[3, 10, 11, 12]
$CaEr_2S_4$	12.87	13.01	3.85	5.18			[3, 10, 11, 12]
$CaTm_2S_4$	12.85	12.98	3.84	5.25			[3, 10, 11, 12]
$CaYb_2S_4$	12.82	12.95	3.83	5.37			[3, 10, 11, 12]
$CaLu_2S_4$	12.82	12.95	3.83	5.41			[3, 10, 11, 12]
CaY_2S_4	12.98	13.11	3.88	3.48		Low-temperature modification; according to ref. [5] the parameters of the CaY_2S_4 lattice at 1050°C are: $a = 12.94$, $b = 13.08$, $c = 3.86$ $D_x = 3.62$, $D_m = 3.61$	[3, 10, 11, 12]
CaR_2S_4 Orthorhombic lattice of the MnY_2S_4 type							
$CaEr_2S_4$	12.77	13.09	3.84				[10]
$CaTm_2S_4$	12.72	13.05	3.84				[10]
$CaYb_2S_4$	12.69	13.01	3.83				[10]
$CaLu_2S_4$	12.69	13.01	3.83				[10]
CaY_2S_4	12.90	13.17	3.87			High-temperature modification. At 1120°C, $D_m = 3.82$ (ref. [12])	[10]
SrR_2S_4 Cubic lattice of the Th_3P_4 type							
$SrLa_2S_4$	8.790			4.83	4.69		[3, 8]
$SrCe_2S_4$	8.718			–	–		[3, 8]
$SrPr_2S_4$	8.682						[3, 8]
$SrNd_2S_4$	8.649						[3, 8]
$SrSm_2S_4$	8.595						[3, 8]
$SrGd_2S_4$	8.551						[3, 8]
$SrDy_2S_4$	8.495					High-temperature modification (800°C) (ref. [13])	[3, 8]
SrR_2S_4 Orthorhombic lattice of the $CaFe_2O_4$ type							
$SrTb_2S_4$	11.96	14.35	4.01	5.16			[11, 14]
$SrDy_2S_4$	11.91	14.29	3.29	5.29			[11, 14]
$SrHo_2S_4$	11.89	14.27	3.98	5.37			[11, 14]
$SrEr_2S_4$	11.84	14.22	3.97	5.47			[11, 14]
$SrTm_2S_4$	11.80	14.17	3.94	5.58			[11, 14]
$SrYb_2S_4$	11.76	14.13	3.90	5.76			[11, 14]
$SrLu_2S_4$	11.71	14.09	3.90	5.84			[11, 14]
SrY_2S_4	11.97	14.34	3.99	5.82			[11, 14]

TABLE 6 (cont'd)

Compound	a (Å)	b (Å)	c (Å)	β	Density (g/cm^3) D_x	D_m	Note	Ref.*
BaR_2S_4 Cubic lattice of the Th_3P_4 type								
$BaLa_2S_4$	8.917				5.09	4.97		[3, 5, 11]
$BaCe_2S_4$	8.864				–	–		[3, 3, 11]
$BaPr_2S_4$	8.817							[3, 5, 11]
$BaNd_2S_4$	8.793						High-temperature phase	[14]
$BaSm_2S_4$	8.720						At 800°C	[13]
BaR_2S_4 Orthorhombic lattice of the $CaFe_2O_4$ type								
$BaNd_2S_4$	12.29	14.78	4.14		4.83	–		[1, 3]
$BaSm_2S_4$	12.24	14.73	4.12		4.98	4.92	Low-temperature modification	[1, 3]
$BaGd_2S_4$	12.21	14.65	4.08		5.19	5.13		[1, 3]
$BaTb_2S_4$	12.16	14.55	4.06		5.32	–		[1, 3]
$BaDy_2S_4$	12.09	14.50	4.04		5.48	5.46		[1, 3]
$BaHo_2S_4$	12.07	14.47	4.02		5.50	–		[1, 3]
$BaEr_2S_4$	11.99	14.42	3.99		5.85	5.82		[1, 3]
$BaTm_2S_4$	11.96	14.40	3.98		5.75	–		[1, 3]
$BaYb_2S_4$	11.91	14.33	3.96		5.94	5.90		[1, 3]
$BaLu_2S_4$	11.90	14.32	3.96		5.99	–		[1, 3]
BaY_2S_4	12.16	14.56	4.06		4.03	4.10	Single crystal	[1, 3]

*References: [1] Patrie et al. (1964a); [2] Patrie and Chavalier (1966); [3] Flahaut et al. (1965); [4] Flahaut (1968); [5] Flahaut et al. (1962a); [6] Ben-Dor and Shilo (1980); [7] Adolphe and Laruelle (1968); [8] Flahaut et al. (1962b); [9] Yim et al. (1973); [10] Patrie and Flahaut (1967); [11] Vovan and Khodadad (1969); [12] Flahaut et al. (1961); [13] Sallavuard and René (1971); [14] Patrie et al. (1964b).

Solid solutions based on R_2Se_3 (R = La–Dy) with the Th_3P_4 type, and also solid solutions based on CaSe (R = Gd, Dy, Er, Yb, Y) with the NaCl type, are found in the CaSe–R_2Se_3 systems. The ternary compounds with the composition CaR_2Se_4 (R = Dy–Lu, Y) have the $CaHo_2Se_4$-type structure (Golabi et al. 1965) (table 7).

The compounds with the composition SrR_2Se_4 and BaR_2Se_4 crystallize in two structure types: Th_3P_4 (for Sr, R = La–Lu, Y; and for Ba, R = La–Nd) and

TABLE 7
Crystallographic data for compounds in R–{Ca, Sr, Ba}–Se systems.

					Density (g/cm³)			
Compound	a (Å)	b (Å)	c (Å)	α	D_x	D_m	Note	Ref.*
CaR_2Se_4 Rhombohedral lattice of the $CaHo_2Se_4$ type								
$CaHo_2Se_4$	8.30			59°02′	5.75	–	$a = 7.197(5)$, $\alpha = 33°08(02)'$ (ref. [2])	[1]
$CaEr_2Se_4$	8.28			58°54′	5.88	5.75		[1]
$CaTm_2Se_4$	8.27			58°42′	5.93	–		[1]
$CaYb_2Se_4$	8.264			58°31′	6.04	5.82		[1]
$CaLu_2Se_4$	8.261			58°24′	6.10	–		[1]
CaY_2Se_4	8.29			59°16′	4.47	4.43		[1]
SrR_2Se_4 Cubic lattice of the Th_3P_4 type								
$SrLa_2Se_4$	9.124							[3]
$SrCe_2Se_4$	9.060							[3]
$SrPr_2Se_4$	9.019							[3]
$SrNd_2Se_4$	8.989							[3]
$SrSm_2Se_4$	8.931							[3]
$SrGd_2Se_4$	8.895							[3]
SrR_2Se_4 Orthorhombic lattice of the $GaFe_2O_4$ type								
$SrTb_2Se_4$	12.50	14.97	4.17		6.14	5.94		[3, 4]
$SrDy_2Se_4$	12.49	14.92	4.15		6.25	6.05		[3, 4]
$SrEr_2Se_4$	12.48	14.81	4.21		6.44	6.24		[3, 4]
$SrYb_2Se_4$	12.47	14.71	4.08		6.65	6.53		[3, 4]
$SrLu_2Se_4$	12.46	14.67	4.06		6.74	6.60		[3, 4]
SrY_2Se_4	12.49	14.88	4.15		5.00	4.75		[3, 4]
BaR_2Se_4 Cubic lattice of the Th_3P_4 type								
$BaLa_2Se_4$	9.258							[5]
$BaCe_2Se_4$	9.186							[5]
$BaPr_2Se_4$	9.150							[5]
$BaNd_2Se_4$	9.120							[5]
BaR_2Se_4 Orthorhombic lattice of the $CaFe_2O_4$ type								
$BaSm_2Se_4$	12.75	14.32	4.26		6.02	6.00		[4]
$BaGd_2Se_4$	12.74	15.20	4.24		6.21	6.11		[4]
$BaDy_2Se_4$	12.73	15.11	4.22		6.77	6.63		[4]
$BaEr_2Se_4$	12.69	14.99	4.18		6.58	6.48		[4]
$BaYb_2Se_4$	12.60	14.87	4.13		6.86	6.83		[4]
$BaLu_2Se_4$	12.59	14.82	4.10		6.97	–		[4]
BaY_2Se_4	12.72	15.10	4.22		5.17	5.23		[4]

*References: [1] Golabi et al. (1965); [2] Souleau et al. (1969); [3] Golabi et al. (1964); [4] Patrie et al. (1964b); [5] Flahaut et al. (1965).

$CaFe_2O_4$ (for Sr, R = Tb–Lu, Y; and for Ba, R = Sm–Lu, Y) (Patrie et al. 1964b, Golabi et al. 1964, Flahaut et al. 1965).

The compounds MgR_2Se_4 (R = Ho–Lu) have the spinel-type structure.

Two types of solid solutions are formed in the CaTe–R_2Te_3 systems: for R = La–Ho, the compounds based on R_2Te_3 have the Th_3P_4-type structure and the compounds based on CaTe have the NaCl-type structure (Pardo and Flahaut 1969). The Th_3P_4-type solutions with a narrow range of homogeneity do not attain the composition CaR_2Te_4. The compounds with composition CaR_2Te_4 (R = Dy–Lu, Y) crystallize in the $CaHo_2Se_4$-type structure and have a wide area of homogeneity (table 8).

The compounds with composition ZnR_2S_4 crystallize in the spinel-type structure (R = Sc), for R = Tm–Lu, the orthorhombic symmetry is observed (Yim et al. 1973).

In the CdS–Gd_2S_3 section of the ternary Cd–Gd–S system (fig. 3) the compounds $CdGd_2S_4$ and $CdGd_4S_7$ exist which melt, respectively, congruently at 1870°C, and incongruently at 1615°C according to the reaction,

$$L + CdGd_2S_4 \rightleftarrows CdGd_4S_7 .$$

The interaction in the La_2S_3–CdS, Ce_2S_3–CdS and Yb_2S_3–CdS systems are of the same character (Aliyev and Azadaliyev 1980, Agayev et al. 1984a).

The compounds with composition CdR_4S_7 (R = Dy–Yb, Y) crystallize in the Th_3P_4-type structure (Flahaut et al. 1961, 1962a, Adolphe et al. 1968, Vovan et al. 1969).

The compounds of composition CdR_2S_4 with R = Dy–Lu crystallize in the Th_3P_4-type structure, in the case of R = Ho–Lu the spinel structural type is proposed (Yim et al. 1973, Flahaut et al. 1965, 1968). According to Yim et al. (1973) the Th_3P_4-type structure has been found only for La, the compounds $CdNd_2S_4$ and $CdPr_2S_4$ crystallize with a trigonal symmetry and the compound $CdGd_2S_4$ crystallizes in monoclinic structure (table 9).

The CdSe–R_2Se_3 (R = La, Gd, Ho) systems are analogous to relative sulphide systems (Aliyev et al. 1984). The compounds $CdPr_2Se_4$ and $CdPr_4Se_7$ which melt,

TABLE 8
Crystallographic data for compounds in R–Ca–Te systems.

Compound	a (Å)	α	Density (g/cm^3) D_x	Note	Ref.*
CaR_2Te_4 Rhombohedral lattice of the $CaHo_2Se_4$ type					
$CaDy_2Te_4$	7.673	33°02′	6.08		[1]
$CaHo_2Te_4$	7.680	32°54′	6.15		[1]
$CaEr_2Te_4$	7.673	32°50′	6.22		[1]
$CaTm_2Te_4$	7.677	32°44′	6.27		[1]
$CaLu_2Te_4$	7.677	32°38′	6.42		[1]
CaY_2Te_4	7.676	32°56′	4.87		[1]

*Reference: [1] Pardo and Flahaut (1969).

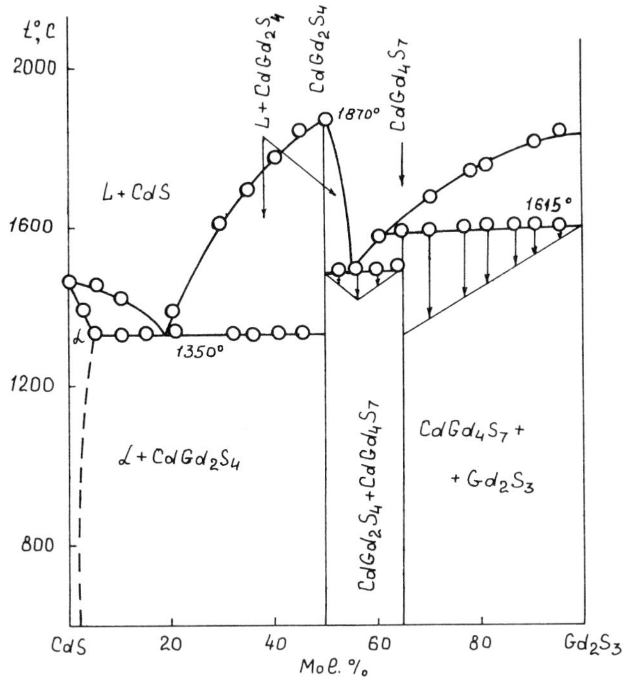

Fig. 3. The phase diagram of the $CdS–Gd_2S_3$ system.

respectively, congruently and according to a peritectic reaction, are found on the quasi-binary section of $CdSe–Pr_2Se_3$ of the ternary Cd–Pr–Se system (fig. 4),

$$L + Pr_2Se_3 \rightleftarrows CdPr_4Se_7 \quad \text{(Agayev et al. 1984b)}.$$

Solubility of Pr_2Se_3 in CdSe is about 3 mol.%. The compounds with composition CdR_2Se_4 crystallize in the following structural types: $MgAl_2O_4$ (R = Dy–Yb, Y), Th_3P_4 (R = La–Tb) (Fujii et al. 1972); the compounds CdR_4Se_7 (R = Ho, Er, Yb) crystallize in the Y_5S_7-type structure (Rustamov et al. 1981) (table 10).

The $CdTe–La_2Te_3$, $CdTe–Gd_2Te_3$ systems exhibit two incongruently melting compounds, CdR_2Te_4 and CdR_4Te_7 (Agayev et al. 1983, Zul'fugarly et al. 1984, Aliyev et al. 1985).

The liquidus of the quasi-binary $CdTe–Dy_2Te_3$ section of the ternary Cd–Dy–Te system consists of three branches: two end-branches correspond to the primary crystallization of CdTe and Dy_2Te_3, the third branch corresponds to the crystallization of the ternary compound $CdDy_2Te_4$, which melts in a peritectic reaction at 1000°C:

$$L + Dy_2Te_3 \rightleftarrows CdDy_2Te_4 .$$

The eutectic in this system is characterized by phase equilibria,

$$L \rightleftarrows \text{solid solution} + CdDy_2Te_4 ,$$

TABLE 9
Crystallographic data for compounds of R–{Zn, Cd}–S systems.

Compound	a (Å)	b (Å)	c (Å)	β	Density (g/cm³) D_x	D_m	Note	Ref.*
ZnR_2S_4 Orthorhombic lattice								
$ZnTm_2S_4$	7.734	13.227	6.263					[1]
$ZnLu_2S_4$	7.658	13.183	6.246					[1]
ZnR_2S_4 Cubic lattice of the $MgAl_2O_4$ type								
$ZnSc_2S_4$	10.483							[1]
CdR_2S_4 Cubic lattice of the Th_3P_4 type								
$CdLa_2S_4$	8.72				5.19	5.12	$a = 8.7033$ (ref. [1])	[2]
$CdCe_2S_4$	8.69				5.27	5.24		[2]
$CdPr_2S_4$	8.66				5.36	5.34		[2]
$CdNd_2S_4$	8.64				5.47	5.40		[2]
$CdSm_2S_4$	8.62				5.63	5.63		[2]
$CdGd_2S_4$	8.60				5.82	5.76	$Cd_{0.5}Gd_{2.3}S_4$: $a = 8.368$ (ref. [3])	[2]
$CdTb_2S_4$	8.58				5.89	5.80	$Cd_{0.4}Tb_{2.4}S_4$: $a = 8.343$ (ref. [3])	[2]
CdR_2S_4 Cubic lattice of the $MgAl_2O_4$ type								
$CdDy_2S_4$	11.26				5.28	5.22		[2]
$CdHo_2S_4$	11.24				5.35	5.28	$a = 11.1674$ (ref. [1]) Electron-microscope study of $CdHo_2S_4$ has shown (ref. [5]) that $CdHo_2S_4$ crystallizes in the spinel structure at 700°C: $a = 11.17$. While at 900–1100°C it occurs in a distorted NaCl structure (rhombohedral lattice, space group R32. The cell parameters in hexagonal axes $a = 4.09$, $c = 20.33$)	[3, 4]
$CdEr_2S_4$	11.19				5.47	5.41	$a = 11.1347$ (ref. [1]) The crystal structure was determined on a single crystal, $a = 11.10$ (ref. [6])	[7]
$CdTm_2S_4$	11.10				5.64	5.63	$a = 11.09$ (ref. [1])	[8, 9]
$CdYb_2S_4$	11.04				5.81	5.79	$a = 11.0684$ (ref. [1])	[8]
$CdLu_2S_4$	11.00				5.91	5.85	$a = 11.45$ (ref. [1])	[7]

CdY_2S_4	11.19				–	–	$a = 11.216$ Å, $D_m = 3.9$ (ref. [12]) Electron-microscope study of CdY_2S_4, with spinel-type structure, revealed the presence of $Cd_2Y_{4/3}S_4$ domains with a defective NaCl-type structure	[8]
$CdSc_2S_4$	10.733				–	–		[1]
CdR_4S_7 Monoclinic lattice of the Y_5S_7 type								
$CdDy_4S_7$	12.84	3.83	11.48	105.65°				[10]
$CdEr_4S_7$	12.58	3.80	11.34	106°			$a = 12.67$, $b = 3.79$, $c = 11.42$, $\beta = 105.70°$ (ref. [11])	[10]
$CdHo_4S_7$	12.60	3.80	11.35	106°			$a = 12.75$, $b = 3.81$, $c = 11.45$, $\beta = 105.60°$ (ref. [11])	[10]
$CdTm_4S_7$	12.79	3.18	11.42	105°				[10]
$CdYb_4S_7$	12.55	3.79	11.29	104°				[10]
CdY_4S_7	12.70	3.83	11.49	106°	4.28	4.15	$a = 12.79$, $b = 3.82$, $c = 11.46$, $\beta = 105.70°$ (ref. [11])	[10]

*References: [1] Yim et al. (1973); [2] Hulliger and Vogt (1966); [3] Ben-Dor and Shilo (1980); [4] Flahaut et al. (1965); [5] Bakkar et al. (1982); [6] Tomas et al. (1978); [7] Fujii et al. (1972); [8] Suchov and Stemple (1964); [9] Pokrzywnicki and Czopnik (1975); [10] Vovan and Khodadad (1969); [11] Adolphe and Laruelle (1968); Tomas et al. (1986).

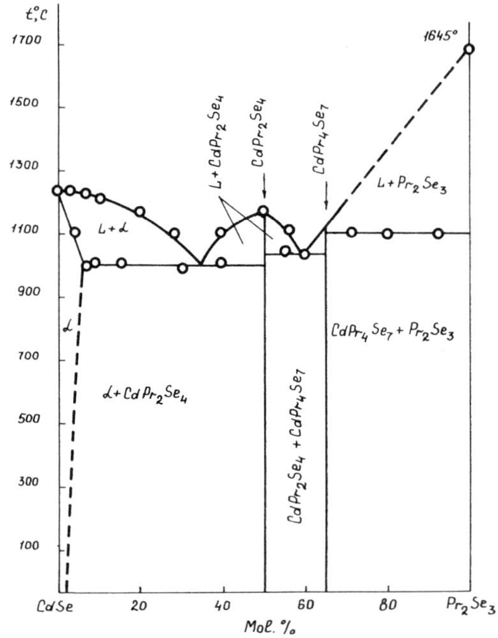

Fig. 4. The phase diagram of the CdSe–Pr_2Se_3 system.

TABLE 10
Crystallographic data for compounds in the R–Cd–Se systems.

Compound	a (Å)	b (Å)	c (Å)	β	Density (g/cm^3)		Note	Ref.*
					D_x	D_m		
CdR_2Se_4 Cubic lattice of the Th_3P_4 type								
$CdLa_2Se_4$	8.94				6.58	6.50		[1]
$CdCe_2Se_4$	8.93				6.63	6.60		[1]
$CdPr_2Se_4$	8.92				6.67	6.64		[1]
$CdNd_2Se_4$	8.90				6.77	6.68		[1]
$CdSm_2Se_4$	8.87				6.80	6.77		[1]
$CdGd_2Se_4$	8.86				7.10	6.94	$a = 8.72$ (ref. [2])	[1]
$CdTb_2Se_4$	8.80				6.94	6.90		[1]
CdR_2Se_4 Cubic lattice of the $MgAl_2O_4$ type								
$CdDy_2Se_4$	11.66				6.28	6.18		[2]
$CdHo_2Se_4$	11.64				6.41	6.32		[3, 4]
$CdEr_2Se_4$	11.60				6.51	6.45		[3, 4]
$CdTm_2Se_4$	11.57				6.59	6.54		[1]
$CdYb_2Se_4$	11.54				6.71	6.60		[2]
$CdLu_2Se_4$	11.50				6.76	6.70		[1]
CdY_2Se_4	11.65				–	–		[1]
CdR_4Se_7 Monoclinic lattice of the Y_5S_7 type								
$CdHo_4Se_7$	13.32	4.02	11.96	105°	7.75	7.85		[5]
$CdEr_4Se_7$	13.30	4.01	11.95	105°	7.20	7.08		[5]
$CdYb_4Se_7$	13.36	4.02	11.97	105°	7.25	7.16		[5]

*References: [1] Hulliger and Vogt (1966); [2] Suchov and Stemple (1964); [3] Fujii et al. (1972); [4] Fujii (1972); [5] Rustamov et al. (1981).

with the eutectic parameters: 33 mol.% Dy_2Te_3 and 997°C (Agayev et al. 1987, Agayev and Kuliyeva 1984).

The quasi-binary CdTe–DyTe section of the ternary Cd–Dy–Te system belongs to the eutectic type with limited solid solutions. The following are the eutectic parameters: 22 mol.% and 877°C. The solubility based on CdTe is 5 mol.% DyTe (Agayev et al. 1987).

Based on $Eu_xHg_{1-x}Te$, the solutions were found in the range $0 \leqslant x \leqslant 0.125$ (Gavaleshko et al. 1985).

References

Adolphe, C., and P. Laruelle, 1968, Bull. Soc. Fr. Miner. Crystallogr. **91**, 219.
Agayev, A., and I. Kuliyeva, 1984, Azerb. Khim. Zh. (5), 61.
Agayev, A., D. Zul'fugarly, S. Aliyev and R. Azadaliyev, 1983, Zh. Neorg. Khim. **28**, 256.
Agayev, A., S. Kuliyeva and U. Kuliyev, 1984a, Azerb. Khim. Zh. (6), 95.
Agayev, A., R. Azadaliyev and U. Aliyeva, 1984b, Azerb. Khim. Zh. (5), 105.
Agayev, A., U. Kuliyeva and P. Rustamov, 1987, Zh. Neorg. Khim. **32**, 2554.
Aliyev, O., and R. Azadaliyev, 1980, Zh. Neorg. Khim. **25**, 3160.
Aliyev, O., A. Agayev, R. Azadaliyev and M. Abdullayeva, 1984, Zh. Neorg. Khim. **29**, 2705.
Aliyev, O., A. Agayev and R. Azadaliyev, 1985, Zh. Neorg. Khim. **30**, 1041.
Bakkar, M., F. Vollebregt and C. Plug, 1982, J. Solid State Chem. **42**, 11.
Ben-Dor, L., and I. Shilo, 1980, J. Solid State Chem. **35**, 278.
Flahaut, J., 1968, in: Progress Science and Technology of Rare Earths, Vol. 3 (Oxford) p. 209.
Flahaut, J., L. Domange and M. Patrie, 1961, Bull. Soc. Chim. Fr. (1), 105.
Flahaut, J., L. Domange and M. Patrie, 1962a, Bull. Soc. Chim. Fr. (1), 159.
Flahaut, J., L. Domange and M. Patrie, 1962b, Bull. Soc. Chim. Fr. (11/12), 2048.
Flahaut, J., M. Guittard, M. Patrie, M.-P. Pardo, S. Golabi and L. Domange, 1965, Acta Crystallogr. **19**, 14.
Fujii, H., 1972, J. Sci. Hiroshima Univ. Ser. A **36**, 67.
Fujii, H., T. Okamoto and T. Kamigaichi, 1972, J. Phys. Soc. Jpn., Suppl. **32**, 1432.
Gavaleshko, N., P. Kav'yuk, V. Lototskii, L. Solonchuk and V. Khomyak, 1985, Izv. Akad. Nauk SSSR, Ser. Neorg. Mater. **21**, 1965.
Golabi, S., J. Flahaut and L. Domange, 1964, C.R. Acad. Sci., Paris, Ser. C **259**, 820.
Golabi, S., J. Flahaut and L. Domange, 1965, C.R. Acad. Sci., Paris, Ser. C **260**, 6385.
Hulliger, F., and O. Vogt, 1966, Phys. Lett. B **21**, 2138.
Pardo, M.-P., and J. Flahaut, 1969, Bull. Soc. Chim. Fr. (1), 6.
Patrie, M., and R. Chevalier, 1966, C.R. Acad. Sci., Paris, Ser. C **263**, 1061.
Patrie, M., and J. Flahaut, 1967, C.R. Acad. Sci., Paris, Ser. C **264**, 395.
Patrie, M., J. Flahaut and L. Domange, 1964a, C.R. Acad. Sci., Paris, Ser. C **258**, 2585.
Patrie, M., S. Golabi, J. Flahaut and L. Domange, 1964b, C.R. Acad. Sci., Paris, Ser. C **259**, 4039.
Pokrzywnicki, S., and A. Czopnik, 1975, Phys. Status Solidi B **70**, K85.
Rustamov, P., O. Aliyev and T. Kurbanov, 1981, V kn: Troiinyye Khal'kogenidy Redkozemel'nykh Elementov. Izd. "Elm", Baku.
Sallavuard, G., and A. René, 1971, C.R. Acad. Sci., Paris, Ser. C **273**, 1428.
Souleau, C., M. Guittard and P. Laruelle, 1969, Bull. Soc. Chim. (1), 9.
Suchov, L., and N. Stemple, 1964, J. Electrochem. Soc. **111**, 191.
Tomas, A., I. Shilo and M. Guittard, 1978, Mater. Res. Bull. **13**, 857.
Tomas, A., M. Guittard, J. Flahaut, M. Guymout, R. Portier and D. Gratias, 1986, Acta Crystallogr. B **42**, 417.
Vovan, T., and P. Khodadad, 1969, Bull. Soc. Chim. Fr. (1), 30.
Vovan, T., and P. Khodadad, 1970, Bull. Soc. Chim. Fr. (8/9), 2888.
Vovan, T., J. Flahaut and L. Domange, 1966, C.R. Acad. Sci., Paris, Ser. C **262**, 278.
Yim, W., A. Fan and E. Stofko, 1973, J. ELectrochem. Soc. **120**, 441.
Zul'fugarly, D., A. Agayev, P. Azadaliyev and Yu. Aliyeva, 1984, Dokl. Akad. Nauk SSSR **40**, 48.

3. The R–M^{III}–X systems

We will now consider the R–M^{III}–X system, with M^{III} designating aluminium, gallium, indium or thallium.

The compounds with the composition $RAlS_3$ were found in the R_2S_3–Al_2S_3 systems (R = La, Ce) (Domange et al. 1960). Based on $LaAlS_3$, the solid solution was found in the range 33.3–60 mol.% Al_2S_3. The composition of the compound with cerium was determined by structural analysis: $Ce_6Al_{10/3}S_{14}$ (Saint–Giniez et al. 1968).

The system La_2S_3–GaS_3 is analogous to the system CeS_2–Ga_2S_3 (Loireau-Lozac'h et al. 1977, 1976, Lozac'h et al. 1972). The compounds $La_6Ga_{10/3}S_{14}$ and $Ce_6Ga_{10/3}S_{14}$ melt conguently at 1150 and 1130°C, respectively; the compounds $LaGaS_3$ and $CeGaS_3$ melt peritectically at 940° and 880°C, respectively. The eutectic parameters in the La_2S_3–Ga_2S_3 system are: 20 mol.% La_2S_3 and 1124°C, 80 mol.% La_2S_3 and 832°C. The Ce_2S_3–Ga_2S_3 is shown in fig. 5.

The ternary compound $NdGaS_3$, congruently melting at 1230°C, was found on the quasi-binary section Nd_2S_3–Ga_2S_3 at the molar relationship of the components 1:1 (Keiserukhskaya et al. 1970, Karayev et al. 1966b). The homogeneity

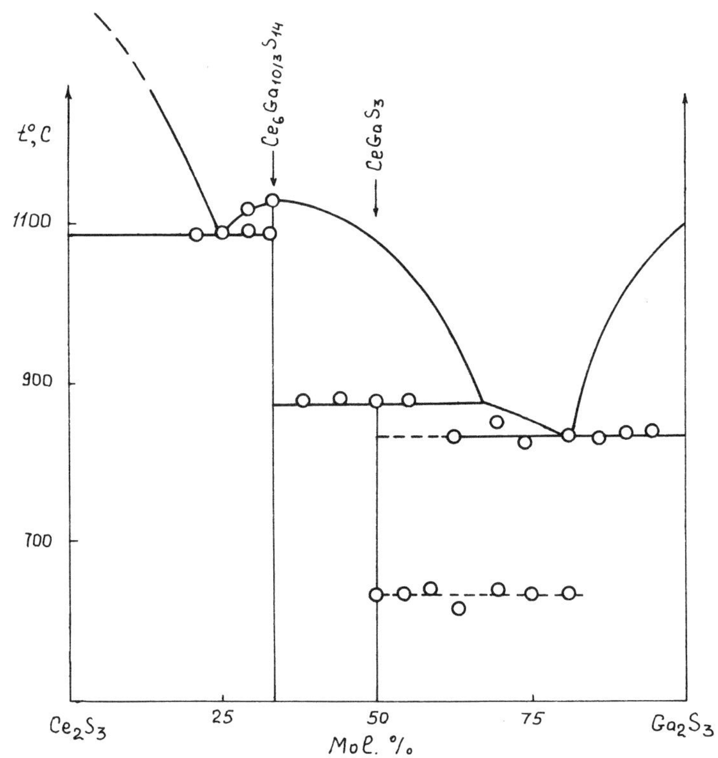

Fig. 5. The phase diagram of the Ce_2S_3–Ga_2S_3 system.

range of the compound is in the range of the compositions from 49.8 to 51 mol.% Ga_2S_3. The solid solution based on Ga_2S_3 was found, as well as the eutectics between $NdGaS_3$ and both binary components with the following eutectic points: 47 mol.% Ga_2S_3 and 1150°C, 73 mol.% Ga_2S_3 and 920°C. The eutectic of $Nd_2S_3 + NdGaS_3$ is shifted towards the ternary compound. According to the data of Aliyev et al. (1987), the compound $Nd_6Ga_{10/3}S_{14}$ melts congruently at 1207°C, the eutectic with Nd_2S_3 has the following parameters: 23 mol.% Ga_2S_3 and 1017°C. Solubility with respect to Ga_2S_3 reaches 1 mol.% at room temperature.

The liquidus line of the Sm_2S_3–Ga_2S_3 system has four branches of primary crystallization: Ga_2S_3, Sm_2S_3, $Sm_6Ga_{10/3}S_{14}$, $SmGaS_3$ (Aliyev et al. 1987). Ternary compounds melt congruently at 1277°C and according to peritectic reaction at 1032°C, respectively,

$$L + Sm_2S_3 \rightleftarrows SmGaS_3 .$$

A continuous series of solid solutions with minimum at 1062°C and 60 mol.% $Nd_6Ga_{10/3}S_{14}$ is formed between the ternary compounds $R_6Ga_{10/3}S_{14}$ (R = Nd, Sm) (Aliyev et al. 1987).

The R_2S_3–Ga_2S_3 systems (R = Dy, Er) are analogous (Loireau-Lozac'h et al. 1977). The only ternary compound R_3GaS_6 (R = Dy, Er) decomposes in a peritectic reaction at 1125°C and 1100°C, respectively. The eutectic points are: 60 mol.% Dy_2S_3 and 950°C, 70 mol.% Er_2S_3 and 960°C, respectively.

The Y_2S_3–Ga_2S_3 system exhibits two ternary compounds, Y_3GaS_6 and $Y_6Ga_{10/3}S_{14}$, which are stable at temperatures up to 1100°C and 1150°C, respectively. The eutectic points are: 66 mol.% Y_2S_3 and 940°C (Loireau-Lozac'h et al. 1977, Lozac'h et al. 1971).

The interaction characteristics in the RS–GaS systems (R = Sm, Eu, Yb) are the same: there is one compound which melts in peritectically, namely $RGaS_2$, and one eutectic (Aliyev et al. 1980). The eutectic composition for R = Sm, Eu is 1.2 mol.% RS. The compound $EuGaS_2$ melts congruently at 1085°C (Nadzhafov et al. 1980). Two incongruently melting compounds, $GaYbS_2$ and $GaYb_4S_7$, are formed in the YbS–GaS system (Rustamov et al. 1977). 0.1 mol.% YbS dissolves in GaS at room temperature, at 960°C this increases up to 1 mol.%. The eutectic composition is 2.5 mol.% YbS. Complex chemical interaction in the SmS–GaS system results in the formation of two peritectically melting compounds:

$$L + \text{solid solution} \rightleftarrows Sm_4GaS_5 \quad \text{at 80 mol.\% SmS and 1090°C} ;$$

$$L + Sm_4GaS_5 \rightleftarrows Sm_3Ga_2S_5 \quad \text{at 60 mol.\% SmS and 1060°C} .$$

The eutectic parameters are: 1.2 mol.% SmS and 935°C. From 3 (at 25°C) to 10 mol.% GdS (1090°C) dissolves in SmS. Based on GaS, the SmS solubility is from 0.2 (25°C) to 0.5 mol.% (935°C) (Aliyev 1980b, Nadzhafov et al. 1978).

The quasi-binary section SmS–Ga_2S_3 is assigned to the types of diagrams with peritectic (1177°C) and limited solid solutions (Aliyeva and Aliyev 1986). Two compounds have been found: $SmGa_2S_4$ melts congruently at 1477°C, and

TABLE 11

Crystallographic data for compounds in the R–{Al, Ga, In, Tl}–S systems.

Compound	a (Å)	b (Å)	c (Å)	γ	Density (g/cm³) D_x	D_m	Note	Ref.*
$R_6Al_{3.33}S_{14}$ Hexagonal lattice of the $Ce_6Al_{3.33}S_{14}$ type. In ref. [3] the "$RAlS_3$" composition is assigned to these compounds								
$La_6Al_{3.33}S_{14}$	10.12(3)		6.07(2)					[1]
$Ce_6Al_{3.33}S_{14}$	10.02(3)		6.08(2)				The crystal structure is determined and the compound composition is specified (ref. [2])	[1]
$Pr_6Al_{3.33}S_{14}$	9.94(3)		6.07(2)					[1]
$Nd_6Al_{3.33}S_{14}$	8.89(3)		6.07(2)					[1]
$Sm_6Al_{3.33}S_{14}$	9.79(4)		6.06(2)					[1]
$Gd_6Al_{3.33}S_{14}$	9.72(4)		6.06(2)					[1]
$Tb_6Al_{3.33}S_{14}$	9.67(3)		6.07(2)					[1]
$Dy_6Al_{3.33}S_{14}$	9.65(3)		6.07(2)					[1]
$Ho_6Al_{3.33}S_{14}$	9.55(3)		6.07(2)					[1]
$Y_6Al_{3.33}S_{14}$	9.58(3)		6.07(2)					[1]
$R_{10/3}Ga_6S_{14}$ Tetragonal lattice of the melilite type								
$La_{10/3}Ga_6S_{14}$	9.30		6.03		4.25	4.17		[4, 5, 6]
$Ce_{10/3}Ga_6S_{14}$	9.29		6.01				$a = 9.96$, $c = 6.08$ (ref. [7])	[4, 5, 6]
								[4, 5, 6]
GaR_3S_6 Orthorhombic lattice (space group $Ccm2_1$)								
$GaDy_3S_6$	10.47	13.22	6.45		5.58			[5, 7]
$GaHo_3S_6$	10.41	13.17	6.43		–			[5, 7]
$GaEr_3S_6$	10.36	13.17	6.40		5.83			[5, 7]
GaY_3S_6	10.45	13.22	6.45		3.76		Single crystal	[5, 7]
$R_6Ga_{3.33}S_{14}$ Hexagonal lattice of the $Ce_6Al_{3.33}S_{14}$ type							The "$RGaS_3$" composition is assigned to these compounds (ref. [3])	
$La_6Ga_{3.33}S_{14}$	10.54(4)		6.08(2)		4.64			[1, 5, 9, 10]
$Ce_6Ga_{3.33}S_{14}$	10.03(4)		6.08(2)		4.77			[1, 5, 9, 10]
$Pr_6Ga_{3.33}S_{14}$	9.94(4)		6.08(2)		–			[1, 5, 9, 10]
$Nd_6Ga_{3.33}S_{14}$	9.90(4)		6.08(2)		–			[1, 5, 9, 10]

$Sm_6Ga_{3.33}S_{14}$	9.78(4)		6.08(2)	–			[1, 5, 9, 10]
$Gd_6Ga_{3.33}S_{14}$	9.72(4)		6.08(2)	–			[1, 5, 9, 10]
$Tb_6Ga_{3.33}S_{14}$	9.67(4)		6.08(2)	–			[1, 5, 9, 10]
$Dy_6Ga_{3.33}S_{14}$	9.68(4)		6.08(2)	–			[1, 5, 9, 10]
$Y_6Ga_{3.33}S_{14}$	9.62(4)		6.04(2)	4.17			[7]
$R_3Ga_5S_{12}$ Hexagonal lattice							
$Gd_3Ga_5S_{12}$	6.729		12.870				[11]
$Yb_3Ga_5S_{12}$	6.431		12.456				[11]
$Y_3Ga_5S_{12}$	6.441		12.447				[11]
$InRS_3$ Orthorhombic lattice (space group Pbam)							
$InLaS_3$	11.80	21.00	3.99	4.71	4.73	$a = 11.81$, $b = 21.27$, $c = 4.039$, $D_x = 4.58$, $D_m = 4.65$ (ref. [12])	[11]
$InCeS_3$	11.78	20.96	3.95	4.78	4.78		[11]
$InPrS_3$	11.79	20.25	3.93	4.91	4.96		[11]
$InNdS_3$	11.78	20.96	3.92	4.97	4.98		[11]
$InSmS_3$	11.76	20.95	3.89	5.04	5.08		[11]
α-$InGdS_3$	10.42	20.26	3.87	–	–		[11]
α-$InYS_3$	11.40	19.80	3.48	5.07	–	β-$InYS_3$ crystallizes in cubic symmetry, $a = 10.65$ (ref. [8])	[11]
RIn_3S_6 Orthorhombic lattice (space group Pbam)							
$LaIn_3S_6$	4.06	11.84	21.35		4.10		[13]
$CeIn_3S_6$	3.99	11.81	21.30		4.38		[13]
$PrIn_3S_6$	3.99	11.78	21.26		4.40		[13]
$NdIn_3S_6$	3.97	11.77	21.25		4.42		[13]
$SmIn_3S_6$	3.95	11.73	21.20		4.47		[13]
$GdIn_3S_6$	3.93	11.71	21.17		4.54		[13]
$TbIn_3S_6$	3.93	11.70	21.15		4.59		[13]
$DyIn_3S_6$	3.92	11.69	21.14		4.67		[13]
$HoIn_3S_6$	3.92	11.68	21.11		4.73		[13]
$ErIn_3S_6$	3.90	11.67	21.03		4.81		[13]
YIn_3S_6	3.94	11.65	21.16		4.38		[13]
R_3InS_6 Orthorhombic lattice of the La_3InS_6 type							
La_3InS_6	16.92	13.95	4.09		4.97	Single crystal, $a = 16.912(9)$, $b = 13.946(5)$, $c = 4.079(2)$, $D_x = 5.1$, $D_m = 5.2$; space group $P2_12_12$ (ref. [14])	[13]
Ce_3InS_6	16.91	13.944	4.081		5.18		[13]

TABLE 11 (cont'd)

Compound	a (Å)	b (Å)	c (Å)	γ	Density (g/cm^3) D_x	D_m	Note	Ref.*
R_3InS_6 (cont'd)								
$PrInS_6$	16.89	13.92	4.07			5.50		[13]
Nd_3InS_6	16.87	13.90	4.04			5.88		[13]
R_3InS_6 Orthorhombic lattice of the Sm_3InS_6 type								
Sm_3InS_6	16.87	13.88	4.03			6.08	The structure is determined on a single crystal (ref. [21]) $a = 16.513(2)$, $b = 13.632(2)$, $c = 3.901(3)$ Å, $D_x = 5.74$, $D_m = 5.6$; space group Pnnm	[13]
Gd_3InS_6	16.86	13.864	4.026			6.10		[13]
Tb_3InS_6	16.84	13.84	4.014			6.14		[13]
Dy_3InS_6	16.82	13.82	4.00			6.26		[13]
Ho_3InS_6	16.80	13.765	3.96			6.34		[13]
Er_3InS_6	16.79	13.74	3.94			6.40		[13]
Tb_3InS_6	16.77	13.72	3.90			6.44		[13]
Yb_3InS_6	16.75	13.70	3.87			6.49		[13]
Lu_3InS_6	16.72	13.67	3.86			6.58		[13]
Sc_3InS_6	16.76	13.74	3.90			6.14		[13]
Y_3InS_6	16.82	13.78	3.98			6.20		[13]
$R_3In_5S_{12}$ Monoclinic lattice								
$Tb_3In_5S_{12}$	11.00	21.26	3.897	96.36°	5.26		The crystal structure was determined on a single crystal (ref. [15]): $a = 10.998$, $b = 21.259$, $c = 3.897$, $\gamma = 96.3°$, $D_x = 5.1$, $D_m = 5.4$	[12]
$Nd_4In_5S_{12}$	11.774(2)	21.254(8)	3.968(1)		–		The crystal structure was determined on a single crystal (ref. [16]); $Nd_4In_5S_{12}$ crystallizes in orthorhombic structure	
$TlRS_2$ Rhombohedral lattice of the α-$NaFeO_2$ type (the cell parameters are given for hexagonal axes)								
$TlSmS_2$	4.13		22.25					[17]
$TlEuS_2$	4.12		22.34					[17]
$TlGdS_2$	4.10		22.34					[17]
$TlTbS_2$	4.07		22.37		–	–		[17]

$TlDyS_2$	4.06		22.35		–	–		[18]
$TlHoS_2$	4.04		22.46		–	–		[17]
$TlErS_2$	4.02		22.47		7.05	6.90		[17]
$TlTmS_2$	4.01		22.52		–	–		[17]
$TlYbS_2$	4.01		22.53		–	–		[17]
$TlLuS_2$	3.98		22.59		–	–		[17]
$TlYS_2$	4.04		22.47		–	–		[17]
$TlRS_2$ Monoclinic lattice								
$TlSmS_2$	7.19	15.83	4.13	110.47°				[19]
$TlEuS_2$	7.16	15.86	4.12	110.12°				[19]
$TlGdS_2$	7.11	15.88	4.10	109.79°	–	–		[19]
$TlTbS_2$	7.09	15.89	4.07	110.21°	–	–		[19]
$TlDyS_2$	7.05	15.88	4.06	110.20°	–	–		[19]
$TlHoS_2$	6.99	15.89	7.04	109.52°	–	–		[19]
$TlErS_2$	6.96	15.94	4.02	109.73°	7.50	7.35	Single crystal	[19]
$TlTmS_2$	6.93	15.93	4.01	109.52°	–	–		[19]
$TlYbS_2$	6.92	15.94	4.01	110.83°	–	–		[19]
$TlLuS_2$	6.90	15.94	3.98	109.57°	–	–		[19]
$TlYS_2$	7.04	15.96	4.04	110.19°	–	–		[19]
TlR_5S_8 Cubic lattice with Th_3P_4 structure type								
$TlLa_5S_8$	8.80				5.63	5.5	$a = 8.806$ (ref. [19])	[20]
$TlPr_5S_8$	8.643				–	–		[19]

*References: [1] Patrie and Guittard (1969); [2] Saint-Giniez et al. (1968); [3] Domange et al. (1960); [4] Lozac'h et al. (1972); [5] Loireau-Lozac'h et al. (1977); [6] Lozac'h et al. (1973); [7] Lozac'h et al. (1971); [8] Karayev et al. (1966a); [9] Sallavuard and Paris (1971); [10] Lozac'h et al. (1973); [11] Rustamov et al. (1981a); [12] Guittard et al. (1978); [13] Aliyev (1980b); [14] Carré et al. (1978); [15] Carré (1977); [16] Guseinov et al. (1979); [17] Kabré et al. (1974); [18] Godzhayev et al. (1986); [19] Carré et al. (1971); [20] Poltmann and Hahn (1971); [21] Messain et al. (1977).

$SmGa_4S_7$ melts incongruently at 1177°C. The compound $SmGa_2S_4$ reacts with SmS according to the eutectic type. The eutectic parameters are: 32 mol.% Ga_2S_3 and 1277°C. The $SmGa_2S_4$–Ga_2S_3 system belongs to the peritectic type. The eutectic parameters are: 81.5 mol.% Ga_2S_3 at 1073°C. Two peritectically melting compounds, $GaYbS_2$ and $GaYb_4S_5$, are encountered in the GaS–YbS section (table 11) (Rustamov et al. 1977).

The compounds of composition La_3InS_6 and $LaInS_3$, decomposing according to a peritectic reaction at 1110 and 1100°C, respectively, are found in the La_2S_3–In_2S_3 system (Guittard et al. 1978). The eutectic parameters are 70 mol.% In_2S_3 and 980°C. The eutectic parameters in the Sm_2S_3–In_2S_3 system are: 80 mol.% In_2S_3 and 1010°C. Sm_3InS_6 melts congruently at 1200°C, and $Sm_3In_5S_{12}$ decomposes peritectally at 1025°C.

Two compounds $TlEr_5S_8$ and $TlErS_2$ are found on the Tl_2S–Er_2S_3 section (Kabré et al. 1974). The eutectic melting point is 452°C.

The selenide systems with aluminium exhibit ternary compounds of compositions $R_6Al_{10/3}Se_{14}$ and $EuAl_2Se_4$ (Patrie and Guittard 1969). The data on the R_2Se_3–Ga_2Se_3 system (R = La–Nd) are inconsistent. According to Efendiyev et al. (1964a,b), the La_2Se–Ga_2Se_3 system exhibits one congruently melting compound with composition $LaGaSe_3$ at 1120°C and two eutectics at 980°C and 860°C. According to the data given by Lozac'h and Guittard (1973), the ternary compound which melts congruently at 1170°C in this system has the composition $La_6Ga_{10/3}Se_{14}$. The eutectic parameters are: 35.7 mol.% La_2Se_3 and 1114°C, 40 mol.% La_2Se_3 and 908°C (fig. 6).

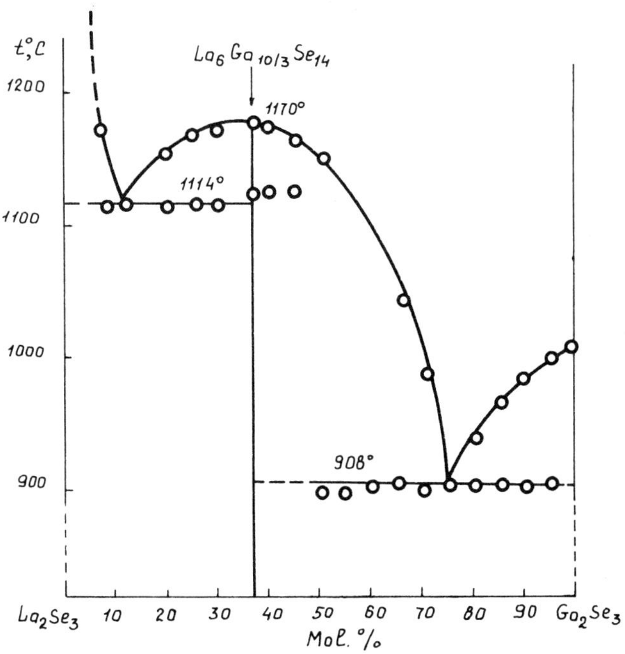

Fig. 6. The phase diagram of the La_2Se_3–Ga_2Se_3 system.

Two congruently melting compounds (at 1185 and 1189°C) with composition $NdGaSe_3$ and $NdGa_4Se_{7.5}$ are formed in the Nd_2Se_3–Ga_2Se_3 system (Efendiyev et al. 1964b). However, according to the data given by Lozac'h et al. (1973), only one compound with composition $Nd_6Ga_{10/3}Se_{14}$, which decomposes in a peritectic reaction at 1226°C, is formed in the system. The eutectic parameters are: 36 mol.% Nd_2Se_3 and 942°C.

The Pr_2Se_3–Ga_2Se_3 system is analogous to the system with neodymium (Nasibov and Karayev 1963). Two compounds $PrGaSe_3$ and $PrGa_4Se_{7.5}$ and two eutectic points at 950 and 900°C are found in this system.

The Gd_2Se_3–Ga_2Se_3 system exhibits the compound $Gd_6Ga_{10/3}Se_{14}$ which decomposes into γ-Gd_2Se_3 and $GdGaSe_3$ in a peritectic reaction at 1076°C (Lozac'h and Guittard 1973).

Two compounds decomposing in a peritectic reaction were found on the quasi-binary section of the YbSe–InSe system (Aliyev 1980a),

$$L + YbSe \rightleftarrows YbIn_4Se_5 \quad \text{at } 830°C\,;$$

$$L + YbIn_4Se_5 \rightleftarrows YbInSe_2 \quad \text{at } 640°C:$$

Between InSe and $YbInSe_2$ a eutectic exists with the following parameters: 27 mol.% YbSe and 520°C. Solubility of YbSe in InSe is ~4 mol.% at 25°C. Cleavage is observed in the concentration range from 54 to 57 mol.% InSe at 600°C.

The liquidus of the YbSe–In_2Se_3 system consists of four branches which correspond with the primary crystallization of YbSe, $YbIn_2Se_4$, and $YbIn_4Se_7$ and the solid solution based on In_2Se_3. The compound $YbIn_2Se_4$ melts congruently at 970°C and forms with YbSe a eutectic with the parameters: 28 mol.% In_2Se_3 and 930°C. The compound $YbIn_2Se_4$ melts congruently at 970°C and with YbSe forms the eutectic with parameters: 28 mol.% In_2Se_3 and 700°C. The compound $YbIn_4Se_7$ melts congruently at 930°C, and with the compounds $YbIn_4Se_7$ and In_2Se_3 it forms eutectics with the parameters: 42 mol.% In_2Se_3 and 850°C and 21 mol.% In_2Se_3 and 800°C. The thermograms of the compositions containing 70–90 mol.% In_2Se_3 exhibit effects at 200, 500 and 650°C which correspond to the temperatures of eutectoid decomposition of the solid solutions based on α, β and γ-In_2Se_3, respectively (Aliyev 1980a).

The compounds with compositions $TlEr_5Se_8$ and $TlErSe_2$ were observed in the Tl_2Se–Er_2Se_3 system. The eutectic point appeared at 384°C (Kabré et al. 1974). The quasi-binary sections TlSe–RSe (R = Dy, Tb) contain one incongruently melting compound $TlRSe_2$ (Guseinov et al. 1981, 1987). For example, the reaction TbSe + L $\rightleftarrows$ $TlTbSe_2$ proceeds at 1292°C. The parameters of the respective eutectic points are 10 mol.% DySe and 327°C, and 10 mol.% TbSe and 350°C. TlSe dissolves 4 mol.% DySe at room temperature (fig. 7).

The YbSe–Tl_2Se system contains one incongruently melting compound $YbTl_4Se_3$

$$L + YbSe \rightleftarrows YbTl_4Se_3\,.$$

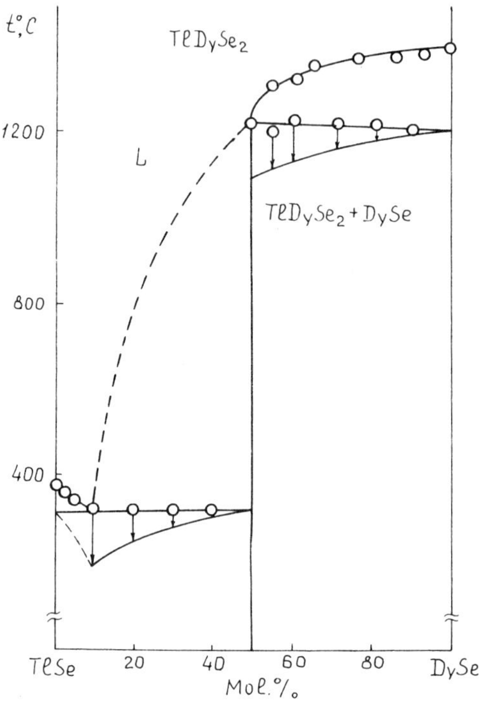

Fig. 7. The phase diagram of the DySe–TlSe system.

The peritectic parameters are: 66.7 mol.% Tl_2Se and 350°C. Between $YbTl_4Se_3$ and Tl_2Se, a eutectic exists with parameters: 15 mol.% YbSe and 300°C (Rustamov et al. 1981b).

The liquidus of the $SmSe–Ga_2Se_3$ state diagram (Aliyeva et al. 1987) consists of three branches which corresponds with the primary crystallization of the phases SmSe, $SmGa_2Se_4$ and Ga_2Se_3. The ternary compound is formed according to the peritectic reaction,

$$L + SmSe \rightleftarrows SmGa_2Se_4 \quad \text{at } 927°C .$$

Between $SmGa_2Se_4$ and Ga_2Se_3, a eutectic exists at 867°C. The polymorphic transformations of Ga_2Se_3 occur at temperatures of 777 and 542°C.

Two congruently melting compounds $EuGa_2Se_4$ (1110°C) and $EuGa_4Se_7$ occur in the $EuSe–Ga_2SE_3$ system. (1060°C) (Aliyev 1980b). Based on Ga_2Se_3, a limited solubility range up to 5 mol.% EuSe exists.

In the InSe–RSe (R–Sm, Er) there are solubility ranges on the basis of InSe. The eutectic parameters are: 8.0 mol.% SmSe and 520°C, 15 mol.% ErSe and 535°C.

One incongruently melting compound with the composition $RGaSe_2$ is formed in each RSe–GaSe (R = Sm, Eu, Yb) system (Aliyev 1980b, Guseinov et al. 1985, 1987) (table 12).

The Yb–Ga–Se constitutional diagram has been established after studying the following sections: $YbSe–Ga_2Se_3$, YbSe–GaSe, $YbGa_2$–GaSe, Yb–GaSe,

Yb_2Se_3–Ga_2Se_3, Yb_2Se_3–GaSe (fig. 8), YbSe–$YbGaSe_3$, Yb_3Se_4–$YbGaSe_3$, Yb-$GaSe_3$–$YbGa_4Se_7$, $YbGa_2Se_4$–$YbGaSe_3$ (Rustamov et al. 1984). The ternary system contains 17 points of non-variant equilibrium, out of which 11 points appear to be ternary eutectic points and 6 points are ternary peritectic points. The YbSe–$YbGaSe_3$ section is quasi-binary. The eutectic parameters are: 30.2 mol.% YbSe and 1057°C. Solid solutions based on $YbGaSe_3$ are formed towards 0.5 mol.% at 25°C. The Yb_3Se_4–$YbGaSe_3$ section is quasi-binary. The eutectic parameters are: 29.8 mol.% Yb_3Se_4, 997°C. The Yb_2Se_3–GaSe section passes through four phase triangles: YbSe–$YbGa_2Se_4$–GaSe, YbSe–$YbGaSe_3$–Yb_3Se_4, Yb_3Se_4–$YbGaSe_5$–Yb_2Se_3, YbSe–$YbGaSe_3$–$YbGa_2Se_4$ (fig. 8). In the 0–35 mol.% GaSe concentration range, the section crosses the secondary ternary system Yb_3Se_4–$YbGaSe_3$–Yb_2Se_3 which shows the formation of a continuous series of solid solutions between Yb_3Se_4 and Yb_2Se_3. The crystallization of the sintered mass terminates at a ternary eutectic temperature of 867°C. Then the section crosses the ternary system YbSe–$YbGaSe_3$–$YbGa_2Se_4$, its components forming a diagram of an eutectic type. The crystallization of the sintered masses terminates at a ternary eutectic temperature of 837°C according to the reaction

$$L \rightleftarrows YbGaSe_3 + YbSe + YbGa_2Se_4 .$$

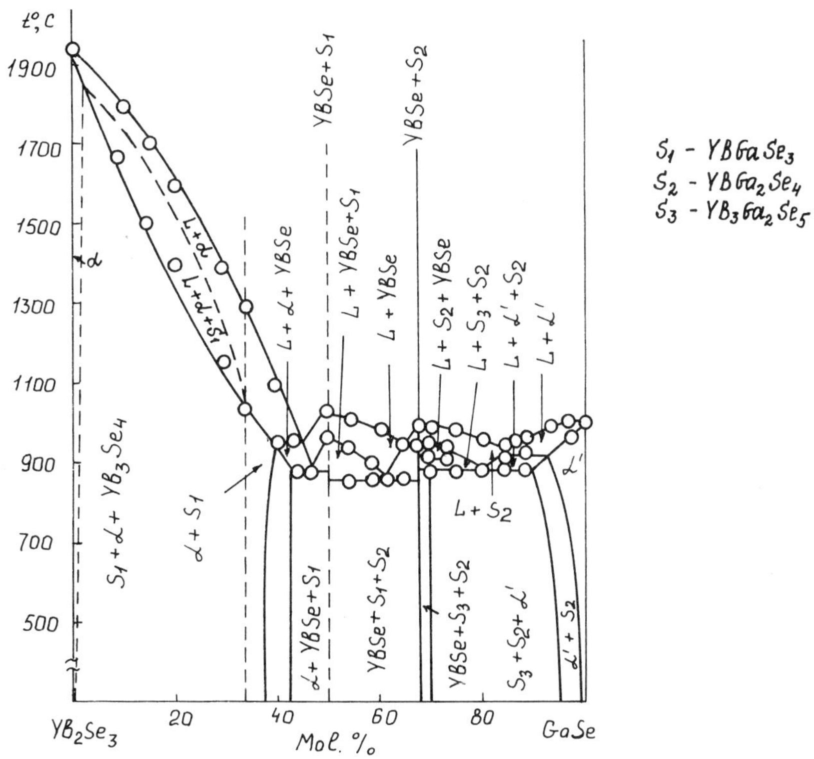

Fig. 8. The phase diagram of the Yb_2Se_3–GaSe system.

TABLE 12
Crystallographic data for compounds of R–{Al, Ga, In, Tl}–Se systems.

Compound	a (Å)	b (Å)	c (Å)	β	Density (g/cm³) D_x	Density (g/cm³) D_m	Note	Ref.*
$R_6Al_{3.33}Se_{14}$ Hexagonal lattice of the $Ce_6Al_{3.33}S_{14}$ type. The compound may be presented in the form of $R_6(Al_{4/3}\square_{2/3})Al_2Se_{14}$ (□ denotes vacancies)								
$La_6Al_{3.33}Se_{14}$	10.43		6.21					[1]
$Ce_6Al_{3.33}Se_{14}$	10.36		6.20					[1]
$Pr_6Al_{3.33}Se_{14}$	10.28		6.20					[1]
$Nd_6Al_{3.33}Se_{14}$	10.23		6.19					[1]
$Sm_6Al_{3.33}Se_{14}$	10.14		6.20					[1]
$Gd_6Al_{3.33}Se_{14}$	10.05		6.19					[1]
$Tb_6Al_{3.33}Se_{14}$	10.03		6.19					[1]
$Dy_6Al_{3.33}Se_{14}$	9.98		6.18					[1]
$Ho_6Al_{3.33}Se_{14}$	9.90		6.18					[1]
$RGaSe_3$ Orthorhombic lattice (space group $Pna2_1$ or Pnam)								
$GdGaSe_3$	16.05	7.10	3.89			6.95		[2]
$DyGaSe_3$	15.93	7.03	3.90		7.16	7.04		[2]
$YbGaSe_3$	5.665	–	–		5.51	5.26	Cubic symmetry with NaCl structure type	[2]
$R_6Ga_{3.33}Se_{14}$ Hexagonal lattice of the $Ce_6Al_{3.33}S_{14}$ type								
$La_6Ga_{3.33}Se_{14}$	10.44		6.20		6.16			[1]
$Ce_6Ga_{3.33}Se_{14}$	10.41		6.18		6.30			[2]
$Pr_6Ga_{3.33}Se_{14}$	10.38		6.17		6.36			[2]
$Nd_6Ga_{3.33}Se_{14}$	10.37		6.16		6.43			[2]
$Sm_6Ga_{3.33}Se_{14}$	10.31		6.15		6.64			[2]
$Gd_6Ga_{3.33}Se_{14}$	10.28		6.14		6.81			[2]
$R_6In_{3.33}Se_{14}$ Hexagonal lattice of the $Ce_6Al_{3.33}S_{14}$ type								
$La_6In_{3.33}Se_{14}$	10.50		6.50					[1]
$InRSe_3$ Hexagonal lattice								
$InLaSe_3$	6.85		4.00		5.80	5.78		[3]
$InCeSe_3$	6.84		3.96		5.82	5.86		[3]
$InPrSe_3$	6.84		3.94		6.20	5.94		[3]
$InNdSe_3$	6.84		3.94		6.24	5.99		[3]
InSmSe	6.83		3.92		6.32	6.08		[3]

$InRSe_3$ Cubic lattice								
$InHoSe_3$	11.382							[3]
$InErSe_3$	11.37							[3]
$InYbSe_3$	11.375						Orthorhombic lattice with $a = 11.28$, $b = 8.12$, $c = 23.84$, space group Fddd, is given for this composition	[3]
$InYSe_3$	11.375							[3]
$TlRSe_2$ Rhombohedral lattice of the α-$NaFeO_2$ type (the cell parameters are given for hexagonal axes)								
$TlPrSe_2$	4.34		23.04					[4]
$TlNdSe_2$	4.32		23.07					[4]
$TlSmSe_2$	4.27		23.10					[4]
$TlGdSe_2$	4.23		23.13					[4]
$TlTbSe_2$	4.21		23.13					[4]
$TlDySe_2$	4.19		23.13				$a = 4.19$, $c = 23.08$, $D_x = 7.44$, $D_m = 7.32$ (ref. [5])	[4]
$TlHoSe_2$	4.17		23.16					[4]
$TlErSe_2$	4.17		23.20					[4]
$TlTmSe_2$	4.15		23.19					[4]
$TlYbSe_2$	4.12		23.25					[4]
$TlLuSe_2$	4.11		23.22					[4]
$TlYSe_2$	4.20		23.26					[4]
$TlRSe_2$ Monoclinic lattice								
$TlSmSe_2$	7.45	16.42	4.29	111°	6.94	7.23		[6]
$TlGdSe_2$	7.38	16.38	4.26	110°	7.13	7.38		[6]
$TlTbSe_2$	7.36	16.35	4.23	110°	7.22	7.52		[6]
$TlDySe_2$	7.29	16.35	4.23	110°	7.36	7.21		[6]
$TlHoSe_2$	7.23	16.31	4.21	110°	7.40	7.45		[6]
$TlErSe_2$	7.22	16.28	4.21	110°	7.56	7.56		[6]
$TlYbSe_2$	7.20	16.21	4.19	110°	7.27	7.23		[6]
$TlYSe_2$	7.27	16.34	4.23	110°	6.34	6.41		[6]
$TlLa_5Se_8$ Cubic lattice of the Th_3P_4 type								
$TlLa_5Se_8$	9.14				6.64	6.5		[7]

*References: [1] Patrie and Guittard (1969); [2] Lozac'h and Guittard (1973); [3] Rustamov et al. (1981a); [4] Kabré et al. (1974); [5] Guseinov et al. (1981); [6] Gamidov and Amanov (1968); [7] Poltmann and Hahn (1971).

Part of the section is crossed by the phase triangle $YbSe–YbGa_2Se_4–YbGaSe_3$. The concentration range 67.5–70 mol.% GaSe contains three phases: YbSe, $YbGaSe_3$, $YbGa_2Se_4$. The sections $YbGa_2Se_4–YbGaSe_3$, $YbSe–YbGa_2$, $YbGaSe_3$–Se are quasi-binary of an eutectic type. The section YbSe–GaSe is quasi-binary. The compound with composition $Yb_3Ga_2Se_5$ is observed at the intersection. The eutectic parameters are 13 mol.% YbSe and 877°C. Solubility of YbSe in GaSe is limited to 2 mol.% at 25°C. According to the data given by Aliyev et al. (1980), no peritectic compound is formed in the system. A GaSe-based solid solution exists at 800°C reaching 5 mol.%, and at 25°C 1.0 mol.% YbSe. The system is eutectic.

The data for the systems with tellurium do not have a systematic character and basically deal with the quasi-binary sections of the related ternary systems.

The ternary system Yb–Ga–Te (Aliyev et al. 1984) contains 13 primary crystallization ranges of separate phases, 23 curves of mono-variant equilibria and 14 non-variant points. The $YbTe–Ga_2Te_3$ section is quasi-binary, of a eutectic type with the formation of the compound $YbGa_2Te_4$ which melts peritectically at 840°C. The eutectic parameters are: 25 mol.% YbTe and 800°C. The Yb–GaTe section crosses two subordinate triangles: $Yb–YbGa_2$–YbTe and $YbGa_2$–YbTe–GaTe. A peritectic reaction occurs at 715°C,

$$L + YbTe \rightleftarrows Yb_3Ga_2Te_5 + YbGa_2 .$$

The GaTe-base solid solution extends towards 7 mol.%, decomposes upon an increase of the ytterbium concentration, and crystallizes in the range 86–93 mol% GaTe together with $Yb_3Ga_2Te_5$. The sections $YbGa_2$–GaTe and $YbGa_2$–YbTe are quasi-binary, of a eutectic type. In the quasi-binary YbTe–GaTe section (fig. 9), the compound of composition $Yb_3Ga_2Te_5$, which melts via a peritectic reaction at 815°C, was observed. The liquidus lines of $Yb_3Ga_2Te_5$ and GaTe cross at the eutectic point at 13 mol.% YbTe and 750°C.

In the Yb–In–Te system, the phase interrelationships were studied in relation to sections $YbTe–In_2Te_s$ (Guseinov et al. 1986, Rustamov et al. 1986). The liquidus of the system YbTe–InTe contains four branches of which corresponds to crystallization of the compounds $YbIn_2Te_4$, $YbIn_4Te_7$, YbTe and InTe. Between $YbIn_4Te_7$ and γ-YbTe, a eutectic exists with the parameters: 18 mol.% YbTe and 627°C. The α-$YbIn_2$-base solid solutions are formed in the 0–5 mol.% concentration range.

The $Sm_2Te_3–Ga_2Te_3$ system exhibits one compound, of composition $SmGaTe_3$ (Karayev et al. 1965).

The liquidus of the system Pr_2Te_3–InTe consists of three branches of primary crystallization: two end-branches correspond to the primary crystallization of Pr_2Te, InTe; whilst the third branch corresponds to that of $InPr_2Te_4$. Based on InTe, solubility attains 1.5 mol.% (Rustamov et al. 1986). In the above-mentioned section, the compound $InPr_2Te_4$ is formed according to the peritectic reaction at 1002°C,

$$L + Pr_2Te_3 \rightleftarrows InPr_2Te_4 .$$

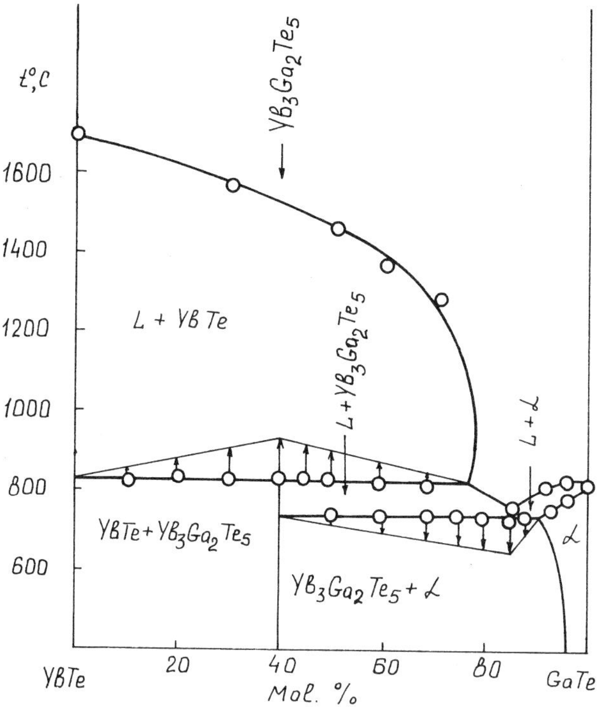

Fig. 9. The phase diagram of the YbTe–GaTe system.

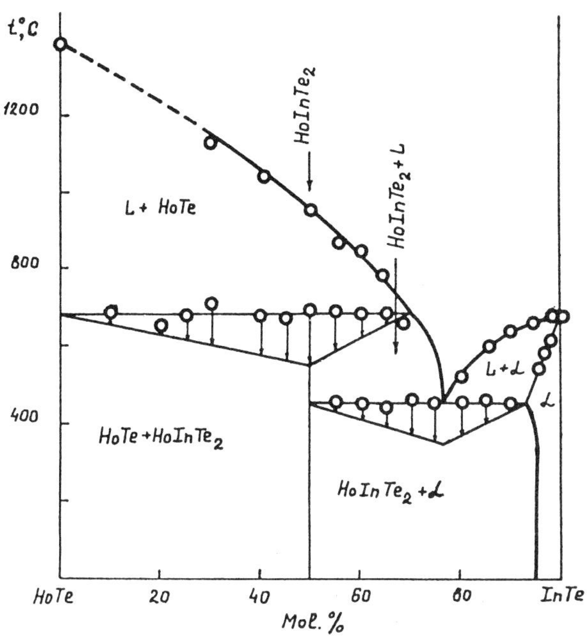

Fig. 10. The phase diagram of the HoTe–InTe system.

TABLE 13
Crystallographic data for compounds in the R–{Ga, In, Al, Tl}–{S, Se, Te} systems.

Compound	a (Å)	b (Å)	c (Å)	β	Density (g/cm³) D_x	Density (g/cm³) D_m	Note	Ref.*
$GaRTe_3$ Hexagonal lattice								
$GaLaTe_3$	10.56		6.48		6.41	6.38		[1]
$GaCeTe_3$	10.54		6.46		6.52	6.44		[1]
$GaPrTe_3$	10.52		6.45		6.58	6.53		[1]
$GaNdTe_3$	10.51		6.45		6.64	6.60		[1]
$GaSmTe_3$	10.49		6.44		6.70	6.71		[1]
$InRTe_3$ Hexagonal lattice								
$InLaTe_3$	6.98		6.42		6.89	6.88		[1]
$InCeTe_3$	6.96		6.18		6.98	6.96		[1]
$InPrTe_3$	6.95		6.16		7.01	7.03		[1]
$InNdTe_3$	6.94		6.16		7.12	7.11		[1]
$InSmTe_3$	6.92		6.15		7.15	7.17		[1]
RM_2X_4 (M = Al, Ga, In; X = S, Se, Te) Orthorhombic lattice. Crystal structure of $EuGa_2S_4$ was determined (ref. [2]). The cell parameters: $a = 20.727(7)$, $b = 20.454(6)$, $c = 12.197(2)$ coincided with the parameters of the $EuGa_2S_4$ cell, which were determined in ref. [3]. The parameters given for the RM_2X_4 composition are the parameters of the subcell								
$EuAl_2S_4$	10.045	10.029	6.020		–	–		[4]
$YbAl_2S_4$	10.18	10.36	6.094		4.32	–		[5]
$EuGa_2S_4$	10.21	10.36	6.094		4.32	4.5		[6]
$SmGa_2S_4$	10.28	10.42	6.16		–	–		[7]
$YbGa_2S_4$	10.26	10.30	6.07		–	4.62		[8]
$EuIn_2S_4$	10.49	10.36	6.50		4.8			[8]
$YbIn_2S_4$	10.40	10.52	6.18			5.07		[8]
$EuAl_2Se_4$	10.73	10.82	6.289		4.75	–		[8]
$EuGa_2Se_4$	10.68	10.81	6.388		5.48			[8]
$SmGa_2Se_4$	21.34	21.60	21.74		–	–		[9]
$YbGa_2Se_4$	10.56	10.70	6.17		5.14	–		[10]
$EuIn_2Se_4$	11.10	10.88	6.70		5.37	–		[5]
$YbIn_2Se_4$	10.80	10.94	6.36		5.38	–		[5]
$EuGa_2Te_4$	11.06	11.00	6.60		–	5.97		[5]

$EuIn_2Te_4$	11.72	11.36	6.92		–	6.20		[5]
$YbIn_2S_4$	20.859(8)	–	3.8743(9)		5.46	–	Hexagonal lattice, crystal structure was determined on a single crystal	[11]
$YbIn_2Te_4$	13.20	–	6.17		–	6.60	Hexagonal lattice	[12]
$TlRTe_2$ Rhombohedral lattice of the α-$NaFeO_2$ type (the cell parameters are given for hexagonal axes)								
$TlPrTe_2$	4.59		24.23		6.76	–		[13, 14]
$TlNdTe_2$	4.56		24.22		6.90	–		[13, 14]
$TlSmTe_2$	4.52		24.24		7.08	–		[13, 14]
$TiGdTe_2$	4.49		24.23		7.26	–		[13, 14]
$TlTbTe_2$	4.48		24.25		7.31	–		[13, 14]
$TlDyTe_2$	4.47		24.24		7.39	–		[13, 14]
$TlHoTe_2$	4.45		24.24		7.48	–		[13, 14]
$TlErTe_2$	4.43		24.24		7.58	7.50		[13, 14]
$TlTmTe_2$	4.43		24.25		7.58	–		[13, 14]
$TlLuTe_2$	4.41		24.26		7.73	–		[13, 14]
$TlYTe_2$	4.47		24.32		6.49	–		[13, 14]
$TlRTe_2$ Monoclinic lattice								
$TlSmTe_2$	8.02	17.68	4.34	111°	7.05	7.15		[15]
$TlGdTe_2$	7.94	17.64	4.34	110°	7.17	7.70		[15]
$TlTbTe_2$	7.85	17.52	4.31	110°	7.37	7.35		[15]
$TlDyTe_2$	7.85	17.52	4.32	110°	7.41	7.52		[15]
$TlHoTe_2$	7.78	17.47	4.28	110°	7.58	7.62		[15]
$TlErTe_2$	7.75	17.46	4.20	110°	7.65	7.45		[15]
$TlYbTe_2$	7.71	17.42	4.27	110°	7.78	7.85		[15]
$TlYTe_2$	7.68	17.51	4.30	110°	6.70	6.75		[15]
TlR_5Te_8 Cubic lattice of the Th_3P_4 type								
$TlLa_5Te_8$	9.69				6.69	6.8		[16]

*References: [1] Rustamov et al. (1981a); [2] Roques et al. (1979); [3] Peters and Baglio (1972); [4] Eholie et al. (1971); [5] Aliyev (1980b); [6] Aliyev et al. (1976); [7] Donohue and Hanlon (1974); [8] Barnier and Guittard (1976); [9] Aliyev et al. (1987); [10] Aliyev et al. (1980); [11] Amirov et al. (1984); [12] Guseinov et al. (1986); [13] Kabré et al. (1972); [14] Kabré et al. (1974); [15] Gamidov and Amanov (1968); [16] Poltmann and Hahn (1971).

The liquidus of the system HoTe–InTe consists of three branches of primary crystallization: HoTe, $HoInTe_2$ and the InTe-base solid solutions. The breadth of the homogeneity range at 25°C is 5 mol.%. The compound of composition HoInTe is formed according to peritectic upon the reaction of the components 1:1 (Agayev et al. 1987),

$$L + HoTe \rightleftarrows HoInTe_2 .$$

The eutectic between $HoInTe_2$ and InTe has the parameters: 67 mol% and 447°C (fig. 10).

The compound of the same formula type is formed in the Tl–Er–Te system (Kabré et al. 1974) (table 13), i.e. $TlErTe_2$.

In the systems R–R′–X, R′ designates the rare earth metals Y and Sc.

Crystalline structures of ternary sulphides of rare earth elements may be conventionally classified in two groups. The first group contains ternary sulphides, isostructural binary sulphides of the structure types NaCl, Th_3P_4, α-Gd_2S_3, δ-Ho_2S_3, etc. As a rule, the ternary sulphides of the second group exhibit peculiar structural types (Eliseev et al. 1982).

In the α-Gd_2S_3, solid solutions crystallize in the α-R_2S_3–α-R'_2S_3 systems (Vovan and Guittard 1979, Carré et al. 1973). The δ-Ho_2S_3-type structure is typical of R, R′–Dy–Tm, Y. In both cases, the continuous series of solid solutions are formed. Solid solutions are observed in the systems R_5S_7–R'_5S_7 (Y_5S_7-type structure). Limited solid solutions are formed in the systems δ-R_2S_3–ε-R'_2S_3 (ε-R'_2S_3 is a distorted structure of corundum: R′ = Tm–Lu), α-R_2S_3-δ-R'_2S_3 (R = La–Sm; R′ = Ho–Tm, Y) (Carré et al. 1973). Thus, addition of R'_2S_3 sulphides to R_2S_3 results in the formation of solid solutions. The smaller the difference between the ionic radii of R and R′, the more readily the formation of solid solutions occurs. An increase in the radii of the rare earth elements results, at first, in the formation of limited solid solutions, and, then, in the appearance of new structure types, such as Er_3ScS_3, $CeTmS_3$ (Rodier and Laruelle 1973, Rodier et al. 1976), $NdYbS_3$ (Carré and Laruelle 1974), etc. (table 14).

The cubic lattice of the Th_3P_4-type is characteristic for many ternary compounds containing bivalent and trivalent cations with the general formula RR'_2S_4, where R = Sm, Eu, Yb, the R′ are trivalent. A series of ternary compounds EuR'_2S_4 (R′ = Ho–Yb, Y, Sc) are synthesized on the basis of bivalent europium. These compounds crystallize in the $CaFe_2O_4$-type structure (Vovan and Khodadad 1969, Hulliger and Vogt 1966).

In the La–Er–Se system, Guittard and Lozac'h (1971) identified a region of existence of polyselenide $La_nEr_{1-n}Se_{1.9}$ ($0 < n < 0.401$), a region of solid solutions based on the NaCl-type structure, a region of solid solutions based on the Th_3P_4-type structure and two compounds: $ErLaSe_3$, with unknown structure, and $LaEr_4Se_7$ with the Y_5S_7-type structure.

The compounds with composition EuR_2Se_4 crystallize in two structure types: Th_3P_4 (R = La–Dy) and $CaFe_2O_4$ (R = Dy–Lu, Y) (Hulliger and Vogt 1966, Souleau and Guittard 1968). The compound $EuDy_2Se_4$ has two polymorphic modifications (Kuz'michyeva et al. 1981) (table 15).

TABLE 14
Crystallographic data for compounds in the R–R′–S systems.

Compound	a (Å)	b (Å)	c (Å)	β	γ	Density (g/cm^3) D_x	Density (g/cm^3) D_m	Note	Ref.*
RR′S$_3$ Orthorhombic lattice of the NdYbS$_3$ type									
NdYbS$_3$	12.545	9.438	3.853			6.42	6.32	The structure was determined on a single crystal	[1]
CeYbS$_3$	12.65	9.46	3.90			–	–		[2]
CeLuS$_3$	12.56	9.46	3.88			–	–		[2]
PrYbS$_3$	12.56	9.45	3.87			–	–		[2]
PrLuS$_3$	12.59	9.47	3.79			–	–		[2]
NdLuS$_3$	12.53	9.42	3.84			–	–		[2]
β-LaYbS$_3$	12.604(5)	9.493(5)	3.914(3)			5.79	–	The structure was determined on a single crystal; low-temperature modification	[3]
RR′S$_3$ Monoclinic lattice of the CeTmS$_3$ type									
CeTmS$_3$	11.09(2)	32.53(3)	3.98(1)		102.95(2)°	5.80	5.69	The structure was determined on a single crystal (ref. [4])	[2]
DyCeS$_3$	11.14	21.46	3.99		103.60°				[2]
YLaS$_3$	11.09	21.36	3.98		102.74°				[2]
HoLaS$_3$	11.12	21.32	3.99		103.88°				[2]
HoCeS$_3$	11.08	21.33	3.98		103.39°				[2]
ErLaS$_3$	11.10	21.20	3.99		103.94°				[2]
ErCeS$_3$	11.11	21.46	3.97		103.37°				[2]
ErPrS$_3$	11.09	21.42	3.98		102.95°				[2]
TmLaS$_3$	11.11	21.59	3.99		102.58°				[2]
TmPrS$_3$	11.06	21.33	3.96		102.95°				[2]
TmNdS$_3$	11.04	21.33	3.95		103.15°				[2]
RR′S$_3$ Orthorhombic lattice of the YScS$_3$ type									
α-LaYbS$_3$	7.06	6.30	9.61					α-LaYbS$_3$ single crystal: $a = 7.356(4)$, $b = 6.698(4)$, $c = 9.867(5)$, $D_x = 5.57$ (ref. [3]); high-temperature modification	[2]
LaLuS$_3$	7.04	6.27	9.61						[2]
YScS$_3$	7.00	6.36	9.46						[5]

TABLE 14 (cont'd)

Compound	a (Å)	b (Å)	c (Å)	β	γ	Density (g/cm^3) D_x	D_m	Note	Ref.*
RR'_2S_4 Cubic lattice of the Th_3P_4 type									
$CeLa_2S_4$	8.703								[6]
$CePr_2S_4$	8.570								[6]
$CeNd_2S_4$	8.555								[6]
$CeSm_2S_4$	8.621								[6]
$EuLa_2S_4$	8.745							Solid solution of the $La_{2.65}Eu_{0.35}S_4$ composition of the Th_3P_4 structure is known ref. [7] $a = 8.761$, $a = 8.739$ (ref. [8])	[6]
$EuCe_2S_4$	8.685							$a = 8.701$ (ref. [9]; $a = 8.679$ (ref. [8])	[6]
$EuNd_2S_4$	8.603							$a = 8.599$ (ref. [9]); $a = 8.607$ (ref. [8])	[6]
$EuSm_2S_4$	8.558							$a = 8.545$ (ref. [9]); $a = 8.551$ (ref. [8])	[6]
$EuGd_2S_4$	8.507							$a = 8.512$ (ref. [9]); $a = 8.500$ (ref. [8])	[6]
$EuPr_2S_4$	8.658							$a = 8.642$ (ref. [8]);	[6]
$EuTb_2S_4$	8.462								[6]
$EuDy_2S_4$	8.440								[6]
$SmLa_2S_4$	8.747								[6]
$SmCe_2S_4$	8.686								[6]
$SmPr_2S_4$	8.644								[6]
$SmNd_2S_4$	8.596								[6]
$SmGd_2S_4$	8.500								[6]
$SmTb_2S_4$	8.454								[6]
$SmDy_2S_4$	8.438								[6]
$SmYb_2S_4$	8.37(2)							1000–1300°C	[10]
$ErLa_2S_4$	8.690								[6]
$ErCe_2S_4$	8.590								[6]
$ErPr_2S_4$	8.530								[6]
$ErNd_2S_4$	8.490								[6]
$YbLa_2S_4$	8.651								[6]
$YbCe_2S_4$	8.589								[6]
$YbPr_2S_4$	8.549								[6]

$YbNd_2S_4$	8.512					[6]
$YbSm_2S_4$	8.465				$a = 8.43(2)$ (ref. [10]); 1000–1300°C	[6]
$YbGd_2S_4$	8.409					[6]
$YbTb_2S_4$	8.369					[6]
$YbDy_2S_4$	8.349					[6]
$YbEu_2S_4$	8.651		8.153		Tetragonal lattice, space group $I\bar{4}2d$ (ref. [11])	[6]
EuR_2S_4 Orthorhombic lattice of the $CaFe_2O_4$ type						
$EuTb_2S_4$	11.878	14.305	3.997			[12]
$EuDy_2S_4$	11.890	14.269	3.795			[12]
$EuHo_2S_4$	11.882	14.217	3.960		$a = 11.86$, $b = 14.23$, $c = 3.96$ (ref. [8])	[12]
$EuEr_2S_4$	11.874	14.172	3.943		$a = 14.151(3)$, $b = 11.861(3)$, $c = 3.939(1)$, $D_x = 6.18$, $D_m = 6.1$ space group Pbnm; the crystal structure was determined on a single crystal (ref. [13])	[12]
$EuYb_2S_4$	11.846	14.086	3.917			[6]
$EuLu_2S_4$	11.836	14.056	3.903			[6]
$EuSc_2S_4$	11.632	13.798	3.838			[12]
RR'_3S_6 Monoclinic lattice						
$ScEr_3S_6$	10.74(2)	11.04(2)	3.82(1)	108.2(2)°	The structure was determined on a single crystal: $Sc_{1.08}Er_{2.98}S_6$ (ref. [15])	[14]
$GdLu_3S_6$	10.915(5)	11.180(5)	3.902(2)	108.73(5)°	The structure was determined on a single crystal	[14]
$LaHo_3S_6$	11.00	11.30	4.00	108.6°		[14]
$LaEr_3S_6$	10.95	11.26	3.98	108.6°		[14]
$LaTm_3S_6$	10.91	11.22	3.96	108.4°		[14]
$LaYb_3S_6$	10.87	11.19	3.94	108.4°		[14]
$LaLu_3S_6$	10.83	11.15	3.92	108.3°		[14]
LaY_3S_6	11.09	11.34	4.00	108.8°		[14]
$CeHo_3S_6$	10.98	11.27	3.98	108.7°		[14]
$CeEr_3S_6$	10.94	11.24	3.96	108.4°		[14]
$CeTm_3S_6$	10.90	11.20	3.94	108.5°		[14]

TABLE 14 (cont'd)

Compound	a (Å)	b (Å)	c (Å)	β	γ	Density (g/cm^3) D_x	D_m	Note	Ref.*
RR'_3S_6 (cont'd)									
$CeYb_3S_6$	10.86	11.16	3.92		108.6°			The structure was determined on a single crystal: $a = 10.93(1)$, $b = 11.20(1)$, $c = 3.940(5)$, $\gamma = 108.8(1)°$	[14]
$CeLu_3S_6$	10.82	11.12	3.90		108.3°				[14]
CeY_3S_6	11.02	11.31	4.00		108.7°				[14]
$PrHo_3S_6$	10.96	11.26	3.98		108.6°				[14]
$PrEr_3S_6$	10.92	11.22	3.96		108.4°				[14]
$PrTm_3S_6$	10.88	11.18	3.94		108.5°				[14]
$PrYb_3S_6$	10.84	11.14	3.92		108.4°				[14]
$PrLu_3S_6$	10.80	11.10	3.90		108.4°				[14]
PrY_3S_6	11.01	11.30	4.00		108.7°				[14]
$NdHo_3S_6$	10.95	11.25	3.97		108.6°				[14]
$NdEr_3S_6$	10.90	11.21	3.95		108.4°				[14]
$NdTm_3S_6$	10.86	11.17	3.92		108.4°				[14]
$NdYb_3S_6$	10.82	11.13	3.90		108.4°				[14]
$NdLu_3S_6$	10.78	11.10	3.87		109.3°				[14]
NdY_3S_6	10.99	11.28	4.00		108.7°				[14]
CeR_4S_7 Monoclinic lattice of the Y_5S_7 type									
$CeTb_4S_7$	12.893	3.767	11.655	105°14′					[6]
$CeDy_4S_7$	12.855	3.840	11.546	105°42′					[6]
$CeHo_4S_7$	12.845	3.831	11.635	105°					[6]
$CeEr_4S_7$	12.814	3.824	11.580	105°28′					[6]
$CeTm_4S_7$	12.777	3.813	11.558	105°18′					[6]
CeY_4S_7	12.827	3.840	11.768	105°22′					[6]
$Sc_2R_3S_7$ Monoclinic lattice of the Y_5S_7 type									
$Sc_2Gd_3S_7$	12.41	3.74	11.37	104.9°					[16]
$Sc_2Tb_3S_7$	12.37	3.72	11.31	104.8°					[16]
$Sc_2Dy_3S_7$	12.36	3.73	11.30	104.8°					[16]
$Sc_2Ho_3S_7$	12.33	3.72	11.28	104.7°					[16]

Compound							
$Sc_{1.8-2.1}$-$Er_{3.2-2.9}S_7$	12.29(1)	3.741(1)	11.26(1)	104.9(2)°			[16]
$R_4R'_{11}S_{22}$ Monoclinic lattice of the $Ce_4Lu_{11}S_{22}$ type							
$La_4Dy_{11}S_{22}$	38.86	11.29	3.93		91°10′		[17]
$La_4Ho_{11}S_{22}$	38.82	11.27	3.92		91°06′		[17]
$La_4Er_{11}S_{22}$	38.79	11.26	3.93		91°04′		[17]
$La_4Tm_{11}S_{22}$	38.76	11.24	3.92		91°03′		[17]
$La_4Y_{11}S_{22}$	38.85	11.28	3.93		91°09′		[17]
$Ce_4Tm_{11}S_{22}$	38.74	11.24	3.92		91°30′		[17]
$Ce_4Lu_{11}S_{22}$	39.60	11.20	3.91		91°30′	$a = 38.65$, $b = 11.23$, $c = 3.91$, $\gamma = 91°04'$	[17]
$Pr_4Ho_{11}S_{22}$	38.65	11.26	3.90		91°01′		[17]
$Pr_4Er_{11}S_{22}$	38.62	11.23	3.91		90°60′		[17]
$Pr_4Tm_{11}S_{22}$	38.62	11.22	3.90		91°04′		[17]
$Pr_4Lu_{11}S_{22}$	38.59	11.2	3.89		91°20′		[17]
$Pr_4Y_{11}S_{22}$	38.67	11.28	3.91		89.09′		[17]
$Nd_4Er_{11}S_{22}$	38.62	11.26	3.89		91°01′		[17]
$Nd_4Tm_{11}S_{22}$	38.59	11.25	3.90		91°20′		[17]
$Nd_4Er_{11}S_{22}$	38.55	11.20	3.90		91°50′		[17]
$Sm_4Er_{11}S_{22}$	38.60	11.24	3.91		91°25′		[17]
$Sm_4Tm_{11}S_{22}$	38.60	11.21	3.90		91°20′	The cell parameters were determined on a single crystal	[17]
$Sm_4Lu_{11}S_{22}$	38.48	11.18	3.88		91°18′		[17]
$R_5La_5S_{14}$ Monoclinic lattice							
$Er_5La_5S_{14}$	29.8(4)	21.9(3)	3.95(4)		122(1)°	$a = 29.71$, $b = 21.83$, $c = 3.94$, $\gamma = 122°$; compound formula $La_{10}Er_9S_{27}$ (ref. [18])	[6]
YbR_2S_4 Orthorhombic lattice of the Yb_3S_4 type							
Yb_3S_4	12.82	12.95	3.82				[19]
$YbTm_2S_4$	12.85	13.00	3.86				[19]
$YbEr_2S_4$	13.00	13.2	3.90				[19]

*References: [1] Carré and Laruelle (1974); [2] Vovan and Khodadad (1971); [3] Rodier and Julien-Pouzol (1983); [4] Rodier (1973); [5] Rodier et al. (1969); [6] Vovan and Khodadad (1969); [7] Amano et al. (1987); [8] Lugscheider et al. (1970); [9] Sato et al. (1980); [10] Lobachevskaya et al. (1987); [11] Eliseev et al. (1980); [12] Hulliger and Vogt (1966); [13] Lemoine et al. (1985); [14] Rodier et al. (1976); [15] Rodier and Laruelle (1973); [16] Rodier and Laruelle (1972); [17] Vovan and Guittard (1979); [18] Carré et al. (1973); [19] Ring and Tecotzky (1964).

TABLE 15
Crystallographic data for compounds in the R–R′–Se systems.

Compound	a (Å)	b (Å)	c (A)	β	Density (g/cm³) D_x	Note	Ref.*
EuR_2Se_4 Cubic lattice of the Th_3P_4 type							
$EuLa_2Se_4$	9.086(5)				6.60		[1]
$EuCe_2Se_4$	9.025(5)				6.77		[1]
$EuPr_2Se_4$	8.987(5)				6.85		[1]
$EuNd_2Se_4$	8.950(5)				7.00		[1]
$EuSm_2Se_4$	8.900(5)				7.24		[1]
$EuDy_2Se_4$	9.2104(1)				–	Cell parameter was determined on a single crystal by Bond's method (ref. [2])	[1]
EuR_2Se_4 Orthorhombic lattice of the $CaFe_2O_4$ type							
$EuTb_2Se_4$	12.452	14.917	4.143				[3]
$EuDy_2Se_4$	12.447	14.865	4.122			$a = 12.50(1)$, $b = 14.78(1)$, $c = 4.17(1)$, $D_x = 6.80$ (ref. [1])	[3]
$EuHo_2Se_4$	12.445	14.826	4.099		–	$a = 12.53(1)$, $b = 14.82(1)$, $c = 4.15(1)$, $D_x = 6.86$ (ref. [1])	[3]
$EuEr_2Se_4$	12.443	14.785	4.092		–	Single-crystal data: $a = 12.49(1)$, $b = 14.82(1)$, $c = 4.15(1)$, $D_x = 6.93$, $D_m = 6.98$ (ref. [1])	[3]
$EuTm_2Se_4$	12.46(1)	14.85(1)	4.13(1)		7.00		[1]
$EuYb_2Se_4$	12.393	14.683	4.068		–	$a = 12.44(1)$, $b = 14.82(1)$, $c = 4.12(1)$, $D_x = 7.11$ (ref. [1])	[3]
$EuLu_2Se_4$	12.378	14.643	4.048		–	$a = 12.40(1)$, $b = 14.82(1)$, $c = 4.12(1)$, $D_x = 7.17$ (ref. [1])	[3]
$EuSc_2Se_4$	12.128	14.243	3.932		–		[3]
β-YbR_2Se_4 Orthorhombic lattice of the Yb_3S_4 type							
$YbHo_2Se_4$	13.2	13.5	4.0				[1]
$YbEr_2Se_4$	13.1	13.5	4.0				[1]
$YbTm_2Se_4$	13.0	13.4	4.0				[1]
YbY_2Se_4	13.5	13.7	4.0				[1]
α-YbR_2Se_4 Rhombohedral lattice of the $CaHo_2Se_4$ type (cell parameters are given for hexagonal axes)							
$YbHo_2Se_4$	4.09(1)		20.33(2)				[1]

$YbEr_2Se_4$	4.07(1)		20.37(2)		[1]
$YbTm_2Se_4$	4.06(1)		20.41(2)		[1]
$YbLu_2Se_4$	4.03(1)		20.47(2)		[1]
YbY_2Se_4	4.12(1)		20.40(2)		[1]
$RE r_4Se_7$ Monoclinic lattice of the Y_5S_7 type					
$LaEr_4Se_7$	13.06	3.97	12.00	105.33°	[4]

*References: [1] Souleau and Guittard (1968); [2] Kuz'michyeva et al. (1981); [3] Hulliger and Vogt (1966); [4] Guittard and Lozac'h (1971).

TABLE 16
Crystallographic data for compounds in the R–R′–Te systems.

Compound	a (Å)	c (Å)	Note	Ref.*
EuR_2Te_4 Rhombohedral lattice of the $EuEr_2Te_4$ type (the cell parameters are given for hexagonal axes)				
$EuSm_2Te_4$	7.727	11.900	Single crystal	[1]
$EuGd_2Te_4$	7.720	11.888	Single crystal	[1]
$EuHo_2Te_4$	7.712	11.872	Single crystal	[1]
$EuEr_2Te_4$	7.710	11.870	Single crystal	[1]
$EuYb_2Te_4$	7.706	11.862	Single crystal	[1]

*Reference: [1] Sadovskaya et al. (1985).

According to Souleau et al. (1969), the compounds of composition YbR_2Se_4 ($R = Ho–Lu, Y$) have two modifications: $CaHo_2Se_4$ and Yb_3S_4.

Ternary tellurides of rare earth elements crystallize in structure types which are derivatives of the NaCl-, Sc_2S_3- and $EuEr_2Te_4$-type structure, as well as in the Th_3P_4-type structure (Eliseev et al. 1980, Sadovskaya et al. 1985) (table 16).

References

Agayev, A., P. Rustamov and Z. Akhmedova, 1987, Zh. Neorg. Khim. **32**, 1030.

Aliyev, O., 1980a, Izv. Akad. Nauk SSSR, Ser. Neorg. Mater. **16**, 616.

Aliyev, O., 1980b, Izv. Akad. Nauk SSSR, Ser. Neorg. Mater. **16**, 1514.

Aliyev, O., T. Kurbanov, P. Rustamov, M. Alidzhanov and S. Salmanov, 1976, Izv. Akad. Nauk SSSR, Ser. Neorg. Mater. **12**, 1944.

Aliyev, O., P. Rustamov, D. Zul'fugarly and A. Agayev, 1980, Azerb. Khim. Zh. (3), 137.

Aliyev, O., P. Rustamov and T. Il'yasov, 1984, Zh. Neorg. Khim. **29**, 1280.

Aliyev, O., O. Aliyeva and P. Rustamov, 1987, Izv. Akad. Nauk SSSR, Ser. Neorg. Mater. **23**, 22.

Aliyeva, O., and O. Aliyev, 1986, Izv. Akad. Nauk SSSR, Ser. Neorg. Mater. **22**, 577.

Aliyeva, O., O. Aliyev and P. Rustamov, 1987, Neorg. Khim. **32**, 252.

Amano, T., B. Beaudry and K. Gschneidner, 1987, J. Less-Common. Met. **127**, 270.

Amirov, A., A. Shnulin, G. Guseinov and Kh. Mamedov, 1984, Kristallografiya **29**, 787.

Barnier, S., and M. Guittard, 1976, C.R. Acad. Sci., Paris, Ser. C **282**, 461.

Carré, D., 1977, Acta Crystallogr. B **33**, 1163.

Carré, D., and P. Laruelle, 1974, Acta Crystallogr. B **30**, 952.

Carré, D., M. Julien-Pouzol, M. Guittard, F. Chopin and J. Flahaut, 1971, C.R. Acad. Sci., Paris, Ser. C **272**, 1721.

Carré, D., J. Flahaut, P. Khodadad, P. Laruelle, N. Rodier and T. Vovan, 1973, J. Solid State Chem. **7**, 321.

Carré, D., M. Guittard and C. Adolphe, 1978, Acta Crystallogr. B **34**, 3499.

Domange, L., J. Flahaut and M. Patrie, 1960, Bull. Soc. Chim. Fr. (1), 229.

Donohue, P., and J. Hanlon, 1974, J. Electrochem. Soc. **121**, 137.

Efendiyev, G., Z. Karayev and I. Nasibov, 1964a, Azerb. Khim. Zh. (5), 103.

Efendiyev, G., Z. Karayev and I. Nasibov, 1964b, Azerb. Khim. Zh. (5), 111.

Eholie, R., O. Gorochov, M. Guittard, A. Mazurier and J. Flahaut, 1971, Bull. Soc. Chim. Fr. (5), 747.

Eliseev, A., G. Kuz'michyeva and A. Belostsockii, 1980, Zh. Neorg. Khim. **25**, 895.

Eliseev, A., O. Sadovskaya and G. Kuz'michyeva, 1982, Izv. Akad. Nauk SSSR, Ser. Neorg. Mater. **18**, 1672.

Gamidov, P., and A. Amanov, 1968, Uch. Zap. Azerb. Univ. Ser. Fiz. Mat. Nauk (6), 76.

Godzhayev, E., G. Zeinalov and S. Mamedov, 1986, Izv. Akad. Nauk SSSR, Ser. Neorg. Mater. **22**, 1907.

Guittard, M., and A.-M. Lozac'h, 1971, Bull. Soc. Chim. Fr. (3), 751.

Guittard, M., D. Carré and T. Kabré, 1978, Mater. Res. Bull. **13**, 279.

Guseinov, G., F. Mamedov, A. Shnulin and Kh. Mamedov, 1979, Dokl. Akad. Nauk SSSR **246**, 1360.

Guseinov, G., V. Aliyev and A. Nadzhafov, 1981, Izv. Akad. Nauk SSSR, Ser. Neorg. Khim. **17**, 802.

Guseinov, G., G. Iskenderov, A. Nadzhafov and E. Kerimova, 1985, Izv. Akad. Nauk SSSR, Ser. Neorg. Khim. **21**, 917.

Guseinov, G., O. Aliyev, A. Agayev and P. Rustamov, 1986, Mater. Res. Bull. **21**, 1155.

Guseinov, G., E. Kerimova, D. Agamaliyev and D. Nadzhafov, 1987, Izv. Akad. Nauk SSSR, Ser. Neorg. Mater. **23**, 1632.

Hulliger, F., and O. Vogt, 1966, Phys. Lett. B **21**, 2138.

Kabré, S., M. Julien-Pouzol and M. Guittard, 1972, C.R. Acad. Sci., Paris, Ser. C **275**, 1367.

Kabré, S., M. Julien-Pouzol and M. Guittard, 1974, Bull. Soc. Chim. Fr. (9/10), 1881.

Karayev, Z., A. Kadymov and N. Murguzov, 1965, Azerb. Khim. Zh. (4), 110.

Karayev, Z., L. Keiserukhskaya, Sh. Aliyeva and A. Gadymov, 1966a, Azerb. Khim. Zh. (1), 112.

Karayev, Z., I. Nasibov and Sh. Aliyeva, 1966b, Izv. Akad. Nauk SSSR, Ser. Neorg. Mater. **11**, 1322.

Keiserukhskaya, L., N. Luzhnaya and Z. Karayev, 1970, Izv. Akad. Nauk SSSR, Ser. Neorg. Mater. **6**, 1869.

Kuz'michyeva, G., A. Eliseev, I. Palyanichko, I. Perepelkin and O. Sadovskaya, 1981, Zh. Neorg. Khim. **26**, 2258.

Lemoine, P., D. Carré and M. Guittard, 1985, Acta Crystallogr. C **41**, 667.

Lobachevskaya, N., V. Bamburov, L. Finkel'shtein and N. Efremova, 1987, Zh. Neorg. Khim. **32**, 1255.

Loireau-Lozac'h, A.-M., M. Guittard and J. Flahaut, 1976, Mater. Res. Bull. **11**, 1489.

Loireau-Lozac'h, A.-M., M. Guittard and J. Flahaut, 1977, Mater. Res. Bull. **12**, 881.

Lozac'h, A.-M., and M. Guittard, 1973, Bull. Soc. Chim. Fr. (1), 6.

Lozac'h, A.-M., S. Jaulmes and M. Guittard, 1971, C.R. Acad. Sci., Paris, Ser. C **272**, 1123.

Lozac'h, A.-M., M. Guittard and J. Flahaut, 1972, C.R. Acad. Sci., Paris, Ser. C **275**, 809.

Lozac'h, A.-M., M. Guittard and J. Flahaut, 1973, Mater. Res. Bull. **8**, 75.

Lugscheider, W., H. Pink, K. Weber and W. Zinn, 1970, Z. Angew. Phys. B **30**, 36.

Messain, D., D. Carré and P. Laruelle, 1977, Acta Crystallogr. B **33**, 2540.

Nadzhafov, A., P. Rustamov, T. Kurbanov, O. Aliyev and M. Ali-Zade, 1978, Izv. Akad. Nauk SSSR, Ser. Neorg. Mater. **14**, 645.

Nadzhafov, A., O. Aliyev and P. Rustamov, 1980, Izv. Akad. Nauk SSSR, Ser. Neorg. Mater. **16**, 605.

Nasibov, I., and Z. Karayev, 1963, Azerb. Khim. Zh. (5), 105.

Patrie, M., and M. Guittard, 1969, C.R. Acad. Sci., Paris, Ser. C **268**, 1136.

Peters, T., and S. Baglio, 1972, J. Electrochem. Soc. **119**, 230.

Poltmann, F., and H. Hahn, 1971, Naturwissenschaften **58**, 544.

Ring, S., and M. Tecotzky, 1964, Inorg. Chem. **3**, 1654.

Rodier, N., 1973, Bull. Soc. Fr. Mineral. Cristallogr. **96**, 350.

Rodier, N., and R. Julien-Pouzol, 1983, Acta Crystallogr. C **39**, 670.

Rodier, N., and P. Laruelle, 1972, Bull. Soc. Fr. Mineral. Cristallogr. **95**, 552.

Rodier, N., and P. Laruelle, 1973, Bull. Soc. Fr. Mineral. Cristallogr. **96**, 30.

Rodier, N., P. Laruelle and J. Flahaut, 1969, C.R. Acad. Sci., Paris, Ser. C **269**, 1391.

Rodier, N., T. Vovan and M. Guittard, 1976, Mater. Res. Bull. **11**, 1209.

Roques, R., R. Rimet, J. Declercq and G. Germain, 1979, Acta Crystallogr. B **35**, 555.

Rustamov, P., O. Aliyev, T. Kurbanov, M. Ali-Zade and A. Nadzhafov, 1977, Izv. Akad. Nauk SSSR, Ser. Neorg. Mater. **13**, 1748.

Rustamov, P., O. Aliyev and T. Kurbanov, 1981a, V kn: Troiinyye Khal'kogenidy redkozemel'nykh Elementovc. Izd. "Elm", Baku.
Rustamov, P., O. Aliyev and E. Miskarli, 1981b, Zh. Neorg. Khim. **26**, 487.
Rustamov, P., O. Aliyev and T. Il'yasov, 1984, Izv. Akad. Nauk SSSR, Ser. Neorg. Mater. **20**, 1801.
Rustamov, P., A. Agayev and Kh. Musayeva, 1986, Dokl. Akad. Nauk Azerb. SSR **46**, 40.
Sadovskaya, O., G. Kuz'michyeva, V. Tolstova, V. Ponomaryev and A. Eliseev, 1985, Zh. Neorg. Khim. **30**, 2512.
Saint-Giniez, D., P. Laruelle and J. Flahaut, 1968, C.R. Acad. Sci., Paris, Ser. C **267**, 1029.
Sallavuard, G., and R.-A. Paris, 1971, C.R. Acad. Sci., Paris, Ser. C **273**, 1428.
Sato, M., G.-Y. Adachi and J. Shiokawa, 1980, J. Solid State Chem. **31**, 337.
Souleau, M., and M. Guittard, 1968, Bull. Soc. Chim. Fr. (9), 3962.
Souleau, M., M. Guittard and P. Laruelle, 1969, Bull. Soc. Chim. Fr. (1), 9.
Vovan, T., and M. Guittard, 1979, Mater. Res. Bull. **14**, 597.
Vovan, T., and P. Khodadad, 1969, Bull. Soc. Chim. Fr. (1), 30.
Vovan, T., and P. Khodadad, 1971, Bull. Soc. Chim. Fr. (10), 3454.

4. The R–M^{IV}–X systems

In the R–M^{IV}–X systems, M^{IV} designates silicon, germanium, tin, lead, zirconium and hafnium.

Compounds of the following compositions are formed in the R_2S_3–SiS_2 systems: R_2SiS_5 (R = La–Nd) (Michelet et al. 1970), $R_4Si_3S_{12}$ (R = Ce–Gd) (Perez and Duale 1969) and $R_6Si_{2.5}S_{14}$ (R = Gd–Dy, Y) (Perez and Duale 1969, Michelet and Flahaut 1969).

Two congruently melting compounds, $La_2Ge_2S_7$ (1071°C) and La_2GeS_5 (1105°C), are found in the La_2S_3–GeS_2 system (Sarkisov et al. 1968, 1970). $La_2Ge_2S_7$ displays a polymorphoric transition, $\beta \rightarrow \alpha$, at 845°C. The eutectic between GeS_2 and $La_2Ge_2S_7$ has the parameters: 7.5 mol.% La_2S_3 and 756°C. There are two more eutectics: between $La_2Ge_2S_7$ and La_2GeS_5 (at 40 mol.% La_2S_3 and 1020°C), between La_2GeS_5 and La_2S_3 (at 57 mol.% La_2S_3 and 1075°C) (fig. 11).

The compound of composition $La_4Ge_3S_{12}$ (Mazurier and Etienne 1974) was not observed in this system.

Two compounds, Sm_2GeS_5 and $Sm_2Ge_2S_7$, were observed in the Sm_2S_3–GeS_2 system with components ratios of 1:1 and 1:3, respectively (Beskrovnaya et al. 1971b, Beskrovnaya and Serebrinnikov 1971).

The compounds Eu_2GeS_4 and $EuGeS_3$ decompose in the EuS–GeS_2 system at 1160 and 786° (Barnier and Guittard 1978). The eutectic parameters are: 75 mol.% EuS and 615°C.

The compound of composition $Gd_4Ge_3S_{12}$ is stable at temperatures up to 1030°C. At 1050°C, it decomposes according to the reaction

$$Gd_4Ge_3S_{12} \rightarrow Gd_{12}Ge_5S_{28} + GeS_2 \qquad \text{(Stepanez et al. 1973a–c)}.$$

At a 1:1 ratio of the components, the R_2S_3–SnS_2 systems produce compounds with the R_2SnS_5 composition with the La_2SnS_5-type structure and the Sm_2SnS_5 composition (Mel'chenko and Serebrennikov 1973, Jaulmes 1974, Julien-Pouzol and Jaulmes 1979).

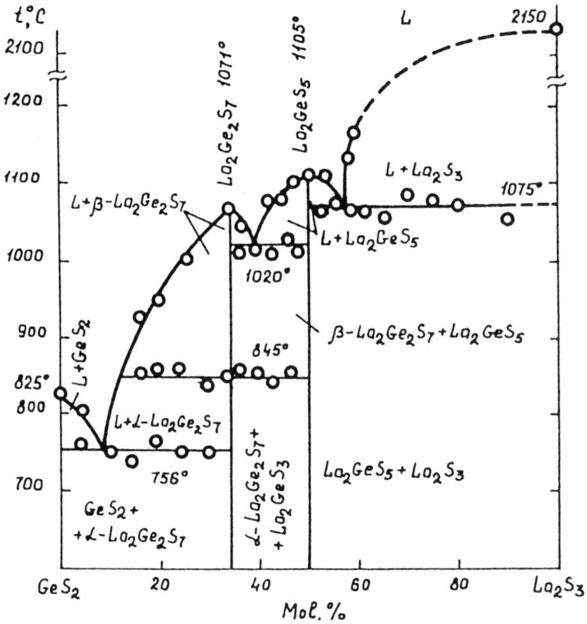

Fig. 11. The phase diagram of the La_2S_3–GeS_2 system.

For a components ratio of 1:1, compounds of composition PbR_2S_4 with the $CaFe_2O_4$-type structure (R = Ho–Lu) and the Th_3P_4-type structure (R = La–Gd) (Patrie et al. 1969, Vovan et al. 1966) are formed in the R_2S_3–PbS systems (table 17).

Numerous authors have studied binary sections of the related ternary systems of the R_2Se_2–$GeSe_2$ type (R = La, Ce, Pr, Nd, Sm, Eu, Gd, Tb, Ho, Yb) (Rustamov et al. 1972, 1973, Aliyeva et al. 1972, Nasibov et al. 1981b–e, Nasibov 1982). Ternary compounds with compositions $R_2Ge_2Se_7$, $Re_2Ge_3Se_9$, R_2GeSe_5 were observed in the systems of the cerium subgroup. In the system with lanthanum, La_2GeSe_5 melts congruently at 1030°C, $La_2Ga_2Se_7$ and $La_2Ge_3Se_9$ were obtained at 925 and 760°C according to the reactions

$$\beta\text{-}La_2GeSe_5 + L \rightleftarrows \beta\text{-}La_2Ge_2Se_7 ,$$

$$\beta\text{-}La_2Ge_2Se_7 + L \rightleftarrows La_2Ge_3Se_9 .$$

Polymorphic transformations were observed at 685 and 605°C for the compounds La_2GeSe_5 and $La_2Ge_2Se_7$, respectively. The eutectic β-La_2GeSe_5 + La_2Se_3 melts at 875°C, while $La_2Ge_3Se_9$ (3 mol.%) + $GeSe_2$ melts at 680°C (Rustamov et al. 1973).

In the system with cerium, the compounds Ce_2GeSe_5 and $Ce_2Ge_2Se_7$ melt congruently at 1075 and 1020°C, the compound $Ce_2Ge_3Se_9$ melts peritectically at 725°C (see fig. 12):

$$\alpha\text{-}Ce_2Ge_2Se_7 + L \rightleftarrows Ce_2Ge_3Se_9 .$$

TABLE 17

Crystallographic data for compounds in the R–(Si, Ge, Sn, Pb)–S systems.

Compound	a (Å)	b (Å)	c (Å)	α	β	γ	Density (g/cm³) D_x	Density (g/cm³) D_m	Note	Ref.*
R_2SiS_5 Monoclinic lattice of the La_2GeS_5 type										
La_2SiS_5	7.857(4)	7.606(4)	12.627(8)			101.55(2)°	4.55			[1, 2]
Ce_2SiS_5	7.798(4)	7.540(4)	12.524(8)			101.60(2)°	–			[1, 2]
Pr_2SiS_5	7.775(4)	7.514(4)	12.489(8)			101.62(2)°	–			[1, 2]
Nd_2SiS_5	7.740(4)	7.480(4)	12.434(8)			101.66(2)°	–			[1, 2]
$R_6Si_{2.5}S_{14}$ Hexagonal lattice of the $Dy_6Ge_{2.5}S_{14}$ type										
$Gd_6Si_{2.5}S_{14}$	9.87		5.71				5.04			[2, 3]
$Tb_6Si_{2.5}S_{14}$	9.82		5.70				5.14			[2, 3]
$Dy_6Si_{2.5}S_{14}$	9.75		5.70				5.28			[2, 3]
$Y_6Si_{2.5}S_{14}$	9.75		5.70				3.73			[2, 3]
$RGeS_3$ Triclinic lattice (space group $P\bar{1}$)										
$EuGeS_3$	8.381	8.453	11.834	111.44°	99.90°	75.39°	4.24		According to ref. [5] in $EuGeS_3$ 100% Eu^{+2}. The crystal structure was determined on a single crystal (ref. [6]): $a = 8.468(3)$, $b = 11.76(1)$, $c = 8.389(7)$, $\alpha = 90.49(11)°$, $\beta = 104.56(4)°$, $\gamma = 69.52(4)°$, space group $P\bar{1}$	[4]
Yb_6GeS_3	8.163	8.395	11.59	110.54°	96.70°	72.12°	–			[7]
R_2GeS_4 Monoclinic lattice of the Sr_2GeS_4 type										
Eu_2GeS_4	6.70	8.22	6.67			108.1°	4.8	–		[4]
Er_2GeS_4	–	–	–			–	6.46	6.39	According to the data of ref. [8], it crystallizes in hexagonal structure	[8]
Yb_2GeS_4	6.632	8.128	6.593			107.9°	–	–		[7]
$R_6Ge_3S_{14}$ Hexagonal lattice of the $Dy_6Ge_{2.5}S_{14}$ type. (In refs. [8, 9] the R_2GeS_5 composition is assigned to these compounds.) Compounds supposedly have the homogeneity region from $R_6(Ge^{+2}\square)Ge_2S_{14}$ to $R_6(Ge^{+4}_{0.5}\square_{1.5})Ge_2S_{14}$										
$La_6Ge_3S_{14}$	10.38		5.81				4.60	–		[1, 3]
$Ce_6Ge_3S_{14}$	10.30		5.80				4.70	–	For $Ce_6Ge_{2.5}S_{14}$ composition: $a = 10.22$, $c = 5.82$, $D_x = 4,63$	[1, 3]

$Pr_6Ge_3S_{14}$	10.19		5.76		4.85		For the $Pr_6Ge_{2.5}S_{14}$ composition: $a = 10.13$, $c = 5.82$, $D_x = 4.74$	[1, 3]
$Nd_6Ge_3S_{14}$	10.18		5.76		4.91		For the $Nd_6Ge_{2.5}S_{14}$ composition: $a = 10.05$, $c = 5.82$, $D_x = 4.89$	[1, 3]
$Sm_6Ge_3S_{14}$	10.03		5.74		–	–	For the $Sm_6Ge_{2.5}S_{14}$ composition: $a = 9.94$, $c = 5.82$, $D_x = 5.12$	[1, 3]
$Gd_6Ge_{2.5}S_{14}$	9.84		5.82		5.36	–		[1, 3]
$Tb_6Ge_{2.5}S_{14}$	9.79		5.82		5.45	–		[1, 3]
$Dy_6Ge_{2.5}S_{14}$	9.73		5.82		5.59	5.54	The crystal structure of $Dy_6(Ge_{0.5}\square_{1.5})Ge_2S_{14}$ (refs. [1, 10]) was determined	[1, 3]
$Ho_6Ge_{2.5}S_{14}$	9.69		5.82		5.68			[10]
$Y_6Ge_{2.5}S_{14}$	9.73		5.82		4.04			[10]
R_2SnS_5 Orthorhombic lattice of the La_2SnS_5 type								
La_2SnS_5	11.26(5)	7.89(3)	3.99(1)				The structure was determined on a single crystal: $a = 11.22$, $b = 7.915$, $c = 3.96$ (ref. [12])	[11]
Ce_2SnS_5	11.24	7.86	3.95					[1]
Pr_2SnS_5	11.17	7.83	3.93					[1]
Nd_2SnS_5	11.15	7.78	3.93					[1]
R_2SnS_5 Orthorhombic lattice of the Sm_2SnS_5 type								
Sm_2SnS_5	11.18(5)	7.76(3)	3.90(1)				The structure was determined on a single crystal	[11]
Gd_2SnS_5	11.16	7.75	3.88					[11]
Tb_2SnS_5	11.15	7.75	3.87					[11]
Dy_2SnS_5	11.14	7.75	3.86					[11]
R_2GeS_5 Monoclinic lattice of the La_2GeS_5 type								
La_2GeS_5	7.893(2)	7.641(2)	12.702(3)	101.39°	4.55	4.45	The structure was determined on a single crystal: $a = 7.887(4)$, $b = 7.675(4)$, $c = 12.720(8)$, $\gamma = 101.40°$, $D_x = 4.55$ (ref. [1])	[13]

TABLE 17 (cont'd)

Compound	a (Å)	b (Å)	c (Å)	α	β	γ	Density (g/cm^3) D_x	D_m	Note	Ref.*
$R_4Ge_3S_{12}$ Rhombohedral lattice (space group R3c) (cell parameters are given for hexagonal axes)										
$La_4Ge_3S_{12}$	19.40		8.10				4.37	4.29	The structure was determined on a single crystal (ref. [15])	[1, 14]
$Ce_4Ge_3S_{12}$	19.36		8.06				4.43	–		[1, 14]
$Pr_4Ge_3S_{12}$	19.30		8.01				4.50	–		[1, 14]
$Nd_4Ge_3S_{12}$	19.24		7.98				4.59	–		[1, 14]
$Sm_4Ge_3S_{12}$	19.19		7.95				4.73	–		[1, 14]
$Gd_4Ge_3S_{12}$	19.09		7.90				4.72	–		[1, 14]
$R_xSn_yS_z$ Orthorhombic lattice										
Yb_2SnS_4	15.20	8.80	11.25				4.76	–	Space group $Pma2_1$	[7]
$Yb_4Sn_2S_9$	4.06	15.60	11.28				5.62	–	Space group Pmcb	[7]
$Yb_3Sn_2S_7$	11.50	12.58	3.95				5.25	–	Space group Pbam	[7]
$Eu_3Sn_2S_7$	11.542(4)	12.690(5)	3.974(1)				5.2	5.2	The structure was determined on a single crystal	[12]
$R_5Sn_3S_{12}$ Orthorhombic lattice (space group $Pm2_1b$)										
$Eu_5Sn_3S_{12}$	3.924(1)	11.509(1)	20.219(1)				5.4	5.2	The structure was determined on a single crystal: $Eu_3^{+2}Eu_2^{+3}Sn_3S_{12}$	[12]
$Yb_5Sn_3S_{12}$	3.90	11.46	20.18				–	5.36		[7]
PbR_2S_4 Cubic lattice of the Th_3P_4 type										
$PbLa_2S_4$	8.767									[16]
$PbCe_2S_4$	8.705									[16]
$PbPr_2S_4$	8.675									[16]
$PbNd_2S_4$	8.632								$a = 8.600$ at 800°C	[16]
$PbSm_2S_4$	8.572								$a = 8.520$ at 720°C	[16]
$PbGd_2S_4$	8.522								$a = 8.485$ at 720°C	[16]
PbR_2S_4 Orthorhombic lattice of the $CaFe_2O_4$ type										
$PbHo_2S_4$	11.89	14.25	4.01							[16]
$PbEr_2S_4$	11.85	14.17	4.00							[16]

$PbTm_2S_4$	11.83	14.10	3.98	[16]
$PbYb_2S_4$	11.78	14.08	3.96	[16]
$PbLu_2S_4$	11.78	14.07	3.96	[16]

*References: [1] Michelet et al. (1975); [2] Michelet et al. (1970); [3] Michelet and Flahaut (1969); [4] Barnier and Guittard [1978]; [5] Finkel'shtein and Samsonova (1987); [6] Bugli et al. (1978); [7] Rustamov et al. (1981a); [8] Gus'kova and Serebrennikov (1973a, b); [9] Stepanez et al. (1973b); [10] Collin et al. (1973); [11] Julien-Pouzol and Jaulmes (1979); [12] Jaulmes and Julien-Pouzol (1977a, b); [13] Mazurier and Etienne (1973); [14] Michelet et al. (1966); [15] Mazurier and Etienne (1974); [16] Patrie et al. (1969).

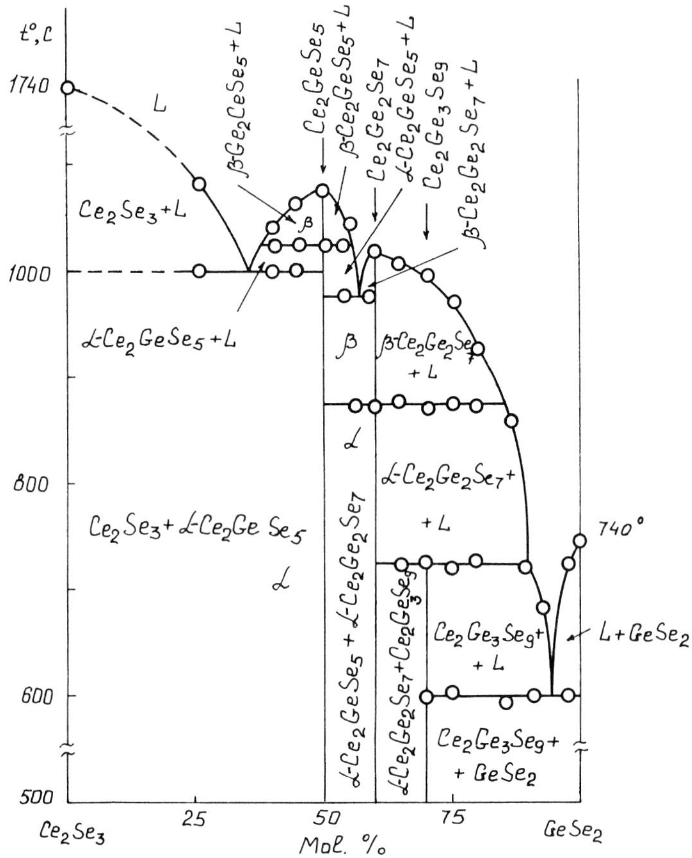

Fig. 12. The phase diagram of the Ce_2Se_3–$GeSe_2$ system.

The compounds Ce_2GeSe_5 and $Ce_2Ge_2Se_7$ undergo polymorphic transition at 1025 and 875°C, respectively. The eutectic α-Ce_2GeSe_5 + Ce_2Se_3 melts at 1000°C. The eutectic parameters between $Ce_2Ge_3Se_9$ and $GeSe_2$ are 3 mol.% Ce_2Se_3 and 600°C.

In the same system with praseodymium, the compound $Pr_2Ge_3Se_9$ melts peritectically at 690°C,

$$\alpha\text{-}Pr_2Ge_2Se_7 + L \rightleftarrows Pr_2Ge_3Se_9 \, .$$

The compounds Pr_2GeSe_5 and $Pr_2Ge_2Se_7$ melt congruently at 1020 and 990°C, undergo polymorphic transition at 880 and 830°C, respectively. The eutectic between β-Pr_2GeSe_5 and Pr_2Se_3 melts at 1000°C, while between β-Pr_2GeSe_5 and $Pr_2Ge_2Se_7$ the eutectic melts at 960°C. The eutectic parameters between $Pr_2Ge_3Se_9$ and $GeSe_2$ are: 3 mol.% Pr_2Se_3 and 550°C (Beskrovnaya et al. 1971a).

The Ho_2Se_3–$GeSe_2$ system (Nasibov 1982), which is shown in fig. 13, contains the compounds Ho_4GeSe_8, Ho_2GeSe_5 and $Ho_2Ge_2Se_7$ which melt congruently at

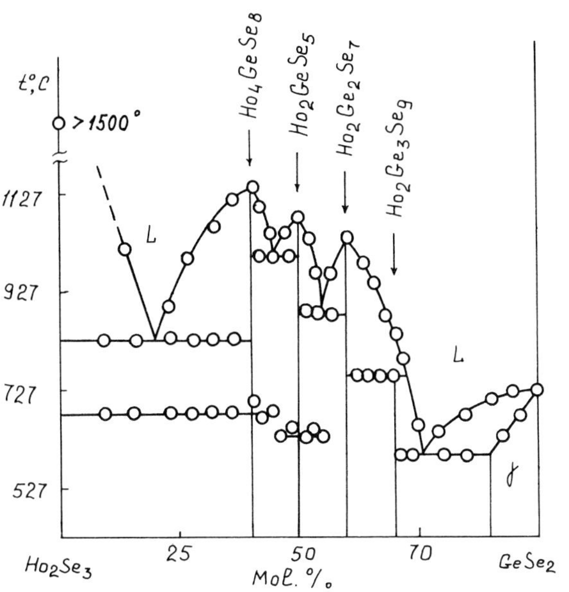

Fig. 13. The phase diagram of the Ho_2Se_3–$GeSe_2$ system.

1157, 1057 and 997°C, respectively. The peritectic compound $Ho_2Ge_3Se_9$ decomposes at 767°C. The compounds Ho_4GeSe_8 and Ho_2GeSe_5 undergo polymorphic transformations at 697 and 648°C, respectively. There are four eutectic points in the system: Ho_2Se_3–Ho_4GeSe_8 (847°C), Ho_4Se_8–Ho_2GeSe_5 (997°C), Ho_2GeSe_5–$Ho_2Ge_2Se_7$ (897°C), $Ho_2Ge_3Se_9$–$GeSe_2$ (597°C). A narrow range of solid solutions is observed on the basis of $GeSe_2$.

The R_2Se_3–$GeSe_2$ systems with R = Gd and Tb are analogous to the Ho_2Se_3–$GeSe_2$ system. There are two incongruently melting compounds in the YbSe–$GeSe_2$ system (Rustamov et al. 1981a): $YbGeSe_3$ and Yb_2GeSe_4. The eutectic parameters are: 20 mol.% YbSe and 530°C.

The quasi-binary sections R_2Se_3–$SnSe_2$ (R = La, Ce, Pr, Nd, Sm, Eu, Tb, Yb) have been studied (Rustamov et al. 1971, 1981b, Nasibov et al. 1972, 1978, Nasibov and Sultanov 1980, Murguzov et al. 1984). A limited field of solid solutions is found in the system with lanthanum (fig. 14). These solid solutions are observed at $SnSe_2$, they decompose with the resulting eutectic having the following parameters: 3 mol.% La_2Se_3 and 610°C. The compound $La_2Sn_3Se_9$ is formed at 650°C according to the peritectic reaction:

$$La_2SnSe_5 + L \rightleftarrows La_2Sn_3Se_9 .$$

The compound La_2SnSe_5 melts congruently at 1115°C. The eutectic observed between La_2SnSe_5 and La_2Sn_3 melts at 1000°C. The type of interaction in the systems with Nd, Pr, Sm, Eu is analogous (table 18).

The congruently melting compounds Ce_2SnSe_5 and $CeSn_3Se_9$ are formed in the system with cerium at the ratio of binary components of 1:1 and 1:3. Eutectics are formed between Ce_2SnSe_5 and Ce_2Se_3, Ce_2SnSe_5 and $Ce_2Sn_3Se_9$ at 910 and

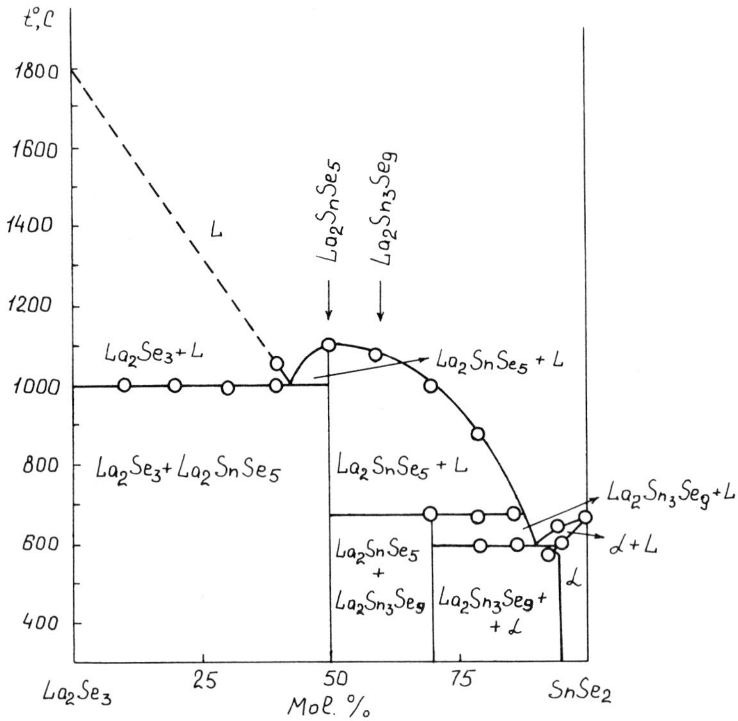

Fig. 14. The phase diagram of the La_2Se_3–$SnSe_2$ system.

800°C, respectively. There is a narrow range of solid solutions at $SnSe_2$. The eutectic observed between $Ce_2Sn_3Se_9$ and $SnSe_2$ melts at 500°C.

The system with terbium exhibits one congruently melting compound, Tb_2SnSe_5, and two incongruently melting compounds,

$$Tb_2SnSe_5 + L \rightleftarrows Tb_2Sn_2Se_7 ,$$

$$Tb_2Sn_2Se_7 + L \rightleftarrows Tb_2Sn_3Se_7 .$$

The eutectic formed between Tb_2SnSe_5 and Tb_2Se_3 melts at 850°C (Nasibov et al. 1978).

In the Yb_2Se_3–$SnSe_2$ system, the compound Yb_2SnSe_5 melts congruently at 1240°C, the compound $Yb_2Sn_2Se_7$ melts peritectically at 850°C. There exists a $SnSe_2$-based solid solution (fig. 15).

The section Ce_2Se_3–SnSe (fig. 16) is a quasi-binary section of the ternary system Ce–Sn–Se and belongs to a peritectic type with limited solid solutions based on SnSe (Murguzov et al. 1986). The parameters of the peritectic point are 70 mol% SnSe and 870 ± 5°C. The ternary compound Ce_2SnSe_4 is formed according to the reaction

$$L + Ce_2Se_3 \rightleftarrows Ce_2SnSe_4 .$$

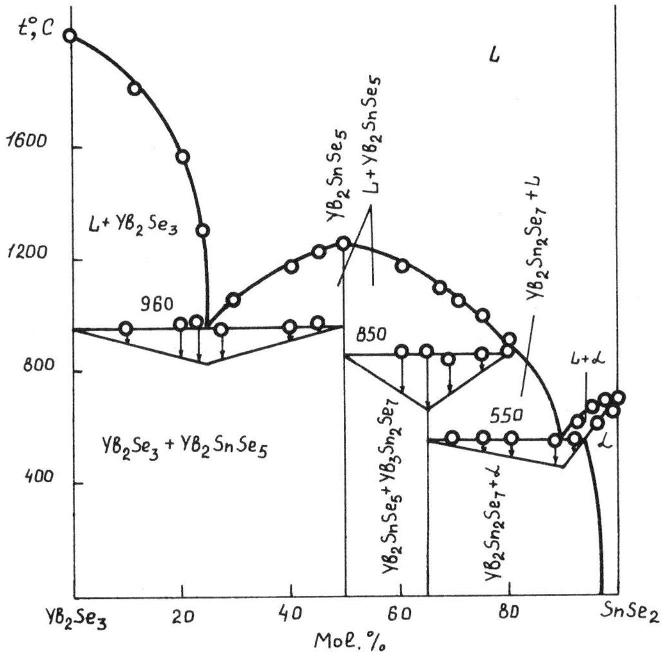

Fig. 15. The phase diagram of the Yb_2Se_3–$SnSe_2$ system.

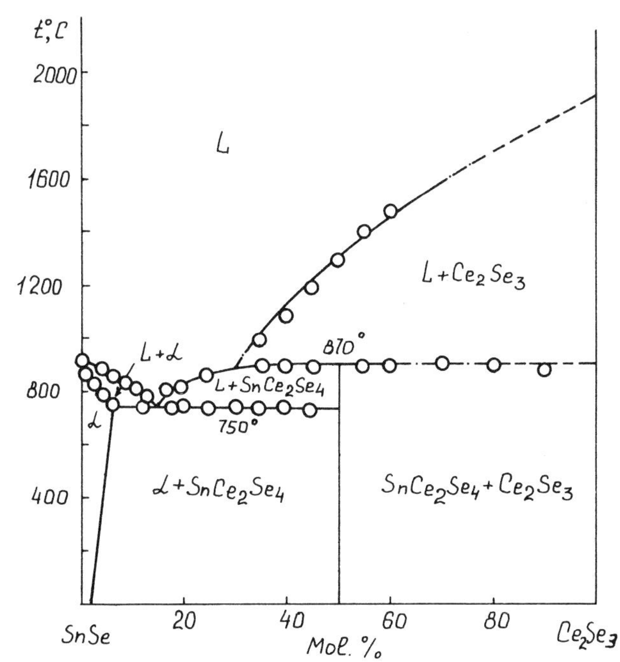

Fig. 16. The phase diagram of the Ce_2Se_3–SnSe system.

TABLE 18

Crystallographic data for compounds in the R–{Ge, Sn, Pb}–Se systems.

Compound	a (Å)	b (Å)	c (Å)	α	β	γ	Density (g/cm^3) D_x	D_m	Note	Ref.*
R_2GeSe_5 Tetragonal lattice										
α-Gd_2GeSe_5	4.15		8.40							[1]
α-Tb_2GeSe_5	4.12		8.36							[2]
$R_2Ge_2Se_7$ Hexagonal lattice										
$Gd_2Ge_2Se_7$	6.79		12.61							[1]
$Tb_2Ge_2Se_7$	6.75		12.59							[2]
$RGeSe_3$ Triclinic lattice										
$YbGeSe_3$	8.34	8.54	11.68	111.08°	99.95°	75.44°		4.01		[3]
R_2GeSe_4 Monoclinic lattice (space group $P2_{1/m}$)										
Yb_2GeSe_4	6.644	8.390	6.840			108°		4.35		[3]
$R_2Ge_3Se_9$ Orthorhombic lattice. In ref. [4], Guittard et al. (1976) did not identify any compound of the $R_2Ge_3Se_9$ (R = La–Sm) composition in the R–Ge–Se system										
$La_2Ge_3Se_9$								5.46		[5]
$Pr_2Ge_3Se_9$								5.05		[6]
$Gd_2Ge_3Se_9$	6.71	11.87	22.34					–		[1]
$Tb_2Ge_3Se_9$	6.66	11.81	22.30					–		[2]
R_2SnSe_4										
Ce_2SnSe_4	11.80	14.05	4.12						Orthorhombic lattice of the $CaFe_2O_4$ type	[7]
Yb_2SnSe_4	12.68	–	14.42						Hexagonal lattice	[8]
R_2SnSe_5 Tetragonal lattice. In ref. [4], Guittard et al. (1976) did not identify any compounds of the R_2SnSe_5 (R = La–Sm) composition in the R–Sn–Se system										
La_2SnSe_5	6.15		8.73				6.29	6.23		[9]
Ce_2SnSe_5	6.12		8.59				6.17	6.16		[9]
Pr_2SnSe_5	6.98		8.47				6.02	6.01		[9]
Nd_2SnSe_5	5.87		8.37				6.40	6.39		[10]
Sm_2SnSe_5	5.87		8.27				5.69	5.69		[10]
Yb_2SnSe_5	5.26		8.04				–	6.85	Single crystal	[8]

Compound					Ref.*
PbR_2Se_4 Cubic lattice of the Th_3P_4 type					
$PbLa_2Se_4$	9.106			$a = 9.066$ (ref. [11])	[12]
$PbCe_2Se_4$	9.045				[12]
$PbPr_2Se_4$	8.996				[12]
$PbNd_2Se_4$	8.968				[13]
$PbSm_2Se_4$	8.909				[12]
$PbGd_2Se_4$	8.857				[12]
$PbTb_2Se_4$	8.737				[12]
$Pb_4R_2Se_7$ Cubic lattice of the Th_3P_y type					
$Pb_4La_2Se_7$	8.923				[11]
PbR_2Se_4 Orthorhombic lattice of the $CaFe_2O_4$ type					
$PbEr_2Se_4$	12.45	14.85	4.12		[11]
$PbTm_2Se_4$	12.59	14.75	4.10		[11]
$PbYb_2Se_4$	12.83	14.64	4.08		[11]
$PbLu_2Se_4$	12.87	14.59	4.07		[11]

*References: [1] Nasibov et al. (1981c); [2] Nasibov et al. (1981a); [3] Rustamov et al. (1981a); [4] Guittard et al. (1976); [5] Rustamov et al. (1973); [6] Aliyeva et al. (1972); [7] Murguzov et al. (1986); [8] Murguzov et al. (1984); [9] Rustamov et al. (1971); [10] Nasibov et al. (1972); [11] Nasibov et al. (1987a); [12] Nasibov et al. (1980).

The compounds Ce_2SnSe_4 and Ce_2Se_3 cocrystallize in the interval of 51–100 mol.% Ce_2Se_3. The system with ytterbium is analogous to the system with cerium (Murguzov et al. 1984).

The eutectic with the following parameters is observed along the section LaSe–SnSe at SnSe: 5.8 mol.% LaSe and 650°C. The solubility of LaSe in SnSe is about 1.6 mol.% at 25°C, and 3.2 mol.% at 650°C. The RSe–SnSe systems (R = Ce, Pr, Nd, Sm, Gd, Tb, Dy, Ho, Er, Yb, Lu) are analogous to the lanthanum system (Gurshumov et al. 1984). A region of solid solutions containing La_2Se_3 in an amount of up to 3 mol.% was found on the basis of PbSe. The system with cerium is analogous to that with lanthanum (Nasibov et al. 1987a, 1979b). The compounds Ce_2PbSe_4 and $Ce_2Pb_4Se_7$ melt congruently at 1100 and 1075°C and undergo polymorphic transformations at 850 and 760°C, respectively. The eutectics which form between β-Ce_2PbSe_4 with Ce_2Se_3, and with β-$Ce_2Pb_4Se_7$ melt at 950 and 1000°C, respectively. The eutectic of PbSe with β-$Ce_2Pb_4Se_7$ melts at 900°C. For the components ratio of 1:1, the Sm_2Se_3–PbSe system (fig. 17) exhibits the compound $PbSm_2Se_4$ which undergoes phase transition at 850°C. The eutectic parameters of β-Sm_2PbSe_4 with Sm_2Se_3 are: 65 mol.% Sm_2Se_3 and 950°C. The compound Sm_2PbSe_4 melts congruently at 1195°C, and $Sm_2Pb_4Se_7$ decomposes according to a peritectic reaction (Shafagatova et al. 1979, Rustamov and Rustamov 1979) (table 18).

The systems R_2Se_3–PbSe (R = Pr, Nd, Lu) are analogous to those with samarium (Nasibov et al. 1979a, 1980, 1981a).

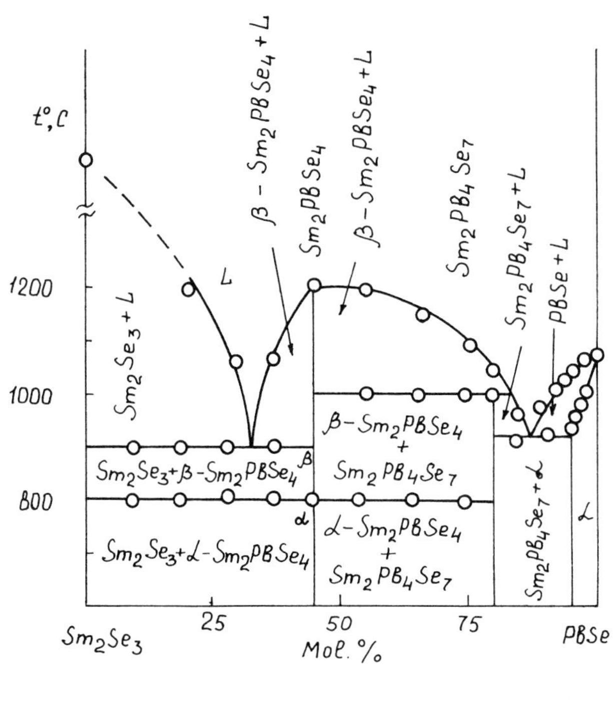

Fig. 17. The phase diagram of the Sm_2Se_3–PbSe system.

In the telluride systems, only some selected binary sections have been studied.

The compound $Sm_5Ge_2Te_7$ is stable in the SmTe–GeTe system up to 927°C. The system's peritectic at GeTe contains 8 mol.% SmTe and melts at 747°C (Aliyev et al. 1986).

The section GeTe–YbTe is quasi-binary and of eutectic type (Mukhtarova et al. 1985). The eutectic parameters are: 47 mol.% YbTe and 650°C. The liquidus of the system contains two branches of primary crystallization of solid solutions based on YbTe and GeTe. Solubility of YbTe in GeTe attains 15 mol.% at 25°C and 25 mol.% at 650°C.

The section Sm_2Te_3–GeTe is a quasi-binary section of the ternary system Sm–Ge–Te. The system's liquidus consists of three branches: Sm_2Te_3, $GeSm_2Te_4$ and the GeTe-base solid solution. The compound $GeSm_2Te_4$ is formed according to a peritectic reaction at 827°C,

$$L + Sm_2Te_3 \rightleftarrows GeSm_2Te_4 \qquad \text{(Aliyev et al. 1986)}.$$

The Eu_2Te_3–PbTe system also belongs to the eutectic type. The eutectic parameters are: 9.8 mol.% Eu_2Te_3, and 750°C (Nasibov et al. 1987c). The constitutional diagram of the Tb_2Te_3–PbTe system can be divided into two comparatively simple diagrams, namely, Tb_2Te_3–Tb_2PbTe_4, which appears to be a eutectic system, and Tb_2PbTe_4–PbTe, which forms peritectic compounds and solid solutions based on PbTe (see fig. 18). Ternary compounds are formed when

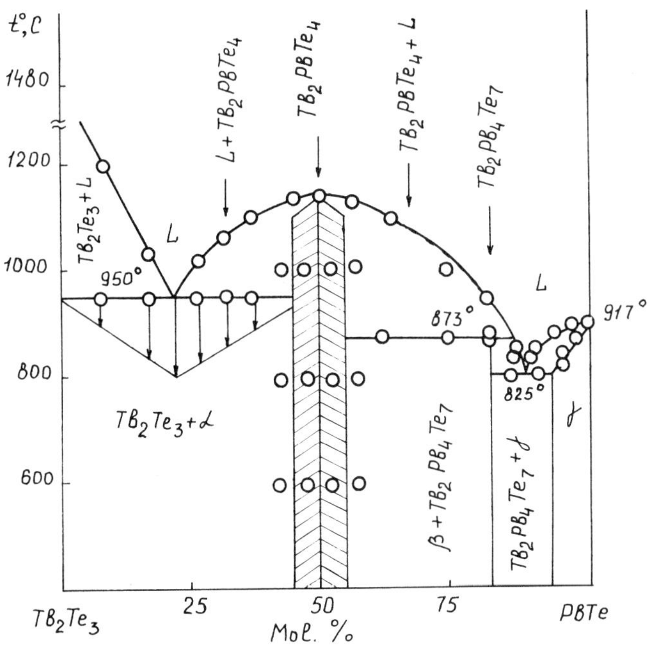

Fig. 18. The phase diagram of the Tb_2Te_3–PbTe system.

TABLE 19
Crystallographic data for compounds in the R–{Zr, Hf}–S (Se) systems.

Compound	a (Å)	b (Å)	c (Å)	Density (g/cm^3) D_x	D_m	Note	Ref.*
$R_2MS(Se)_5$	Orthorhombic lattice						
La_2ZrS_5	11.4864(5)	7.3894(3)	8.2167(5)	5.065	5.05		[1]
Sm_2ZrS_5	11.491(1)	7.2848(5)	7.9092(9)	6.468	6.45		[1]
Ho_2ZrS_5	11.4781(8)	7.211(1)	7.7154(5)	–	–		[1]
Er_2ZrS_5	11.4664(7)	7.1958(4)	7.6810(6)	–	–		[1]
Y_2ZrS_5	11.4958(6)	7.2262(2)	7.7294(4)	–	–		[1]
Ce_2HfS_5	11.4177(6)	7.3389(4)	8.1277(5)	–	–		[1]
Sm_2HfS_5	11.4340(6)	7.2724(4)	7.8964(4)	–	–		[1]
Ho_2HfS_5	11.415(3)	7.204(2)	7.777(2)	–	–		[1]
Er_2HfS_5	11.442(2)	7.1932(8)	7.681(20	–	–		[1]
Y_2HfS_5	11.4585(3)	7.2207(2)	7.7215(3)	–	–	The crystal structure was determined on a single crystal: $a = 11.4585(3)$, $b = 7.7215(3)$, $c = 7.2207(2)$ (ref. [2])	[1]
La_2ZrSe_5	12.058(3)	7.719(2)	8.505(2)	–	–		[1]
Sm_2ZrSe_5	12.026(2)	7.607(1)	8.2031(1)	–	–		[1]
Gd_2ZrSe_5	12.064(8)	7.6028(5)	8.1379(7)	–	–		[1]
Tb_2ZrSe_5	12.020(4)	7.604(3)	8.138(3)	–	–		[1]
La_2HfSe_5	11.919(1)	7.680(1)	8.524(1)				[1]
Ce_2HfSe_5	11.886(1)	7.6540(7)	8.4292(7)				[1]

*References: [1] Donohue and Jeitschko (1974); [2] Jeitschko and Donohue (1975).

the ratios of binary components are 1:1 and 1:4: the compound Tb_2PbTe_4 melts congruently at 1125°C and the compound $Tb_2Pb_4Te_7$ melts in the peritectic reaction

$$Tb_2PbTe_4 + L \rightleftarrows Tb_2Pb_4Te_7 .$$

The eutectic between Tb_2PbTe_4 and Tb_2Te_3 melts at 950°C, whereas the one between $Tb_2Pb_4Te_7$ and PbTe melts at 825°C (Nasibov et al. 1987b).

Crystallographic data for the compound with the composition $R–M^{IV}–X$ (M^{IV} = Zr, Hf; X = S, Se) are given in table 19.

References

Aliyev, O., T. Kurbanov and Z. Mukhtarova, 1986, Zh. Neorg. Khim. **31**, 2628.
Aliyeva, M., I. Nasibov and P. Rustamov, 1972, Dokl. Akad. Nauk Azerb. SSR **28**, 21.
Barnier, S., and M. Guittard, 1978, C.R. Acad. Sci., Paris, Ser. C **286**, 205.
Beskrovnaya, R., and V. Serebrennikov, 1971, Vestn. Tomsk. Univ. Ser. Khim. **204**(5/6), 302.
Beskrovnaya, R., I. Kamayeva and V. Serebrennikov, 1971a, Izv. Akad. Nauk SSSR, Ser. Neorg. Mater. **7**, 326.
Beskrovnaya, R., G. Kustova and V. Serebrennikov, 1971b, Izv. Akad. Nauk SSSR, Ser. Neorg. Mater. **7**, 1078.
Bugli, G., D. Carré and S. Barnier, 1978, Acta Crystallogr. B **34**, 3186.
Collin, G., J. Étienne and P. Laruelle, 1973, Bull. Soc. Fr. Mineral. Crystallogr. **96**, 12.
Donohue, P., and W. Jeitschko, 1974, Mater. Res. Bull. **9**, 1333.
Finkel'shtein, L., and N. Samsonova, 1987, Zh. Neorg. Khim. **32**, 2595.
Guittard, M., M. Julien-Pouzol and S. Jaulmes, 1976, Mater. Res. Bull. **11**, 1073.
Gurshumov, A., B. Kuliyev, A. Akhmedov, D. Safarov and V. Lazaryev, 1984, Izv. Akad. Nauk SSSR, Ser. Neorg. Mater. **20**, 1090.
Gus'kova, V., and V. Serebrennikov, 1973a, Vestn. Tomsk. Univ. Ser. Khim. **240**, 420.
Gus'kova, V., and V. Serebrennikov, 1973b, Vestn. Tomsk. Univ. Ser. Khim. **249**. 137.
Jaulmes, S., 1974, Acta Crystallogr. B **30**, 2283.
Jaulmes, S., and M. Julien-Pouzol, 1977a, Acta Crystallogr. B **33**, 1191.
Jaulmes, S., and M. Julien-Pouzol, 1977b, Acta Crystallogr. B **33**, 3898.
Jeitschko, W., and P. Donohue, 1975, Acta Crystallogr. B **31**, 1890.
Julien-Pouzol, M., and S. Jaulmes, 1979, Acta Crystallogr. B **35**, 2672.
Mazurier, A., and J. Étienne, 1973, Acta Crystallogr. B **29**, 817.
Mazurier, A., and J. Étienne, 1974, Acta Crystallogr. B **30**, 759.
Mel'chenko, G., and V. Serebrennikov, 1973, Zh. Neorg. Khim. **18**, 2572.
Michelet, A., and J. Flahaut, 1969, C.R. Acad. Sci., Paris, Ser. C **268**, 326.
Michelet, A., P. Laruelle and J. Flahaut, 1966, C.R. Acad. Sci., Paris, Ser. C **262**, 753.
Michelet, A., G. Perez, J. Étienne and M. Darriet-Duale, 1970, C.R. Acad. Sci., Paris, Ser. C **271**, 513.
Michelet, A., A. Mazurier, G. Collin, P. Laruelle and J. Flahaut, 1975, J. Solid State Chem. **13**, 65.
Mukhtarova, Z., T. Kurbanov and O. Aliyev, 1985, Zh. Neorg. Khim. **30**, 1332.
Murguzov, M., A. Gurshumov and Kh. Kadyrly, 1984, Zh. Neorg. Khim. **29**, 2696.
Murguzov, M., A. Gurshumov and B. Gadyrov, 1986, Zh. Neorg. Khim. **31**, 1906.
Nasibov, I., 1982, Izv. Akad. Nauk SSSR, Ser. Neorg. Mater. **18**, 1972.
Nasibov, I., and T. Sultanov, 1980, Izv. Akad. Nauk SSSR, Ser. Neorg. Khim. **16**, 422.
Nasibov, I., T. Sultanov, P. Rustamov, Z. Karayev and A. Nazarov, 1972, Azerb. Khim. Zh. (1), 99.
Nasibov, I., T. Sultanov, P. Rustamov and T. Dzhali-zade, 1978, Izv. Akad. Nauk SSSR, Ser. Neorg. Mater. **14**, 1265.

Nasibov, I., T. Sultanov, P. Rustamov and G. Shafagatova, 1979a, Izv. Akad. Nauk SSSR, Ser. Neorg. Mater. **15**, 1535.
Nasibov, I., T. Sultanov and G. Shafagatova, 1979b, Zh. Neorg. Khim. **24**, 2574.
Nasibov, I., T. Sultanov, A. Rustamov, P. Rustamov and G. Shafagatova, 1980, Izv. Akad. Nauk SSSR, Ser. Neorg. Mater. **16**, 426.
Nasibov, I., T. Sultanov and B. Mardakhayev, 1981a, Zh. Neorg. Khim. **26**, 2263.
Nasibov, I., T. Sultanov and T. Dzhalil-Zade, 1981b, Zh. Neorg. Khim. **17**, 1341.
Nasibov, I., T. Sultanov and P. Sultanov, 1981c, Izv. Akad. Nauk SSSR, Ser. Neorg. Mater. **17**, 1983.
Nasibov, I., T. Sultanov and P. Rustamov, 1981d, Zh. Neorg. Khim. **26**, 2524.
Nasibov, I., T. Sultanov and T. Dzhalil-Zade, 1981e, Zh. Neorg. Khim. **26**, 2529.
Nasibov, I., T. Sultanov, G. Shafagatova and I. Abilov, 1987a, Izv. Akad. Nauk SSSR, Ser. Neorg. Mater. **23**, 403.
Nasibov, I., V. Valiyev, A. Khalilov, T. Sultanov and T. Dzhalil-Zade, 1987b, Zh. Neorg. Khim. **32**, 1011.
Nasibov, I., T. Sultanov, V. Valiyev and O. Alidzhanova, 1987c, Izv. Akad. Nauk SSSR, Ser. Neorg. Mater. **23**, 515.
Patrie, M., M. Guittard and M.-P. Pardo, 1969, Bull. Soc. Chim. Fr. (11), 3832.
Perez, G., and M. Duale, 1969, C.R. Acad. Sci., Paris, Ser. C **269**, 984.
Rustamov, A., and P. Rustamov, 1979, Zh. Neorg. Khim. **24**, 2197.
Rustamov, P., Z. Karayev, I. Nasibov, T. Sultanov and R. Gamidov, 1971, Azerb. Khim. Zh. (3), 97.
Rustamov, P., I. Nasibov and M. Aliyeva, 1972, Izv. Akad. Nauk SSSR, Ser. Neorg. Mater. **6**, 1155.
Rustamov, P., I. Nasibov and M. Aliyeva, 1973, Izv. Akad. Nauk SSSR, Ser. Neorg. Mater. **9**, 1900.
Rustamov, P., O. Aliyev, A. Abilgasymova and T. Kurbanov, 1981a, Izv. Akad. Nauk SSSR, Ser. Neorg. Mater. **17**, 1571.
Rustamov, P., O. Aliyev and T. Kurbanov, 1981b, V kn: Troiinyye Khal'kogenidy Redkozemel'nykh Elementov. Izd. "Elm", Baku.
Sarkisov, E., R. Lidin and Yu. Khodzhainov, 1968, Izv. Akad. Nauk SSSR, Ser. Neorg. Mater. **4**, 2033.
Sarkisov, E., R. Lidin and V. Shum, 1970, Izv. Akad. Nauk SSSR, Ser. Neorg. Mater. **6**, 2054.
Shafagatova, G., I. Nasibov, T. Sultanov, A. Rustamov and P. Rustamov, 1979, Zh. Neorg. Khim. **24**, 2197.
Stepanez, M., V. Serebrennikov, I. Kamayeva and L. Vedeneyeva, 1973a, Vestn. Tomsk. Univ. Ser. Khim. **249**, 133.
Stepanez, M., V. Serebrennikov and V. Agafonnikov, 1973b, Vestn. Tomsk. Univ. Ser. Khim. **240**, 95.
Stepanez, M., V. Serebrennikov, I. Kamayeva and L. Vedeneyeva, 1973c, Vestn. Tomsk. Univ. Ser. Khim. **249**, 129.
Vovan, T., J. Flahaut and L. Domange, 1966, C.R. Acad. Sci., Paris, Ser. C **262**, 278.

5. The R–M^V–X systems

The R–M^V–X systems contain some elements of group V; phosphorus, arsenic, antimony, bismuth and tantalum.

The compounds with the compositions RPS (R = Pr–Er, Y), Eu_4P_2S and RPS_4 are known to be observed in the R = P–S systems (Hulliger 1979, 1968, Wibbelmann and Brockner 1984, Yampol'skaya and Serebrennikov 1971).

The section As_2S_3–La_2S_3 of the ternary system La–As–S shows one eutectic at 700°C, one peritectic at 450°C and a ternary compound $LaAsS_3$, at equimolar ratio of the components (Serebrennikov and Dashiyev 1969).

In the As_2S_3–Sm_2S_3 system, one ternary compound, $SmAsS_3$, is formed by a peritectic reaction at 767°C,

$$L + Sm_2S_3 \rightleftarrows SmAsS_3 .$$

Up to 0.8 mol.% Sm_2S_3 dissolves in As_2S_3 (Il'yasov et al. 1986a). The R_2S_3–As_2S_3 systems (R = La, Pr, Sm, Yb) are analogous (Rcheushvili et al. 1987).

In the As_2S_3–SmS system (fig. 19), the compound $SmAs_4S_7$ melts congruently at 577°C, whereas $SmAs_2A_4$ and $Sm_3As_4S_9$ melt peritectically,

$$L + Sm_3As_4S_9 \rightleftarrows SmAs_2S_4 \quad \text{at } 827°C,$$

$$L + SmS \rightleftarrows Sm_3As_4S_9 \quad \text{at } 1047°C.$$

Up to 0.5 mol.% SmS dissolves in As_2S_3 at 25°C (Il'yasov et al. 1986a).

On the quasi-binary section As_2S_3–YbS, the compounds $YbAs_4S_7$ and $Yb_3As_4S_9$ melt congruently and the compound $YbAs_2S_4$ melts according to a peritectic reaction at 647°C,

$$L + Yb_3As_4S_9 \rightleftarrows YbAs_2S_4 \quad \text{(Rustamov et al. 1985)}.$$

The summarized results of the physicochemical analyses were used to plot the projection of the liquidus surface of the ternary Yb–As–S system (Rustamov et al. 1987). The system displays the following ternary compounds: $YbAs_4S_7$, $YbAs_2S_4$, $YbAsS_3$ and $Yb_3As_4S_9$. Triangulation has shown that the system can be divided into 12 subordinate triangles formed by singular secants: Yb_4As–Yb–YbS, Yb_4As_3–YbAs–YbS, YbS–YbAs–As_4S_4, As_4S_4–YbS–$Yb_3As_4S_9$, As_4S_4–As_2S_3–$YbAs_2S_7$, YbAs–As_4S_4–As, As_4S_4–YbS–$Yb_3As_4S_9$, As_2S_3–Yb_2S_3–$YbAs_4S_7$, $YbAs_4S_7$–$Yb_3As_4S_9$–Yb_2S_3, $Yb_3As_4S_9$–Yb_2S_3–Yb_3S_4, $Yb_3As_4S_9$–

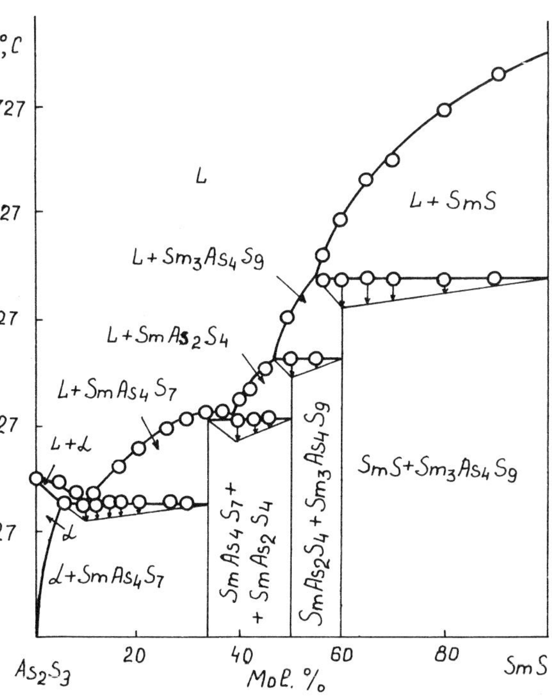

Fig. 19. The phase diagram of the SmS–As_2S_3 system.

YbS–Yb_3S_4, As_2S_3–Yb_2S_3–S. To construct the liquidus surface, the corresponding quasi- and non-quasi-binary sections, e.g., As_2S_3–Yb_2S_3, have been studied (Il'yasov et al. 1984).

The EuS–Sb_2S_3 system (fig. 20) appears to be a quasi-binary section of the ternary system Eu–Sb–S (Khasayev et al. 1982). The system liquidus consists of five branches; the two terminal lines correspond to the primary crystallization of EuS and Sb_2S_3, all the other lines correspond to ternary compounds: $Eu_3Sb_4S_9$, $EuSb_2S_4$ and $EuSb_4S_7$. The first and the third melt congruently at 830 and 750°C. The compound $EuSb_2S_4$ melts in the peritectic reaction at 650°C,

$$L + Eu_3Sb_4S_9 \rightleftarrows EuSb_2S_4 .$$

The eutectic composition contain 18.41 and 69 mol.% EuS. The Sb_2S_3-base solid solutions are formed up to 2 mol.% EuS.

The compounds of the compositions $RBiS_3$ and $EuBi_2S_4$ are formed when R_2S_3 reacts with Bi_2S_3 and EuS reacts with Bi_2S_3 in a 1:1 ratio of the components (table 20) (Rustamov et al. 1977a).

In the systems with selenium (R–P–Se), the compounds $RPSe_4$ (R = La–Sm) are observed upon the interaction of the elements or of the binary components R_2Se_3 and P_2Se_5. The compound with composition Eu_4P_2Se is also known (Yampol'skaya and Serebrennikov 1973).

In the systems with arsenic (R–As–Se), the compounds with the composition RAsSe (R = La, Gd) and Eu_4As_2Se were identified (Hulliger 1968, 1979).

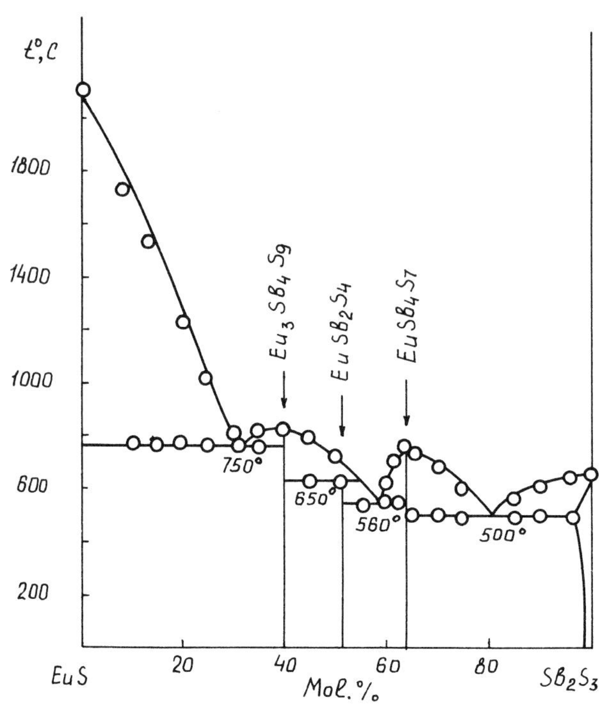

Fig. 20. The phase diagram of the EuS–Sb_2S_3 system.

TABLE 20
Crystallographic data for compounds in the R–{P, As, Sb, Bi, Ta}–S systems.

Compound	a (Å)	b (Å)	c (Å)	β	γ	Density (g/cm^3) D_x	Density (g/cm^3) D_m	Note	Ref.*
RPS Orthorhombic lattice (space group $P22_12$)									
GdPS	5.41	5.37	16.76					The structure derived from the PbFCl type structure: $a = \sqrt{2}\, a_0$; $c = 2c_0$, where a_0 and c_0 are the PbFCl parameters	[1]
RPS_4 Tetragonal lattice									
$PrPS_4$	10.914(5)		19.361(8)					The structure was determined on a single crystal	[2]
R_4M_2S Cubic lattice of the anti-Th_3P_4 type									
Eu_4P_2S	9.056(1)								[3]
Eu_4As_2S	9.199(2)								[3]
RAsS Monoclinic lattice									
CeAsS	4.047	5.616	17.450		135.85°			The crystal structure of CeAsS was determined	[4]
$RAsS_4$ Orthorhombic lattice									
$LaAsS_4$	8.31(1)	8.14(1)	16.47(3)						[5]
$NdAsS_4$	8.31(1)	8.07(1)	16.27(3)						[5]
R_3AsS_3 Cubic lattice									
Ce_3AsS_3	8.730					5.70	5.90		[6]
RSb_2S_4 Orthorhombic lattice (space group Pbam)									
$EuSb_2S_4$	11.2	13.9	3.4					The structure resembles $FeSb_2S_4$ $a = 11.22$, $b = 13.86$, $c = 3.98$ (ref. [9]). According to ref. [10] there is 90% Eu^{+2} in $EuSb_2S_4$	[7, 8]
$R_8Sb_2S_{15}$ Tetragonal lattice (space group $I4_1cd$)									
$Ce_8Sb_2S_{15}$	15.66		19.68						[11]
$Pr_8Sb_2S_{15}$	15.626		19.659					The structure was determined on a single crystal	[11]
$R_3Sb_4S_9$ Orthorhombic lattice (space group Pbnm)									
$Sm_3Sb_4S_9$	17.10	24.68	4.02					100% Eu^{+2} in $Eu_3Sb_4S_9$ (ref. [10])	[9, 12]
$Eu_3Sb_4S_9$	17.08	24.60	4.00						[9, 12]

TABLE 20 (cont'd)

Compound	a (Å)	b (Å)	c (Å)	β	γ	Density (g/cm³) D_x	Density (g/cm³) D_m	Note	Ref.*
RSb_4S_7 Orthorhombic lattice (space group Pnmm)									
$EuSb_4S_7$	17.4	25.02	4.19					According to ref. [10], there is 80% Eu^{+2} in $EuSb_4S_7$	[9, 13]
$RBiS_3$ Orthorhombic lattice of the Sb_2S_3 type									
$LaBiS_3$	11.38	11.41	3.99						[14]
$CeBiS_3$	8.73	–	–					Cubic lattice	[14]
$PrBiS_3$	11.20	11.28	3.96						[14]
$NdBiS_3$	11.40	11.27	3.94						[14]
$GdBiS_3$	11.10	11.24	3.90						[14]
RBi_2S_4 Orthorhombic lattice. The structure resembles $FeSb_2S_4$									
$EuBi_2S_4$	11.22	14.06	3.59						[7, 8]
$R_xBi_2S_4$ Hexagonal lattice (space group $P6_3/m$)									
$Eu_{1.1}Bi_2S_4$	24.82		4.080			6.39	6.2		[15]
$R_{1.25}Bi_{3.78}S_8$									
$Ce_{1.25}Bi_{3.78}S_8$	16.55(1)	4.053(2)	21.52(1)					Crystal structure was determined on a single crystal $Ce(Ce_{0.25}\square_{0.75})Bi_3(Bi_{0.78}\square_{0.22})[S_8]$	[16]
$RTaS_3$ Monoclinic lattice of the $YbTaS_3$ type									
$YbTaS_3$	5.6	14.2	11.8	91°					[17]

*References: [1] Hulliger (1968); [2] Wibbelmann and Brockner (1984); [3] Hulliger (1979); [4] Sfez and Adolphe (1972); [5] Angapova et al. (1973); [6] Céolin and Khodadad (1971); [7] Rustamov et al. (1977a); [8] Godzhayev et al. (1977); [9] Khasayev et al. (1982); [10] Finkel'shtein and Samsonova (1987); [11] Guseinov et al. (1981); [12] Rustamov et al. (1981b); [13] Aliyev et al. (1978); [14] Rustamov et al. (1981a); [15] Lemoine et al. (1986); [16] Ceolin et al. (1977); [17] Donohue (1975).

Various sections of the ternary system Yb–As–Se have been studied by Rustamov and Il'yasov (1984), Il'yasov et al. (1986c). The section As_2Se_3–YbSe shows formation of the compounds $YbAs_4Se_7$ and $Yb_3As_4Se_9$, melting congruently at 725 and 880°C, respectively, as well as a peritectic reaction at 730°C for the compound $YbAs_2Se_4$,

$$L + Yb_3As_4Se_9 \rightleftarrows YbAs_2Se_4 \,.$$

The eutectics have melting points of 320, 600, 320°C, respectively. The sections AsSe–YbSe (fig. 21), AsSe–$YbAs_4Se_7$, AsSe–$Yb_3As_4Se_9$ are quasi-binary. The parameters of the eutectic between AsSe and YbSe are: 11 mol.% YbSe and 257°C. A liquid immiscibility region was detected in the 42–63 mol.% YbSe concentration range. The temperature of the monotectic horizontal is 997°C. The sections YbSe–$YbAs_4Se_7$ and AsSe–$Yb_3As_4Se_9$ are triangulated sections of the ternary system (Il'yasov et al. 1986c). Solid solution based on AsSe extends to 0.6 mol.%, and the one based on YbSe reaches 1.5 mol.%. The eutectic parameters are: 19 mol.% YbSe and 267°C. The eutectic parameters along the other triangulated section are: 15 mol.% $YbAs_4Se_7$, and 267°C. Solubility on the basis of $YbAs_4Se_7$ is 2 mol.%. The section AsSe–SmSe is a quasi-binary section of the ternary system As–Sm–Se (Il'yasov et al. 1986b). The constitutional diagram belongs to a peritectic type with liquid immiscibility, the immiscibility region being 19.5–61 mol.% SmSe. A monotectic process proceeds at 727°C according to the scheme

$$L_2 \rightleftarrows L_1 + SmSe \text{ (solid solution)} \,.$$

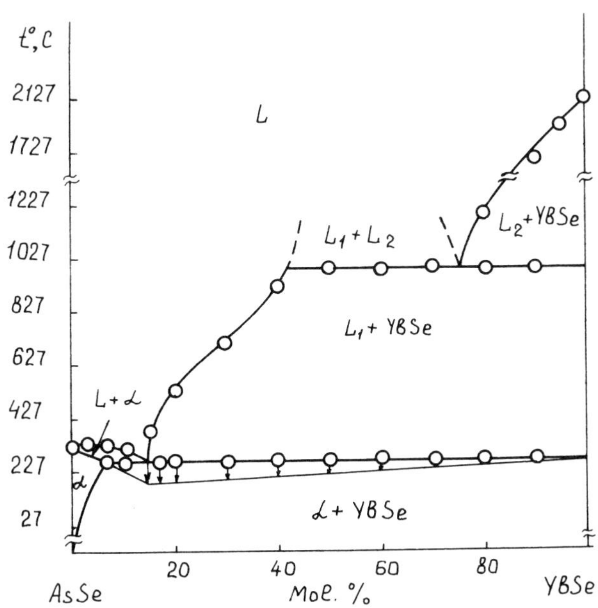

Fig. 21. The phase diagram of the YbSe–AsSe system.

At 627°C, a peritectic reaction occurs:

$$L_1 + SmSe \rightleftarrows AsSe \text{ (solid solution)} .$$

Solubility on the basis of AsSe is 3 mol.%, on the basis of SmSe it is 2 mol.%.

In the quasi-binary section La_2Se_3–Sb_2Se_3 (Aliyev et al. 1977), the peritectic compound is formed at 650°C,

$$L + La_2Se_3 \rightleftarrows LaSbSe_3 .$$

A region of liquid immiscibility is observed in the concentration range from 62 to 87 mol.% Sb_2Se_3. The eutectic coordinates are: 12 mol.% La_2Se_3 and 526°C.

In the Ce_2Se_3–Sb_2Se_3 system, the peritectic compound $CeSbSe_3$ is also formed at a components ratio of 1:1 and at a temperature of 590°C (table 21). The eutectic parameters between CeSbSe and Sb_2Se_3 are: 17 mol.% Ce_2Se_3 and 510°C. Solubility on the basis of Ce_2Se_3 extends up to 5 mol.% (Aliyev et al. 1977).

TABLE 21
Crystallographic data for compounds in the R–{P, As, Sb, Bi}–Se systems.

Compound	a (Å)	b (Å)	c (Å)	Note	Ref.*
R_4M_2Se Cubic lattice of the anti-Th_3P_4 type					
Eu_4P_2Se	9.126(3)				[1]
Eu_4As_2Se	9.289(1)				[1]
RAs_xSe_y Orthorhombic lattice					
$YbAs_2Se_4$	11.34	13.04	3.98		[2]
$YbAs_4Se_7$	6.85	23.52	4.06		[2]
RAsSe Orthorhombic lattice					
GdAsSe	3.99	3.95	8.77	Slight structural distortion of the PbFCl type	[3]
$RSbSe_3$ Orthorhombic lattice of the Sb_2S_3 type					
$LaSbSe_3$	11.64(1)	11.82(1)	3.985(5)		[4]
RSb_2Se_4 Orthorhombic lattice					
$EuSb_2Se_4$	11.4	14.2	3.76	$a = 11.37$, $b = 11.35$, $c = 4.0$ (ref. [6]). The structure resembles $FeSb_2S_4$ (ref. [7])	[5–7]
$R_3Sb_4Se_9$ Orthorhombic lattice (space group Pnam)					
$Eu_3Sb_4Se_9$	17.174(2)	24.846(3)	4.169		[8]
$Sm_3Sb_4Se_9$	17.44	24.92	4.08		[8]
$RBiSe_3$ Orthorhombic lattice					
$LaBiSe_3$	11.41	11.45	4.01		[4]
$CeBiSe_3$	8.81	–	–	Cubic lattice	[4]
$PrBiSe_3$	11.30	11.42	4.00		[4]
$NdBiSe_3$	11.30	11.90	3.98		[4]
$NdBiSe_3$	11.42	11.98	3.97		[4]
RBi_2Se_4 Orthorhombic lattice. The structure resembles $FeSb_2S_4$					
$EuBi_2Se_4$	11.26	14.16	3.72		[5, 7]

*References: [1] Hulliger (1979); [2] Rustamov and Il'yasov (1984); [3] Hulliger (1968); [4] Rustamov et al. (1981a); [5] Godzhayev et al. (1977); [6] Aliyev and Rustamov (1978); [7] Rustamov et al. (1977a); [8] Rustamov et al (1981b).

The R_2Se_3–Sb_2Se_3 systems (R = Pr, Nd, Sm, Gd) are analogous to the system with cerium (Rustamov et al. 1981a). Upon transition from La to Gd, the solubility range on the basis of R_2Se_3 increases from 4 mol.% La_2Se_3 to 6 mol.% Gd_2Se_3 (Rustamov et al. 1977b), as shown in fig. 22, for the Nd_2Se_3–Sb_2Se_3 system.

When EuSe reacts with Bi_2Se_3 in the 1:1 ratio, the compound $EuBi_2Se_4$ is formed (Rustamov et al. 1981a, Godzhayev et al. 1977).

The Eu–As–Te system is known to exhibit the compound Eu_4As_2Te, resulting from a direct interaction of binary components (Hulliger 1979).

In the section La_2Te_3–Sb_2Te_3, the compound $LaSbTe_3$ melts by a peritectic reaction at 650°C,

$$L + La_2Te_3 \rightleftarrows LaSbTe_3$$ (Geidarova et al. 1982, Geidarova and Rustamov 1985).

The eutectic parameters between $LaSbTe_3$ and Sb_2Te_3 are: 75 mol.% Sb_2Te_3 and 540°C. 4 mol.% La_2Te_3 dissolves in Sb_2Te_3 at 25°C.

Analogous interaction of the components, resulting in formation of the ternary compound $RSbTe_3$, is observed in the R_2Te_3–Sb_2Te_3 systems (R = Pr, Nd, Sm, Dy, Er, Y). Narrow homogeneity ranges are observed in these systems on the basis of Sb_2Te_3.

The EuTe–Sb_2Te_3 system exhibits the compound $EuSb_2Te_4$ congruently melting at 670°C and the compound $EuSb_4Te_7$ incongruently melting at 620°C. The latter forms eutectics with $EuSb_2Te_4$ (570°C) and Sb_2Te_3 (530°C) (Aliyev and Rustamov 1978).

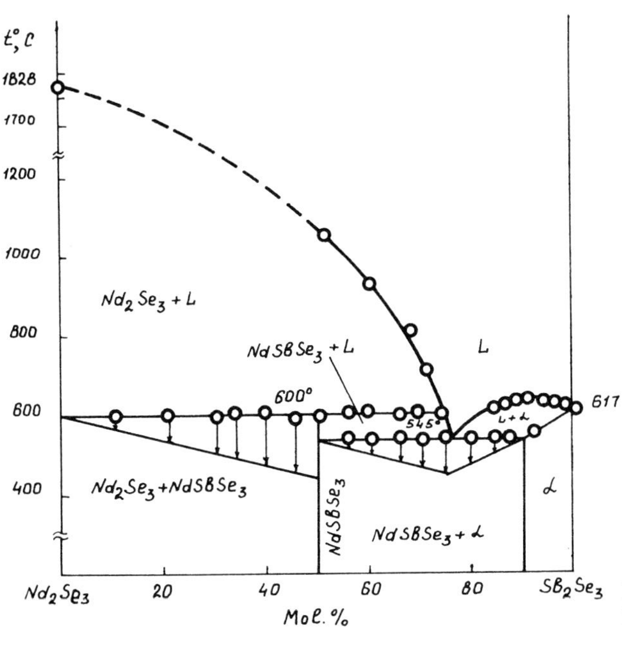

Fig. 22. The phase diagram of the Nd_2Se_3–Sb_2Se_3 system.

The eutectic parameters on the quasi-binary sections GdTa–Sb_2Te_3 and TbTe–Sb_2Te_3 are: 10 mol% GdTe and 853°C, and 15 mol.% TbTe and 858°C, respectively (Rustamov and Geidarova 1984). A range of solid solutions up to 5 mol.% GdTe(TbTe) is observed on the basis of Sb_2Te_3.

The Yb–Sb–Te system has been studied by Rustamov et al. (1981b, 1982a) and (Aliyev et al. 1985). The plot of the liquidus surface diagram shows that the sections $YbSb_2$–Sb_2Te_3 (fig. 23), Yb_4Sb_3–Sb_2Te_3, Sb_2Te_3–Yb, $YbSb_4Te_7$–Yb appear to be quasi-binary, whereas the section YbSb–Sb_2Te_3 is a complex, non-quasi-binary one. The system exhibits 14 fields of primary crystallization of the phases, and 27 curves of monovariant equilibria. The liquidus of YbTe–Sb_2Te_3 system consists of four branches: the two terminal branches correspond to the primary crystallization of YbTe and a solid solution based on Sb_2Te_3, and the other two correspond to the formation of the compound $YbSb_4Te_7$, which melts congruently at 670°C, and the compound $YbSb_4Te_7$, which melts according to peritectic at 650°C,

$$L + YbTe \rightleftarrows YbSb_2Te_4 .$$

The eutectics have the composition 23 and 46 mol.% YbTe. The compound $YbSb_2Te_4$ (table 22) has two modifications: a low-temperature modification (α) and a high-temperature modification (β). The transition $\alpha \rightarrow \beta$ occurs at 420°C. Solubility of YbTe in Sb_2Te_3 increases from 6 mol.% at 25°C to 8 mol% at 600°C.

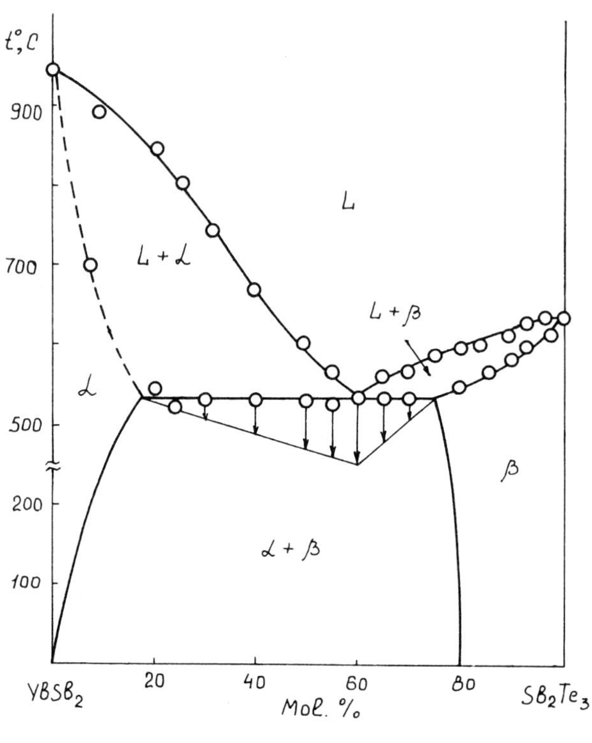

Fig. 23. The phase diagram of the $YbSb_2$–Sb_2Te_3 system.

TABLE 22
Crystallographic data for compounds in the R–{As, Sb, Bi}–Te systems.

Compound	a (Å)	b (Å)	c (Å)	Density (g/cm³) D_x	D_m	Note	Ref.*
R_4As_2Te Cubic lattice of the anti-Th_3P_4 type							
Eu_4As_2Te	9.453(3)						[1]
$RSbTe_3$ Orthorhombic lattice of the Bi_2Te_2S type (the cell parameters are given for hexagonal axes)							
$LaSbTe_3$	4.24		30.40			$a = 4.42$, $c = 29.42$ (ref. [3])	[2]
$NdSbTe_3$	4.32		29.40				[2]
$GdSbTe_3$	4.16		30.24				[2]
$EuSbTe_3$	4.16		30.39				[2]
$SmSbTe_3$	4.46		30.20				[2]
$DySbTe_3$	4.50		30.38				[2]
$YSbTe_3$	4.47		30.32				[2]
RSb_xTe_{2-x} Orthorhombic lattice. The structure is intermediate between $LaTe_2$ of the Cu_2Sb type and $LaSb_2$ (space group Cmca)							
$LaSb_{0.2}Te_{1.8}$	4.503	4.451	9.168				[4]
$LaSb_{0.3}Te_{1.7}$	4.493	4.414	9.232				[4]
$LaSb_{0.5}Te_{1.5}$	4.477	4.398	9.263				[4]
$LaSb_1Te_1$	4.440	4.388	9.330				[4]
$LaSb_{1.5}Te_{0.5}$	4.383	4.383	9.438			Tetragonal lattice	[4]
RSb_2Te_4 Cubic lattice							
$EuSb_2Te_4$	8.84					Structure of the Th_3P_4 type	[5, 6]
α-$YbSb_2Te_4$	10.40						[7]
R_4Sb_2Te Cubic lattice of the anti-Th_3P_4 type							
Eu_4Sb_2Te	9.814(5)						[1]
RSb_4Te_7 Cubic lattice							
$YbSb_4Te_7$	10.625						[6]
$RBiTe_3$ Rhombohedral lattice of the Bi_2Te_2S type (the cell parameters are given for hexagonal axes)							
$LaBiTe_3$	4.39		30.20				[6]
$CeBiTe_3$	4.19		31.6				[6]
$PrBiTe_3$	4.20		31.06	–	–		[6]
$NdBiTe_3$	4.18		31.1				[6]
$GdBiTe_3$	4.16		31.2				[6]
$DyBiTe_3$	4.14		31.02				[6]
$ErBiTe_3$	4.12		31.10				[6]
$YBiTe_3$	4.46		31.69	6.70	6.78		[8]
R_4Bi_2Te Cubic lattice of the anti-Th_3P_4 type							
Eu_4Bi_2Te	9.926(2)						[1]
Sm_4Bi_2Te	9.899(3)						[1]
Yb_4Bi_2Te	9.571(2)						[1]
RBi_2Te_4 Cubic lattice							
$EuBi_2Te_4$	8.90					Structure of the Th_3P_4 type	[9, 10]
$YbBi_2Te_4$	10.48						[9, 10]
$YbBi_4Te_7$ Cubic lattice							
$YbBi_4Te_7$	10.64						[9, 10]

*References: [1] Hulliger (1979); [2] Geidarova and Rustamov (1985); [3] Geidarova et al. (1982); [4] Wang et al. (1967); [5] Godzhayev et al. (1977); [6] Rustamov et al. (1981a); [7] Rustamov et al. (1977a); [8] Rustamov et al. (1978); [9] Rustamov et al. (1981b); [10] Aliyev and Maksudova (1986).

The sections NdTe–Bi, NdBi–NdTe, Bi_2Te_3–Nd, NdBi–Te, Bi_2Te_3–Nd_3Te_4, Bi_2Te_3–Nd_2Te_3 have been studied using a physicochemical analysis (Sadygov et al. 1986, Sadygov and Rustamov 1983, Geidarova and Rustamov 1985). In the ternary system Nd–Bi–Te, the quasi-binary sections appear to be Bi_2Te_3–NdTe, NdBi–NdTe, Bi_2Te_3–Nd_3Te_4, Bi_2Te_3–Nd_2Te_3, NdTe–Bi. The system triangulates into six secondary ternary systems: Nd–NdBi–NdTe, NdBi–Bi–NdTe, Bi–NdTe–Bi_2Te_3, Bi_2Te_3–NdTe–Nd_3Te_4, Bi_2Te_3–Nd_3Te_4–Nd_2Te_3, Bi_2Te_3–Nd_2Te_3–Te. The liquidus of the system consists of the 18 fields of primary crystallization, the field NdTe being the largest one. The monovariant curves intersect in 18 non-variant points, 6 of them being eutectic and 12 being peritectic.

The compound $DyBiTe_3$ is formed at the quasi-binary section Bi_2Te_3–Dy_2Te_3 according to a peritectic reaction at 970°C,

$$L + Dy_2Te_3 \rightleftarrows DyBiTe_3 \qquad \text{(Sadygov et al. 1987)}.$$

The eutectic parameters between the ternary compound and Bi_2Te_3 are: 80 mol.% Bi_2Te_3 and 552°C. The analogous sections are observed in the systems with erbium and gadolinium.

The systems R_3Te_4–Bi_2Te_3 (R = La, Dy) are analogous, quasi-binary, of eutectic type (Rustamov et al. 1982b, Sadygov and Rustamov 1983). The eutectic parameters are: 25 mol.% Dy_3Te_4 and 557°C.

The sections DyTe–Bi, DyTe–Bi_2Te_3 are quasi-binary, of eutectic type (Sadygov and Rustamov 1983, Murguzov et al. 1984). The eutectic parameters in the first section are: 3 mol.% DyTe and 262°C.

The section EuTe–Bi_2Te_3 exhibits the ternary compounds $EuBi_2Te_4$, which melts congruently, and $EuBi_4Te_7$, which melts incongruently.

The ternary system Yb–Bi–Te has been studied by Aliyev and Rustamov (1978) and Aliyev and Maksudova (1986). The liquidus of the YbTe–Bi_2Te_3 system consists of four branches; the two terminal lines correspond to the primary crystallization of YbTe and a Bi_2Te_3-base solid solution, the other two lines correspond to the formation of the compound $YbBi_4Te_7$, which melts congruently at 647°C, and the compound $YbBi_2Te_4$, which melts incongruently at 600°C. The eutectic parameters between $YbBi_4Te_7$ and Bi_2Te_3 are: 24 mol.% YbTe and 550°C. The compound $YbBi_2Te_4$ has two modifications (table 22): a low-temperature (α) and a high-temperature modification (β), the transition temperature of the latter being 400°C.

The liquidus of the Yb_4Bi_3–Bi_2Te_3 system is quasi-binary and of the eutectic type, it consists of the branches of the primary crystallization of the Bi_2Te_3- and $YbBi_2$-base solid solutions (Aliyev and Maksudova 1986). The branches of the primary crystallization intersect at 40 mol.% $YbBi_2$ and 550°C where their cocrystallization starts according to the following eutectic scheme,

$$L \rightleftarrows \text{solid solutions based on } Bi_2Te_3 + YbBi_2 .$$

The section $YbBi_2$–Bi_2Te_3 is also quasi-binary, of the eutectic type. In addition to the compounds described in the systems R–Bi–Te, the compound of the composition R_4Bi_2Te (R = Sm, Eu, Yb) is also known (Hulliger 1979).

Table 20 gives the crystallographic data for the compound $YbTaS_3$.

References

Aliyev, O., and T. Maksudova, 1986, Zh. Neorg. Khim. **31**, 2639.
Aliyev, O., and P. Rustamov, 1978, Zh. Neorg. Khim. **23**, 2800.
Aliyev, O., E. Magerramov and P. Rustamov, 1977, Zh. Neorg. Khim. **22**, 2836.
Aliyev, O., P. Rustamov, G. Guseinov and M. Guseinov, 1978, Izv. Akad. Nauk SSSR, Ser. Neorg. Mater. **14**, 1346.
Aliyev, O., T. Maksudova and P. Rustamov, 1985, Izv. Akad. Nauk SSSR, Ser. Neorg. Mater. **21**, 1479.
Angapova, L., I. Kamayeva and V. Serebrennikov, 1973, Vestn. Tomsk. Univ. Ser. Khim. **240**, 81.
Céolin, R., and P. Khodadad, 1971, C.R. Acad. Sci., Paris, Ser. C **272**, 1769.
Céolin, R., P. Toffoli, P. Khodadad and N. Rodier, 1977, Acta Crystallogr. B **33**, 2804.
Donohue, P., 1975, J. Solid State Chem. **12**, 80.
Finkel'shtein, L., and N. Samsonova, 1987, Zh. Neorg. Khim. **32**, 2595.
Geidarova, E., and P. Rustamov, 1985, Zh. Neorg. Khim. **30**, 2992.
Geidarova, E., P. Rustamov and I. Aliyev, 1982, Zh. Neorg. Khim. **27**, 1880.
Godzhayev, E., P. Rustamov, M. Guseinov and O. Aliyev, 1977, Izv. Akad. Nauk SSSR, Ser. Neorg. Mater. **13**, 624.
Guseinov, G., F. Mamedov, I. Amiraslanov and Kh. Mamedov, 1981, Kristallografiya **26**, 831.
Hulliger, F., 1968, Nature **219**, 373.
Hulliger, F., 1979, Mater. Res. Bull. **14**, 259.
Il'yasov, T., A. Mamedov and P. Rustamov, 1984, Zh. Neorg. Khim. **29**, 2094.
Il'yasov, T., A. Mamedov and P. Rustamov, 1986a, Zh. Neorg. Khim. **31**, 2075.
Il'yasov, T., L. Mamedova and Z. Guliyeva, 1986b, Zh. Neorg. Khim. **31**, 2706.
Il'yasov, T., P. Rustamov and L. Mamedova, 1986c, Zh. Neorg. Khim. **31**, 2437.
Khasayev, D., O. Aliyev and M. Guseinov, 1982, Izv. Akad. Nauk SSSR, Ser. Neorg. Mater. **18**, 331.
Lemoine, P., D. Carré and M. Guittard, 1986, Acta Crystallogr. C **42**, 259.
Murguzov, M., A. Gurshumov, Kh. Kadyrly, M. Karayeva and S. Bagirova, 1984, Zh. Neorg. Khim. **29**, 2965.
Rcheushvili, V., I. Bakhtadze and Ye. Nanobashvili, 1987, Soob. Akad. Nauk GSSR **126**, 545.
Rustamov, P., and E. Geidarova, 1984, Zh. Neorg. Khim. **29**, 2982.
Rustamov, P., and T. Il'yasov, 1984, Zh. Neorg. Khim. **29**, 2975.
Rustamov, P., E. Godzhayev, M. Guseinov and O. Aliyev, 1977a, Zh. Neorg. Khim. **22**, 540.
Rustamov, P., E. Magerramov, O. Aliyev, T. Guliyev and M. Alidzhanov, 1977b, Azerb. Khim. Zh. (1), 93.
Rustamov, P., F. Sadygov, Z. Melikova and M. Safarov, 1978, Zh. Neorg. Khim. **23**, 849.
Rustamov, P., O. Aliyev and T. Kurbanov, 1981a, Troiinyye Khal'kogenidi Redkozemel'nykh Elementov. Izd. "Elm", Baku.
Rustamov, P., O. Aliyev and T. Maksudova, 1981b, Izv. Akad. Nauk SSSR, Ser. Neorg. Mater. **17**, 966.
Rustamov, P., D. Khasayev and O. Aliyev, 1981c, Izv. Akad. Nauk SSSR, Ser. Neorg. Mater. **17**, 1976.
Rustamov, P., T. Maksudova and O. Aliyev, 1982a, Zh. Neorg. Mater. **27**, 1792.
Rustamov, P., F. Sadygov and Kh. Abilov, 1982b, Izv. Akad. Nauk SSSR, Ser. Neorg. Mater. **18**, 1123.
Rustamov, P., T. Il'yasov and M. Mamedov, 1985, Zh. Neorg. Khim. **30**, 769.
Rustamov, P., T. Il'yasov, A. Mamedov and F. Sadydov, 1987, Zh. Neorg. Khim. **32**, 158.
Sadygov, F., and P. Rustamov, 1983, Zh. Neorg. Khim. **28**, 2701.
Sadygov, F., P. Rustamov and T. Il'yasov, 1986, Zh. Neorg. Khim. **31**, 3106.
Sadygov, F., P. Rustamov, T. Il'yasov and V. Cherstvova, 1987, Izv. Akad. Nauk SSSR, Ser. Neorg. Mater. **23**, 545.
Serebrennikov, V., and M. Dashiyev, 1969, Izv. Akad. Nauk SSSR, Ser. Neorg. Mater. **5**, 2210.
Sfez, G., and C. Adolphe, 1972, Bull. Soc. Fr. Mineral. Crystallogr. **95**, 553.
Wang, R., H. Steinfink and A. Raman, 1967, Inorg. Chem. **6**, 1298.

Wibbelmann, C., and W. Brockner, 1984, Z. Naturforsch. A **39**, 190.
Yampol'skaya, V., and V. Serebrennikov, 1971, Vestn. Tomsk. Univ. Ser. Khim. **204**, 410.
Yampol'skaya, V., and V. Serebrennikov, 1973, Vestn. Tomsk. Univ. Ser. Khim. **249**, 181.

6. The R–M^{VI}–X systems

This section deals with the ternary compounds formed by rare earth elements and chalcogenides with the elements of groups VI through VIII: chromium, manganese, iron, cobalt, nickel and molybdenum.

The only phase, $RCrS_3$, with a very narrow homogeneity range is formed in the R_2S_3–Cr_2S_3 systems (R = La, Nd, Gd, Dy, Y) (Rustamov et al. 1977, Kurbanov et al. 1980). The temperature of congruent melting is, for the following compounds: $NdCrS_3$ at 1505°C, $GdCrS_3$ at 1520 ± 50°C, $DyCrS_3$ at 1600°C, $YCrS_3$ at 1620°C. Two eutectics were observed: between R_2S_3 and $RCrS_3$, and between $RCrS_3$ and Cr_2S_3 (fig. 24). The eutectic parameters for the Dy_2S_3–Cr_2S_3 system are: 60 mol.% Cr_2S_3 and 1400°C, 73 mol.% Cr_2S_3 and 1275 ± 50°C.

Two congruently melting compounds, $CrLa_2S_4$ and $CrLa_4S_7$, are observed at the section La_2S_3–CrS. The liquidus of the system consists of the curves of primary crystallization of CrS, $CrLa_2S_4$, $CrLa_4S_7$ and La_2S_3. Eutectics are observed between CrS and $CrLa_2S_4$, $CrLa_2S_4$ and $CrLa_4S_7$, $CrLa_4S_7$ and La_2S_3. The phase with the CrR_4S_7 composition was observed in all R_2S_3–CrS (R = La–Tm, Y) systems (Kurbanov et al. 1980). Two crystal classes for the CrR_2S_4 compounds are known: monoclinic for R = Pr–Dy, and orthorhombic for R = Ho–Yb, Y, having the MnY_2S_4-type structure (Patrie et al. 1964, Patrie and Chevalier 1966). Some authors report a doubling of the lattice period along the

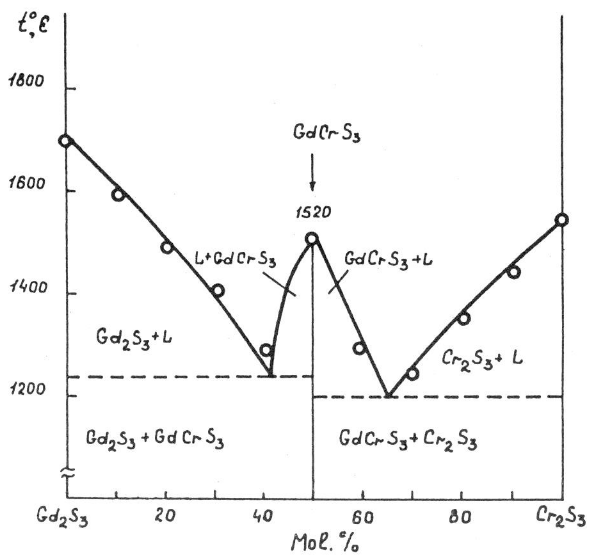

Fig. 24. The phase diagram of the Gd_2S_3–Cr_2S_3 system.

c-axis (Tomas et al. 1976b, Tomas and Guittard 1980). The melting temperatures of some compounds CrR_2S_4 are known (Kurbanov et al. 1980). The compound of the composition R_3CrS_6 melts incongruently, while RCr_3S_6 melts congruently (Rustamov et al. 1984, Takahashi et al. 1974).

The composition R_2MS_4 was initially assigned to the phase $R_{32.66}M_{11}S_{60}$ (M = Cr, Mn, Fe) (Patrie et al. 1968).

The following ternary compounds are observed in the ErS–CrS system: $CrEr_2S_4$, $CrEr_4S_7$, $Cr_2Er_6S_{11}$, $Cr_2Er_8S_{14}$, $Cr_3Er_8S_{15}$ (table 23) (Tomas and Guittard 1980, Tomas 1980).

The compound MnR_4S_7 (R = Dy–Yb, Y) has the Y_5S_7-type structure (Flahaut et al. 1961, 1962a, Adolphe and Laruelle 1968, Tomas et al. 1984). Compounds of this composition have structural types: R = Tb–Tm, Y with the MnY_2S_4-type structure (Patrie et al. 1964, Patrie and Chevalier 1966, Tomas and Guittard 1980), and R = Tm–Lu, Sc with the spinel type (Longo and Raccah 1967, Patrie et al. 1964, Patrie and Guittard 1966, Guittard et al. 1964). The compound $MnTm_2S_4$ is stable only at high temperature.

Various phases have been observed in the $MnS–Y_2S_3$ system using electron microscopy (Bakker and Hyde 1978): MnY_2S_4, MnY_4S_7, $Mn_2Y_6S_{11}$, $Mn_4Y_{10}S_{19}$, $Mn_3Y_{10}S_{18}$, $Mn_4Y_{14}S_{25}$. The structures of the phases are formed from the cell MnS (the NaCl type) by means of twinning.

The data on the composition of isostructural compounds with a hexagonal lattice are contradictory, Takahashi et al. (1971) reports the composition as either $LaMS_3$(M = Fe, Co, Ni) or $La_6M_{10/3}S_{14}$ of the $Ce_6Al_{10/3}S_{14}$ type structure. On the other hand Collin et al. (1968) assigned the same structure type to the compounds FeR_4S_7 (R = La–Nd), $NiLa_4S_7$, CoR_4S_7 (R = La, Ce).

According to Mössbauer spectroscopy and X-ray studies, the structure of the compounds $FeYb_2S_4$ and $FeLu_2S_4$ appears to be of an intermediate type which is between the spinel and the NaCl type (Tomas et al. 1980).

The compounds with the compositions $La_2Fe_{2-x}\square_xS_5$ (□ designates vacancies, $x = 0$, 0.3, 0.24) are known (Besrest and Collin 1977). The vacancies of iron result in the appearance of a superstructure.

The information about selenide systems is not complete. The compounds of composition $RCrSe_3$ belong to two different crystal classes: R = Gd–Lu, Y (monoclinic) and R = La–Dy (orthorhombic).

The compound of the composition CrR_2Se_4 belongs to the monoclinic symmetry (Rustamov et al. 1978), Kurbanov et al. 1980).

The Mn–R–Se system exhibits ternary compound $MnRSe_4$ (R = Yb, Lu) with the spinel-type structure (table 24) (Flahaut 1968).

Table 24 gives the data for the ternaray telluride compound $EuCr_2Te_4$.

TABLE 23
Crystallographic data for compounds in the R–{Cr, Mn, Fe, Co, Ni, Mo}–S systems.

Compound	a (Å)	b (Å)	c (Å)	α	β	γ	Density (g/cm³) D_x	D_m	Note	Ref.*
$RCrS_3$ Orthorhombic lattice										
$GdCrS_3$	13.8	16.54	3.95						Single crystal	[1]
$DyCrS_3$	13.75	16.43	3.94						Single crystal	[1]
$RCrS_3$ Monoclinic lattice (space group $P2_1/a$)										
$LaCrS_3$	6.00(1)	17.12(2)	11.05(2)		95.3(1)°		4.22	4.45	Single crystal	[2, 3, 4, 5]
$GdCrS_3$	5.86	10.69	21.41		91.1°		5.29	5.08		[2, 3, 4, 5]
$DyCrS_3$	5.71	10.71	21.28		–		5.56	5.32	Orthorhombic lattice, space group Pcab	[2, 3, 4, 5]
$HoCrS_3$	5.66	10.61	21.27		95.1°		5.72	5.42		[2, 3, 4, 5]
$ErCrS_3$	5.79	10.68	21.26		95.6°		–	–		[2, 3, 4, 5]
$YCrS_3$	5.75	10.65	21.24		95.3°		4.44	4.24		[2, 3, 4, 5]
$RCrS_3$ Triclinic lattice										
$LaCrS_3$	5.94	17.2	66.2	90.3°	95.3°	90.0°	4.24	4.53	The structure was determined on a single crystal	[6]
CrR_2S_4 Monoclinic lattice										
$CrPr_2S_4$	10.92	3.0	5.90		92°					[1]
CrR_2S_4 Orthorhombic lattice										
$CrHo_2S_4$	12.632(6)	12.563(7)	7.584(6)						$a = 12.51$, $b = 12.51$, $c = 3.76$, $D_m = 5.40$ (ref. [8])	[7]
$CrEr_2S_4$	12.555(6)	12.481(6)	7.536(4)						The structure was determined on a single crystal (ref. [7, 9]). According to ref. [8]: $a = 12.46$, $b = 12.46$, $c = 3.74$, $D_m = 5.52$	[7]
$CrTm_2S_4$	12.509(7)	12.397(8)	7.507(6)						$a = 12.44$, $b = 12.44$, $c = 3.59$, $D_m = 5.59$ (ref. [8])	[7]
$CrYb_2S_4$	12.62(1)	12.50(1)	7.59(1)						$a = 12{,}41$, $b = 12.41$, $c = 3.72$, $D_m = 5.73$ (ref. [8])	[7]
CrY_2S_4	12.58(1)	12.50(1)	7.59(1)						$a = 12.53$, $b = 12.53$, $c = 3.77$, $D_m = 3.66$ (ref. [8]). $a = 12.52(3)$, $b = 12.52(3)$, $c = 3.77(1)$, $D_x = 4.03$, $D_m = 4.04$ (ref. [10])	[7]

CrR_4S_7 Monoclinic lattice of the Y_5S_7 type									
$CrHo_4S_7$	12.56	3.77	11.32	105.75°		–	–		[11]
$CrEr_4S_7$	12.50	3.76	11.29	105.65°		–	–		[9]
$CrTm_4S_7$	12.46	3.74	11.26	105.55°		–	–		[12]
CrY_4S_7	12.61	3.79	11.41	106°		3.99	3.91		[10]
RCr_2S_4 Hexagonal lattice (space group P6)									
$EuCr_2S_4$	21.422		3.453						[13]
$YbCr_2S_4$	22.32		3.39						[13]
$Cr_xEr_yS_z$									
$Cr_2Er_6S_{11}$	12.471	34.156	3.751		–			Orthorhombic lattice, the structure was determined on a single crystal (ref. [14])	[9]
$Cr_2Er_8S_{14}$	12.40	42.8	3.75		90°			Monoclinic lattice, space group Bb	[9]
$Cr_3Er_8S_{15}$	12.46	47.26	3.76		90°			Monoclinic lattice, space group Bb	[9]
RCr_3S_6 Orthorhombic lattice, space group P2nm or Pmnm									
$LaCr_3S_6$	13.12	11.27	10.70				4.08		[15]
$CeCr_3S_6$	13.08	11.23	10.68				4.19		[15]
$PrCr_3S_6$	13.05	11.20	10.65				4.30		[15]
$NdCr_3S_6$	12.99	11.19	10.64				4.48		[15]
$SmCr_3S_6$	12.90	11.00	10.63				4.50		[15]
$GdCr_3S_6$	12.88	10.95	10.60				4.57	$a = 12.81$, $b = 10.97$, $c = 10.65$, $D_x = 4.530$, $D_m = 4.15$; single-crystal data (ref. [16])	[15]
$TbCr_3S_6$	12.76	10.95	10.60				4.60		[15]
$DyCr_3S_6$	12.74	10.95	10.60				4.69	$a = 12.78$, $b = 10.95$, $c = 10.52$, $D_x = 4.609$, $D_m = 4.26$; single-crystal data (ref. [16])	[17]
$HoCr_3S_6$	12.76	10.90	10.50				4.88	$a = 12.75$, $b = 10.94$, $c = 10.51$, $D_x = 4.651$, $D_m = 4.26$; single-crystal data (ref. [16])	[15]
$ErCr_3S_6$	12.73	10.87	10.50				4.96	$a = 12.74$, $b = 10.88$, $c = 10.46$, $D_x = 4.723$, $D_m = 4.52$; single-crystal data (ref. [16])	[15]
$TmCr_3S_6$	12.72	10.84	10.48			5.00			[15]
$YbCr_3S_6$	12.65	10.82	10.40			4.44			[15]

TABLE 23 (cont'd)

Compound	a (Å)	b (Å)	c (Å)	α	β	γ	Density (g/cm³) D_x	Density (g/cm³) D_m	Note	Ref.*
RCr_3S_6 (cont'd)										
$LuCr_3S_6$	12.60	10.76	10.35				5.24			[15]
YCr_3S_6	12.80	10.93	10.52				4.71		$a = 12.75$, $b = 10.93$, $c = 3.977$, $D_x = 3.977$, $D_m = 3.85$; single-crystal data (ref. [16])	[15]
R_3CrS_6 Orthorhombic lattice										
La_3CrS_6	13.56	16.85	4.26					4.94		[15]
Ce_3CrS_6	13.52	16.80	4.20					5.06		[15]
Pr_3CrS_6	13.48	16.77	4.16					5.16		[15]
Nd_3CrS_6	13.46	16.74	4.16					5.25		[15]
Sm_3CrS_6	13.37	16.66	4.10					5.40		[15]
Gd_3CrS_6	13.26	16.46	3.90					5.53		[15]
Tb_3CrS_6	13.25	16.44	3.86					5.77		[15]
Dy_3CrS_6	13.23	16.44	3.76					5.80		[15]
Ho_3CrS_6	13.15	16.40	3.70					5.36		[15]
Er_3CrS_6	13.12	16.37	3.64					5.72		[15]
Tm_3CrS_6	13.10	16.21	3.54					5.99		[15]
Yb_3CrS_6	12.94	16.11	3.52					5.60		[15]
Lu_3CrS_6	12.90	16.08	3.50					6.14		[15]
Y_3CrS_6	13.18	16.41	3.74					5.21		[15]
MnR_2S_4 Orthorhombic lattice of the MnY_2S_4 type										
$MnTb_2S_4$	12.63	12.76	3.78					5.47		[8]
$MnDy_2S_4$	12.61	12.72	3.77					5.58		[8]
$MnHo_2S_4$	12.54	12.66	3.76					5.71		[8]
$MnEr_2S_4$	12.52	12.63	3.75					5.80		[8]
MnY_2S_4	12.62	12.75	3.78					3.94	There exists a non-stoichiometric phase with a cubic lattice, $a = 5.375$ (ref. [18]) $a = 12.64$, $b = 12.71$, $c = 3.76$, $D_x = 3.62$, $D_m = 3.64$ (ref. [19]) The compound was studied in ref. [9]	[8]

$Mn_4Y_{10}S_{19}$	12.62	60.3	3.78					Electron microscopy	[18]
$Mn_2Y_6S_{12}$	12.62	35	3.78						[18]
$Mn_3Y_{10}S_{11}$	12.62	56.9	3.78						[9]
$Mn_4Y_{14}S_{25}$	12.62	78.9	3.78					Electron microscopy	[18]
MnR_4S_7 Hexagonal lattice of the $Ce_6Al_{3.33}S_{14}$ type									
$MnLa_4S_7$	10.330		5.751						[20]
$MnCe_4S_7$	10.217		5.660						[20]
MnR_4S_7 Monoclinic lattice of the Y_5S_7 type. The compounds of the MnR_4S_7 composition are also studied (ref. [11])									
$MnDy_4S_7$	12.679	3.804	11.448	105.55°		5.79	5.81		[21]
$MnHo_4S_7$	12.637	3.786	11.436	105.48°					[21]
$MnEr_4S_7$	12.577	3.768	11.395	105.31°		6.05	6.03	The structure was determined on a single crystal of the $Mn_{0.4}Er_{4.6}S_7$ composition: $a = 12.573(4)$, $b = 11.390(4)$, $c = 3.777(4)$, $\beta = 105.45°$ (ref. [22])	[21]
$MnTm_4S_7$	12.519	3.755	11.357	105.40°					[21]
$MnYb_4S_7$	12.473	3.745	11.315	105.20°					[21]
MnY_4S_7	12.636	3.790	11.443	105.45°				$a = 12.67$, $b = 3.79$, $fc = 11.43$, $\beta = 105.5°$, $D_x = 3.89$, $D_m = 4.00$ (ref. [19])	[21]
$R_{32.66}Mn_{11}S_{60}$ Monoclinic lattice									
$La_{32.66}Mn_{11}S_{60}$	16.650	10.906	14.082		102.58°			The structure was determined on a single crystal in hexagonal approximation: $a = 8.12$, $b = 7.26$, space group $P6_3/mcm$ for compounds of the $La_6Mn_2S_{10}$ composition	[23]
FeR_2S_4 Orthorhombic lattice of the MnY_2S_4 type									
$FeYb_2S_4$	12.38	12.49	3.70					Low-temperature modification $Fe_{0.76}Yb_{2.16}S_4$ has a structure of the spinel type	[24]

TABLE 23 (cont'd)

Compound	a (Å)	b (Å)	c (Å)	α	β	γ	Density (g/cm^3) D_x	D_m	Note	Ref.*
FeR_4S_7 Hexagonal lattice (space group $P6_3$)										
$FeLa_4S_7$	10.311		5.749							[20]
$FeCe_4S_7$	10.202		5.657							[20]
$FePr_4S_7$	10.152		5.553							[20]
FeR_4S_7 Monoclinic lattice of the Y_5S_7 type. These compounds were studied in refs. [10, 11, 19, 25]										
$FeHo_4S_7$	12.570	3.778	11.341		105.70°		–	–		[21]
$FeEr_4S_7$	12.521	3.756	11.317		105.61°		6.15	6.16		[21]
$FeTm_4S_7$	12.499	3.746	11.298		105.52°		–	–		[21]
$FeYb_4S_7$	12.448	3.732	11.279		105.31°		–	–		[21]
FeY_4S_7	12.614	3.788	11.368		105.78°		4.04	4.08		[21]
$R_2Fe_2S_5$ Orthorhombic lattice										
$La_2Fe_2S_5$	3.997(t)	16.485(5)	11.394(3)						The structure was determined on a single crystal, space group $Cmc2_1$	[26]
$Pr_2Fe_2S_5$	3.907	16.18	11.11							[27]
$La_2Fe_{1.87}S_5$	3.995	49.508	11.308						The structure was determined on a single crystal, space group $Cmc2_1$	[26]
$La_2Fe_{1.76}S_5$	4.001	32.936	11.291						The structure was determined on a single crystal, space group $Pmc2_1$	[28]
$R_{32.66}Fe_{11}S_{60}$ Monoclinic lattice										
$La_{32.66}Fe_{11}S_{60}$	16.545	10.845	13.983			102.91°			The structure was determined on a single crystal for the hexagonal approximation	[29]
$Pr_{32.66}Fe_{11}S_{60}$	7.920	–	7.115						Hexagonal lattice, space group $P6_3/mcm$	[29]

R_4CoS_7 Hexagonal lattice (space group $P6_3$)					
La_4CoS_7	10.285	5.747		For the phase of the La_4CoS_7 composition, the tetragonal lattice of the K_2NiF_4-type with $a_0 = 4.104$, $c_0 = 16.377$ is given (refs. [23, 29]) superstructural reflections have been found: $a = 4a_0$, $c = c_0$	[23]
Ce_4CoS_7	10.193	5.645			[23]
$R_{32.66}Co_{11}S_{60}$ Monoclinic lattice					
$La_{32.66}Co_{11}S_{60}$	7.959	7.149		Parameters are given for the hexagonal approximation	[27]
R_4NiS_7 Tetragonal lattice of the K_2NiF_4 type					
La_4NiS_7	4.081(1)	16.334(5)		The structure was determined on a single crystal. Superstructure reflections have been found: $a' = 4a$, $c' = c$. There exists a phase with a hexagonal lattice: $a = 10.264$, $c = 5.744$, space group $P6_3$ (ref. [20])	[29]
Ce_4NiS_7	4.040(1)	16.175(5)			[29]
Pr_4NiS_7	4.017(1)	16.175(5)			[29]
Nd_4NiS_7	3.995(1)	16.037(5)			[29]
RMo_6S_8 Rhombohedral lattice					
$CeMo_6S_8$	6.504		88.76°	Single crystal	[30]
$EuMo_6S_8$	6.5531(2)		88.931(5)°	The structure was determined on a single crystal	[31]

*References: [1] Kurbanov et al. (1980); [2] Takahashi et al. (1973); [3] Takahashi et al. (1971); [4] Dismukes et al. (1970); [5] Kurbanov et al. (1977); [6] Kato and Kawada (1977); [7] Tomas et al. (1976a, b); [8] Patrie and Chevalier (1966); [9] Tomas and Guittard (1980); [10] Flahaut et al. (1962b); [11] Adolphe and Laruelle (1968); [12] Flahaut et al. (1961); [13] Omloo and Iellinek (1968); [14] Tomas (1980); [15] Rustamov et al. (1984); [16] Takahashi et al. (1974); [17] Guseinov et al. (1984); [18] Bakker and Hyde (1978); [19] Flahaut et al. (1962a); [20] Collin et al. (1968); [21] Adolphe (1965); [22] Tomas et al. (1984); [23] Collin and Laruelle (1974); [24] Tomas and Guittard (1977); [25] Adolphe et al (1964); [26] Besrest and Collin (1977); [27] Dy et al. (1984); [28] Besrest and Collin (1978); [29] Collin and Flahaut (1974); [30] Peña et al. (1987); [31] Peña et al. (1986).

TABLE 24
Crystallographic data for compounds in the R–Cr–{Se, Te} systems.

Compound	a (Å)	b (Å)	c (Å)	β	Density (g/cm^3) D_m	Note	Ref.*
$RCrSe_3$ Orthorhombic lattice of the $CeCrSe_3$ type							
$LaCrSe_3$	8.11	13.79	3.96		–	The structure was determined on a single crystal	[1, 2]
$CeCrSe_3$	8.08	13.74	3.95		–		[1, 2]
$PrCrSe_3$	8.05	13.70	3.93		–		[1, 2]
$NdCrSe_3$	8.01	13.61	3.91		–		[1, 2]
$TbCrSe_3$	8.07	13.54	3.88		–		[1, 2]
$DyCrSe_3$	7.97	13.50	3.86		–		[1, 2]
$GdCrSe_3$	14.04	16.74	3.97		5.14		[3]
CrR_2Se_4							
$CrPr_2Se_4$	11.26	3.06	4.96	94°		Monoclinic lattice similar to $CrPr_2S_4$ it resembles vacant structures of the $M_3\square X_4$ type with an orderly arrangement of vacancies (ref. [4])	[3]
CrY_2Se_4	7.27	11.70	6.00	–		Orthorhombic lattice	[5]
RCr_2Se_4 Hexagonal lattice (space group P6)							
$EuCr_2Se_4$	22.380		3.627				[6, 7, 8]
$YbCr_2Se_4$	22.40		3.50				[6, 7, 8]
R_3CrSe_6 Orthorhombic lattice							
Gd_3CrSe_6	14.02	16.81	3.97			$a = 13.975(9)$, $b = 16.770(19)$, $c = 3.950(3)$. The crystal structure was determined on a single crystal	[9]
RCr_2Te_4 Hexagonal lattice (space group P6)							
$EuCr_2Te_4$	22.62		3.86				[5, 6, 8]

*References: [1] Nguyen et al. (1969); [2] Nguyen et al. (1971); [3] Kurbanov et al. (1980); [4] Rustamov et al. (1978); [5] Rustamov and Aliyev (1981); [6] Omloo and Iellinek (1968); [7] Pink and Gobel (1973); [8] Omloo et al. (1971); [9] Guseinov et al. (1981).

References

Adolphe, C., 1965, Ann. Chim. (Paris) **10**, 271.
Adolphe, C., and P. Laruelle, 1968, Bull. Soc. Fr. Mineral. Crystallogr. **91**, 219.
Adolphe, C., M. Guittard and P. Laruelle, 1964, C.R. Acad. Sci., Paris, Ser. C **258**, 4773.
Bakker, M., and B. Hyde, 1978, Philos. Mag. A **38**, 615.
Besrest, F., and G. Collin, 1977, J. Solid State Chem. **21**, 161.

Besrest, F., and G. Collin, 1978, J. Solid State Chem. **24**, 301.
Collin, G., and J. Flahaut, 1974, J. Solid State Chem. **9**, 352.
Collin, G., and P. Laruelle, 1974, Acta Crystallogr. B **30**, 1134.
Collin, G., F. Rouyer and J. Loriers, 1968, C.R. Acad. Sci., Paris, Ser. C **266**, 689.
Dismukes, J., R. Smith and J. White, 1970, Z. Kristallogr. **132**(4–6), 272.
Nguyen, Huy Dung, J. Étienne and P. Laruelle, 1969, C.R. Acad. Sci., Paris, Ser. C **269**, 120.
Nguyen, Huy Dung, J. Étienne and P. Laruelle, 1971, Bull. Soc. Chim. Fr. (7), 2433.
Dy, Y. Tang, Z. Li and D. Yang, 1984, Microstructures and Properties Ceramic Materials Proc., Vol. 1, China-US, Bilateral Semin. Inogr. Mater. Res. Shanghai, May 17–21, 1983, Beijing, Amsterdam, p. 202.
Flahaut, J., 1968, in: Progress Science and Technology of Rare Earths, Vol. 3 (Oxford) p. 209.
Flahaut, J., L. Domange and M. Patrie, 1961, Bull. Soc. Chim. Fr. (1), 105.
Flahaut, J., L. Domange and M. Patrie, 1962a, Bull. Soc. Chim. Fr. (1), 159.
Flahaut, J., L. Domange and M. Patrie, 1962b, Bull. Soc. Chim. Fr. (1), 157.
Guittard, M., C. Souleau and H. Farsam, 1964, C.R. Acad. Sci., Paris, Ser. C **259**, 2847.
Guseinov, G., I. Amiraslanov, Yu. Yusifov and Kh. Mamedov, 1984, Dokl. Akad. Nauk Azerb. SSR **40**, 40.
Guseinov, T., V. Gasymov, I. Aliyev and Kh. Mamedov, 1981, Izv. Akad. Nauk SSSR, Ser. Neorg. Mater. **17**, 805.
Kato, K., and I. Kawada, 1977, Acta Crystallogr. B **33**, 3437.
Kurbanov, T., P. Rustamov, O. Aliyev and I. Aliyev, 1977, Izv. Akad. Nauk SSSR, Ser. Neorg. Mater. **13**, 742.
Kurbanov, T., P. Rustamov, O. Aliyev, L. Abdulayeva, A. Einullayev and I. Aliyev, 1980, Izv. Akad. Nauk SSSR, Ser. Neorg. Mater. **16**, 611.
Longo, J., and P. Raccah, 1967, Mater. Res. Bull. **2**, 541.
Omloo, W., J. Bommerson, H. Heikens, H. Risselada and M. Vellinga, 1971, Phys. Status Solidi (a) **5**, 349.
Omloo, W.P., and F. Iellinek, 1968, Reckel. Trav. Chim. **87**, 545.
Patrie, M., and R. Chevalier, 1966, C.R. Acad. Sci., Paris, Ser. C **263**, 1061.
Patrie, M., J. Flahaut and L. Domange, 1964, C.R. Acad. Sci., Paris, Ser. C **258**, 2585.
Patrie, M., Nguyen Huy Dung and J. Flahaut, 1968, C.R. Acad. Sci., Paris, Ser. C **266**, 1575.
Peña, O., R. Horin, C. Geantet, P. Gougeon, J. Padion and M. Sergent, 1986, J. Solid State Chem. **63**, 62.
Peña, O., R. Horin, C. Geantet, M. Sergent, A. Dinia, S. Quezel, L. Regnault, M. Bonnet and J. Rossat-Mignod, 1987, J. Magn. & Magn. Mater. **63–64**, 40.
Pink, H., and H. Gobel, 1973, Z. Anorg. Allg. Chem. **402**, 312.
Rustamov, P., and O. Aliyev, 1981, V kn: Redkozemel'nyye Poluprovodniki, Izd. "Elm", Baku, p. 93.
Rustamov, P., T. Kurbanov, O. Aliyev, A. Kamarzin, V. Sokolov and I. Aliyev, 1977, Zh. Neorg. Khim. **22**, 1948.
Rustamov, P., G. Guseinov, T. Kurbanov, A. Einullayev and O. Aliyev, 1978, Izv. Akad. Nauk SSSR, Ser. Neorg. Mater. **14**, 649.
Rustamov, P., T. Kurbanov, O. Aliyev and I. Aliyev, 1984, Izv. Akad. Nauk SSSR, Ser. Neorg. Mater. **20**, 1919.
Takahashi, T., T. Oka, O. Yamada and K. Ametani, 1971, Mater. Res. Bull. **6**, 173.
Takahashi, T., S. Ocaka and O. Yamada, 1973, J. Phys. Chem. Solids **34**, 1131.
Takahashi, T., K. Ametani and O. Yamada, 1974, J. Cryst. Growth **24–25**, 151.
Tomas, A., 1980, Acta Crystallogr. B **36**, 1987.
Tomas, A., and M. Guittard, 1977, Mater. Res. Bull. **12**, 1043.
Tomas, A., and M. Guittard, 1980, Mater. Res. Bull. **15**, 1547.
Tomas, A., R. Chevalier, P. Laruelle and B. Bachet, 1976a, Acta Crystallogr. B **32**, 3287.
Tomas, A., M. Guittard, R. Chevalier and J. Flahaut, 1976b, C.R. Acad. Sci., Paris, Ser. C **282**, 587.
Tomas, A., L. Brossard and M. Guittard, 1980, J. Solid State Chem. **34**, 11.
Tomas, A., M. Robert, C. Adolphe and M. Guittard, 1984, Mater. Res. Bull. **19**, 1643.

Handbook on the Physics and Chemistry of Rare Earths, Vol. 13
edited by K.A. Gschneidner Jr. and L. Eyring

Chapter 90

THE SYSTEMS R_2O_3–M_2O_3–M′O

[R is In, Sc, Y or one of the lanthanides, M is Fe, Ga, or Al, and M′ is one of the divalent cation elements]

Noboru KIMIZUKA and Eiji TAKAYAMA-MUROMACHI
National Institute for Research in Inorganic Materials, 1-1, Namiki, Tsukuba-shi, Ibaraki-ken, 305 Japan

Kiiti SIRATORI*
Department of Physics, Faculty of Science, Osaka University, 1-1, Machikaneyama, Toyonaka-shi, Osaka-fu, 560 Japan

Contents

*Visiting research officer of the National Institute for Research in Inorganic Materials.

List of symbols

a_i	activity of the i-th component
$\boldsymbol{a}, \boldsymbol{b}$	unit cell vectors in the c-plane
$\boldsymbol{a}^*, \boldsymbol{b}^*$	reciprocal vectors in the c-plane
A, B, C	triangular lattice points
a, c	lattice parameters
c, h	symbols, used to designate two types of oxygen layers
d, d'	R–O and shortest M(M′)–O distances in $RMM'O_4$, respectively
e	electronic charge
E	electromotive force (Volt)
E_Q	quadrupole splitting of a Mössbauer spectrum
F	Faraday constant
$f(O_2)$	oxygen fugacity
G^0	standard Gibbs free energy
H^0	standard enthalpy
H_{cool}	external magnetic field applied during the cooling of specimens
H_{meas}	external magnetic field applied during measurement
I, I_z	nuclear spin, z-component of it, respectively
I.S.	isomer shift of a Mössbauer spectrum
J	exchange constant
k_B	Boltzmann constant
l_z	quantum number of 3d orbital
l, p	numbers of negative and positive signs in a Zhdanov's notation, respectively
M	magnetization
m, n	integral numbers, used to designate a chemical formula, $(RMO_3)_n(M'O)_m$
n	mole number
N	total number of signs in a Zhdanov's notation
p	pressure
$P(O_2)$	oxygen partial pressure
q	order parameter of spins of transition metal ions on both sides of the layer of the rare earth element
$\boldsymbol{q}$	wave vector in the reciprocal space
r	ionic radius
$\boldsymbol{r}$	position vector in the real space
s	order parameter of spins of transition metal ions on adjacent c-planes

S	Seebeck coefficient	Z	number of formula units in a unit cell
S^0	standard entropy	γ	critical exponent for magnetic susceptibility
t	tolerance factor		
T	temperature	Γ	width of a Mössbauer absorption with a Lorentzian profile
T_N	Néel temperature		
T, U	layers having the composition (M, M′)O or RO_2, respectively	κ	inverse of the correlation length of spins
U	activation energy	ν	critical exponent for correlation length of spins
v	velocity, a scale of γ-ray energy in Mössbauer spectroscopy		
		ρ	electrical resistivity
w	hopping frequency of carriers	σ	electrical conductivity
x	deviation from the stoichiometric composition	χ	magnetic susceptibility
		ω	angular frequency
x, y, z	fractional atomic coordinates		

1. Introduction and scope

Since the 1970s, phase diagrams of ternary systems R_2O_3–M_2O_3–M′O (R: Y or lanthanides, M: trivalent cation elements and M′: divalent cation elements) have been systematically studied and new homologous compounds RFe_2O_4 and $R_2Fe_3O_7$ with hexagonal layered structures have been synthesized. Phase relationships in the systems R_2O_3–Fe_2O_3–Fe (M = M′ = Fe in the above notation) at 1200°C are classified into four groups with respect to the assemblages of $RFeO_3$ (perovskites), $R_3Fe_5O_{12}$ (garnets), RFe_2O_4 and $R_2Fe_3O_7$:

Type A: R = La, Pr and Nd. Only $RFeO_3$ exists stably.

Type B: R = Sm, Eu, Gd, Tb and Dy. Two phases, $RFeO_3$ and $R_3Fe_5O_{12}$, exist stably.

Type C: R = Y, Ho, Er and Tm. Three phases, $RFeO_3$, $R_3Fe_5O_{12}$ and RFe_2O_4, exist stably.

Type D: R = Yb and Lu. Four phases, $RFeO_3$, $R_3Fe_5O_{12}$, RFe_2O_4 and $R_2Fe_3O_7$, exist stably.

Further investigations revealed that there is a series of higher-order compounds that can be expressed as $(YbFeO_3)_nFeO$. The constituents of the compounds extend to R = In, Y or lanthanides of Ho–Lu, M = Fe, Ga or Al and M′ = Mg, Mn, Co, Cu, Fe(II), Zn or Cd. Another series of compounds, the formula of which can be written as $RMO_3(M'O)_m$ (R: Sc, In, Y or Er–Lu, M: Fe, Ga or Al, and M′ = Zn) with a layered structure were also synthesized. Thus, the existence of a large number of homologous compounds with a general formula $(RMO_3)_n(M'O)_m$ has been disclosed in the ternary systems. RFe_2O_4 and $R_2Fe_3O_7$ are members of this series with $n = 1$, $m = 1$ and $n = 2$, $m = 1$, respectively.

Section 2 of the present chapter offers details of the phase relationships in the ternary systems and sect. 3 describes data of the thermochemical properties for $RFeO_3$, $R_3Fe_5O_{12}$, RFe_2O_4 and $R_2Fe_3O_7$.

The crystal structures of $YbFe_2O_4$ and $Yb_2Fe_3O_7$ were analyzed by means of

X-ray diffraction of single crystals. The $(RMO_3)_n(M'O)_m$-type compounds with higher n or m have been examined by powder X-ray diffraction and electron microscopy. The common characteristics of their structures are:

(i) Both cations and anions form two-dimensional triangular lattices,

(ii) The M and M′ ions are coordinated by five oxygen ions, forming a trigonal bipyramid,

(iii) The R ion is located in an octahedron of the oxygen ions, the corners of which are vertices of the above-mentioned trigonal bipyramid.

Details of the crystal structures, tables of lattice constants and atomic positions, as well as the general rules of constructing the $(RMO_3)_n(M'O)_m$-type structure are presented in sect. 4. Hexagonal $YAlO_3$ and würtzite-type ZnO are argued as the end members of the series, with $n = \infty$, $m = 0$ and $n = 0$, $m = \infty$, respectively. Atomic states of 3d transition metal ions in trigonal bipyramidal coordination, as well as conditions of the appearance of the present series of compounds in relation to other types of compounds, are also discussed in sect. 4.

Sections 5 and 6 are devoted to the electronic and magnetic properties, respectively. The physical properties of this series are believed to be determined mainly by the layers of transition metal ions. They are characterized by a triangular net perpendicular to the c-axis. The frame of the crystal lattice is, on the other hand, determined by the rare earth elements. The character of the R ions is reflected in the properties indirectly but definitely.

The physical properties of this series have been investigated most extensively in the Fe compound. Based upon the ionic-crystal model, RFe_2O_4 is considered to contain an equal number of Fe(II) and Fe(III) in equivalent crystal sites, as in Fe_3O_4. Coexistence of Fe(II) and Fe(III) ions, or charge fluctuation of Fe ions, results in a rather good electrical conductivity, $\sim 10^{-1}$ $(\Omega\,cm)^{-1}$ at room temperature. Electrical conductivity shows a marked anisotropy. Conductivity within a c-plane is about two orders of magnitude higher than that along the c-axis at room temperature. The stoichiometric compound YFe_2O_4 shows a two-step phase transformation at about 220 K with a conductivity gap of 10^2 $(\Omega\,cm)^{-1}$ at each step, accompanied by a crystal deformation. This phase transformation is compared with the Verwey transition in Fe_3O_4 or Ti_4O_7.

In the compounds containing Fe, exchange interactions have appreciable magnitude. Here also the two-dimensional character of the crystal appears. The main exchange coupling is antiferromagnetic between nearest-neighbor sites of the triangular lattice. These series of compounds offer examples of frustrated spin systems, the ground states of which are infinitely degenerate. Much interest has been stimulated in recent years in these compounds. The magnetic properties of $RFeM'O_4$ (M′: divalent transition metal ions except Fe) are similar to spin glasses or metamagnets. Qualitatively, no difference is observed in the magnetic properties of RFe_2O_4 and $RFeM'O_4$, although the difference is rather large quantitatively. The difference can be explained by the strong uniaxial magnetic anisotropy of the Fe(II) ions. Magnetic properties are the subject of discussion in sect. 6.

The crystal chemical and physical properties of RMO_3 (perovskite) and $R_3M_5O_{12}$ (garnet) in the system R_2O_3–M_2O_3 are very interesting and important subjects. We will not, however, describe them, since Khattak and Wang (1979) have already reviewed them comprehensively.

2. Phase relationships

The systems R–O, R_2O_3–Fe_2O_3 and M_2O_3–M′O and M′O–R_2O_3 which are the subsystems for the R_2O_3–M_2O_3–M′O systems have been reviewed so far. Holley et al. (1968) and Eyring (1979) reviewed comprehensively the systems R–O and summarized detailed thermochemical properties of R_2O_3 and their lower and higher oxides. Roth (1964) critically reviewed the binary systems, R_2O_3–Fe_2O_3, R_2O_3–Ga_2O_3, R_2O_3–Cr_2O_3, R_2O_3–Al_2O_3 and R_2O_3–R'_2O_3 (R, R′: Y, Sc or lanthanides) and showed the stability ranges of RMO_3 (perovskite) and $R_3M_5O_{12}$ (garnet). We will describe these binary systems, only when it is necessary to understand the ternary systems. The binary system Fe_2O_3–Fe at elevated temperatures was thoroughly investigated by Darken and Gurry (1945, 1946). Smiltens (1957) reported the equilibrium oxygen partial pressure between Fe_3O_4 and Fe_2O_3 in the temperature range from 1081 to 1458°C. Muan and Osborn (1965) presented a comprehensive review article on the system Fe–O and summarized the relationships between temperature, composition and oxygen fugacity.

2.1. *Experimental methods*

There are three established methods that enable us to determine the phase relationships of the oxide systems at elevated temperatures:

(i) the classical quenching method;

(ii) the thermogravimetric method with controlling oxygen fugacities $f(O_2)$ in the atmosphere. (In this chapter, the oxygen fugacity will be considered to be, approximately, the oxygen partial pressure so we will use $P(O_2)$ (atm) instead of $f(O_2)$ hereafter.); and

(iii) the solid-state electrogalvanic cell method.

Since they have been described in many standard textbooks (e.g., Nafziger et al. 1971, Sato 1971) we do not describe each of the methods in detail.

The classical quenching method gives us the relationships in phases that were rapidly cooled from the established temperature, and both the thermogravimetric method, with controlled $P(O_2)$ in the atmosphere, and the solid-state electrogalvanic cell method provide the in situ data at elevated temperatures and make possible the determination of equilibrium oxygen partial pressures or electromotive force of the oxidation/reduction system.

In the following part, we briefly describe the starting compounds, temperature and $P(O_2)$ controls, e.m.f. measurements which are fundamental experimental conditions for evaluating the reliability of the phase relationships, $P(O_2)$, ΔG^0, ΔH^0 and ΔS^0 reported in the systems R_2O_3–M_2O_3–M′O at elevated temperatures.

(a) *The starting compounds*

The purities of the starting compounds used for determining the phase relationships in the system R_2O_3–M_2O_3–M′O cited here were above 99.9% unless otherwise stated.

(b) *Temperature and oxygen fugacity controls, and e.m.f. measurements*

Thanks to the development of techniques in electronic controlling systems and of ceramic forming techniques at high temperature, we can expect temperature control with high precision and a high quality of solid-state electrolytes. So, the phase relationships which are reviewed in the present section were determined at the established temperatures ±2°C.

In order to control the oxygen partial pressure in the atmosphere, a buffer system, either CO_2–CO, H_2O–H_2 or CO_2–H_2, was used. In the phase relationships in which $P(O_2)$ was controlled by means of mixed-gas system, $P(O_2)$ measured with oxygen sensors was compared with $P(O_2)$ calculated from the mixing ratios of CO_2/CO, H_2O/H_2 or CO_2/H_2. And when they were identical within experimental error, the phase relationships reported are given. It is almost meaningless to use the atmospheres of mixed gas systems which are not in equilibrium in the atmosphere for determining equilibrium oxygen partial pressure values of the solid phases, even if it is measured with a $P(O_2)$ sensor (Takayama and Kimizuka 1979).

Stabilized ZrO_2 was used, in general, as electrolyte. The transference number of the electrolyte stabilized with Y_2O_3, MgO or CaO is considered unity in the oxygen partial pressure and temperature ranges discussed in this review.

(c) *Determination of equilibrium states*

In the case of the classical quenching method, equilibrium was considered to have been attained when X-ray powder diffraction for the heated samples showed no further change.

Both for the gravimetric method and electrogalvanic method, each of the reactions was approached from both sides, oxidation and reduction, and when the sample weights or e.m.f. were identical within experimental error, the systems were considered to have been in equilibrium states.

2.2. *The phase relationships in the systems R_2O_3–Fe_2O_3–Fe (R: Y and the lanthanides) determined by means of the gravimetric method with controlled oxygen partial pressure*

In this section, we will discuss the phase relationships determined by thermogravimetry while controlling $P(O_2)$ from 0 to 15.5 in terms of $-\log P(O_2)$. R_2O_3 is stable in this $P(O_2)$ range and a nonstoichiometric range was not detected, except for Eu_2O_3, Pr_2O_3 and Tb_2O_3. The composition of europium oxide ranges from $Eu_2O_{2.98}$ to $Eu_2O_{3.00}$ at 1200°C.

The details of the relationships between the compositions of praseodymium oxide, and terbium oxide and $-\log P(O_2)$ will be described later.

The phase relationships in the systems R_2O_3–Fe_2O_3–Fe at 1200°C are classified into four groups with respect to the assemblages of $RFeO_3$, $R_3Fe_5O_{12}$, RFe_2O_4 and $R_2Fe_3O_7$.

2.2.1. *R = La, Pr and Nd (type A)*

The system La_2O_3–Fe_2O_3 in air in the temperature range of 1250–1900°C was originally reported by Moruzzi and Shafer (1960). $LaFeO_3$ and $LaFe_{12}O_{19}$ appear as stable binary phases and the latter decomposes to Fe_2O_3 and $LaFeO_3$ below 1380°C and to magnetite and $LaFeO_3$ above 1420°C. Figures 1 and 2 show the phase relationships in the systems La_2O_3–Fe_2O_3–Fe at 1204 (Kimizuka and Katsura 1974), 1100 and 1000°C (Sugihara, 1978) and Nd_2O_3–Fe_2O_3–Fe at 1200, 1100 and 1000°C (Sugihara 1978), respectively. There are higher oxides $PrO_{1.6+x}$ and $PrO_{1.71+x}$ in addition to $PrO_{1.5}$ in the system Pr–O (Eyring 1979) and the relation between x in $PrO_{1.50+x}$ and $-\log P(O_2)$ at 1200, 1100 and 1000°C are shown in fig. 3a. The phase relationships in the system Pr–Fe–O at 1200, 1100 and 1000°C are shown in figs. 3b,c (Sugihara 1978). Using a classical quenching method, Tretyakov et al. (1976b) investigated the system Pr–Fe–O at 1300 K, and reported only one binary phase, $PrFeO_3$. There is only one phase $RFeO_3$ in the binary system R_2O_3–Fe_2O_3 (R: La, Pr or Nd) and no ternary phase exists. The phase assemblages in the system Pr_2O_3–Fe_2O_3–Fe are the same at 1200 and at 1100°C, however, there is an additional $PrO_{1.71+x}$ phase at 1000°C. No significant deviations from the stoichiometric composition were found in the perovskites, $RFeO_3$ (R = La, Pr and Nd).

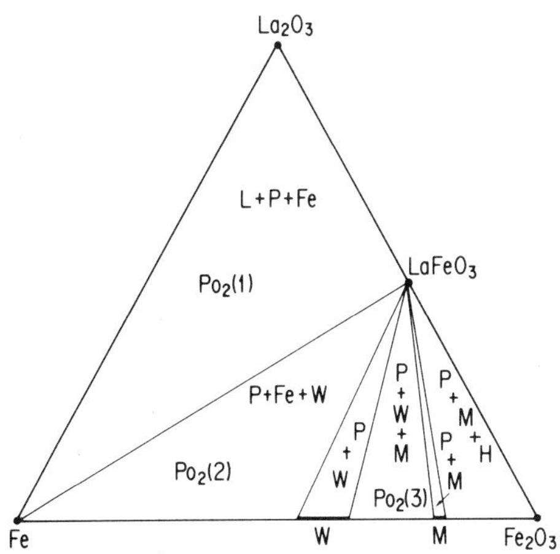

Fig. 1. The phase relationships in the systems La_2O_3–Fe_2O_3–Fe at 1204, 1100 and 1000°C. L: La_2O_3, P: $LaFeO_3$, Fe: γ-iron, W: wüstite, M: magnetite and H: hematite, $-\log P(O_2)(1) = 13.57$ (1204°C), 15.13 (1100°C) and 17.02 (1000°C), $-\log P(O_2)(2) = 11.92$ (1204°C), 13.34 (1100°C) and 14.93 (1000°C), $-\log P(O_2)(3) = 9.12$ (1204°C), 10.72 (1100°C) and 13.36 (1000°C), where $P(O_2)(1)$, $P(O_2)(2)$ and $P(O_2)(3)$ are the equilibrium oxygen partial pressures over the three phases L + P + Fe, P + Fe + W, and P + W + M, respectively. (Kimizuka and Katsura 1974a, b, Sugihara 1978.)

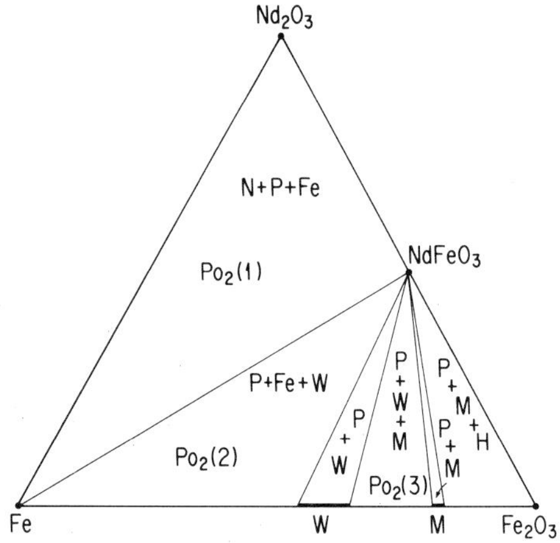

Fig. 2. The phase relationships in the system Nd_2O_3–Fe_2O_3–Fe at 1200, 1100 and 1000°C. N: Nd_2O_3, P: $NdFeO_3$, Fe: γ-iron, W: wüstite, M: magnetite and H: hematite, $-\log P(O_2)(1) = 12.99$ (1200°C), 14.54 (1100°C) and 16.40 (1000°C), $-\log P(O_2)(2) = 11.92$ (1200°C), 13.34 (1100°C) and 14.93 (1000°C) and $-\log P(O_2)(3) = 9.14$ (1200°C), 10.72 (1100°C) and 13.36 (1000°C), where $P(O_2)(1)$, $P(O_2)(2)$ and $P(O_2)(3)$ are the equilibrium oxygen partial pressures over the three phases N + P + Fe, P + Fe + W, and P + W + M, respectively (Sugihara 1978).

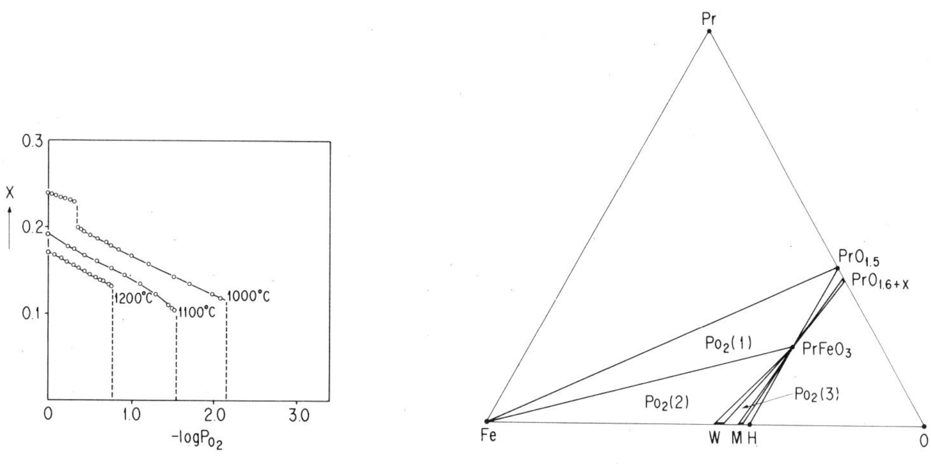

Fig. 3. (a) The relations between x in $PrO_{1.5+x}$ and $-\log P(O_2)$ at 1200, 1100 and 1000°C. (Sugihara 1978.) (b) The phase relationships in the system $PrO_{1.5}$–$PrO_{1.6+x}$–Fe–O at 1200, and 1100°C. W: wüstite, M: magnetite and H: hematite, $-\log P(O_2)(1) = 13.07$ (1200°C) and 14.62 (1100°C), $-\log P(O_2)(2) = 11.92$ (1200°C) and 13.34 (1100°C) and $-\log P(O_2)(3) = 9.14$ (1200°C) and 10.72 (1100°C). (Sugihara 1978.) (c) The phase relationships in the system $PrO_{1.5}$–$PrO_{1.71+x}$–Fe–O at 1000°C. W: wüstite, M: magnetite and H: hematite, $-\log P(O_2)(1) = 16.56$, $-\log P(O_2)(2) = 14.93$ and $-\log P(O_2)(3) = 13.36$. (Sugihara 1978.)

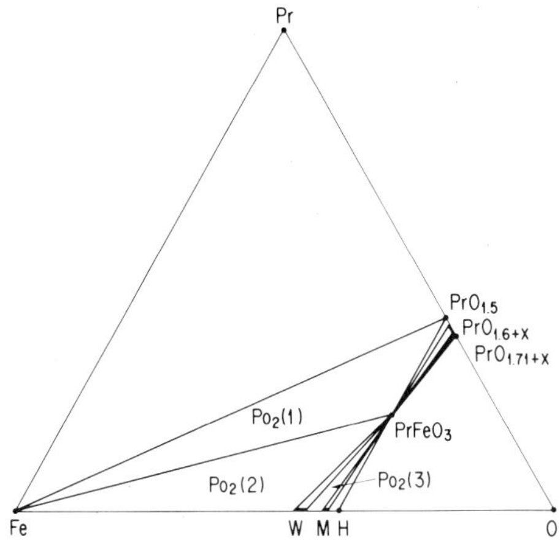

Fig. 3(c).

2.2.2. *The system Ce_2O_3–Fe–O (type A)*

The phase relationships for the system Ce_2O_3–Fe–O at 1200°C are shown in fig. 4 (Kitayama et al. 1985). The system CeO_2–Ce_2O_3 at elevated temperatures has been studied by means of thermogravimetric method controlling $P(O_2)$ with CO_2–CO, H_2O–H_2 or CO_2–H_2 mixture system, and solid-state galvanic cell method (Eyring 1979). The relation between the composition and the structure in the CeO_2–Ce_2O_3 system has already been reviewed by Eyring (1979). The component, composition, $-\log P(O_2)$ and activity in the system Ce_2O_3–Fe–O at various temperatures are summarized in table 1a. There is only one binary phase, $CeFeO_3$, having the perovskite structure in the oxygen partial pressure range from 7.25 to 12.47 in terms of $-\log P(O_2)$ at 1200°C. The composition ranges

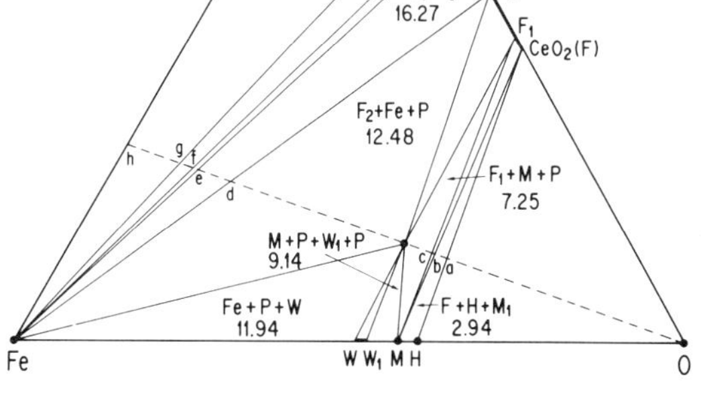

Fig. 4. The phase relationships in the system Ce_2O_3–Fe–O at 1200°C. The numerical values in the three solid regions are equilibrium oxygen partial pressures in terms of $-\log P(O_2)$. The symbols are the same as those in table 1a. (Kitayama et al. 1985.)

Table 1A
Composition and activity ($=a_i$) in the system Ce_2O_3–Fe–O (Kitayama et al. 1985).

T (°C)	Component	Composition	Symbol	$-\log P(O_2)$	$\log a_i$
1000	CeO_2	$CeO_{2.00}$	F	0–4.0[a]	0
		$CeO_{1.99}$	F_1	10.72	0.010
		$CeO_{1.87}$	F_2	15.68	0.101
	Ce_2O_3	$Ce_2O_{3.00}$	R		0
	$CeFeO_3$	$CeFeO_{3.00}$	P	10.72–15.68	0
1100	CeO_2	$CeO_{2.00}$	F	0–3.2[a]	0
		$CeO_{1.99}$	F_1	8.63	0.013
		$CeO_{1.86}$	F_2	13.96	0.162
	Ce_2O_3	$Ce_2O_{3.00}$	R		0
	$CeFeO_3$	$CeFeO_{3.00}$	P	8.63–13.96	0
1153	CeO_2	$CeO_{2.00}$	F	0–3.0[a]	0
		$CeO_{1.72}$	F_3	16.91	0.563
	Ce_3O_5	$Ce_3O_{5.10}$	C_1	16.91	−0.0108
		$Ce_3O_{4.96}$	C	17.65	0
	Ce_2O_3	$Ce_2O_{3.00}$	R	17.65–	0
1200	CeO_2	$CeO_{2.00}$	F	0–2.7[a]	0
		$CeO_{1.99}$	F_1	7.25	0.0134
		$CeO_{1.84}$	F_2	12.48	0.191
		$CeO_{1.72}$	F_3	16.27	0.600
	Ce_3O_5	$Ce_3O_{5.09}$	C_1	16.27	−0.0054
		$Ce_3O_{4.96}$	C	16.70	0
	Ce_2O_3	$Ce_2O_{3.00}$	R	16.70–	0
	$CeFeO_3$	$CeFeO_{3.00}$	P	7.25–12.48	0
1249	CeO_2	$CeO_{2.00}$	F	0–2.5[a]	0
		$CeO_{1.71}$	F_3	15.55	0.637
	Ce_3O_5	$Ce_3O_{5.07}$	C_1	15.55	0.0029
		$Ce_3O_{4.95}$	C	16.00	0
	Ce_2O_3	$Ce_2O_{3.00}$	R	16.00–	0
1310	CeO_2	$CeO_{2.00}$	F	0–2.5[a]	0
		$CeO_{1.70}$	F_3	14.87	0.654
	Ce_3O_5	$Ce_3O_{5.05}$	C_1	14.87	−0.0002
		$Ce_3O_{4.95}$	C	15.10	0
	Ce_2O_3	$Ce_2O_{3.00}$	R	15.10–	0
1330	CeO_2	$CeO_{2.00}$	F	0–2.3[a]	0
		$CeO_{1.69}$	F_3	14.57	0.703
	Ce_3O_5	$Ce_3O_{5.00}$	C_1	14.57	−0.00234
		$Ce_3O_{4.94}$	C	14.83	0
	Ce_2O_3	$Ce_2O_{3.00}$	R	14.83–	0

[a]Estimated value by extrapolation.

TABLE 1B
The standard Gibbs free energies of reaction in the system CeO_2–Fe–O (Kitayama et al. 1985).

Reaction	T (°C)	$-\log P(O_2)$	$-\Delta G^0$ (kJ)	$-\Delta G^{0\,a}$ (kJ)
$\frac{3}{2}Ce_2O_3 + \frac{1}{4}O_2 \rightarrow Ce_3O_5$	1153	17.65	120.5	
	1200	16.70	117.7	
	1249	16.00	116.6	
	1310	15.10	114.4	
	1330	14.83	113.8	
$\frac{1}{3}Ce_3O_5 + \frac{1}{6}O_2 \rightarrow CeO_2$	1153	16.91	61.7	
	1200	16.27	59.6	
	1249	15.55	57.0	
	1310	14.87	55.3	
	1330	14.57	54.5	
$\frac{1}{3}Fe_3O_4 + CeO_2 \rightarrow CeFeO_3 + \frac{1}{6}O_2$	1000	10.72	−43.3	−41.5
	1100	8.63	−36.3	−37.0
	1200	7.25	−33.7	−32.4
$CeO_2 + Fe + \frac{1}{2}O_2 \rightarrow CeFeO_3$	1000	15.68	193.5	190.8
	1100	13.96	187.7	183.2
	1200	12.48	181.4	175.6

[a]Tretyakov et al. (1976a).

from $CeFeO_{2.99}$ to $CeFeO_{3.00}$. In table 1b, the chemical reactions, $-\log P(O_2)$ and Gibbs free-energy changes of reactions in the system Ce_2O_3–Fe–O are summarized. Tretyakov et al. (1976a) investigated, with the e.m.f. method, the phase equilibria and the thermodynamics of co-existing phases in the system Ce–Fe–O at 1300 K, and showed that only one ternary phase, $CeFeO_3$, with the perovskite structure existed.

2.2.3. $R =$ *Sm, Eu, Gd, Tb and Dy* (*type B*)

The phase relationships in the systems Sm_2O_3–Fe_2O_3–Fe (Kitayama and Katsura 1976), Eu_2O_3–Fe_2O_3–Fe (Sugihara et al. 1975), Gd_2O_3–Fe_2O_3–Fe and Dy_2O_3–Fe_2O_3–Fe (Sugihara 1978) at 1200, 1100 and 1000°C are shown in figs. 5–8.

There are two phases, $RFeO_3$ (perovskite) and $R_3Fe_5O_{12}$ (garnet), in the binary system R_2O_3–Fe_2O_3 and no ternary phase exists at 1000–1200°C. The composition of the perovskites ranges from, $SmFeO_{2.982}$ to $SmFeO_{3.000}$, $EuFeO_{2.976}$ to $EuFeO_{3.000}$, $TbFeO_{2.97}$ to $TbFeO_{3.00}$ and $DyFeO_{2.98}$ to $DyFeO_{3.00}$, respectively, at 1200°C. No significant deviation from the stoichiometric composition was observed in $GdFeO_3$. In table 2, components, compositions and activities in the system Sm_2O_3–Fe_2O_3–Fe at 1200°C are shown.

Terbium oxide has a homogeneity range as expressed by $TbO_{1.5+x}$. The relation between x in $TbO_{1.5+x}$ and $-\log P(O_2)$ at 1200, 1100 and 1000°C are shown in fig. 9a. The phase relationships in the system Tb–Fe–O at 1200, 1100 and 1000°C (Sugihara 1978) are shown in fig. 9b. McCarthy and Fischer (1972) and Tretyakov

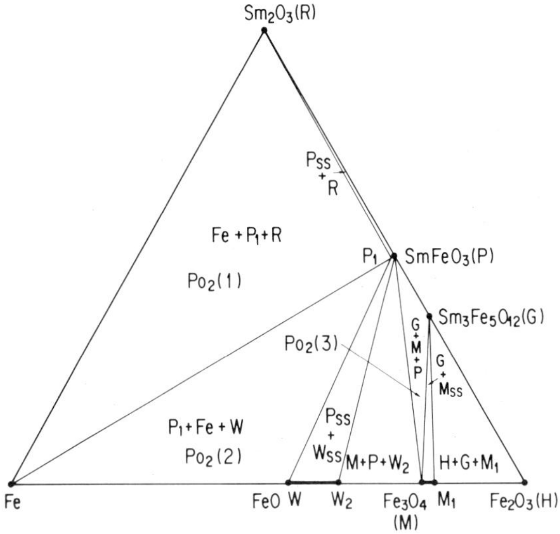

Fig. 5. The phase relationships in the system Sm_2O_3–Fe_2O_3–Fe at 1200, 1100 and 1000°C. R: Sm_2O_3, P: perovskite phase, Fe: γ-iron, W: wüstite, M: magnetite, H: hematite $-\log P(O_2)(1) = 12.68$ (1200°C), 14.23 (1100°C) and 16.11 (1000°C), $-\log P(O_2)(2) = 11.94$ (1200°C), 13.34 (1100°C) and 14.93 (1000°C), $-\log P(O_2)(3) = 3.72$ (1200°C), 5.75 (1100°C) and 8.44 (1000°C), where $P(O_2)(1)$, $P(O_2)(2)$ and $P(O_2)(3)$ are the equilibrium oxygen partial pressures over the three phases, R + P_1 + Fe, P_1 + Fe + W and M + P + G. The letters R, P, G and M represent stoichiometric compositions of Sm_2O_3, $SmFeO_3$, $Sm_3Fe_5O_{12}$ and Fe_3O_4, respectively. M_1 means the end-member of the magnetite solid solution with a chemical composition. $Fe_{2.957}O_4$, P_{ss}, W_{ss} and M_{ss} mean the solid solution of $SmFeO_3$ from P to P_1, of FeO from W to W_2, and Fe_3O_4 from M to M_1, respectively. The other letters as in table 2. (Kitayama and Katsura 1976.)

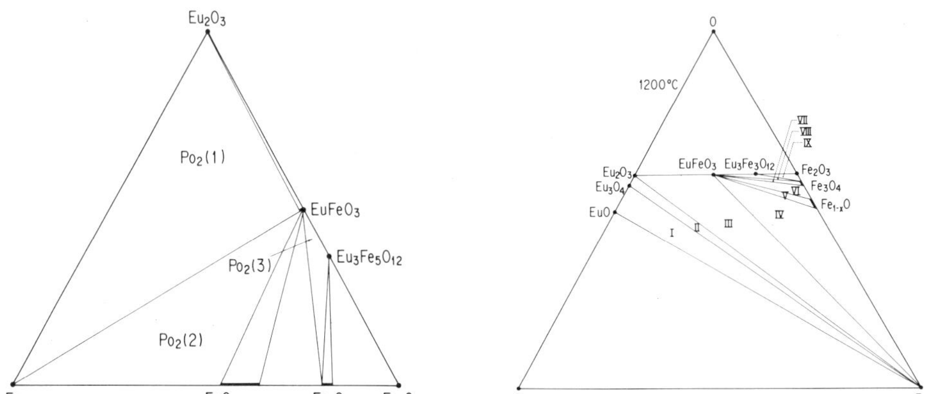

(a) (b)

Fig. 6. (a) The phase relationships in the system Eu_2O_3–Fe_2O_3–Fe at 1200, 1100 and 1000°C. $-\log P(O_2)(1) = 12.52$ (1200°C), 14.08 (1100°C) and 15.99 (1000°C), $-\log P(O_2)(2) = 11.92$ (1200°C), 13.34(1100°C) and 14.93 (1000°C) and $-\log P(O_2)(3) = 4.33$ (1200°C), 6.29 (1100°C) and 9.14 (1000°C), where $P(O_2)(1)$, $P(O_2)(2)$ and $P(O_2)(3)$ are the equilibrium oxygen partial pressures over the three phases. Eu_2O_3 + Fe + $EuFeO_3$, $EuFeO_3$ + Fe + W, and $EuFeO_3$ + $Eu_3Fe_5O_{12}$ + M, respectively. (Sugihara et al. 1975, Sugihara 1978.) (b) The phase relationships in the system Eu_2O_3–EuO–Fe_2O_3–Fe at 1200°C. The meaning of the numerical symbols is given in table 3. (McCarthy and Fischer 1972.)

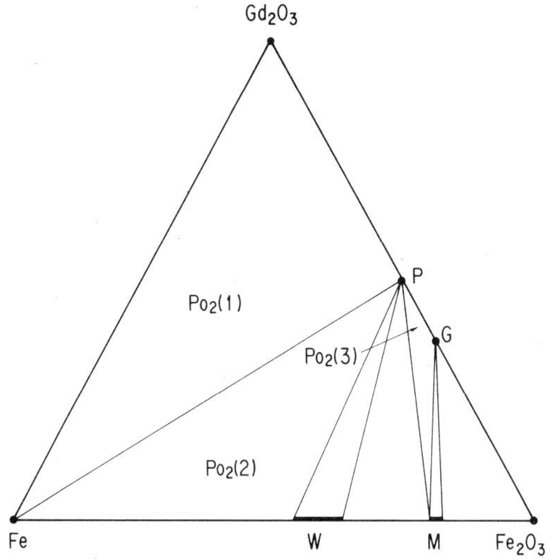

Fig. 7. The phase relationships in the system Gd_2O_3–Fe_2O_3–Fe at 1200, 1100 and 1000°C. P: perovskite phase, Fe: γ-iron, W: wüstite, M: magnetite, $-\log P(O_2)(1) = 12.49$ (1200°C), 14.04 (1100°C) and 15.91 (1000°C), $-\log P(O_2)(2) = 11.94$ (1200°C), 13.34 (1100°C) and 14.93 (1000°C) and $-\log P(O_2)(3) = 5.12$ (1200°C), 6.45 (1100°C) and 9.48 (1000°C), where $P(O_2)(1)$, $P(O_2)(2)$ and $P(O_2)(3)$ are the equilibrium oxygen partial pressures over the three phases, Gd_2O_3 + P + Fe, P + Fe + W and P + G + M, respectively. (Sugihara 1978.)

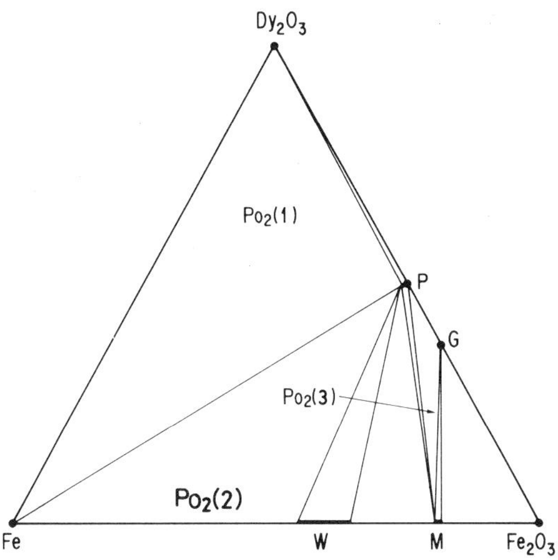

Fig. 8. The phase relationships in the system Dy_2O_3–Fe_2O_3–Fe at 1200, 1100 and 1000°C. $-\log P(O_2)(1) = 12.11$ (1200°C), 13.65 (1100°C) and 15.56 (1000°C). $-\log P(O_2)(2) = 11.92$ (1200°C), 13.34 (1100°C) and 14.93 (1000°C). $-\log P(O_2(3) = 5.35$ (1200°C), 7.44 (1100°C) and 9.84 (1000°C), where $P(O_2)(1)$, $P(O_2)(2)$ and $P(O_2)(3)$ are the equilibrium oxygen partial pressures over the three phases Dy_2O_3 + P + Fe, P + Fe + W, and P + G + M. (Sugihara 1978.)

TABLE 2
Composition and activity ($=a_i$) in the system Sm_2O_3–Fe_2O_3–Fe at 1200°C (Kitayama and Katsura 1976).

Component	Composition	Symbol	$-\log P(O_2)$	$\log a_i$
FeO	$FeO_{1.049}$	W	11.94	0
	$FeO_{1.125}$	W_1	10.00	−0.0805
	$FeO_{1.166}$	W_2	9.14	−0.150
$SmFeO_3$	$SmFeO_{2.982}$	P_1	12.68	0
	$SmFeO_{3.000}$	P	9.14	0.0512
			0.68	0.0512
$Sm_3Fe_5O_{12}$	$Sm_3Fe_5O_{12.00}$	G	3.72	0

et al. (1976c) reported the phase relationships in the system Eu_2O_3–EuO–Fe_2O_3–Fe at 1200°C which were determined by means of classical quenching methods. The phase relationships of the subsystem Eu_2O_3–Fe_2O_3–Fe in their phase diagram (fig. 6b) are identical with those in fig. 6a which was obtained by thermogravimetry. The phase assemblages indicated with Roman numerals in fig. 6b are listed in table 3.

2.2.4. *R = Y, Ho, Er and Tm (type C)*

Figures 10–13 are the phase relationships in the systems Y_2O_3–Fe_2O_3–Fe (Kimizuka and Katsura 1975a), Ho_2O_3–Fe_2O_3–Fe (Katsura et al. 1978), Er_2O_3–

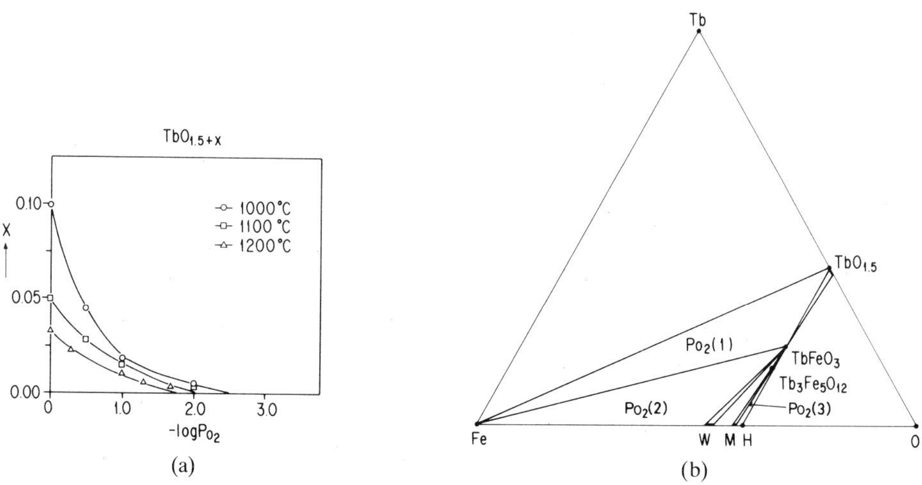

Fig. 9. (a) The relations between x in $TbO_{1.5+x}$ and $-\log P(O_2)$ at 1200, 1100 and 1000°C. (Sugihara 1978.) (b) The phase relationships in the system $TbO_{1.5}$–$TbO_{1.5+x}$–Fe–O at 1200, 1100 and 1000°C, $-\log P(O_2)(1) = 12.22$ (1200°C), 13.83 (1100°C) and 15.70 (1000°C), $-\log P(O_2)(2) = 11.94$ (1200°C), 13.34 (1100°C) and 14.93 (1000°C) and $-\log P(O_2)(3) = 5.83$ (1200°C), 7.52 (1100°C) and 9.87 (1000°C), where $P(O_2)(1)$, $P(O_2)(2)$ and $P(O_2)(3)$ are equilibrium oxygen partial pressures over the three phases, Tb_2O_3 + Fe + $TbFeO_3$, $TbFeO_3$ + W + Fe and $TbFeO_3$ + $Tb_3Fe_5O_{12}$ and M. (Sugihara 1978.)

TABLE 3
Phase assemblages in fig. 6b.

No.	Assemblage
I	$EuO + Eu_3O_4 + Fe$
II	$Eu_3O_4 + Eu_2O_3 + Fe$
III	$Eu_2O_3 + EuFeO_3 + Fe$
IV	$EuFeO_3 + Fe + Fe_{1-x}O$
V	$EuFeO_3 + Fe_{1-x}O$
VI	$EuFeO_3 + Fe_{1-x}O + Fe_3O_4$
VII	$EuFeO_3 + Fe_3O_4$
VIII	$EuFeO_3 + Fe_3O_4 + Eu_3Fe_5O_{12}$
IX	$Eu_3Fe_5O_{12} + Fe_3O_4 + Fe_2O_3$

Fe_2O_3–Fe (Kitayama and Katsura 1976) and Tm_2O_3–Fe_2O_3–Fe (Katsura et al. 1978), respectively. There are two phases, $RFeO_3$ (perovskite) and $R_3Fe_5O_{12}$ (garnet) in the binary systems R_2O_3–Fe_2O_3, and one ternary phase RFe_2O_4 [$(RFeO_3)(FeO)$] in the systems R_2O_3–Fe_2O_3–Fe at 1200°C. No significant deviation from the stoichiometric composition was observed in $YFeO_3$, while the compositions of the other perovskites range from $HoFeO_{2.955}$ to $HoFeO_{3.000}$, $ErFe_{2.974}$ to $ErFeO_{3.000}$ and $TmFeO_{2.953}$ to $TmFeO_{3.000}$ at 1200°C. RFe_2O_4 is stable with a variable composition. Each of the compositional ranges is seen in

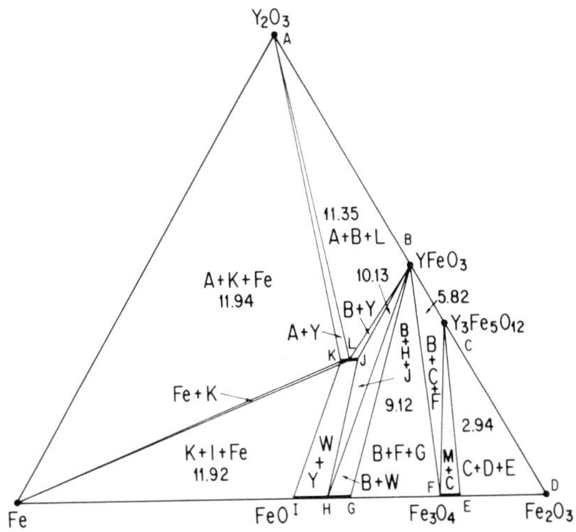

Fig. 10. The phase relationships in the Y_2O_3–Fe_2O_3–Fe at 1200°C. The numbers mean values of $-\log P(O_2)$ in fields equilibrated with three solids phases. The labels A, B, C, D, F and J represent stoichiometric compositions of the Y_2O_3, $YFeO_3$, $Y_3Fe_5O_{12}$, Fe_2O_3, Fe_3O_4 and YFe_2O_4 phases, respectively. M means the magnetite solid solution from E to F, W means the partial solid solution of wüstite from G to H. Y means the solid solution of YFe_2O_4 from K to J. The chemical composition of each letter is as follows: E, $Fe_{2.957}O_4$; G, $Fe_{0.862}O$; H, $Fe_{0.894}O$; I, $Fe_{0.953}O$; K, $YFe_2O_{3.905}$; L, $YFe_2O_{3.944}$. The composition of the point K should have a very small interval, but the authors could not determine its variation. (Kimizuka and Katsura 1975a.)

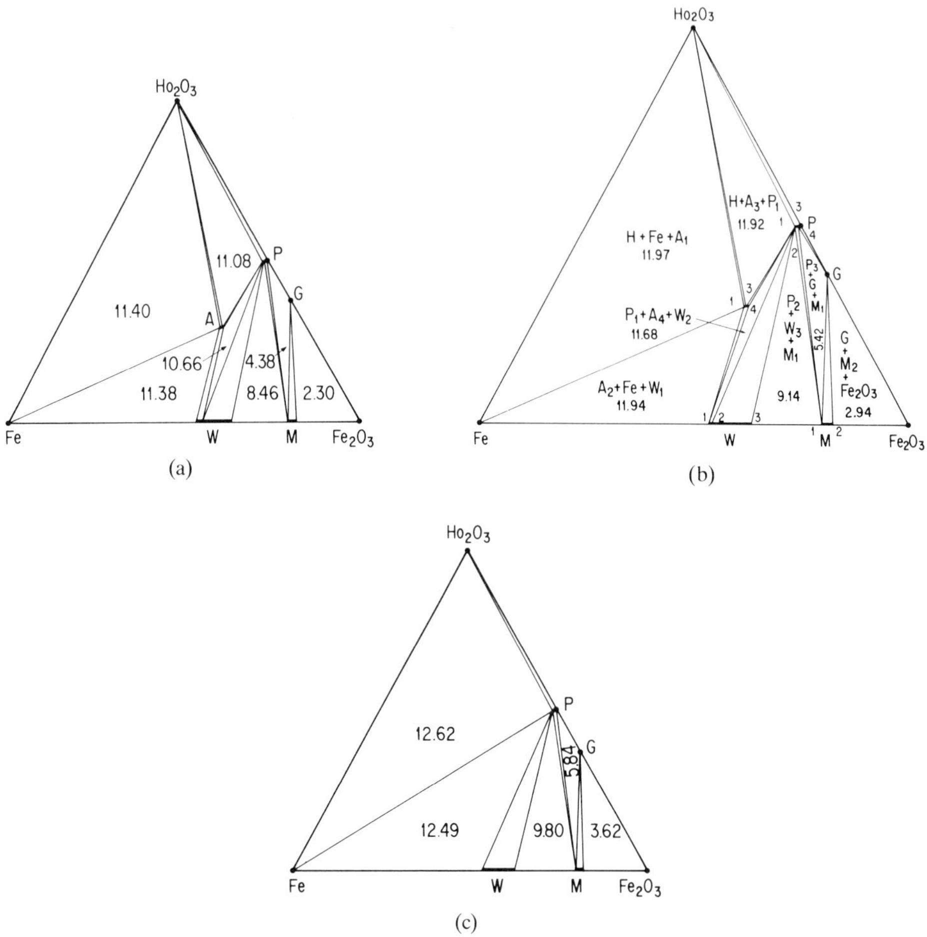

Fig. 11. (a)–(c). The phase relationships in the system Ho_2O_3–Fe_2O_3–Fe at 1250, 1200 and 1150°C. The numbers mean values of $-\log P(O_2)$ in fields equilibrated with three solid phases. The composition of each symbol is given in table 4. (Katsura et al. 1978.)

each figure. No ternary phase $HoFe_2O_4$ exists at 1100°C as shown in fig. 11c. Tables 4–6 supply the values for chemical composition, $P(O_2)$ and activity in the system R_2O_3–Fe_2O_3–Fe (R: Ho, Er and Tm).

2.2.5. *R = Yb and Lu (type D)*

Figures 14 and 15 show the phase relationships in the systems Yb_2O_3–Fe_2O_3–Fe (Kimizuka and Katsura 1975b) and Lu_2O_3–Fe_2O_3–Fe (Sekine and Katsura 1976) at 1200°C, respectively. There are two binary phases, $RFeO_3$ (perovskite) and $R_3Fe_5O_{12}$ (garnet) and two ternary phases, RFe_2O_4 and $R_2Fe_3O_7$ $[(RFeO_3)_2(FeO)]$ at 1200°C. The deviations from the stoichiometric composition

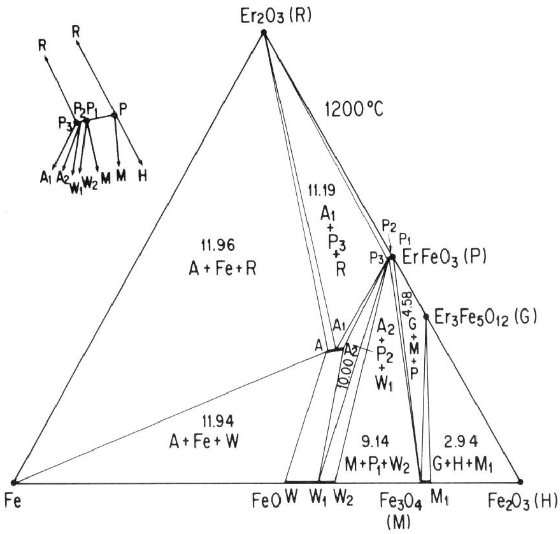

Fig. 12. The phase relationships in the system Er_2O_3–Fe_2O_3–Fe at 1200°C. The numbers mean values of $-\log P(O_2)$ in fields equilibrated with three solid phases. The letters R, P, G and M represent stoichiometric compositions of Er_2O_3, $ErFeO_3$, $Er_3Fe_5O_{12}$ and Fe_3O_4, respectively. M_1 means the end-member of the magnetite solid solution with a chemical composition of $Fe_{2.957}O4$. The other letters as in table 5. (Kitayama and Katsura 1976.)

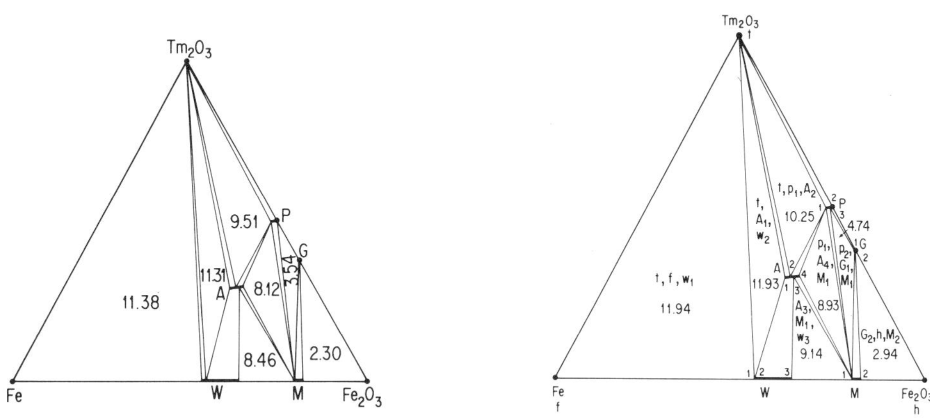

(a) (b)

Fig. 13. (a) The phase relationships in the system Tm_2O_3–Fe_2O_3–Fe at 1250°C. (Katsura et al. 1978.) (b) The phase relationships in the system Tm_2O_3–Fe_2O_3–Fe at 1200°C. (Katsura et al. 1978.) (c) The phase relationships in the system Tm_2O_3–Fe_2O_3–Fe at 1150°C. The composition of each symbol is given in table 6. (Katsura et al. 1978.)

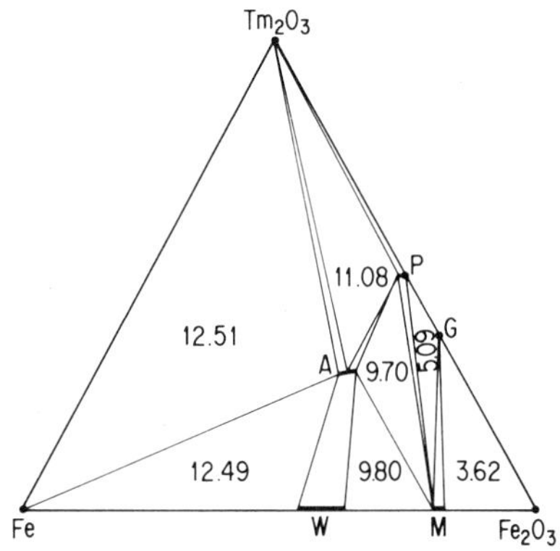

(c)

Fig. 13. (continued).

TABLE 4

Composition and activity ($=a_i$) in the system Ho_2O_3–Fe_2O_3–Fe (Katsura et al. 1978). Symbols in parentheses correspond to those used in fig. 11.

Composition	$-\log P(O_2)$	$\log a_i$
1150°C		
$FeO_{1.048}$	12.49	0
$FeO_{1.159}$	9.80	−0.1394
$HoFeO_{2.965}$	12.62	0
$HoFeO_{3.000}$		0
$Ho_3Fe_5O_{12.00}$	5.84	0
1200°C		
$FeO_{1.048}(W_1)$	11.94	0
$FeO_{1.070}(W_2)$	11.68	−0.0069
$HoFe_2O_{3.869}(A_1)$	11.97	0
$HoFe_2O_{3.879}(A_3)$	11.92	0.0032
$HoFe_2O_{3.927}(A_4)$	11.68	0.0148
$HoFeO_{2.955}(P_1)$	11.92	0
$HoFeO_{3.000}(P_4)$		0
$Ho_3Fe_5O_{12}(G)$	5.42	0
1250°C		
$FeO_{1.050}$	11.38	0
$FeO_{1.077}$	10.66	−0.0192
$HoFe_2O_{3.867}$	11.40	0
$HoFe_2O_{3.888}$	11.08	0.0196
$HoFe_2O_{3.916}$	10.66	0.0402
$HoFeO_{2.981}$	11.08	0
$HoFeO_{3.000}$		0
$Ho_3Fe_5O_{12.00}$	4.38	0

TABLE 5

Composition and activity ($=a_i$) in the system Er_2O_3–Fe_2O_3–Fe at 1200°C (Kitayama and Katsura 1976).

Component	Composition	Symbol	$-\log P(O_2)$	$\log a_i$
FeO	$FeO_{1.049}$	W	11.94	0
	$FeO_{1.125}$	W_1	10.00	−0.0805
	$FeO_{1.168}$	W_2	9.14	−0.150
$ErFe_2O_4$	$ErFe_2O_{3.903}$	A	11.96	0
	$ErFe_2O_{3.942}$	A_1	11.19	0.0350
	$ErFe_2O_{3.995}$	A_2	10.00	0.0478
$ErFeO_3$	$ErFeO_{2.973}$	P_3	11.19	0
	$ErFeO_{2.977}$	P_2	10.00	0.0213
	$ErFeO_{2.980}$	P_1	9.14	0.0240
	$ErFeO_{3.000}$	P	3.64	0.0509
$Er_3Fe_5O_{12}$	$Er_3Fe_5O_{12.00}$	G	4.58	0

TABLE 6

Composition and activity ($=a_i$) in the system Tm_2O_3–Fe_2O_3–Fe (Katsura et al. 1978). The symbols in parentheses correspond to those used in fig. 13.

Composition	$-\log P(O_2)$	$\log a_i$
1150°C		
$FeO_{1.048}$	12.49	0
$FeO_{1.159}$	9.80	−0.1394
$TmFe_2O_{3.904}$	12.51	0
$TmFe_2O_{3.954}$	11.08	0.0507
$TmFe_2O_{4.003}$	9.70	0.0654
$TmFeO_{2.980}$	11.08	0
$TmFeO_{3.000}$		0
$Tm_3Fe_5O_{12.00}$	5.09	0
1200°C		
$FeO_{1.048}(W_1)$	11.94	0
$FeO_{1.052}(W_2)$	11.93	0.0003
$FeO_{1.165}(W_3)$	9.14	−0.1435
$TmFe_2O_{3.896}(A_1)$	11.93	0
$TmFe_2O_{3.966}(A_2)$	10.25	0.0579
$TmFe_2O_{4.022}(A_4)$	8.93	0.0620
$TmFeO_{2.977}(P_1)$	10.25	0
$TmFeO_{3.000}(P_3)$		0
$Tm_3Fe_5O_{11.98}(G_1)$	4.74	0
1250°C		
$FeO_{1.050}$	11.38	0
$FeO_{1.054}$	11.31	−0.0017
$FeO_{1.172}$	8.46	−0.1566
$TmFe_2O_{3.879}$	11.31	0
$TmFe_2O_{3.953}$	9.51	0.0756
$TmFe_2O_{4.010}$	8.12	0.0885
$TmFeO_{2.977}$	9.51	0
$TmFeO_{3.000}$		0
$Tm_3Fe_5O_{12.00}$	3.54	0

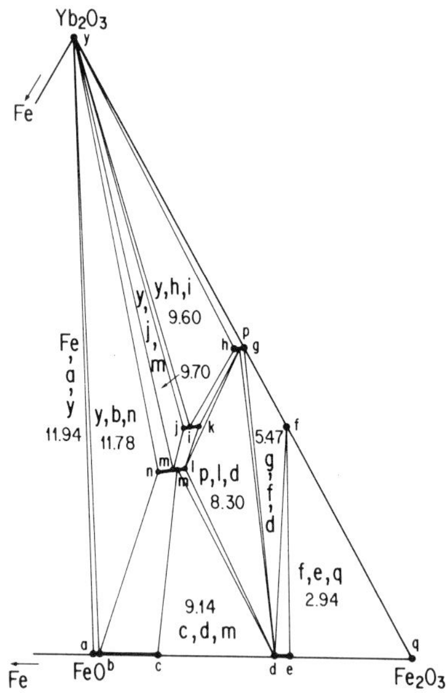

Fig. 14. The phase relationships in the system Yb_2O_3–Fe_2O_3–Fe at 1200°C. The numbers are values of $-\log P(O_2)$ in fields equilibrated with three solid phases. The composition at each letter is: (a) $Fe_{0.953}O$, (b) $Fe_{0.947}O$, (c) $Fe_{0.858}O$, (d) $Fe_{3.000}O_4$, (e) $Fe_{2.957}O$, (f) $Yb_3Fe_5O_{12.000}$, (g) $YbFeO_{3.000}$, (h) $YbFeO_{2.973}$, (i) $Yb_2Fe_3O_{6.885}$, (j) $Yb_2Fe_3O_{6.884}$, (k) $Yb_2Fe_3O_{7.000}$, (l) $YbFe_2O_{4.052}$, (m) $YbFe_2O_{4.002}$, (n) $YbFe_2O_{3.929}$, (p) $YbFeO_{2.989}$ (see table 7). (Kimizuka and Katsura 1975b.)

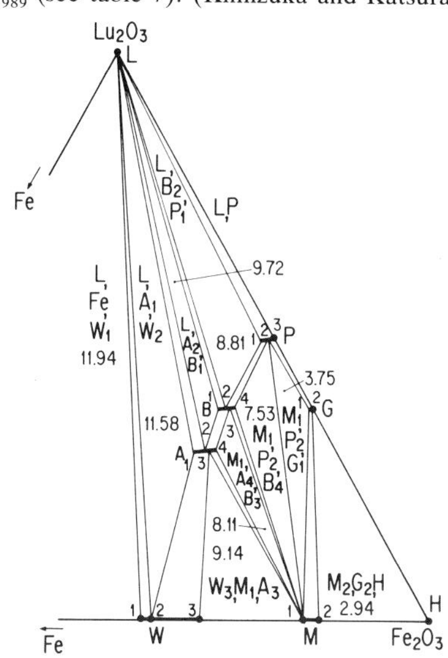

Fig. 15. The phase relationships in the system Lu_2O_3–Fe_2O_3–Fe at 1200°C. The numbers are the values of $-\log P(O_2)$ in fields equilibrated with three solid phases. The composition corresponding to each letter is given in table 8. A boundary line M_1–P_2 by which the area $G_1 + M_1 + P_2$ is separated from the area $B_4 + M_1 + P_2$ is strictly drawn, composed of two boundary lines: one is M_1–P_2, and the other M_1–P_2'. The composition of P_2' is very slightly shifted to the direction of stoichiometric perovskite P. The small area for M_1 + (solid solution from P_2 to P_2') was recognized experimentally. (Sekine and Katsura 1976.)

of the perovskite phases range from $YbFeO_{2.973}$ to $YbFeO_{3.000}$ and from $LuFeO_{2.958}$ to $LuFeO_{3.000}$. Both RFe_2O_4 and $R_2Fe_3O_7$ have nonstoichiometric ranges. Tables 7 and 8 supply the values for chemical composition, $P(O_2)$ and activity in the system R_2O_3–Fe_2O_3–Fe (R: Yb and Lu).

A series of homologous compounds $(YbFeO_3)_n(FeO)$ ($n = 1$, 2, 3 and 4) were prepared in a carbon dioxide gas atmosphere at 1650°C (Kimizuka et al. 1976). The crystal structure models for each compound are given in the next section. In summary, there exists $RFeO_3$ in the system (Y and La–Lu)$_2O_3$–Fe_2O_3–Fe,

TABLE 7

Numerical values for calculating the activity ($=a_i$) in four solid solutions in the system Yb_2O_3–Fe_2O_3–Fe at 1200°C (Kimizuka and Katsura 1975b).

Solid solution	Limit of		A[c]	B[c]	Composition range
	$-\log P(O_2)(1)$	$-\log P(O_2)(2)$			
FeO	11.94	9.14	0.5388	0.0410	$FeO_{1.0495}$ to $FeO_{1.166}$
$YbFe_2O_4$	11.78	8.30	0.3437	0.0352	$YbFe_2O_{3.929}$ to $YbFe_2O_{4.052}$
$Yb_2Fe_3O_7$	9.70	8.2[a]	0.0190	0.0100	$Yb_2Fe_3O_{6.884}$ to $Yb_2Fe_3O_{7.000}$
$YbFeO_3$	9.60	7.40[b]	0.0911	0.0123	$YbFeO_{2.973}$ to $YbFeO_{3.000}$

[a]An exact limit could not be determined.
[b]Nearly stoichiometric composition was maintained at $\log P(O_2)$ ranging from −7.40 to 0.00.
[c]A and B are two parameters for calculating activity of each component. The activities were calculated based upon the following equation:

$$\log a_i = -\tfrac{1}{2}\int_{\log P(O_2)(1)}^{\log P(O_2)(2)} [A + B \log P(O_2)]\, \mathrm{d} \log P(O_2)$$

TABLE 8

Composition and activity ($=a_i$) in the system Lu_2O_3–Fe_2O_3–Fe at 1200°C (Sekine and Katsura 1976).

Component	Composition	Symbol	$-\log P(O_2)$	$\log a_i$
FeO	$FeO_{1.049}$	W_1	11.94	0
	$FeO_{1.064}$	W_2	11.58	−0.0107
	$FeO_{1.165}$	W_3	9.14	−0.1435
$LuFe_2O_4$	$LuFe_2O_{3.935}$	A_1	11.58	0
	$LuFe_2O_{3.982}$	A_2	9.72	0.0397
	$LuFe_2O_{4.000}$	A_3	9.14	0.0437
	$LuFe_2O_{4.018}$	A_4	8.11	0.0414
$Lu_2Fe_3O_7$	$Lu_2Fe_3O_{6.879}$	B_1	9.72	0
	$Lu_2Fe_3O_{6.944}$	B_2	8.81	0.0436
	$Lu_2Fe_3O_{6.984}$	B_3	8.11	0.0540
	$Lu_2Fe_3O_{7.000}$	B_4	7.53	0.0527
$LuFeO_3$	$LuFeO_{2.958}$	P_1	8.81	0
	$LuFeO_{2.985}$	P_2	7.53	0.0184
	$LuFeO_{2.999}$	P_3	3.75	0.0649
	$LuFeO_{3.000}$	P_4	0.00	0.0650
$Lu_3Fe_5O_{12}$	$Lu_3Fe_5O_{11.96}$	G_1	3.75	0
	$Lu_3Fe_5O_{12.000}$	G_2	0.00	

$R_3Fe_5O_{12}$ in the system (Y and Sm–Lu)$_2O_3$–Fe_2O_3–Fe, RFe_2O_4 in the system (Y and Ho–Lu)$_2O_3$–Fe_2O_3–Fe and $R_2Fe_3O_7$ in the system (Yb and Lu)$_2O_3$–Fe_2O_3–Fe at 1200°C and no $HoFe_2O_4$ exists stably at 1100°C.

2.3. *The phase relationships in the systems Yb_2O_3–M_2O_3–$M'O$ and In_2O_3–M_2O_3–$M'O$ (M: Fe, Ga or Cr, and M': Co, Ni, Cu or Zn)*

In the preceding subsection, we discussed the phase relationships in the system M = M′ = Fe. In the present section, we describe the systems where M or M′ is substituted by other transition metal elements. Since thermogravimetric methods can not be applied to such systems, the phase relationships shown here were determined by the quenching method. We also refer to the systems including In_2O_3 instead of rare earth oxides.

The phase relationships in the systems Yb_2O_3–M_2O_3–M′O and In_2O_3–M_2O_3–M′O at elevated temperatures were determined in air or in Pt tubes. $YbFeO_3$, $YbCrO_3$ and $Yb_3M_5O_{12}$ are stable binary phases in the systems Yb_2O_3–M_2O_3 (M = Fe or Ga), and $M_2M'O_4$ with a spinel structure is a stable binary phase in the systems M_2O_3–M′O at the temperature range mentioned below.

2.3.1. *The systems Yb_2O_3–Fe_2O_3–$M'O$ (M′: Co, Ni, Cu or Zn) (Kimizuka and Takayama 1982b)*

(i) The phase relationships in the Yb_2O_3–Fe_2O_3–CoO at 1350 and 1300°C are shown in figs. 16a and 16b. No binary phase exists in the system CoO–Yb_2O_3 and there is only one ternary phase, $YbFeCoO_4$ with a $YbFe_2O_4$ type of structure at 1350°C. $YbFeCoO_4$ was decomposed to $YbFeO_3$ and CoO, and no ternary phase exists at 1300°C.

(ii) The phase relationships in the system Yb_2O_3–Fe_2O_3–NiO at 1300 and 1200°C are shown in fig. 17. No ternary phase exists in the system Yb_2O_3–Fe_2O_3–NiO at 1300 and 1200°C.

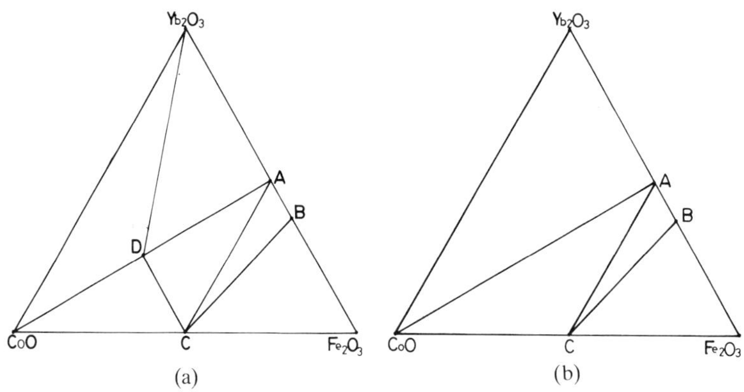

Fig. 16. (a) The phase relationships in the system Yb_2O_3–Fe_2O_3–CoO at 1350°C. A: $YbFeO_3$, B: $Yb_3Fe_5O_{12}$, C: Fe_2CoO_4 and D: $YbFeCoO_4$. (Kimizuka and Takayama 1982b.) (b) The phase relationships in the system Yb_2O_3–Fe_2O_3–CoO at 1300°C. A: $YbFeO_3$, B: $Yb_3Fe_5O_{12}$, C: Fe_2CoO_4. (Kimizuka and Takayama 1982b.)

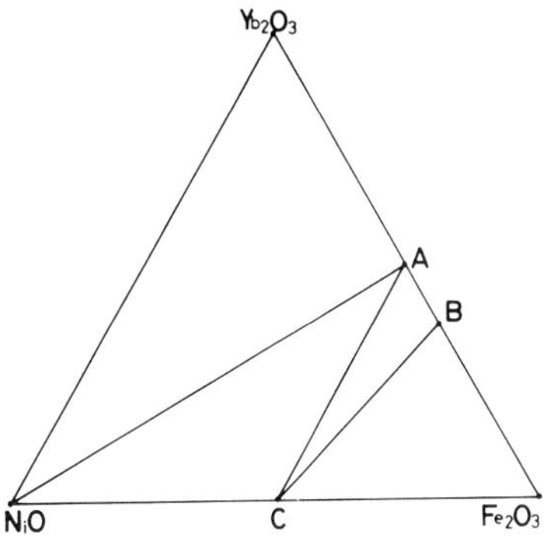

Fig. 17. The phase relationships in the system Yb_2O_3–Fe_2O_3–NiO at 1300 and 1200°C. A: $YbFeO_3$, B: $Yb_3Fe_5O_{12}$ and C: Fe_2NiO_4. (Kimizuka and Takayama 1982b.)

(iii) The phase relationships in the system Yb_2O_3–Fe_2O_3–CuO at 1000°C are shown in fig. 18. There is one phase $Yb_2Cu_2O_5$ with $In_2Cu_2O_5$ type of structure in the CuO–Yb_2O_3 (Bergerhoff and Kasper 1968) and one ternary phase $YbFeCuO_4$ with a $YbFe_2O_4$ type of structure in the system Yb_2O_3–Fe_2O_3–CuO at 1000°C.

(iv) The phase relationships in the system Yb_2O_3–Fe_2O_3–ZnO at 1300°C are shown in fig. 19. There is no binary phase in the system ZnO–Yb_2O_3 but there is a ternary phase, $YbFeZnO_4$ with the $YbFe_2O_4$-type structure in the system Yb_2O_3–Fe_2O_3–ZnO.

2.3.2. *The systems Yb_2O_3–Ga_2O_3–M'O (M': Co, Ni, Cu or Zn) (Kimizuka and Takayama 1982c)*

(i) The phase relationships in the Yb_2O_3–Ga_2O_3–CoO at 1350 and 1200°C are shown in fig. 20. There is a binary phase $Yb_3Ga_5O_{12}$ (garnet) in the system Yb_2O_3–Ga_2O_3. No $YbGaO_3$ (perovskite) phase exists. Marezio et al. (1966) reported that $YbGaO_3$ (distorted perovskite structure) was synthesized at 1100°C under 70 kbar and it was unstable under one atmosphere. There is no binary phase in the system CoO–Yb_2O_3. There is a ternary phase $YbGaCuO_4$ with a $YbFe_2O_4$–type structure in the system Yb_2O_3–Ga_2O_3–CoO.

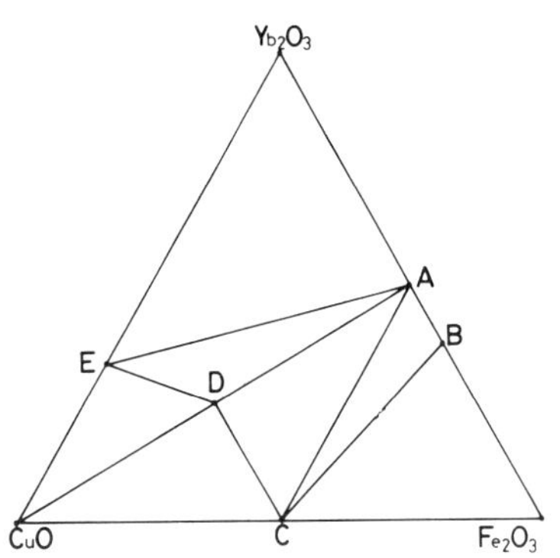

Fig. 18. The phase relationships in the system Yb_2O_3–Fe_2O_3–CuO at 1000°C. A: $YbFeO_3$, B: $Yb_3Fe_5O_{12}$, C: Fe_2CuO_4, D: $YbFeCuO_4$ and E: $Yb_2Cu_2O_5$. (Kimizuka and Takayama 1982b.)

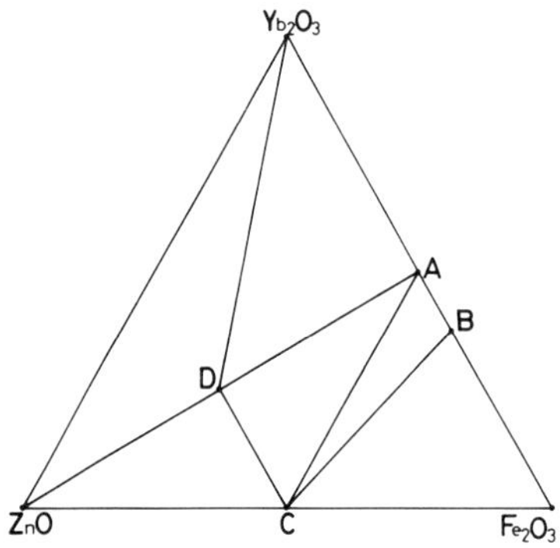

Fig. 19. The phase relationships in the system Yb_2O_3–Fe_2O_3–ZnO at 1350°C. A: $YbFeO_3$, B: $Yb_3Fe_5O_{12}$, C: Fe_2ZnO_4 and D: $YbFeZnO_4$. (Kimizuka and Takayama 1982b.)

The $YbGaCoO_4$ which exists at 1350°C decomposes to Yb_2O_3, CoO and $Yb_3Fe_5O_{12}$. No ternary stable phase exists in the system Yb_2O_3–Ga_2O_3–CoO at 1200°C.

(ii) The phase relationships in the Yb_2O_3–Ga_2O_3–NiO at 1300 and 1200°C are shown in fig. 21. There is no ternary stable phase in the system Yb_2O_3–Ga_2O_3–NiO.

(iii) The phase relationships in the Yb_2O_3–Ga_2O_3–CuO at 1000°C are shown in fig. 22. There is a stable ternary $YbGaCuO_4$ with a $YbFe_2O_4$-type structure in this ternary system.

(iv) The phase relationships in the Yb_2O_3–Ga_2O_3–ZnO at 1300 and 1200°C are shown in fig. 23. There is a ternary stable phase $YbGaZnO_4$ with a $YbFe_2O_4$-type structure. At 1200°C, $YbGaZnO_4$ decomposes to Yb_2O_3, $Yb_3Ga_5O_{12}$ and ZnO. No ternary stable phase exists.

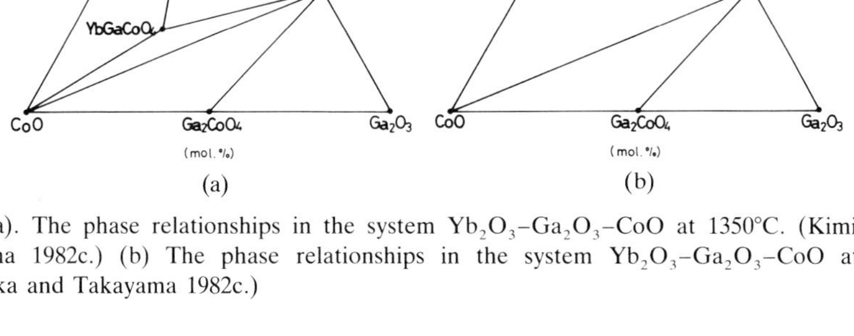

Fig. 20(a). The phase relationships in the system Yb_2O_3–Ga_2O_3–CoO at 1350°C. (Kimizuka and Takayama 1982c.) (b) The phase relationships in the system Yb_2O_3–Ga_2O_3–CoO at 1200°C. (Kimizuka and Takayama 1982c.)

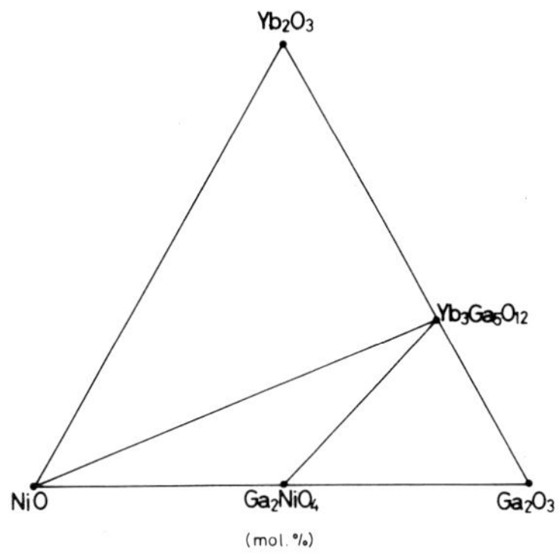

Fig. 21. The phase relationships in the system Yb_2O_3–Ga_2O_3–NiO at 1300 and 1200°C. (Kimizuka and Takayama 1982c.)

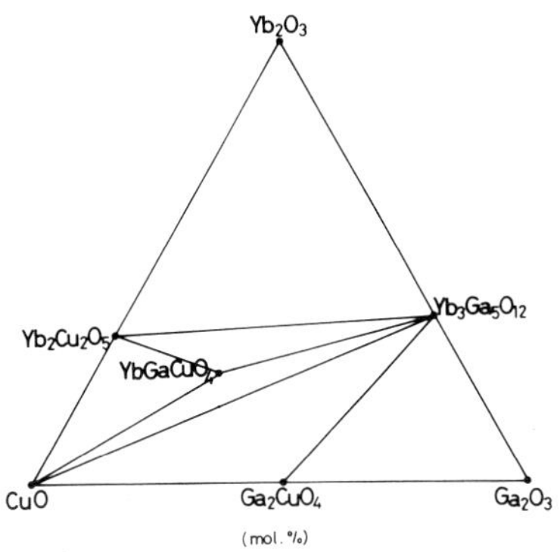

Fig. 22. The phase relationships in the system Yb_2O_3–Ga_2O_3–CuO at 1000°C. (Kimizuka and Takayama 1982c.)

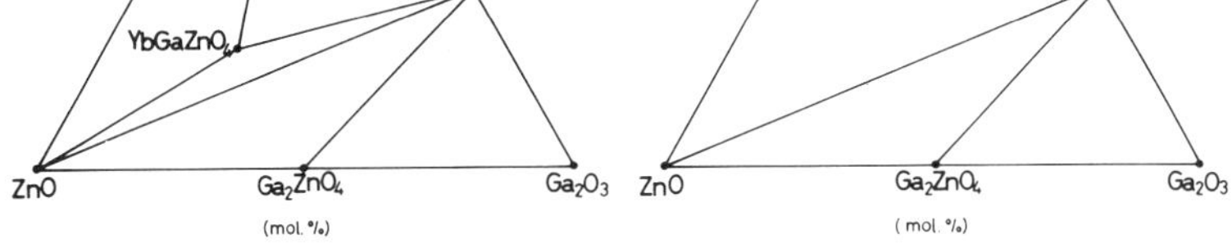

Fig. 23. (a) The phase relationships in the system Yb_2O_3–Ga_2O_3–ZnO at 1350°C. (Kimizuka and Takayama 1982c.) (b) The phase relationships in the system Yb_2O_3–Ga_2O_3–ZnO at 1200°C. (Kimizuka and Takayama 1982c.)

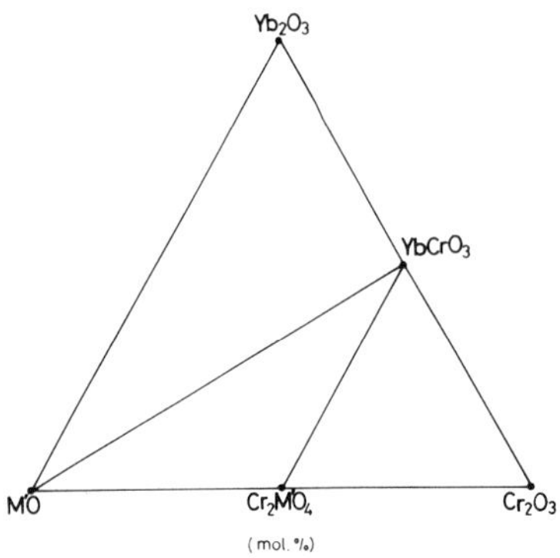

Fig. 24. The phase relationships in the system Yb_2O_3–Cr_2O_3–M′O (M′: Co, Ni or Zn) at 1300°C. (Kimizuka and Takayama 1982c.)

2.3.3. *The systems Yb_2O_3–Cr_2O_3–M′O (M′: Co, Ni, Cu or Zn) (Kimizuka and Takayama 1982c)*

(i) The phase relationships in the system Yb_2O_3–Cr_2O_3–M′O (M′: Co, Ni or Zn) are shown in fig. 24. There is a stable binary phase $YbCrO_3$ (distorted perovskite) in the system Yb_2O_3–Cr_2O_3 at 1300 and 1200°C. No $Yb_3Cr_5O_{12}$ exists. No ternary phase exists in the system Yb_2O_3–Cr_2O_3–M′O at 1300 and 1200°C.

(iii) The phase relationships in the Yb_2O_3–Cr_2O_3–CuO system at 1000°C are shown in fig. 25. There is a stable binary phase $YbCrO_3$ in the system Yb_2O_3–Cr_2O_3. No ternary phase exists in the Yb_2O_3–Cr_2O_3–CuO at 1000°C.

2.3.4. *The systems In_2O_3–M_2O_3–M′O (Kimizuka and Takayama 1984)*

(i) The phase relationships in the In_2O_3–M_2O_3–CuO at 1000°C (M: Fe or Ga) are shown in figs. 26 and 27. There are three ternary phases, $InFeO_3(CuO)$, $(InFeO_3)_2CuO$ and $(InFeO_3)_3CuO$ that are isostructural with $(YbFeO_3)_nFeO$ ($n = 1$, 2 or 3) in the system In_2O_3–Fe_2O_3–CuO. In the phase relationships in the system In_2O_3–Ga_2O_3–CuO at 1000°C there are two ternary phases, $InGaO_3(CuO)$ and $(InGaO_3)_2CuO$, that are isostructural with $(YbFeO_3)_nFeO$ ($n = 1$ or 2). The crystal strutures of $(YbFeO_3)_nFeO$ will be described later.

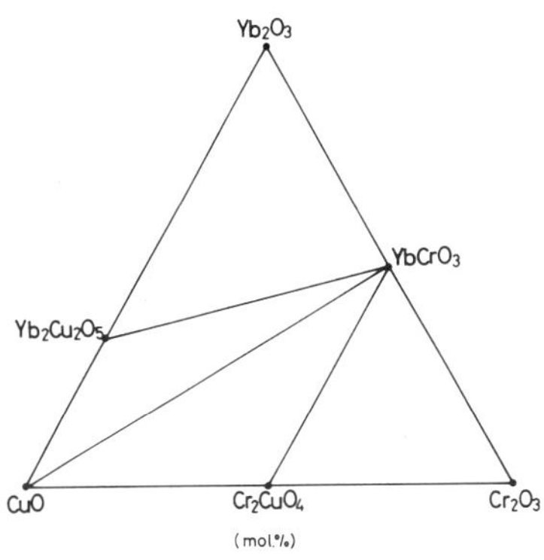

Fig. 25. The phase relationships in the system Yb_2O_3–Cr_2O_3–CuO at 1000°C. (Kimizuka and Takayama 1982c.)

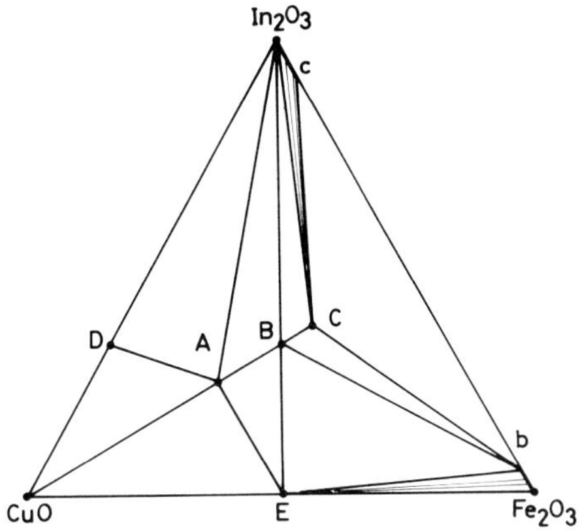

Fig. 26. The phase relationships in the system In_2O_3–Fe_2O_3–CuO at 1000°C. (A) $InFeCuO_4$, (B) $In_2Fe_2CuO_7$, (C) $In_3Fe_3CuO_{10}$, (D) $In_2Cu_2O_5$, (E) Fe_2CuO_4, (b) $In_2O_3 : Fe_2O_3 = 0.05 : 0.95$ (in mole), (c) $In_2O_3 : Fe_2O_3 = 0.925 : 0.075$ (in mole). (Kimizuka and Takayama 1984.)

(ii) The phase relationships in the systems In_2O_3–M_2O_3–CoO at 1300°C (M: Fe or Ga) are shown in figs. 28 and 29. There is no binary phase in the system CoO–In_2O_3 at 1300°C. In the system In_2O_3–Fe_2O_3–CoO, there is a solid solution from $InFeCoO_4$ ($a = 8.638(3)$ Å) to Fe_2CoO_4 with a spinel structure. The dependence of the lattice constant of the spinel phase upon the constituent cation is shown in fig. 30, which shows a good linearity in the whole compositional range. In the system In_2O_3–Ga_2O_3–CoO, there is a ternary phase, $InGaO_3(CoO)$ having a $YbFe_2O_4$–type structure and rather extensive solubility of $InGaCoO_4$ in the spinel Ga_2CoO_4 is observed. The solid solution ranges from Ga_2CoO_4 to $In_{0.4}Ga_{0.6}CoO_4$ and the lattice constant varies from $a = 8.323(1)$ Å (Ga_2CoO_4) to $a = 8.426(1)$ Å ($In_{0.4}Ga_{0.6}CoO_4$).

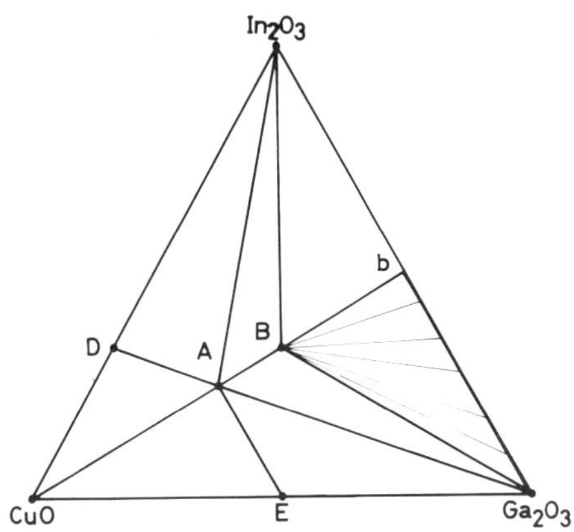

Fig. 27. The phase relationships in the system In_2O_3–Ga_2O_3–CuO at 1000°C. (A) $InGaCuO_4$, (B) $In_2Ga_2CuO_7$, (D) $In_2Cu_2O_5$, (E) Ga_2CuO_4, (b) $In_2O_3 : Ga_2O_3 = 1 : 1$ (in mole). (Kimizuka and Takayama 1984.)

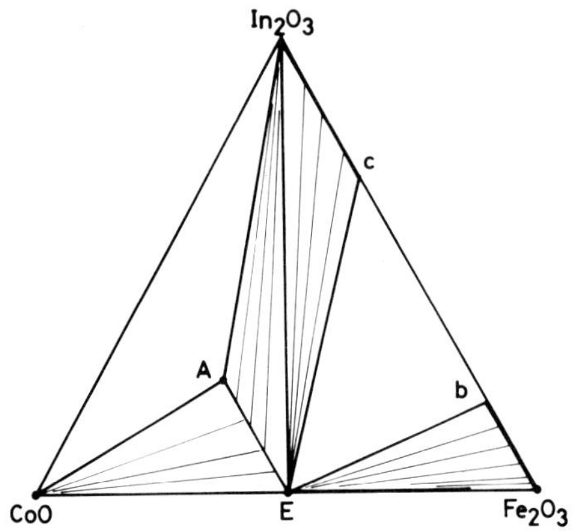

Fig. 28. The phase relationships in the system In_2O_3–Fe_2O_3–CoO at 1300°C. (A) $InFeCoO_4$ (spinel), (E) Fe_2CoO_4, (b) $In_2O_3 : Fe_2O_3 = 0.25 : 0.75$ (in mole), (c) $In_2O_3 : Fe_2O_3 = 0.60 : 0.40$ (in mole). (Kimizuka and Takayama 1984.)

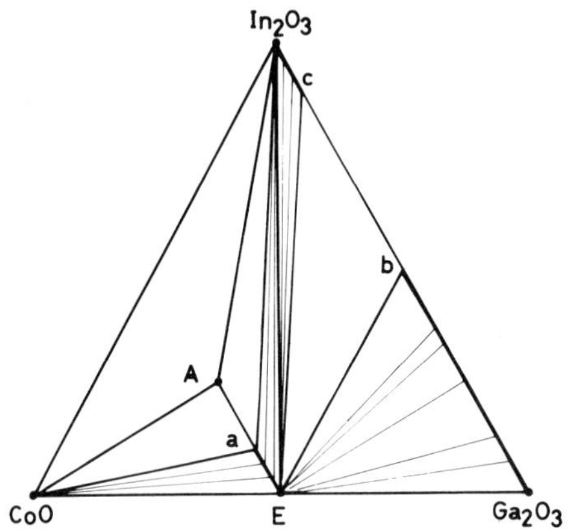

Fig. 29. The phase relationships in the system In_2O_3–Ga_2O_3–CoO at 1300°C. (A) $InGaCoO_4$ ($YbFe_2O_4$ type), (E) Ga_2CoO_4, (a) $In_2O_3 : Ga_2O_3 : CoO = 0.10 : 0.40 : 0.50$ (in mole), (b) $In_2O_3 : Ga_2O_3 = 1 : 1$ (in mole). (Kimizuka and Takayama 1984.)

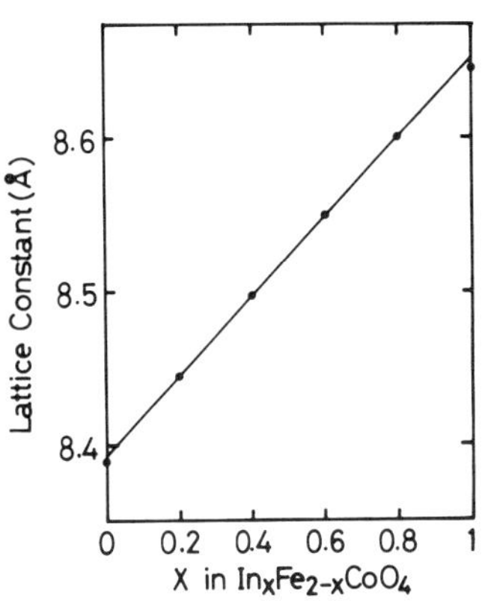

Fig. 30. The relation between the lattice constant and x in $In_{2-x}Fe_xCoO_4$. (Kimizuka and Takayama 1984.)

2.4. *The phase relationships in the system Eu_2O_3–Fe_2O_3–SrO (Drofenik et al. 1974)*

The phase relationships in the system Eu_2O_3–Fe_2O_3–SrO at 1300°C were determined by means of a classical quenching method.

When the size of the M′ ion becomes large compared with the transition metal ions, phases with different crystal structures appear in the systems. As an example, we show the above phase relationships in fig. 31. The starting compounds were reagent grade $SrCO_3$, Fe_2O_3 and Eu_2O_3. There are three homologous phases. $(EuFeO_3)_nSrO$ ($n = 1$, 2 and 3). These phases are not isostrutural with the $(RMO_3)_n(M'O)_m$ compounds mentioned above but they have $(KNiF_3)_nKF$-type structures (see next section) in which both Eu and Sr are surrounded by 9 oxygen ions, while Fe occupies an octahedral site.

2.5. *Single-crystal growth*

A few papers have been published so far on single-crystal growth of $(RMO_3)_n(M'O)_m$-type compounds. They are restricted to $RMM'O_4$, i.e., the case $n = m = 1$.*

The earliest report was made by Schmitz-DuMont and Kasper (1965) on $InAlCuO_4$. To make single crystals of $Cu_2In_2O_5$, they heated a mixture of PbF_2, CuO and In_2O_3 in an alumina crucible. After slow cooling, they found dark blue hexagonal platelets, single crystals of $InAlCuO_4$. This was the first discovery of the compound with $n = m = 1$ (see sect. 4.1). In other attempts, they used mixture of PbO and PbF_2 as a flux agent. The starting material was a mixture of

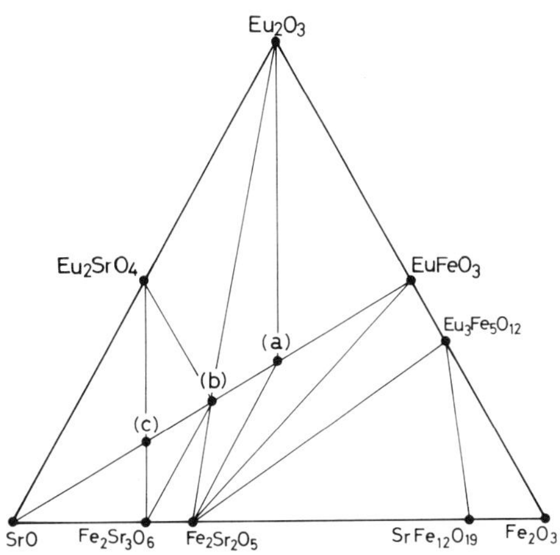

Fig. 31. The phase relationships in the system Eu_2O_3–Fe_2O_3–SrO at 1300°C (Drofenik et al. 1974). (a) $(EuFeO_3)_2SrO$; (b) $(EuFeO_3)SrO$; (c) $EuFeO_3(SrO)_2$.

*Single-crystal growth was reported on $LuMnO_3$ (Yakel et al. 1963) by the flux method with Bi_2O_3. This compound can be considered as one of the end members of the series with $n = \infty$ (see section 4.2).

In_2O_3, Al_2O_3 and CuO in molar ratio 1 : 1 : 2. The mixture was heated to 1000°C in an alumina crucible and then cooled to 800°C and maintained for 24 h. The flux agent was dissolved with ammonium tartarate.

Ten years later, after the discovery of Fe-containing compounds, the floating-zone method was applied to YFe_2O_4 with appreciable success (Shindo et al. 1976). Several grains of single crystal were grown with the growth direction perpendicular to *c*-axis and the largest size of the grain was the order of several mm. It was found that the compound has an incongruent melting point. The same method was applied to $YbFe_2O_4$, $LuFe_2O_4$, $LuFeCoO_4$ and $YbFeMgO_4$ (Iida 1988).

The oxygen partial pressure of the atmosphere should be controlled carefully during the crystal growth of Fe^{2+}-containing compounds. Mixtures of CO and CO_2 were used for them, while single crystals of $LuFeCoO_4$ and $YbFeMgO_4$ were grown in air. In the case of $YbFeMgO_4$, the nominal composition should be slightly different from the stoichiometric one (Yb : Fe : Mg = 1 : 1.12 : 0.88), since the melting point is high (about 1600°C).

One point to be made here is the homogeneity and stoichiometry of the crystal thus obtained. Since there is a very large temperature gradient in a zone-melting furnace, the crystals grown are believed to be inhomogeneous in composition. In particular, oxygen nonstoichiometry is a serious problem. (Note that the oxygen nonstoichiometry of the compound depends on temperature and oxygen fugacity.) Annealing after crystal growth seems rather ineffective in making a stoichiometric and homogeneous single crystal, at least for YFe_2O_4, where nonstoichiometry affects physical properties drastically (see sect. 5.1). All single crystals of YFe_2O_4 obtained so far are oxygen deficient.

3. Thermochemistry. The Gibbs free-energy changes, enthalpy changes and entropy changes of $RFeO_3$, $R_3Fe_5O_{12}$, RFe_2O_4 and $R_2Fe_3O_7$

3.1. *$RFeO_3$ (perovskite) (R: Y and lanthanides)*

As shown in the preceding section, $RFeO_3$ (R: La, Pr, Nd, Sm, Eu, Gd, Tb and Dy) is in equilibrium with R_2O_3 and Fe metal at 1000–1200°C. Therefore, we can calculate the Gibbs free-energy change for the following chemical equation,

$$\tfrac{1}{2}R_2O_3 + Fe + \tfrac{3}{4}O_2 \rightarrow RFeO_3\,, \qquad (1)$$

using

$$\Delta G^0(1) = \tfrac{3}{4}RT \ln P(O_2)\,. \qquad (2)$$

$P(O_2)$ values for each chemical reaction are given in the preceding section.

Mizusaki et al. (1982) reported the decomposition pressure of oxygen of the $LaFeO_3$ phase using the CO_2–CO system as a reducing atmosphere in a process for measuring electrical conductivity and the Seebeck coefficient of $LaFeO_3$ (fig. 32). ΔH^0 (kJ/mol) = -499.6 ± 4.2, ΔS^0 (J/deg mol) = -136.8 ± 4.2 are esti-

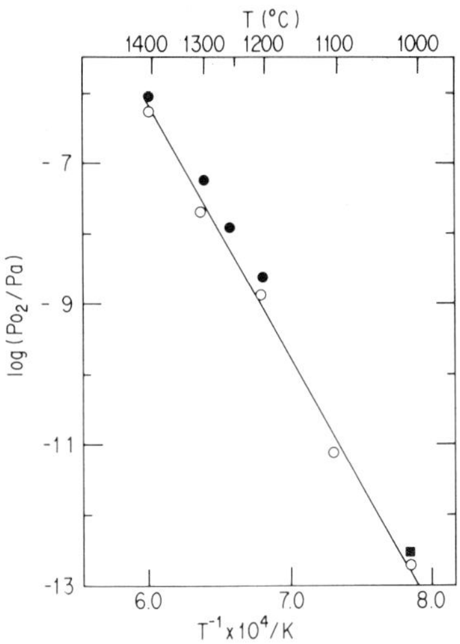

Fig. 32. Decomposition $P(O_2)$ of $LaFeO_3$ versus $1/T$: ○ (Mizusaki et al. 1982), ● (Katsura et al. 1975), ■ (Nakamura et al. 1979.)

mated from fig. 32. Nakamura et al. (1979) measured the same parameters of $LaFeO_3$ using the CO_2–H_2 system at 1000 and 1200°C. Their results are as follows: $\Delta G^0(1273) = -309.7 \pm 1.3$ kJ/mol, $\Delta G^0(1473) = -288.3 \pm 1.3$ kJ/mol. Katsura et al. (1975) measured equilibrium oxygen partial pressures of $RFeO_3$ (R = La, Nd, Sm, Eu, Gd, Tb and Dy) at $T = 1473$–1673 K. On the other hand, we can also measure the Gibbs free-energy change with the Coulombic titration method using the cell system (I) in fig. 33. Piekarczyk et al. (1978) measured the dissociation oxygen partial pressures and the Gibbs free-energy change of $YFeO_3$ phase by means of a solid-state galvanic cell method in the temperature range of 900–1250°C (fig. 34). Their results show ΔG^0 (kJ/mol) $= -421.3 + 0.1148T \pm 1.0$. At 1200°C, ΔG^0 (kJ/mol) $= -252.2 \pm 0.7$ is calculated from their equation. Tretyakov et al. (1977) also reported the e.m.f. of the $RFeO_3$ (R = La, Pr, and Eu) cell in the temperature range of 950–1200°C by means of a solid-state galvanic cell. And Sreedharan and Chandrasekharaiah (1986) reported the e.m.f. values of $LaFeO_3$ at $T = 1094$–1299 K.

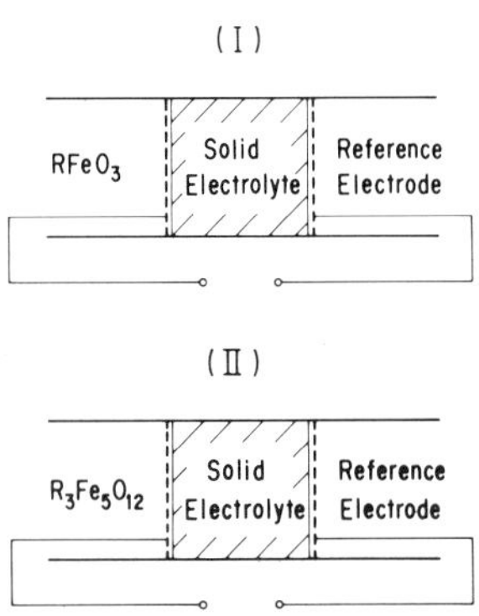

Fig. 33. Solid-state galvanic cell systems.

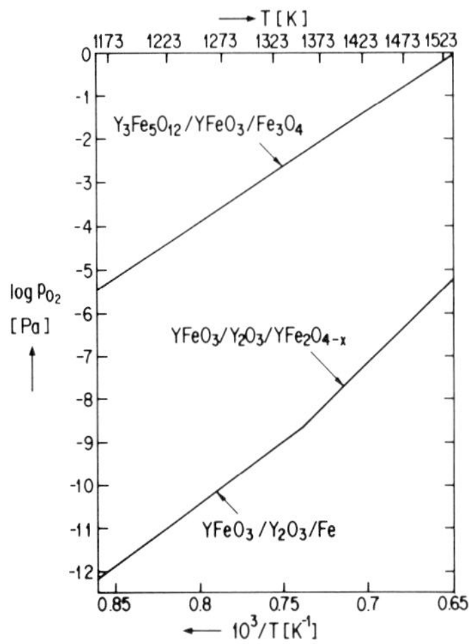

Fig. 34. Temperature dependence of the oxygen dissociation pressure of $Y_3Fe_5O_{12}$ and $YFeO_3$. (Piekarczyk et al. 1978.)

The values for $P(O_2)$, E or ΔG^0 which we referred to above are summarized in table 9. As shown in the preceding section, $RFeO_3$ (R = Y, Ho–Tm) is decomposed to R_2O_3 and Fe via RFe_2O_4, and $RFeO_3$ (R = Yb and Lu) is to R_2O_3 and Fe via both RFe_2O_4 and $R_2Fe_3O_7$ at 1200°C. So we can calculate ΔG^0 values for $RFeO_3$ (R = Y and Ho–Lu) as shown in tables 10–15. Kimizuka et al. (1983) reevaluated ΔG^0 for $RFeO_3$ (R = Y and lanthanides), which are shown in table 16.

One of the interesting characteristics in the $RFeO_3$ is that the $RFeO_3$ (R = La–Dy) is directly converted to iron metal and R_2O_3, or vice versa, without taking the Fe(II) ion state. We can say that the iron in these perovskite phases is in the very stable state for the reducive atmosphere.

From the temperature dependence of ΔG^0, both ΔH^0 and ΔS^0 were calculated, which are shown in table 17. Schmaltzried and Navrotzky (1975) estimated the

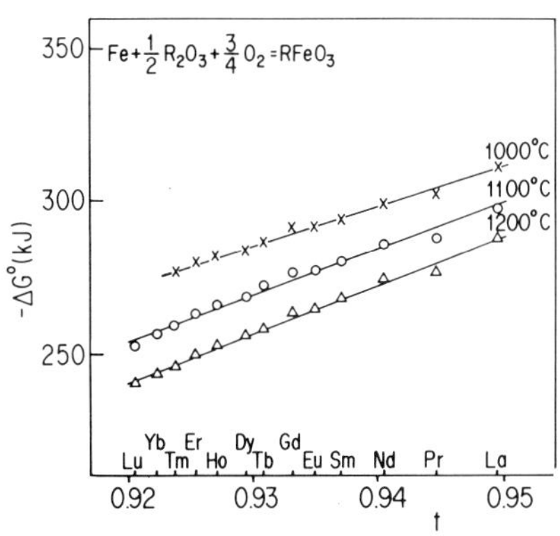

Fig. 35. The relation between $-\Delta G^0$ and the tolerance factor of $RFeO_3$. (Katsura et al. 1975.)

TABLE 9

ΔG^0 or $P(O_2)$ for the chemical reaction: $\frac{1}{2}R_2O_3 + Fe + \frac{3}{4}O_2 \rightarrow RFeO_3$.

$RFeO_3$	Method*	Atmosphere	Preparation of $RFeO_3$	Temperature range (K)	ΔG^0(kJ/mol) or $P(O_2)$	Ref.**
$LaFeO_3$	G.A.	CO–H_2	Mechanical mixing	1473–1673	$\Delta G^0 = -447.7 + 0.1096 \times T \pm 1.0$	[1]
	e.m.f. (ZrO_2–Y_2O_3)		Co-precipitation	1223–1473	$\log P(O_2) = A - B/T$, $A = 8.29$, $B = 32041$	[2]
	e.m.f. (ZrO_2–CaO)		Mechanical mixing	1094–1299	$\Delta G^0 = -414.62 + 0.0834 \times T \pm 1.9$	[3]
	Seebeck coefficient	CO_2–CO	Co-precipitation	1273–1673	See fig. 32	[4]
	G.A.	CO_2–H_2	Mechanical mixing	1273–1473	$\Delta G^0 = -447.8 + 0.1076 \times T \pm 1.0$	[5]
$PrFeO_3$	e.m.f. (ZrO_2–Y_2O_3)		Co-precipitation	1223–1473	$\log P(O_2) = A - B/T$, $A = 9.21$, $B = 32485$	[2]
$NdFeO_3$	G.A.	CO_2–H_2	Mechanical mixing	1473–1620	$\Delta G^0 = -447.7 + 0.1184 \times T \pm 1.0$	[1]
$SmFeO_3$	G.A.	CO_2–H_2	Mechanical mixing	1473–1570	$\Delta G^0 = -447.7 + 0.1230 \times T \pm 1.0$	[1]
$EuFeO_3$	G.A.	CO_2–H_2	Mechanical mixing	1473–1570	$\Delta G^0 = -447.7 + 0.1251 \times T \pm 1.0$	[1]
	e.m.f. (ZrO_2–Y_2O_3)		Co-precipitation	1223–1473	$\log P(O_2) = A - B/T$, $A = 9.04$, $B = 31547$	[2]
$GdFeO_3$	G.A.	CO_2–H_2	Mechanical mixing	1473–1570	$\Delta G^0 = -447.7 + 0.1255 \times T \pm 1.0$	[1]
$YFeO_3$	e.m.f. (ZrO_2–CaO)		Mechanical mixing	1173–1373	$\Delta G^0 = -421.3 + 0.1148 \times T \pm 1.0$	[6]

*G.A. means gravimetric analysis; and e.m.f. is electromotive force analysis.

**References: [1] Katsura et al. (1975); [2] Tretyakov et al. (1977); [3] Sreedharan and Chandrasekharaiah (1986); [4] Mizusaki et al. (1982); [5] Nakamura et al. (1979); [6] Piekarczyk et al. (1978).

TABLE 10
The Gibbs free energies of formation of $YFeO_3$, $YFe_2O_{4-x_0}$ and $Y_3Fe_5O_{12}$ at 1200°C (Kimizuka and Katsura 1975a).

Chemical reaction	$-\log Po_2$	$-\Delta G^0$ (kJ)	$-\log a_i$ (other than unity)
$2Fe + \frac{1}{2}Y_2O_3 + \frac{1}{4}(5-2x_0)O_2 \rightarrow YFe_2O_{4-x_0}$ ($x_0 = 0.095$)	11.94	405.0 ± 0.8	
$Fe + \frac{1}{2}Y_2O_3 + \frac{3}{4}O_2 \rightarrow YFeO_3$	11.35	250.2 ± 0.8	Fe: 0.358
$5Fe + \frac{3}{2}Y_2O_3 + \frac{15}{4}O_2 \rightarrow Y_3Fe_5O_{12}$		1208.8 ± 2.5	
$3YFeO_3 + \frac{2}{3}Fe_3O_4 + \frac{1}{6}O_2 \rightarrow Y_3Fe_5O_{12}$	5.82	27.2 ± 2.5	
$3Fe + 2O_2 \rightarrow Fe_3O_4$	9.14	646.4 ± 1.3	Fe: 1.545
$YFe_2O_{4-x_0} + \frac{1}{2}x_0O_2 \rightarrow FeO + YFeO_3$ ($x_0 = 0.095$)	10.13	13.0 ± 0.8	FeO: 0.070; $YFe_2O_{4-x_0}$: 0.043
$Fe + \frac{1}{2}O_2 \rightarrow FeO$	11.92	168.2 ± 0.8	

TABLE 11
The standard Gibbs free energy for the Ho–Fe–O compounds (Katsura et al. 1978).

Chemical reaction	$-\Delta G^0$ (kJ) 1150°C	1200°C	1250°C
$Fe + \frac{1}{2}Ho_2O_3 + \frac{3}{4}O_2 \rightarrow HoFeO_3$	257.7 ± 0.4	253.1 ± 0.8	247.7 ± 0.8
$Fe + \frac{1}{2}O_2 \rightarrow FeO$	170.3 ± 0.4	168.2 ± 0.4	166.1 ± 0.4
$2Fe + \frac{1}{2}Ho_2O_3 + \frac{5}{4}O_2 \rightarrow HoFe_2O_4$		422.2 ± 0.4	415.5 ± 0.4
$\frac{1}{2}HoFe_2O_4 + \frac{1}{4}Ho_2O_3 + \frac{1}{8}O_2 \rightarrow HoFeO_3$		41.8 ± 0.4	40.2 ± 0.4
$FeO + HoFeO_3 \rightarrow HoFe_2O_4$		0.63 ± 0.20	1.84 ± 0.20
$3FeO + \frac{1}{2}O_2 \rightarrow Fe_3O_4$	144.8 ± 0.4	141.4 ± 0.4	137.0 ± 0.4
$3HoFeO_3 + \frac{2}{3}Fe_3O_4 + \frac{1}{6}O_2 \rightarrow Ho_3Fe_5O_{12}$	26.4 ± 0.4	25.5 ± 0.4	23.4 ± 0.4
$5Fe + \frac{3}{2}Ho_2O_3 + \frac{15}{4}O_2 \rightarrow Ho_3Fe_5O_{12}$	1237.2 ± 3.3	1215.5 ± 3.4	1189.9 ± 3.4

TABLE 12
The standard Gibbs free energy for the Er–Fe–O compounds at 1200°C (Kitayama and Katsura 1976).

Chemical reaction	$-\Delta G^0$ (kJ)
$2Fe + \frac{1}{2}Er_2O_3 + \frac{5}{4}O_2 \rightarrow ErFe_2O_4$	421.8 ± 0.4
$Fe + \frac{1}{2}Er_2O_3 + \frac{3}{4}O_2 \rightarrow ErFeO_3$	249.8 ± 1.3
$5Fe + \frac{3}{2}Er_2O_3 + \frac{15}{4}O_2 \rightarrow Er_3Fe_5O_{12}$	1197.5 ± 2.9
$3Fe + 2O_2 \rightarrow Fe_3O_4$	646.4 ± 1.3
$\frac{1}{4}Er_2O_3 + \frac{1}{2}ErFe_2O_4 + \frac{1}{8}O_2 \rightarrow ErFeO_3$	38.9 ± 0.8
$3ErFeO_3 + \frac{2}{3}Fe_3O_4 + \frac{1}{6}O_2 = Er_3Fe_5O_{12}$	17.2 ± 1.3

TABLE 13
The standard Gibbs free energy for the Tm–Fe–O compounds (Katsura et al. 1978).

Chemical reaction	$-\Delta G^0$ (kJ)		
	1150°C	1120°C	1250°C
$2Fe + \frac{1}{2}Tm_2O_3 + \frac{5}{4}O_2 \rightarrow TmFe_2O_4$	425.9 ± 0.4	420.9 ± 0.8	414.2 ± 0.8
$2FeO + \frac{1}{2}Tm_2O_3 + \frac{1}{4}O_2 \rightarrow TmFe_2O_4$		84.1 ± 0.4	82.4 ± 0.4
$\frac{1}{2}TmFe_2O_4 + \frac{1}{4}Tm_2O_3 + \frac{1}{8}O_2 \rightarrow TmFeO_3$	37.0 ± 0.2	35.3 ± 0.2	33.6 ± 0.2
$TmFe_2O_4 + \frac{1}{6}O_2 \rightarrow \frac{1}{3}Fe_3O_4 + TmFeO_3$	42.3 ± 0.4	40.2 ± 0.4	36.8 ± 0.4
$3TmFeO_3 + \frac{2}{3}Fe_3O_4 + \frac{1}{6}O_2 \rightarrow Tm_3Fe_5O_{12}$	23.0 ± 0.4	22.2 ± 0.4	17.2 ± 0.4
$Fe + \frac{1}{2}Tm_2O_3 + \frac{3}{4}O_2 \rightarrow TmFeO_3$	249.8 ± 1.3	245.6 ± 1.3	240.2 ± 1.3
$5Fe + \frac{3}{2}Tm_2O_3 + \frac{15}{4}O_2 \rightarrow Tm_3Fe_5O_{12}$	1210.0 ± 3.4	1190.4 ± 3.8	1161.1 ± 3.8

TABLE 14
The standard Gibbs free energy for the Yb–Fe–O compounds at 1200°C (Kimizuka and Katsura 1975b).

Chemical reaction	$-\Delta G^0$ (kJ)
$2Fe + \frac{1}{2}Yb_2O_3 + \frac{5}{4}O_2 \rightarrow YbFe_2O_4$	420.0 ± 0.4
$3Fe + Yb_2O_3 + 2O_2 \rightarrow Yb_2Fe_3O_7$	662.7 ± 0.4
$Fe + \frac{1}{2}Yb_2O_3 + \frac{3}{4}O_2 \rightarrow YbFeO_3$	243.4 ± 1.4
$5Fe + \frac{3}{2}Yb_2O_3 + \frac{15}{4}O_2 \rightarrow Yb_3Fe_5O_{12}$	1185.8 ± 2.9
$\frac{1}{2}Yb_2O_3 + 2FeO + \frac{1}{4}O_2 \rightarrow YbFe_2O_4$	83.3 ± 0.4
$\frac{3}{2}YbFe_2O_4 + \frac{1}{4}Yb_2O_3 + \frac{1}{8}O_2 \rightarrow Yb_2Fe_3O_7$	32.7 ± 0.4
$\frac{1}{6}Yb_2O_3 + \frac{1}{3}Yb_2Fe_3O_7 + \frac{1}{12}O_2 \rightarrow YbFeO_3$	22.5 ± 0.4
$YbFe_2O_4 + \frac{1}{6}O_2 \rightarrow YbFeO_3 + \frac{1}{3}Fe_3O_4$	38.9 ± 0.4
$3YbFeO_3 + \frac{2}{3}Fe_3O_4 + \frac{1}{6}O_2 \rightarrow Yb_3Fe_5O_{12}$	24.7 ± 1.3

TABLE 15
The standard Gibbs free energy for the Lu–Fe–O compounds at 1200°C (Sekine and Katsura 1976).

Chemical reaction	$-\Delta G^0$ (kJ)
$2Fe + \frac{1}{2}Lu_2O_3 + \frac{5}{4}O_2 \rightarrow LuFe_2O_4$	419.7 ± 0.8
$3Fe + Lu_2O_3 + 2O_2 \rightarrow Lu_2Fe_3O_7$	661.9 ± 0.8
$Fe + \frac{1}{2}Lu_2O_3 + \frac{3}{4}O_2 \rightarrow LuFeO_3$	240.6 ± 1.3
$5Fe + \frac{3}{2}Lu_2O_3 + \frac{15}{4}O_2 \rightarrow Lu_3Fe_5O_{12}$	1167.8 ± 3.8
$2FeO + \frac{1}{2}Lu_2O_3 + \frac{1}{4}O_2 \rightarrow LuFe_2O_4$	82.4 ± 0.4
$\frac{3}{2}LuFe_2O_4 + \frac{1}{4}Lu_2O_3 + \frac{1}{8}O_2 \rightarrow Lu_2Fe_3O_7$	32.6 ± 0.4
$\frac{1}{3}Lu_2Fe_3O_7 + \frac{1}{6}Lu_2O_3 + \frac{1}{12}O_2 \rightarrow LuFeO_3$	20.3 ± 0.4
$2LuFe_2O_4 + \frac{1}{6}O_2 \rightarrow Lu_2Fe_3O_7 + \frac{1}{3}Fe_3O_4$	37.0 ± 0.4
$\frac{1}{2}Lu_2Fe_3O_7 + \frac{1}{12}O_2 \rightarrow LuFeO_3 + \frac{1}{6}Fe_3O_4$	17.5 ± 0.4
$3LuFeO_3 + \frac{2}{3}Fe_3O_4 + \frac{1}{6}O_2 \rightarrow Lu_3Fe_5O_{12}$	14.9 ± 0.8

TABLE 16

The Gibbs free-energy change of the chemical reaction: $Fe + \frac{1}{2}R_2O_3 + \frac{3}{4}O_2 \rightarrow RFeO_3$ (Kimizuka et al. 1983).

$RFeO_3$	$-\Delta G^0$ (kJ/mol) ± 1.3									
	1000°C	1100°C	1150°C	1200°C	1224°C	1250°C	1275°C	1297°C	1347°C	1400°C
$LaFeO_3$	311.1	298.3		288.3		282.3		276.6		265.2
$PrFeO_3$	302.7	288.2		276.4						
$NdFeO_3$	298.6	286.7		274.8		268.4		262.8	257.6	
$SmFeO_3$	294.5	280.5		268.2		262.1		256.4		
$EuFeO_3$	292.3	277.6		264.8	262.4	259.4		253.4		
$GdFeO_3$	290.8	276.8		264.2	261.2	258.1		251.7		
$TbFeO_3$	287.0	272.5	263.8	258.5	257.2	253.9	249.2			
$DyFeO_3$	284.4	269.1	260.9	256.1	254.4	250.6	245.5			
$HoFeO_3$	282.6	266.5	257.9	253.0		247.8				
$ErFeO_3$	280.8	263.6		249.8						
$TmFeO_3$	277.1	259.7	250.0	245.7		240.2				
$YbFeO_3$		256.9		243.4						
$LuFeO_3$		253.8	244.2	240.7		234.7				
$YFeO_3$				250.2						

entropy change accompanying the oxygen production in the decomposition of the oxides. According to one of their conclusions, the entropy change for the series of the present chemical reactions should be $44 \times n$ ($n = \frac{3}{4}$ for the present case) = 33 e.u., which seems to be in good agreement with the results obtained.

Madelung energies for $RFeO_3$ were calculated (Nakamura 1981, Kimizuka et al. 1983) based on the point-charge model using the atomic positions which were determined by Marezio et al. (1970). They show the same dependence upon the constituent cations as the Gibbs free-energy change does in table 16. The relation

TABLE 17

Enthalpy change and entropy change for the reaction $Fe + \frac{1}{2}R_2O_3 + \frac{3}{4}O_2 \rightarrow RFeO_3$ (Kimizuka et al. 1983).

$RFeO_3$	$-\Delta H^0$ (kJ/mol)	$-\Delta S^0$ (J/K mol)
$LaFeO_3$	456.1 ± 12.6	113.0 ± 4
$PrFeO_3$	460.2 ± 12.6	125.5 ± 4
$NdFeO_3$	456.1 ± 12.6	121.3 ± 4
$SmFeO_3$	456.1 ± 12.6	125.5 ± 4
$EuFeO_3$	456.1 ± 12.6	129.7 ± 4
$GdFeO_3$	456.1 ± 12.6	129.7 ± 4
$TbFeO_3$	451.9 ± 12.6	129.7 ± 4
$DyFeO_3$	447.7 ± 12.6	129.7 ± 4
$HoFeO_3$	447.7 ± 12.6	133.9 ± 4
$ErFeO_3$	447.7 ± 12.6	133.9 ± 4
$TmFeO_3$	439.3 ± 12.6	129.7 ± 4
$YbFeO_3$	435.1 ± 12.6	129.7 ± 4
$LuFeO_3$	431.0 ± 12.6	129.7 ± 4
$YFeO_3$		

between $\Delta G^0(1)$ and the tolerance factor which was defined by Goldschmidt (1926) is shown in fig. 35 (Katsura et al. 1975). We can see a good linearity in this relation.

3.2. $R_3Fe_5O_{12}$ (*garnet*) (*R*: *Y and Sm–Lu*)

As shown in the preceding section, $R_3Fe_5O_{12}$ is in equilibrium with $RFeO_3$ and Fe_3O_4 in the temperature range 1000–1200°C. Therefore, we can calculate the Gibbs free-energy change for the following chemical equation,

$$3RFeO_3 + \tfrac{2}{3}Fe_3O_4 + \tfrac{1}{6}O_2 \rightarrow R_3Fe_5O_{12}\,, \tag{3}$$

using

$$\Delta G^0(3) = \tfrac{1}{6}RT \ln P(O_2)\,. \tag{4}$$

$P(O_2)$ values for each chemical reaction are given in the preceding subsection.

We can also measure the Gibbs free-energy change with the Coulombic titration for the cell system II in fig. 33b. Piekarczyket et al. (1978) measured dissociation oxygen partial pressure and the Gibbs free-energy change of $Y_3Fe_5O_{12}$ by means of a solid-state galvanic cell method in the temperature range of 900–1250°C. Their results show ΔG^0 (kJ/mol) $= -84.1 + 0.0387T \pm 0.2$. At 1200°C, ΔG^0 (kJ/mol) $= -27.1 \pm 0.2$ is calculated from the above equation (see fig. 34). Since the reaction rate of the above chemical reaction is much smaller than that of the chemical reaction of eq. (1), we can say that the experimental errors in determining $P(O_2)$ are greater than that of $RFeO_3$. We summarize the $\Delta G^0(3)$ in table 18a.

The Gibbs free-energy change for the chemical reaction

$$3RFeO_3 + Fe_2O_3 \rightarrow R_3Fe_5O_{12}$$

TABLE 18A
The Gibbs free-energy change of the chemical reaction: $3RFeO_3 + \tfrac{2}{3}Fe_3O_4 + \tfrac{1}{6}O_2 \rightarrow R_3Fe_5O_{12}$ (Kimizuka et al. 1983).

	$-\Delta G^0$ (kJ/mol) ± 2.5				
$R_3Fe_5O_{12}$	1000°C	1100°C	1150°C	1200°C	1250°C
$Sm_3Fe_5O_{12}$	34.3	25.2		13.4	
$Eu_3Fe_5O_{12}$	37.1	26.2		20.3	
$Gd_3Fe_5O_{12}$	38.5	28.2		24.1	
$Tb_3Fe_5O_{12}$	40.1	32.9		27.4	
$Dy_3Fe_5O_{12}$	40.0	32.6		25.2	
$Ho_3Fe_5O_{12}$	39.0	31.1	26.5	25.9	21.3
$Er_3Fe_5O_{12}$	38.2	29.0			
$Tm_3Fe_5O_{12}$	35.2	26.7	23.1	23.1	18.5
$Yb_3Fe_5O_{12}$		24.1			
$Lu_3Fe_5O_{12}$		23.1		14.1	
$Y_3Fe_5O_{12}$				27.2	

TABLE 18B

The Gibbs free-energy change of the chemical reaction: $5Fe + \frac{3}{2}R_2O_3 + \frac{15}{4}O_2 \rightarrow R_3Fe_5O_{12}$ (Kimizuka et al. 1983).

$R_3Fe_5O_{12}$	$-\Delta G^0$ (kJ/mol) ± 2.5				
	1000°C	1100°C	1150°C	1200°C	1250°C
$Sm_3Fe_5O_{12}$	1391.2	1320.5		1248.9	
$Eu_3Fe_5O_{12}$	1387.8	1313.8		1246.0	
$Gd_3Fe_5O_{12}$	1384.5	1312.1		1250.2	
$Tb_3Fe_5O_{12}$	1374.9	1303.7		1236.4	
$Dy_3Fe_5O_{12}$	1366.9	1293.3		1226.8	
$Ho_3Fe_5O_{12}$	1363.6	1284.1	1237.2	1207.1	1189.9
$Er_3Fe_5O_{12}$	1353.9	1273.2		1197.5	
$Tm_3Fe_5O_{12}$	1351.0	1259.4	1210.0	1190.4	1161.1
$Yb_3Fe_5O_{12}$		1248.5		1185.8	
$Lu_3Fe_5O_{12}$		1238.1		1167.8	
$Y_3Fe_5O_{12}$				1208.8	

versus R is shown in fig. 36. There exists no $R_3Fe_5O_{12}$ (R = La, Ce(III) or Nd) as mentioned above. It is thought that R = La, Ce(III) and Nd are too large, and Sc and In are too small to form the $R_3Fe_5O_{12}$ crystal structure under one atmosphere. So it could be reasonable that there should exist a most suitable size of the R ion for forming the garnet-type of structure. From fig. 36, we can guess that R around Tb may be the most suitable size for constructing the $R_3Fe_5O_{12}$ structure. The calculations of the Madelung energies for $R_3Fe_5O_{12}$ also suggested the same conclusion (Kimizuka et al. 1983).

3.3. *RFe_2O_4 (R: Y and Ho–Lu) and $R_2Fe_3O_7$ (R: Yb and Lu)*

As shown in the preceding section, there are stable ternary phases, RFe_2O_4 and $R_2Fe_3O_7$, in rather limited ranges of oxygen partial pressure at 1200°C. However,

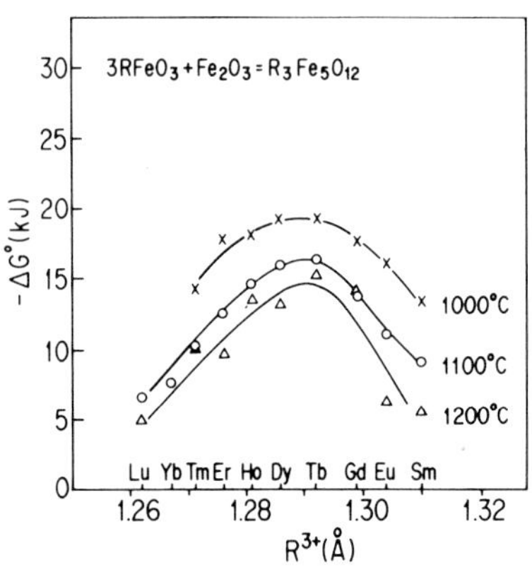

Fig. 36. The relation between $-\Delta G^0$ and the constituent R in the garnet. (Sugihara 1978.)

each of RFe_2O_4 and $R_2Fe_3O_7$ has a compositional range from the stoichiometric composition as in $RFeO_3$. Each of the activity values for $RFeO_3$, RFe_2O_4 ($YFe_2O_{3.905}$ in the case of the system Y_2O_3–Fe_2O_3–Fe) and $R_2Fe_3O_7$ was calculated, based upon the Gibbs–Duhem equation using the experimental relation between the oxygen composition and $-\log P(O_2)$. In tables 4–8, we show the chemical reactions which take part in RFe_2O_4 and $R_2Fe_3O_7$ and the activity for each component. Finally, Gibbs free-energy changes for RFe_2O_4 and $R_2Fe_3O_7$ which were re-evaluated by Kimizuka et al. (1983) are shown in tables 19 and 20.

TABLE 19A
The Gibbs free-energy change of the chemical reaction: $RFeO_3 + FeO \rightarrow RFe_2O_4$ (Kimizuka et al. 1983).

	$-\Delta G^0$ (kJ) ± 2.0				
RFe_2O_4	1000°C	1100°C	1150°C	1200°C	1250°C
$HoFe_2O_4$				0.8	1.7
$ErFe_2O_4$		0.8		2.9	
$TmFe_2O_4$	1.3	5.4	5.9	7.1	8.8
$YbFe_2O_4$		8.4		8.4	
$LuFe_2O_4$		9.6	10.9	11.3	12.6

TABLE 19B
The Gibbs free-energy change of the chemical reaction: $2Fe + \frac{1}{2}R_2O_3 + \frac{5}{4}O_2 \rightarrow RFe_2O_4$ (Kimizuka et al. 1983).

	$-\Delta G^0$ (kJ) ± 2.0				
RFe_2O_4	1000°C	1100°C	1150°C	1200°C	1250°C
$HoFe_2O_4$				422.2	415.5
$ErFe_2O_4$		440.2		421.8	
$TmFe_2O_4$	465.7	438.9	425.9	420.9	414.4
$YbFe_2O_4$		439.7		420.1	
$LuFe_2O_4$		437.2	424.3	419.7	412.3
$YFe_2O_{3.905}$				405.0	

TABLE 20A
The Gibbs free-energy change of the chemical reaction: $2RFeO_3 + FeO \rightarrow R_2Fe_3O_7$ (Kimizuka et al. 1983).

	$-\Delta G^0$ (kJ) ± 2.0			
$R_2Fe_3O_7$	1100°C	1150°C	1200°C	1250°C
$Yb_2Fe_3O_7$	8.8		9.2	
$Lu_2Fe_3O_7$	11.7	12.6	13.0	14.3

TABLE 20B
The Gibbs free-energy change of the chemical reaction: $3Fe + R_2O_3 + 2O_2 \rightarrow R_2Fe_3O_7$ (Kimizuka et al. 1983).

	$-\Delta G^0$ (kJ) ± 2.0			
$R_2Fe_3O_7$	1100°C	1150°C	1200°C	1250°C
$Yb_2Fe_3O_7$	697.1		662.8	
$Lu_2Fe_3O_7$	697.1	670.3	661.9	648.5

4. Crystal chemistry of the $(RMO_3)_n(M'O)_m$-type compounds

In this section, we will discuss the crystal chemistry of the series of $(RMO_3)_n(M'O)_m$-type compounds (hereafter denoted simply as (n, m)-type compounds). $YbFe_2O_4$ and $Yb_2Fe_3O_7$ described in the preceding section are members of the series with $n = 1$, $m = 1$ and $n = 2$, $m = 1$, respectively. So far, many other compounds with higher n or m, although one of two is unity, have been known. It will be shown from the structural viewpoint that $YAlO_3$ found by Bertaut and Mareschal (1963) is an end member of the series ($n = 1$, $m = 0$ or $n = \infty$, $m = 0$) and würtzite ZnO can be considered as an opposite end member with $n = 0$, $m = 1$ (or $n = 0$, $m = \infty$).

4.1. *Crystal structure of the (1, 1)- and (2, 1)-type compounds*

The first report on the (1, 1)-type compounds of the present series appeared in 1965: Schmitz-DuMont and Kasper prepared a series of oxides $(InMO_3)(M'O)$ (M: Al, Ga; M′: Mg, Mn, Co, Cu, Zn, Cd). They were estimated to belong to the space group R3m. After ten years, the structure of the (1, 1)-type compounds was determined independently by two groups, by means of single-crystal X-ray diffraction. A Japanese group (Kato et al. 1975) investigated $YbFe_2O_4$ $[(YbFeO_3)(FeO)]$ and a French group (Malaman et al. 1975) investigated $Yb_{0.5}Eu_{0.5}Fe_2O_4$ $[(Yb_{0.5}Eu_{0.5}FeO_3)(FeO)]$. In 1980 and 1983, atomic positions of three more compounds were reported: $InAlCuO_4$ and $InFe_{1.75}Si_{0.25}O_4$ by Gérardin et al. (1980) and YFe_2O_4 by Kimizuka et al. (1983). These compounds crystallize in trigonal $R\bar{3}m$ (not R3m) with atoms in positions shown in table 21 where the crystal data are expressed on the hexagonal system rather than the rhombohedral one. (Throughout this chapter, we use the hexagonal system to discuss crystal structure.)

In the $(RMO_3)(M'O)$-type [(1, 1)-type] structure, each element, R, M + M′ or O, composes triangular planar nets and stacking of these nets along the c-axis of a hexagonal cell forms a three-dimensional lattice, as is shown in fig. 37 for $YbFe_2O_4$. The structure can be expressed as –U–T–T–U–T–T–U–T–T– by using two kinds of layers U and T, whose compositions are RO_2 and (M, M′)O, respectively. A triangular net of rare earth elements R is adjoined by two nets of

TABLE 21
Atom positions for (1, 1)-type compounds: $YbFe_2O_4$ (column labelled A), $Yb_{0.5}Eu_{0.5}Fe_2O_4$ (B), $InAlCuO_4$ (C), $InFe_{1.75}Si_{0.25}O_4$ (D) and YFe_2O_4 (E).

Atom	Wyckoff notation	x	y	z-coordinate				
				A [1]	B [2]	C [3]	D [3]	E [4]
R	3(a)	0	0	0	0	0	0	0
M (M′)	6(c)	0	0	0.2150	0.2141	0.2128	0.2168	0.2144
O(1)	6(c)	0	0	0.2925	0.2914	0.2894	0.2925	0.2924
O(2)	6(c)	0	0	0.1292	0.1295	0.1302	0.1285	0.1282

References: [1] Kato et al. (1975); [2] Malaman et al. (1975); [3] Gérardin et al. (1980); [4] Kimizuka et al. (1983).

oxygen as shown in fig. 38 for $Yb_{0.5}Eu_{0.5}Fe_2O_4$, constructing a U block. The R ion is coordinated by six oxygen ions octahedrally. A T layer is composed of triangular nets of transition metals, M + M′, and of oxygen on an almost identical c-plane. M and M′ ions are believed to distribute randomly in an equivalent crystallographic site. They are coordinated by five oxygen ions, by three in the same T layer and by two in the adjacent T or U layer, forming a trigonal bipyramid. Note that the unit cell consists of three times the repeating of a unit, –U–T–T–, whose composition is $RMM'O_4$, the chemical formula. If we take a monoclinic primitive cell, it contains one formula unit. The M (M′) atom is not necessarily located at the center of the trigonal bipyramid: in each of the double T layers of fig. 37, the M (M′) atom is generally displaced along [0 0 1] toward the

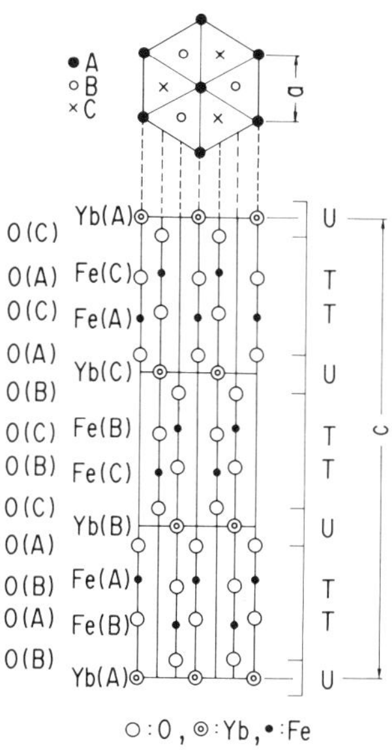

Fig. 37. The crystal structure of $YbFe_2O_4$.

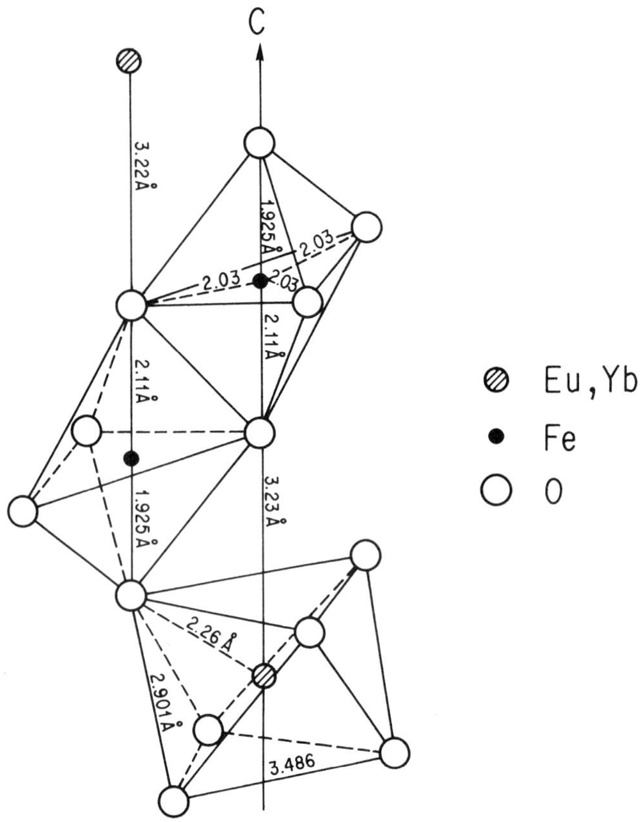

Fig. 38. Cation coordinations in (Yb, Eu)Fe_2O_4. (Malaman et al. 1975.)

adjacent U block, while O atoms shift toward the adjacent T layer. This trigonal bipyramidal coordination of M (M′) ions is an earmark of the present series of compounds. One metal site in the magnetoplumbite structure has this coordination and it was supposed to cause the strong magnetic anisotropy in hexagonal Ba ferrites (Went et al. 1952).

Interatomic distances in the (1, 1)-type compounds are listed in table 22. The O–O distance within a triangular net [O(1)–6O(1) in the table], which is equal to the *a*-cell dimension, is considerably longer than twice the ionic radius of O^{2-} [2.8 Å after Shannon and Prewitt (1969)], whereas interlayer O–O distances [O(1)–3O(1), O(1)–3O(2) and O(2)–3O(2)] are in rough accordance with it. This fact indicates that an oxygen ion is in contact with oxygen ions in the adjacent layers but is apart from those in the same layer.

The structure of the (2, 1)-type compounds was independently determined, for $Yb_2Fe_3O_7$ [$(YbFeO_3)_2(FeO)$], also by Kato et al. (1976) and by Malaman et al. (1976). It is schematically represented in fig. 39 and the atomic positions are listed in table 23. $Yb_2Fe_3O_7$ crystallizes in the space group $P6_3/mmc$.

TABLE 22
Interatomic distances (in Å) in: $YbFe_2O_4$ (column labelled A), $Yb_{0.5}Eu_{0.5}Fe_2O_4$ (B), $InAlCuO_4$ (C), $InFe_{1.75}Si_{0.25}O_4$ (D) and YFe_2O_4 (E).

Bond	Interatomic distance				
	A [1]	B [2]	C [3]	D [3]	E [4]
R–6O(1)	2.241	2.268	2.189	2.188	2.269
M(M′)–1O(1)	1.944	1.925	1.86	1.987	1.934
M(M′)–3O(2)	2.013	2.03	1.92	1.933	2.043
M(M′)–1O(2)	2.149	2.11	2.01	2.32	2.137
O(1)–6O(1)	3.455	3.486	3.308	3.303	3.516
O(1)–3O(1)	2.856	2.901	2.87	2.870	2.870
O(1)–3O(2)	2.980	2.97	2.84	2.989	2.967
O(2)–3O(2)	2.74	2.73	2.61	2.768	2.785

References: [1] Kato et al. (1975); [2] Malaman et al. (1975); [3] Gérardin et al. (1980); [4] Kimizuka et al. (1983).

Apparently, the (2, 1)-type structure is closely related to the (1, 1)-type. It can be expressed by the use of U and T layers as –U–T–T–U–T–U–T–T–U–T–. Both anions and cations occupy the triangular lattice points and the coordinations of cations are the same: with the R ion at an octahedral site and the M or M′ ion at a trigonal bipyramidal site, as shown in fig. 40. It should be noted, however, that single T layers appear in the stacking sequence of the (2, 1)-type structure, besides double T layers found in the (1, 1)-type structure. According to the

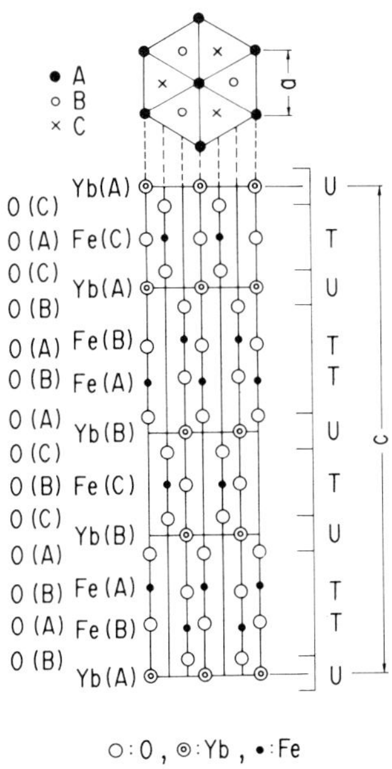

Fig. 39. The crystal structure of $Yb_2Fe_3O_7$.

TABLE 23
Atom positions in $Yb_2Fe_3O_7$ after Kato et al. (1976) and Malaman et al. (1976).

Atom	Wyckoff notation	x	y	z Kato et al.	z Malaman et al.
Yb	4(f)	$\frac{1}{3}$	$\frac{2}{3}$	0.1480	0.1482
Fe(1)	2(b)	0	0	$\frac{1}{4}$	$\frac{1}{4}$
Fe(2)	4(f)	$\frac{2}{3}$	$\frac{1}{3}$	0.4568	0.4574
O(1)	4(e)	0	0	0.317	0.320
O(2)	4(f)	$\frac{2}{3}$	$\frac{1}{3}$	0.389	0.385
O(3)	4(f)	$\frac{1}{3}$	$\frac{2}{3}$	0.467	0.465
O(4)	2(c)	$\frac{1}{3}$	$\frac{2}{3}$	$\frac{1}{4}$	$\frac{1}{4}$

symmetry of the structure, a metal ion in a single T layer should be located at the central position of a trigonal bipyramid. In addition, the unit of stacking, –U–T–T–U–T– is repeated twice, instead of three times, in a unit cell. As will be seen below, the following rule is kept in the present series:

The crystal symmetry expected for a (n, m)-type compound is $R\bar{3}m$ and the unit of stacking is repeated three times (i.e., $Z = 3$) when $n + m$ is even whereas the symmetry expected is $P6_3/mmc$ and the unit is repeated twice (i.e., $Z = 2$) when $n + m$ is odd.

The interatomic distances in the $Yb_2Fe_3O_7$ are shown in table 24. The average Fe(1)–O distance (1.97 Å as calculated from the distance by Kato et al. (1976)) is close to the ideal one (1.98 Å) obtained from the ionic radii of O^{2-} and Fe^{3+} in the five-fold coordination (Shannon 1976). Moreover, the average Fe(2)–O distance (2.04 Å) is in good agreement with the average Fe–O distance in $YbFe_2O_4$ (2.03 Å) where Fe^{2+}–Fe^{3+} mixed ions occupy the Fe sites. These facts

TABLE 24
Interatomic distances in $Yb_2Fe_3O_7$.

Bond	Interatomic distance	Bond	Interatomic distance
Yb–3O(1)	2.23 (2.20)*	O(1)–3O(2)	2.85 (2.73)
Yb–3O(2)	2.26 (2.21)	O(1)–3O(4)	2.78 (2.83)
Yb–1O(3)	3.27 (3.21)	O(2)–3O(3)	2.99 (3.03)
Yb–1O(4)	2.90 (2.896)	O(3)–3O(3)	2.75 (2.82)
Fe(1)–2O(1)	1.92 (1.99)		
Fe(1)–3O(4)	2.01 (2.00)		
Fe(2)–1O(2)	1.93 (2.06)		
Fe(2)–3O(3)	2.03 (2.01)		
Fe(2)–1O(3)	2.17 (2.20)		

*Values without parentheses from Kato et al. (1976); those in parentheses are from Malaman et al. (1976).

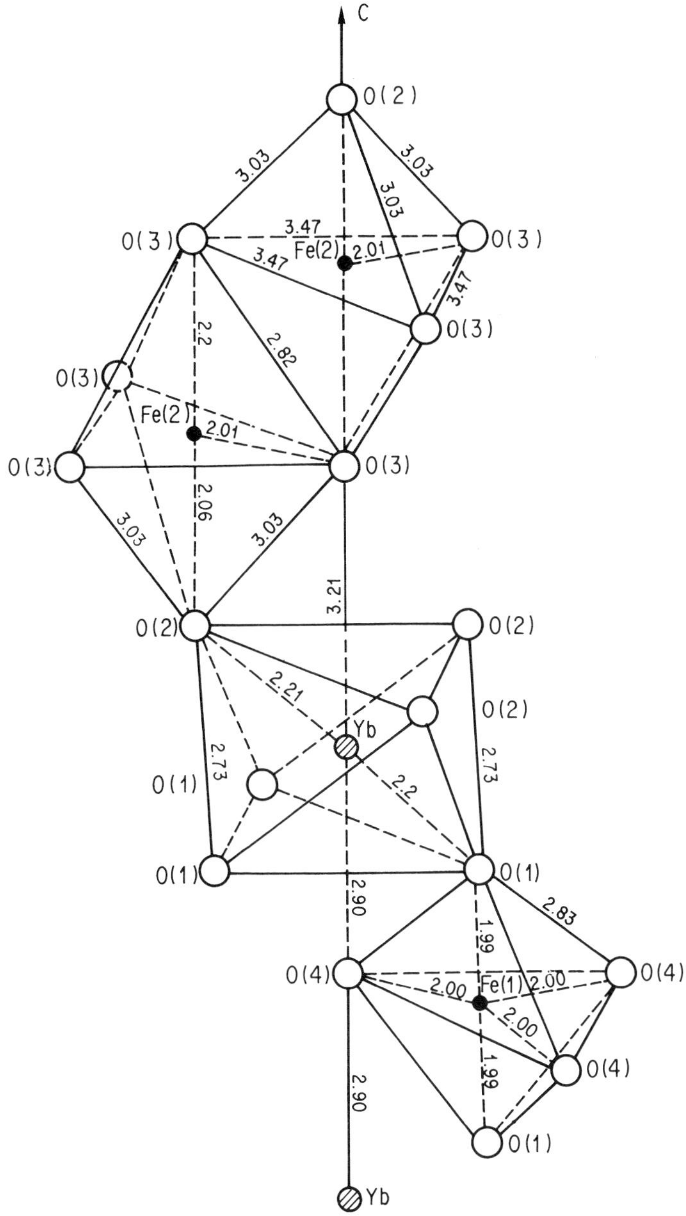

Fig. 40. Cation coordinations in $Yb_2Fe_3O_7$ (Malaman et al. 1976). Interatomic distances are given in Å.

suggest that the Fe(1) sites (in the single T layers) are occupied by Fe^{3+} while the Fe(2) sites (in the double T layers) are occupied by Fe^{2+}–Fe^{3+} mixed ions (Malaman et al. 1976, Kato et al. 1976). The interlayer O–O distances in the $Yb_2Fe_3O_7$ are, as in the (1, 1)-type compounds, close to twice the O^{2-} radius. Oxygen ions are packed nearly in contact with each other along the [0 0 1] direction.

4.2. *Construction of the crystal lattice of the* (n, m)*-type compounds*

Although single-crystal X-ray analyses have been applied to only a few compounds of the present series, the compounds having higher n or m were investigated by powder X-ray and electron microscopic analyses and the basic features of their structures have been elucidated. Kimizuka and his collaborators synthesized $Yb_3Fe_4O_{10}$ and $Yb_4Fe_5O_{13}$ [(3, 1)- and (4, 1)-type compounds] and indicated that the unit stacking sequences of their structures are –U–T–T–U–T–U–T– and –U–T–T–U–T–U–T–U–T–, respectively (Kimizuka et al. 1976, Matsui et al. 1979, Matsui 1980). These sequences suggest the limiting case of $n = \infty$, i.e., an (1, 0)-compound RMO_3 constructed by a sequence of –U–T–U–T–. Indeed, Bertaut and Mareschal have reported a series of compounds $RAlO_3$ (R: Y, Eu, Gd, Tb, Dy, Ho and Er) already in 1963. The structure was determined for $YAlO_3$: the unit cell is hexagonal, space group $P6_3/mmc$, $a = 3.678$, $c = 10.52$ Å and $Z = 2$ (Bertaut and Mareschal 1963). The Al coordination is trigonal bipyramidal and the R ion occupies an octahedral site. From the structure shown in fig. 41, it is easy to see that $RAlO_3$ is really the end member of the present homologous series. Bertaut and his collaborators reported that $RMnO_3$ also crystallizes in the same structure with a little modification (Bertaut and Mareschal 1963, Yakel et al. 1963).

On the other side of the present series, Kimizuka et al. prepared a series of compounds $(RMO_3)(ZnO)_m$ (R: Sc, In, Er, Tm, Yb, Lu; M: Fe, Ga, Al; $m = 2$–11) and determined their stacking sequences based on the high-resolution lattice images and powder X-ray diffractions (Kimizuka et al. 1988, Kimizuka and

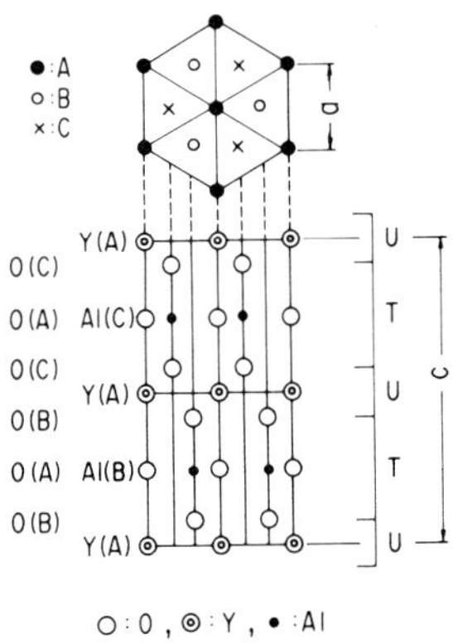

Fig. 41. The crystal structure of $YAlO_3$.

Mohri 1989). The symmetry of the crystal follows the rule mentioned before, i.e., hexagonal for odd $n+m$, while rhombohedral for even $n+m$. The metal positions determined from the lattice images indicated that these compounds are also composed of two kinds of layers U and T, i.e., they belong to $(1, m)$-type compounds. The unit stacking sequences are rather simple, –U–T–T–T–T– in $(RMO_3)(ZnO)_3$, for instance, or generally $-U-(T-)_{m+1}$ in $(1, m)$-type compound. Figure 42 shows the lattice image of $(InGaO_3)(ZnO)_5$. One can see six lines of weaker dark dots separated by a single line of stronger dark dots: the former corresponds to Zn–Ga mixed ions in T layers while the latter indicates In ions in a U layer.

On the other hand, there was a report by Kasper (1967) of oxides $In_2O_3(ZnO)_m$ ($m = 2$, 3, 4, 5, 7) in the binary system ZnO–In_2O_3. The phases crystallized rhombohedrally ($m = 3$, 5, 7) or hexagonally ($m = 2$, 4) and were suggested to consist of würtzite-type ZnO slabs separated by a In_2O_3 slab (Kasper 1967). Note that the symmetry of the compound is consistent with the above rule. In 1988, Cannard and Tilley reinvestigated the same system using powder X-ray diffraction and electron microscopy and found additional compounds with $m = 6$, 9, 11, although they were not isolated as single phases. Although they proposed a similar structure model to that by Kasper, detailed stacking of metal and oxygen atom planes was not determined. Comparing the powder X-ray patterns of $In_2O_3(ZnO)_m$ and $(RMO_3)(ZnO)_m$, it seems that they are isostructural and that the former is the special case with R = M = In. Indeed, ten lines of weaker dark dots were observed separated by a single line of stronger dark dots in the high-resolution structure images of $In_2O_3(ZnO)_9$ projected along [1 0 0] (Matsui

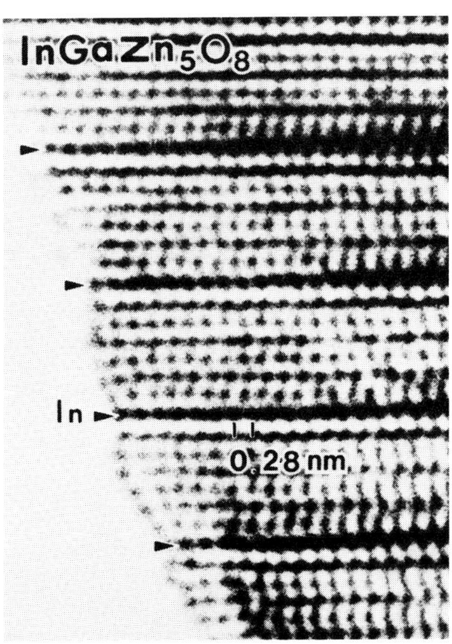

Fig. 42. Lattice image of $(InGaO_3)(ZnO)_5$ projected along [1 0 0]: by courtesy of Dr Y. Matsui.

1988). The former corresponds to T layers including In–Zn mixed ions and the latter corresponds to a U layer composed of InO_6 octahedra.

When *m* tends to infinity, the crystal will be constructed by a stacking of only T layers. Although no binary compounds with an ideal (0, 1)-type structure are known, the würtzite-type one (Wyckoff 1964) is closely related to it. If metal ions in an ideal (0, 1)-type structure are displaced from trigonal bipyramidal position to tetrahedral position along [0 0 1], the resulting structure is würtzite type. It should, however, be noted that the symmetry of the structure is lowered by such a displacement to $P6_3mc$ from $P6_3/mmc$ expected for the ideal (0, 1)-type structure. If we disregard this point, würtzite-type ZnO can be considered as another end member of the present series (see sect. 4.3).

It is not so difficult to find out the sequence using U and T for the general (*n*, *m*)-type structure, taking into account the empirical fact that U layers are not stacked in succession. To understand the stacking of triangular nets of cations and anions, however, notation of U and T is not sufficient. Three kinds of triangular nets should be distinguished, those are ordinarily denoted as A, B and C. See figs. 37, 39 or 41. A layer of octahedrally coordinated cations, a U block, is constructed by a stacking of three different triangular lattices: O(C)–R(A)–O(B), etc., and a T layer is composed of two triangular nets on nearly the same *c*-plane: O(A) + M(M′)(B), etc. There are six kinds of U blocks and six kinds of T layers. If we assume the following rule, the crystal structure of the present series of compounds can be understood systematically:
(i) Triangular nets are stacked along [0 0 1] so that cations and anions in adjacent layers may be located as near each other as possible.

Let us assume a U block of O(C)–R(A)–O(B) at the origin. Following this block, a T layer will come and it should be O(A) + M(M′)(B), since the position of anions nearest to R is A, just upon the R^{3+} net, and that of cations is B, just upon the O^{2-} net. The second T layer on the first T should be O(B) + M(M′)(A), by the same principle. Thus, in the (1, 0)-, (1, 1)- and (2, 1)-compounds, triangular nets of cations and oxygen ions are stacked as shown in figs. 41, 37 and 39, respectively. It is easy to see that the structure of any compound in this series can be constructed by using the above rule. Some examples are shown in fig. 43.

Other methods of description of the stacking of triangular nets are also possible and sometimes useful. If we consider only oxygen layers ignoring cations, they can be classified into two types: one is denoted by h if two neighboring layers are equivalent (i.e., both A, both B or both C), another by c if neighboring layers are different (Pauling 1960). The condition (i) is interpreted as:
(ii) An oxygen layer within a T block belongs to h, whereas two oxygen layers in a U block belong to c, i.e., an R ion occupies an octahedral site between c layers and an M(M′) ion occupies a trigonal bipyramidal site in an h layer.

Since U blocks are not stacked in succession, a pair of c layers (cc layer) is always sandwiched by h layers as . . . hcch . . . (a sequence such as . . . ccc . . . does not appear). From condition (ii), it can be understood that a (*n*, *m*)-type structure contains 2*n* times c layers (*n* times cc layers) and (*m* + *n*) times h layers per unit formula. If either *n* or *m* is unity (all compounds found so far satisfy this),

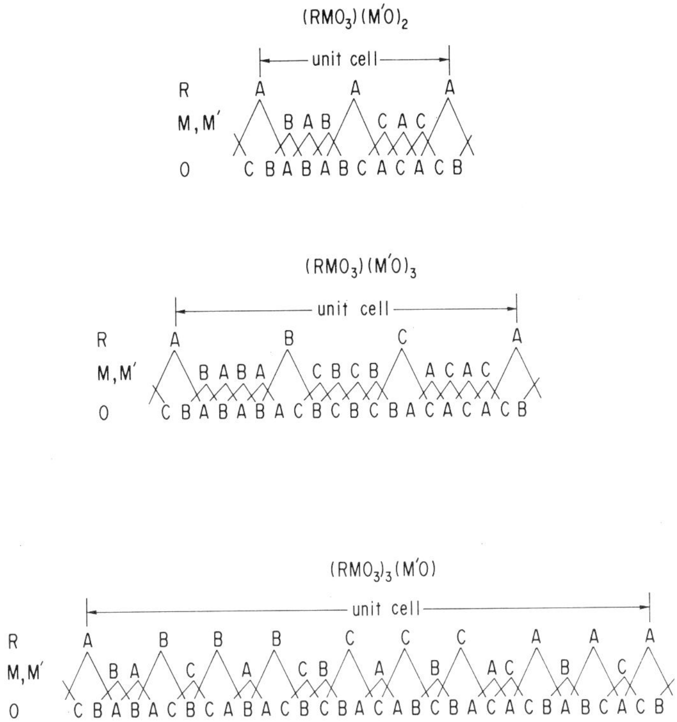

Fig. 43. Atom positions in the (1, 2)-, (1, 3)- and (3, 1)-type structures.

the stacking sequence of oxygen layers using h and c can be determined unequivocally. As an example, consider the (2, 1)-type compound. The stacking of oxygen layers becomes hcchcch for one unit formula and a unit cell is formed by twice that sequence (fig. 44). The earlier notation using A, B and C can be obtained from this sequence starting arbitrarily with AB . . . and continuing until the layer sequence repeats (Wells 1975a). The notation using h and c is, thereby, equivalent with that using A, B and C. In addition, representation by U and T blocks may be obtained easily by substituting cc with U and h with T.

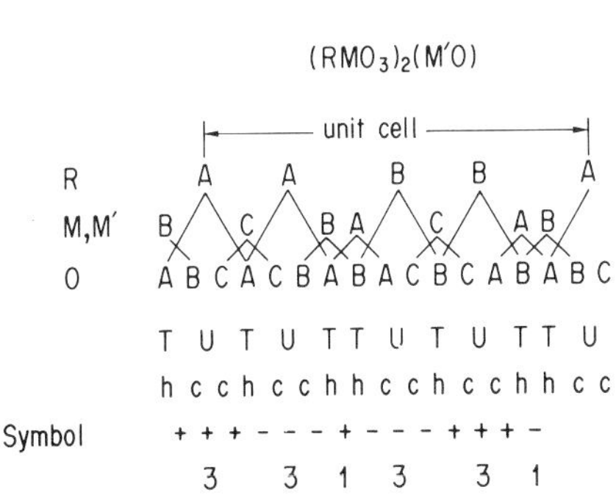

Fig. 44. Notations for stacking sequence of layers in the (2, 1)-type structure.

The highest symmetry expected for the structure of a (n, m)-type compound is identical with that of the closely packed oxygen lattice, ignoring metal ions, and every (n, m)-type compound found so far is, indeed, believed to have the highest symmetry. To see the symmetry, Zhdanov's symbol (Zhdanov 1945, Patterson and Kasper 1967) is useful. In a sequence of oxygen layers expressed by A, B and C, a change $A \rightarrow B \rightarrow C \rightarrow A$ is called positive (+), while $A \rightarrow C \rightarrow B \rightarrow A$ is called negative (−) [Hägg's notation (Hägg 1943)]. A succession of positive or negative changes is abbreviated by a number, so that 22, for instance, stands for ++−−. As shown in fig. 44, the unit sequence of the (2, 1)-type structure is 331331 and this sequence repeats infinitely in a crystal. Let p and l be the total numbers of positive and negative signs in a unit sequence and $N = p + l$. The lattice will be hexagonal with layer-number N if $p - l = 0 \pmod 3$ while it is rhombohedral with layer-number 3N if $p - l = \pm 1 \pmod 3$. More detailed symmetry of the stacked oxygen layers can be determined from the notation (see Patterson and Kasper 1967). In the present series of compounds, the space group is $P6_3/mmc$ for the hexagonal crystal while it is $R\bar{3}m$ when rhombohedral. For the case of the (2, 1)-type structure, $p - l = 0$. Therefore, it has a hexagonal cell with $P6_3/mmc$. The oxygen sequences of all (n, m)-type compounds found so far are summarized in table 25.

4.3. *The lattice parameters of the (n, m)-type compounds*

The lattice parameters of the (n, m)-type compounds are listed in table 26. There is a strong correlation between a-dimension and the size of the R ion, i.e., it increases with a larger R ion. If we simply assume that the oxygen ion is in

TABLE 25
Stacking of oxygen layers in $(RMO_3)_n(M'O)_m$ compounds.

n	m	Symbol		M*	Space group
1	0	hcc	33	6	$P6_3/mmc$
1	1	hcch	31	12	$R\bar{3}m$
2	1	hcchcch	331331	14	$P6_3/mmc$
3	1	hcchcchcch	3331	30	$R\bar{3}m$
4	1	hcchcchcchcch	3333133331	26	$P6_3/mmc$
1	2	hcchh	311311	10	$P6_3/mmc$
1	3	hcchhh	3111	18	$R\bar{3}m$
1	4	hcchhhh	3111131111	14	$P6_3/mmc$
1	5	hcchhhhh	311111	24	$R\bar{3}m$
1	6	hcchhhhhh	31111113111111	18	$P6_3/mmc$
1	7	hcchhhhhhh	31111111	30	$R\bar{3}m$
1	8	hcchhhhhhhh	311111111311111111	22	$P6_3/mmc$
1	9	hcchhhhhhhhh	3111111111	36	$R\bar{3}m$
1	10	hcchhhhhhhhhh	3111111111131111111111	26	$P6_3/mmc$
1	11	hcchhhhhhhhhhh	311111111111	42	$R\bar{3}m$

*Number of oxygen layers in a unit cell.

TABLE 26

Lattice constants of the $(RMO_3)_n(M'O)_m$ compounds.

Compound	*a* (Å)	*c* (Å)	Ref.**
(1, 0)-type (P6_3/mmc)			
$YAlO_3$	3.68	10.52	[1]
$EuAlO_3$	3.76	10.52	[1]
$GdAlO_3$	3.73	10.51	[1]
$TbAlO_3$	3.71	10.51	[1]
$DyAlO_3$	3.70	10.50	[1]
$HoAlO_3$	3.67	10.51	[1]
$ErAlO_3$	3.66	10.50	[1]
(1, 1)-type (R$\bar{3}$m)			
$ScAlMgO_4$	3.236	25.15	[2]
$ScAlMnO_4$	3.260	25.98	[2]
$ScAlCoO_4$	3.247	25.19	[2]
$ScAlCuO_4$	3.277	24.17	[3]
$ScAlZnO_4$	3.242	25.54	[4]
$ScGaMgO_4$	3.272	25.62	[2]
$ScGaCuO_4$	3.313	24.65	[3]
$ScGaZnO_4$	3.259	25.91	[3]
$InAlMgO_4$	3.296	25.65	[5]
	3.294	25.63	[6]
	3.290	25.66	[2]
$InAlMnO_4$	3.319	26.21	[5]
	3.308	26.56	[6]
	3.319	26.21	[2]
$InAlCoO_4$	3.302	25.73	[5]
	3.301	25.56	[6]
	3.301	25.74	[2]
$InAlCuO_4$	3.312	24.34	[5]
	3.308	24.36	[6]
	3.315	24.36	[3]
$InAlZnO_4$	3.309	26.33	[4]
$InAlCdO_4$	3.32	27.6	[5]
	3.321	27.50	[2]
$InFeMnO_4$	3.356	26.01	[6]
	3.356	26.40	[2]
$InFe_2O_4$	3.33	26.2	[7]
	3.337	26.11	[6]
	3.339	26.08	[3]
$InFe_{1.75}Si_{0.25}O_4$	3.303	26.26	[6]
$InFeCuO_4$	3.358	24.96	[6]
	3.374	24.84	[8]
$InFeZnO_4$	3.326	26.10	[6]
	3.321	26.09	[4]
$InGaMgO_4$	3.304	25.89	[5]
	3.304	25.65	[6]
	3.304	25.81	[3]
$InGaMnO_4$	3.338	26.52	[5]
	3.311	26.55	[6]
	3.329	26.52	[3]
(1, 1)-type (R$\bar{3}$m) (cont'd)			
$InGaFeO_4$	3.313	26.17	[2]
$InGaCoO_4$	3.311	25.93	[5]
	3.312	25.83	[6]
	3.309	25.86	[8]
$InGaCuO_4$	3.355	24.82	[5]
	3.348	24.79	[6]
	3.350	24.82	[8]
$InGaZnO_4$	3.298	26.13	[5]
	3.295	26.07	[3]
$In_{0.5}Lu_{0.5}Fe_2O_4$	3.382	25.58	[7]
$In_{0.8}Lu_{0.2}Fe_2O_4$	3.358	25.90	[7]
$YFeMnO_4$	3.496	25.46	[9]
YFe_2O_4	3.518	24.81	[6, 11*]
	3.516	24.79	[9, 12*]
$YFeZnO_4$	3.489	25.20	[2]
$Y_{0.5}Dy_{0.5}Fe_2O_4$	3.53	24.76	[7]
$YGaMnO_4$	3.467	25.29	[15]
$HoFe_2O_4$	3.518	24.81	[7, 11*]
	3.519	24.81	[9, 14*]
$HoGaMnO_4$	3.491	25.14	[15]
$ErFeMgO_4$	3.521	25.69	[9]
$ErFeMnO_4$	3.480	25.55	[9]
$ErFe_2O_4$	3.492	24.93	[7, 11*]
	3.497	24.94	[9, 14*]
$ErGaMgO_4$	3.433	25.12	[15]
$ErGaMnO_4$	3.474	25.30	[15]
$TmAlMnO_4$	3.459	24.91	[15]
$TmFeMgO_4$	3.503	25.55	[9]
$TmFeMnO_4$	3.469	25.56	[9]
$TmFe_2O_4$	3.472	25.01	[7, 11*]
	3.473	25.01	[9, 14*]
$TmFeCoO_4$	3.456	25.17	[9]
$TmFeCuO_4$	3.497	24.08	[9]
$TmFeZnO_4$	3.450	25.25	[15]
$TmGaMgO_4$	3.417	25.10	[15]
$TmGaMnO_4$	3.456	25.69	[15]
$TmGaFeO_4$	3.466	25.07	[2]
$TmGaCoO_4$	3.437	25.09	[15]
$TmGaCuO_4$	3.473	24.16	[15]
$TmGaZnO_4$	3.430	25.07	[15]
$YbAlMnO_4$	3.447	24.97	[15]
$YbFeMgO_4$	3.425	25.11	[7]
	3.429	25.16	[9]
$YbFeMnO_4$	3.457	25.59	[7]
	3.457	25.62	[9]
$YbFe_2O_4$	3.452	25.11	[7, 11*]
	3.455	25.11	[9, 14*]
$Yb_{0.5}Eu_{0.5}Fe_2O_4$	3.486	24.92	[16, 11*]

TABLE 26 (cont'd)

Compound	a (Å)	c (Å)	Ref.**
(1, 1)-type ($R\bar{3}m$) (cont'd)			
$YbFeCoO_4$	3.435	25.20	[7]
	3.430	25.20	[9]
$YbFeCuO_4$	3.485	23.97	[7]
	3.481	24.10	[9]
$YbFeZnO_4$	3.430	25.30	[7]
	3.425	25.28	[9]
$YbGaMgO_4$	3.423	24.61	[7]
	3.399	25.13	[15]
$YbGaMnO_4$	3.442	25.61	[7]
	3.444	25.70	[15]
$YbGaFeO_4$	3.447	25.18	[2]
$YbGaCoO_4$	3.418	25.08	[7]
	3.417	25.08	[15]
$YbGaCuO_4$	3.459	24.16	[7]
	3.460	24.17	[15]
$YbGaZnO_4$	3.414	25.15	[7]
	3.415	25.09	[15]
$LuAlMnO_4$	3.425	25.02	[15]
$LuFeMgO_4$	3.415	25.25	[7]
	3.428	25.31	[9]
$LuFeMnO_4$	3.449	25.65	[7]
	3.445	25.63	[9]
$LuFe_2O_4$	3.433	25.25	[7, 11*]
	3.437	25.25	[9, 14*]
$LuFeCoO_4$	3.415	25.27	[7]
	3.421	25.31	[9]
$LuFeCuO_4$	3.469	23.99	[7]
	3.465	24.18	[9]
$LuFeZnO_4$	3.416	25.42	[7]
	3.411	25.39	[9]
$LuGaMgO_4$	3.418	24.51	[7]
	3.386	25.23	[15]
$LuGaMnO_4$	3.431	25.54	[7]
	3.427	25.82	[15]
$LuGaFeO_4$	3.430	25.31	[2]
$LuGaCoO_4$	3.398	25.20	[7]
	3.400	25.19	[15]
$LuGaCuO_4$	3.443	24.22	[7]
	3.441	24.28	[15]
$LuGaZnO_4$	3.402	25.28	[7]
	3.400	25.25	[15]
(2, 1)-type ($P6_3/mmc$)			
$Sc_2Ga_2CuO_7$	3.303	28.12	[3]
$In_2Fe_2CuO_7$	3.352	28.87	[8]
$In_2FeGaCuO_7$	3.342	28.82	[8]
$In_2Ga_2MnO_7$	3.333	29.69	[3]
$In_2Ga_2CuO_7$	3.332	28.70	[8]
$In_2Ga_2ZnO_7$	3.308	29.48	[3]
(2, 1)-type ($P6_3/mmc$) (cont'd)			
$Yb_2Fe_3O_7$	3.470	28.45	[16]
	3.472	28.33	[17]
$Lu_2Fe_3O_7$	3.455	28.44	[18, 11*]
	3.452	28.42	[17]
(3, 1)-type ($R\bar{3}m$)			
$In_3Fe_3CuO_{10}$	3.343	61.81	[8]
$Yb_3Fe_4O_{10}$	3.490	60.79	[19]
(4, 1)-type ($P6_3/mmc$)			
$Yb_4Fe_5O_{13}$	3.503	53.03	[19]
(1, 2)-type ($P6_3/mmc$)			
$ScAlZn_2O_5$	3.245	22.24	[4]
$ScFeZn_2O_5$	3.278	22.51	[4]
$ScGaZn_2O_5$	3.362	22.50	[4]
$InAlZn_2O_5$	3.295	22.56	[4]
$InFeZn_2O_5$	3.309	22.57	[4]
$InGaZn_2O_5$	3.292	22.52	[4]
$In_2Zn_2O_5$	3.376	23.15	[20]
$TmGaZn_2O_5$	3.392	21.93	[2]
$YbFeZn_2O_5$	3.391	22.05	[2]
$YbGaZn_2O_5$	3.378	21.98	[2]
$LuFeZn_2O_5$	3.377	22.13	[2]
$LuGaZn_2O_5$	3.365	22.05	[2]
(1, 3)-type ($R\bar{3}m$)			
$ScAlZn_3O_6$	3.242	41.12	[4]
$ScFeZn_3O_6$	3.274	41.59	[4]
$ScGaZn_3O_6$	3.263	41.54	[4]
$InAlZn_3O_6$	3.281	41.35	[4]
$InFeZn_3O_6$	3.300	41.68	[4]
$InGaZn_3O_6$	3.288	41.56	[4]
$In_2Zn_3O_6$	3.355	42.52	[20]
$TmGaZn_3O_6$	3.357	40.70	[2]
$YbFeZn_3O_6$	3.368	40.84	[2]
$YbGaZn_3O_6$	3.352	40.76	[2]
$LuFeZn_3O_6$	3.358	40.99	[2]
$LuGaZn_3O_6$	3.344	40.88	[2]
(1, 4)-type ($P6_3/mmc$)			
$ScAlZn_4O_7$	3.243	32.55	[4]
$ScFeZn_4O_7$	3.271	32.92	[4]
$ScGaZn_4O_7$	3.262	32.89	[4]
$InAlZn_4O_7$	3.277	32.72	[4]
$InFeZn_4O_7$	3.294	32.99	[4]
$InGaZn_4O_7$	3.284	32.89	[4]
$In_2Zn_4O_7$	3.339	33.52	[20]
$ErGaZn_4O_7$	3.310	32.51	[2]

TABLE 26 (cont'd)

Compound	a (Å)	c (Å)	Ref.**
(1, 4)-type ($P6_3/mmc$) (cont'd)			
$TmGaZn_4O_7$	3.331	32.38	[2]
$YbFeZn_4O_7$	3.347	32.45	[2]
$YbGaZn_4O_7$	3.334	32.38	[2]
$LuFeZn_4O_7$	3.327	32.53	[2]
$LuGaZn_4O_7$	3.328	32.45	[2]
(1, 5)-type ($R\bar{3}m$)			
$ScAlZn_5O_8$	3.245	56.46	[4]
$ScFeZn_5O_8$	3.267	57.21	[4]
$ScGaZn_5O_8$	3.261	57.10	[4]
$InAlZn_5O_8$	3.272	56.80	[4]
$InFeZn_5O_8$	3.288	57.28	[4]
$InGaZn_5O_8$	3.280	57.14	[4]
$In_2Zn_5O_8$	3.327	58.11	[20]
$ErGaZn_5O_8$	3.298	56.61	[2]
$TmFeZn_5O_8$	3.315	56.61	[2]
$TmGaZn_5O_8$	3.314	56.50	[2]
$YbFeZn_5O_8$	3.328	56.50	[2]
$YbGaZn_5O_8$	3.322	56.39	[2]
$LuFeZn_5O_8$	3.327	56.35	[2]
$LuGaZn_5O_8$	3.320	56.44	[2]
(1, 6)-type ($P6_3/mmc$)			
$ScAlZn_6O_9$	3.246	42.93	[4]
$ScFeZn_6O_9$	3.264	43.62	[4]
$ScGaZn_6O_9$	3.258	43.24	[4]
$InFeZn_6O_9$	3.283	43.36	[4]
$InGaZn_6O_9$	3.275	43.26	[4]
$TmFeZn_6O_9$	3.316	42.87	[2]
$TmGaZn_6O_9$	3.309	42.84	[2]
$YbFeZn_6O_9$	3.310	42.92	[2]
$YbGaZn_6O_9$	3.305	42.80	[2]
$LuAlZn_6O_9$	3.313	42.82	[2]
$LuFeZn_6O_9$	3.317	42.96	[2]
$LuGaZn_6O_9$	3.313	42.80	[2]
(1, 7)-type ($R\bar{3}m$)			
$ScAlZn_7O_{10}$	3.247	72.19	[4]
(1, 7)-type ($R\bar{3}m$) (cont'd)			
$ScFeZn_7O_{10}$	3.263	72.81	[4]
$ScGaZn_7O_{10}$	3.258	72.74	[4]
$InAlZn_7O_{10}$	3.263	72.24	[4]
$InFeZn_7O_{10}$	3.279	72.85	[4]
$InGaZn_7O_{10}$	3.274	72.74	[4]
$In_2Zn_7O_{10}$	3.313	73.62	[20]
$TmFeZn_7O_{10}$	3.306	72.57	[2]
$TmGaZn_7O_{10}$	3.300	72.11	[2]
$YbFeZn_7O_{10}$	3.307	72.19	[2]
$YbGaZn_7O_{10}$	3.300	72.07	[2]
$LuAlZn_7O_{10}$	3.288	71.54	[2]
$LuFeZn_7O_{10}$	3.310	72.22	[2]
$LuGaZn_7O_{10}$	3.307	72.01	[2]
(1, 8)-type ($P6_3/mmc$)			
$InFeZn_8O_{11}$	3.276	53.75	[4]
$TmFeZn_8O_{11}$	3.301	53.42	[2]
$LuGaZn_8O_{11}$	3.301	53.17	[2]
(1, 9)-type ($R\bar{3}m$)			
$ScFeZn_9O_{12}$	3.263	88.51	[4]
$InFeZn_9O_{12}$	3.274	88.41	[4]
$ErFeZn_9O_{12}$	3.288	87.88	[2]
$TmFeZn_9O_{12}$	3.294	87.77	[2]
$TmGaZn_9O_{12}$	3.292	87.70	[2]
$YbFeZn_9O_{12}$	3.296	87.79	[2]
$YbGaZn_9O_{12}$	3.296	87.66	[2]
$YbAlZn_9O_{12}$	3.280	87.11	[2]
$LuAlZn_9O_{12}$	3.279	87.36	[2]
$LuFeZn_9O_{12}$	3.297	87.84	[2]
(1, 10)-type ($P6_3/mmc$)			
$InFeZn_{10}O_{13}$	3.272	64.02	[2]
$TmFeZn_{10}O_{13}$	3.285	63.74	[2]
(1, 11)-type ($R\bar{3}m$)			
$YbAlZn_{11}O_{14}$	3.273	102.60	[2]

*Original paper in which an incorrect crystal system was assumed.

**References: [1] Bertaut and Mareschal (1963); [2] Kimizuka and Mohri (1989); [3] Kimizuka and Mohri (1985); [4] Kimizuka et al. (1988); [5] Schmitz-DuMont and Kasper (1965); [6] Gérardin et al. (1980); [7] Zoumalan et al. (1975); [8] Kimizuka and Takayama (1984); [9] Kimizuka and Takayama (1981); [11] Evrard et al. (1974a); [12] Kimizuka and Katsura (1975a); [13] Kimizuka and Takayama (1982b); [14] Kimizuka et al. (1974b); [15] Kimizuka and Takayama (1982a); [16] Evrard et al. (1974b); [17] Kimizuka et al. (1974a); [18] Gérardin et al. (1982); [19] Kimizuka et al. (1976); [20] Kasper (1967).

contact with oxygen ions of adjacent layers and apply this to a RO_6 octahedra, the a-dimension can be estimated using the ionic radii of the R ion and the oxygen ion, $r_{R^{3+}}$ and $r_{O^{2-}}$ in octahedral coordination (Shannon 1976),

$$a = 2\sqrt{r_{R^{3+}}(r_{R^{3+}} + 2r_{O^{2-}})}\,.$$

In the (1, 0) compounds $RAlO_3$, the a-dimension can be estimated satisfactorily by this equation, i.e., the experimental values vary from 3.76 (R = Eu) to 3.66 Å (R = Er) while the corresponding calculated ones are 3.77 and 3.62 Å, respectively. Note that the distance between Al and the three planar oxygen ions in $YAlO_3$, 2.12 Å (Bertaut and Mareschal 1963), is much larger than the sum of the respective ionic radii [1.88 Å after Shannon (1976)]. In $RAlO_3$, the a-dimension seems to be determined essentially by the size of the R ion. For a compound having higher m, the experimental a-dimension is, however, usually smaller than the estimated one. Its a-dimension is dependent not only on the size of the R ion but also on that of the M(M′) ion. In contrast to the a-dimension, the c-dimension tends to decrease for a larger R ion (Gérardin et al. 1980, Kimizuka and Takayama 1981, 1982a). Comparing the trigonal bipyramids formed by Fe and oxygen ions in $YbFe_2O_4$, $Yb_{0.5}Eu_{0.5}Fe_2O_4$ and YFe_2O_4 (table 22), the distance between Fe and the planar oxygen ions (Fe–3O(2)) increases with a larger R ion due to increase of the a-dimension, while the distances between Fe and the axial oxygen ions [Fe-1O(1) and Fe-1O(2)] decrease (probably in order to conserve the average Fe–O distance), resulting in a decrease of the c-dimension. This tendency of a- and c-dimensions is found not only in the (1, 1)-type compounds but in other series of compounds as well. Although a larger M (M′) ion usually tends to enlarge the c-dimension, as pointed out first by Schmitz-DuMont and Kasper (1965) for $(InMO_3)(M'O)$, it is notable that the c-dimension of a Cu-containing compound is considerably smaller.

The cell dimensions of the $(RMO_3)(ZnO)_m$ compounds are worthy of special mention. Kasper (1967) plotted the a-dimension of $In_2O_3(ZnO)_m$ against the O : Zn ratio and found that it decreased with increasing O : Zn ratio and the extrapolated value at O : Zn = 1 ($m = \infty$) was almost identical with that of würtzite ZnO: $a = 3.250$, $c = 5.207$ Å and $Z = 2$ (Wyckoff 1964). In the case of $(ScAlO_3)(ZnO)_m$, the a-dimension of $ScAlZnO_4$ (3.242 Å) is, in contrast, smaller than that of würtzite ZnO, due to the small Sc and Al ions and it increases slightly with increasing m, resulting in 3.247 Å for $ScAlZn_7O_{10}$. Kasper (1967) calculated the thickness of a ZnO slab from the difference between the c-dimensions of the compounds containing different numbers of ZnO slabs, obtaining a value of 2.59 Å which close to half the c-dimension of würtzite ZnO (2.603 Å). By nearly the same method, Kimizuka et al. (1988) estimated the c-dimensions of $InFeO_3(ZnO)_m$ (m = 1–9). In the stacking sequence expressed by $-T-U-T-(T-)_{m-1}$, they assumed the thickness of a composite T–U–T layer to be 8.698 Å and that of one of remaining T layers to be 2.597 Å. The calculated c-dimension using these values was in agreement with the experimental value within 0.02 Å. The value 2.597 Å also close to the thickness of a ZnO layer in the würtzite structure (Kimizuka et al. 1988). These facts correspond to a würtzite-type oxide being an end member of the present series.

4.4. *Atomic levels of 3d states*

Only a few experiments have been reported on the atomic levels in the present series of compounds. In the first report, Schmitz-DuMont and Kasper (1965) measured optical absorption of $InAlCuO_4$, $InAlCoO_4$, etc., from 4000 to 24 000 cm^{-1}. As they considered the crystal symmetry R3m, instead of $R\bar{3}m$, they assumed that the transition metal ions were coordinated octahedrally and attempted to interpret absorption peaks by d–d transitions in 3d metal ions. They found that the d-level splitting, attributed to the cubic crystalline field by them, was small: the lowest energy absorption appeared at 4700 cm^{-1} in Co-containing compounds and at 5400 cm^{-1} in $InAlZnO_4$, etc., containing 1% Ni. The absorption band in Cu-containing compounds lay at 9800 and 11 800 cm^{-1}.

In 1986, Blasse et al. (1986) measured the luminescence of $ScGaZnO_4$ and found a broad blue emission band with a peak at 21 300 cm^{-1} at room temperature but with double peak at 24 400 and 22 200 cm^{-1} at 5 K. The peak of the excitation spectrum was at 37 000 cm^{-1} (300 K) or 38 500 cm^{-1} (5 K). They ascribed these spectra to T layers composed of Zn, Ga and O, and explained the double peak at 5 K by a disordered distribution of Zn and Ga. In the In-containing compounds, they observed red luminescence only below 250 K, with a peak at 16 700 cm^{-1} ($InGaZnO_4$ at 5 K) or 18 200 cm^{-1} ($InGaMgO_4$ at 5 K). Because of the similarity in these two compounds, they considered that the luminescence of the In-containing compounds is due to U blocks composed of In and O.

As was pointed out at the beginning, a particular structural feature of the present series of compounds is the presence of a T layer, transition metals are coordinated trigonal bipyramidally. The energies of the crystal-field states in the trigonal bipyramidal coordination have been evaluated for the various d^n configurations (see, e.g., Wood, 1972). Because of the axial symmetry of the site, the z-component of the orbital angular momentum, l_z, is a rather good quantum number of the orbital and the 3d-levels split into three, d_{z^2} ($l_z = 0$), d_{yz} and d_{zx} ($l_z = \pm 1$), and d_{xy} and $d_{x^2-y^2}$ ($l_z = \pm 2$). The quantization axis, the z axis, is along the c-axis, of course. The antibonding 3d orbital(d_{z^2}) extending along the c-axis has the highest energy level while d_{xy} and $d_{x^2-y^2}$ orbitals extending on x–y plane has the second highest level. The electron cloud of the d_{yz} or d_{zx} orbitals extends to the intermediate direction between anions and this state is the ground state. Figure 45 shows 3d orbitals corresponding to $l_z = 0$, $l_z = \pm 1$ and $l_z = \pm 2$.

If the Hund rule of maximum spin is fulfilled, the above-mentioned 3d-levels are consistent with the experimental fact that the magnetic anisotropy of Fe^{2+} ions is so large that they can be regarded, along the c-axis, as Ising spins. On the other hand, the Co^{2+} ions do not seem so anisotropic magnetically (see sect. 6.2.4). Another evidence of the level scheme is the absence of the (1, 1)-type

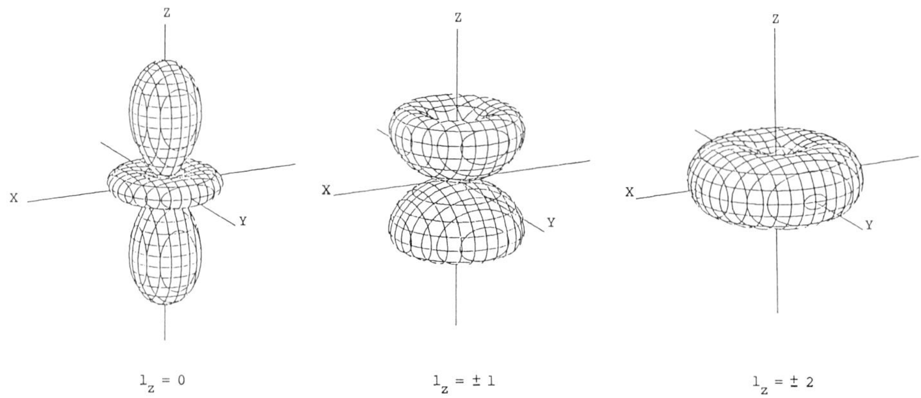

Fig. 45. 3d orbitals with $l_z = 0$, $l_z = \pm 1$ and $l_z = \pm 2$: by courtesy of Prof. A. Yanase.

compounds containing Ni^{2+} and Cr^{3+}, $3d^8$ and $3d^3$ ions, in this coordination except for a very small amount of substitution. They are very stable in an octahedral coordination in which the lowest one-electron orbital state is degenerate triply and 3 or 8 (=5 + 3) electrons fill up this subshell to form a singlet ground state. This is to be compared with the present site having a doubly degenerate orbital state. If this energy difference is large enough, the whole crystal will change to another form such as spinel (see below). According to the results of Schmitz-DuMont and Kasper (1965), the overall splitting of the 3d-level is about $10\,000\ \mathrm{cm}^{-1}$.

Strictly speaking, anions and cations in a T layer are not on an identical plane except in the cases where they are forced to be by the symmetry of the crystal. Unequal positions along the *c*-axis result in a term of the crystalline field proportional to z, and makes the electric dipole transition between 3d-levels possible. Then, this term, proportional to z, mixes the first excited level, d_{xy} and $d_{x^2-y^2}$ ($l_z = \pm 2$), with the ground state. By this mixing of the state, the ground state changes gradually to a $d\gamma$ orbital, the ground doublet in the cubic crystalline field of a tetrahedral coordination as in the würtzite structure. The würtzite-type oxide is a natural end member of the present series.

4.5. *Structure types of AB_2O_4 oxides*

It has been known that AB_2O_4 oxides crystallize in many types of structures, depending on the sizes and the site preferences of metal ions. Table 27 shows cation coordinations in those structures including the present (1, 1)-type (Wells 1975b, Kimizuka and Mohri 1989). Figures 46a–d indicate the structure types found in the R_2O_3–M_2O_3–M′O system where R = Sc–La; M = Al, Ga, Cr, and Fe; M′ = Ni–Ba (Kimizuka and Mohri 1989). Four types of structures appear in fig. 46, the K_2NiF_4-, (1, 1)-, spinel-, and $CaFe_2O_4$-type structures. In a K_2NiF_4-type structure, where both R and M′ ions occupy nine-fold sites while the M ion is

TABLE 27
Crystal structures of compounds AB_2O_4 (Kimizuka and Mohri 1989).

	CN of B				
CN of A	4	5	6	8	9 and/or 10
4	Phenacite		Olivine Spinel	K_2WO_4	β-K_2SO_4
6		(1, 1)-type	Sr_2PbO_4, Ca_2IrO_4, etc.		K_2NiF_4
8			$CaFe_2O_4$ $CaTi_2O_4$		

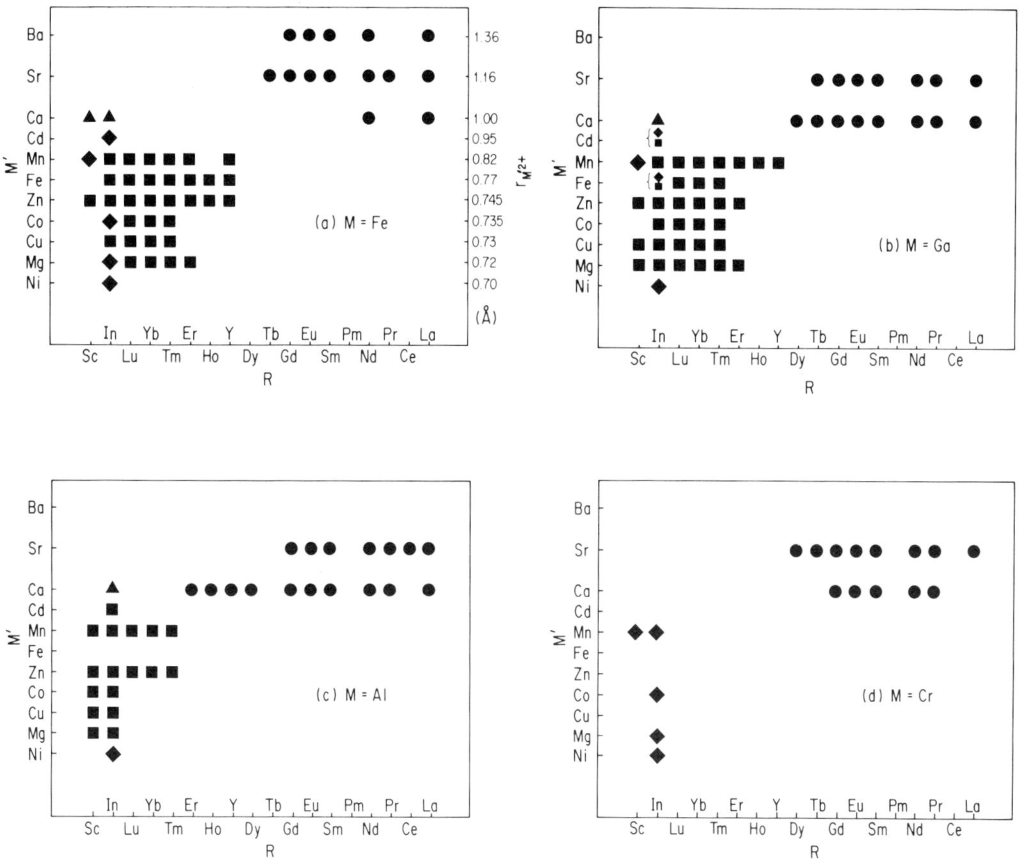

Fig. 46. (a)–(d). Structure types of RMM′O_4 oxides in the R_2O_3–M_2O_3–M′O system (●: K_2NiF_4 type, ■: (1, 1) type, ◆: spinel type, ▲: $CaFe_2O_4$ type). (Kimizuha and Mohri 1989.)

at an octahedral site (table 27), the relation between the ionic radii of cations should be $r_{R^{3+}} \approx r_{M'^{2+}} \gg r_{M^{3+}}$. Similarly, we expect $r_{R^{3+}} \approx r_{M'^{2+}} > r_{M^{3+}}$ for a spinel-type structure if the R and M′ ions are at octahedral sites and the M ion is at a tetrahedral site, $r_{M'^{2+}} > r_{R^{3+}} \approx r_{M^{3+}}$ for a $CaFe_2O_4$-type, and $r_{R^{3+}} > r_{M^{3+}} \approx r_{M'^{2+}}$ for a (1, 1)-type. As shown in fig. 46, these conditions are qualitatively satisfied. Zoumalan et al. (1975) substituted Yb in $YbFe_2O_4$ by Tb, Eu and La. From the limiting compositions, $Yb_{0.3}Tb_{0.7}Fe_2O_4$, $Yb_{0.5}Eu_{0.5}O_4$ and $Yb_{0.8}La_{0.2}Fe_2O_4$, they gave the maximum size of the R ion in RFe_2O_4, as 0.902 Å. This value is close to the ionic radii of the Y ion in octahedral coordination [0.900 Å after Shannon (1976)] and as shown in fig. 46a, the largest R element in (1, 1)-type RFe_2O_4 is indeed Y.

Gérardin et al. (1980) discussed the geometric conditions for adoption of the (1, 1)-type structure more quantitatively. They gave the condition

$$0.842 < d'/d < 0.888 ,$$

where d' is the shortest cation–oxygen distance in the trigonal bipyramidal site, which can be calculated by

$$d' = \tfrac{1}{2}(r_{M^{3+}} + r_{M'^{2+}}) + r_{O^{2-}} ,$$

by using the ionic radii in tetrahedral coordination obtained by Shannon (1976) while d is the R–O distance which is close to the sum of Shannon's ionic radii of ions in octahedral sites. Although this simple condition is valid in general terms, it is not a strict one. For instance, acceptable R elements are Sc, In, Lu and Yb when M = Ga and M′ = Mg, according to the above condition (Gérardin et al. 1980). The experimental results in fig. 46b, however, show that the larger elements Tm and Er can form the (1, 1)-type structure.

The structure type depends not only on sizes of metal ions but on their site preferences as well. It should be emphasized that no (1, 1)-type compounds containing Ni^{2+} or Cr^{3+} have been found. For $3d^3$ or $3d^8$ configurations, the crystal-field stabilization energy in an octahedral coordination is much larger than that in a trigonal bipyramidal one (Wood 1972). The strong octahedral site preference of Ni^{2+} ($3d^8$) or Cr^{3+} ($3d^3$) is a possible reason for the absence of the (1, 1)-type compounds containing them (Gérardin et al. 1980, Kimizuka and Takayama 1982b).

5. Electronic structure and Verwey transition

The main concern in this section is the iron-containing compound, RFe_2O_4. If we assume that the compound is purely ionic, there are equal numbers of Fe^{2+} and Fe^{3+} ions in an equivalent site of the crystal. Since an electron transfer from 2+ to 3+ Fe ions results in a new but nearly degenerate state, rather high electrical conductivity is expected for these oxides. This view is a limit of a strong correlation of electrons due to the Coulomb interaction. From a purely itinerant-electron point of view, the problem is a narrow 3d band having the Fermi level within the band. Low electrical conductivity compared with ordinary metals will result from the narrowness of the band but the state of all Fe atoms is identical without charge fluctuation. This is a result of the neglect of correlation. The real situation lies between these two limits. Each Fe atom fluctuates between two charged states but they cannot be considered as purely 2+ or 3+. Here, RFe_2O_4 is similar to Fe_3O_4 or Ti_4O_7, which are considered to have equal numbers of same element, Fe or Ti, with differently charged states. These oxides exhibit characteristic properties, rather high electrical conductivity for oxides, a phase transition due to the charge ordering (Verwey transition), etc.

In this section, the electronic properties of RFe_2O_4 are discussed. Topics are the Verwey-type phase transition in stoichiometric YFe_2O_4 and its derivatives, transport properties, direct determination of charge fluctuation frequency by Mössbauer spectroscopy, and so on.

5.1. *Verwey-type phase transition*

Half a century ago, a phase transition in Fe_3O_4 at around 120 K was attributed to the ordering of differently charged Fe ions in a crystallographically equivalent site, the B site in a spinel, by Verwey (Verwey and Haayman 1941). This sort of transition is called a Verwey transition. This transition is a little different from the metal–insulator transition in V_2O_3, VO_2, etc., since there is only one atomic state in the latter from the ionic point of view. Two successive phase transitions were found later in Ti_4O_7 (Bartholomew and Frankl 1969), and assigned to the same sort of transition (Schlenker and Marezio 1980). These transitions are very sensitive to the stoichiometry, compression or substitution of cations by other elements, indicating a delicate balance of the transfer integral and the correlation energy.

A low-temperature transition in stoichiometric YFe_2O_4 was discovered by magnetic measurements (Nakagawa et al. 1979) first and then confirmed by Mössbauer spectroscopy (Tanaka et al. 1979), and electrical conductivity and specific heat measurement (Tanaka et al. 1982). Figure 47 shows the temperature

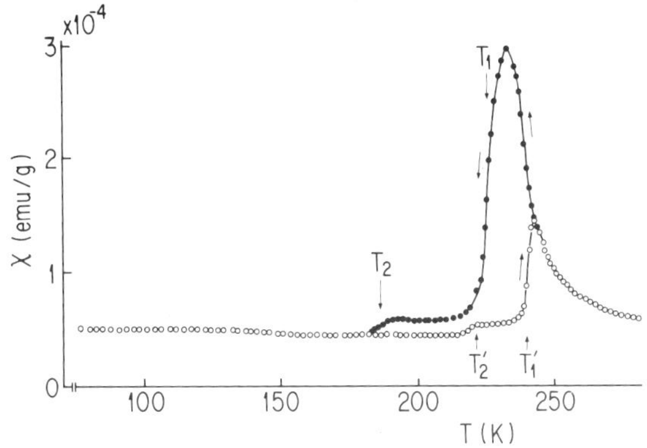

Fig. 47. Temperature dependence of the magnetic susceptibility of stoichiometric YFe_2O_4. (Nakagawa et al. 1979.)

dependence of magnetic susceptibility of polycrystalline specimens of the stoichiometric YFe_2O_4. There are two sharp successive transitions with large thermal hysteresis. This curve differs sharply from those of non-stoichiometric (oxygen deficient) YFe_2O_4 or of compounds with R other than Y, as will be discussed in the next section in detail. One point which should be made here is the fact that at all temperatures magnetization is proportional to the external field at least up to 15 kOe, i.e., no remanent magnetization appears in stoichiometric YFe_2O_4. Figure 47 can be interpreted as consisting of thermomagnetic curves in a constant external magnetic field. By X-ray powder diffraction, Nakagawa et al. (1979) disclosed that the crystal distorts from rhombohedral to monoclinic at the higher transition point and further to triclinic at the lower transition. Splitting of diffraction peaks is shown in fig. 48 and the unit-cell dimension is tabulated in table 28, after Nakagawa et al. (1979).

The character of the transition is made clearer by electrical measurements. The electrical resistivity of YFe_2O_4 was investigated both in its stoichiometric and non-stoichiometric form, and the results are shown in fig. 49, after Tanaka et al. (1982). This is a plot of the electrical resistivity against inverse temperature for several polycrystalline specimens annealed at 1200°C in an atmosphere of controlled oxygen partial pressure, a mixture of H_2 and CO_2, the ratio of which is shown in the figure. The composition of the specimens was estimated by thermogravimetry as $x = 0.00$, 0.03, 0.04 and 0.05, from top to bottom. Here x is the amount of oxygen deficiency: YFe_2O_{4-x}. When oxygen is deficient, electrical conductivity decreases with decreasing temperature following an Arrhenius equation: $\rho \propto \exp(-U/kT)$, where U, the activation energy, is 0.21 eV above 270 K and is 0.29 eV below 220 K. Note that antiferromagnetic spin ordering develops in this temperature range (see sect. 6), although there is no sharp anomaly in the conductivity. In the case of less oxygen-deficient crystals, however, there appears a transition where resistivity increases rapidly with decreasing temperature. The transition shows a large thermal hysteresis as is seen in magnetic susceptibility (fig. 47). This is a characteristic of a first-order phase transition. The transition point moves to higher temperatures in the more stoichiometric crystals and the

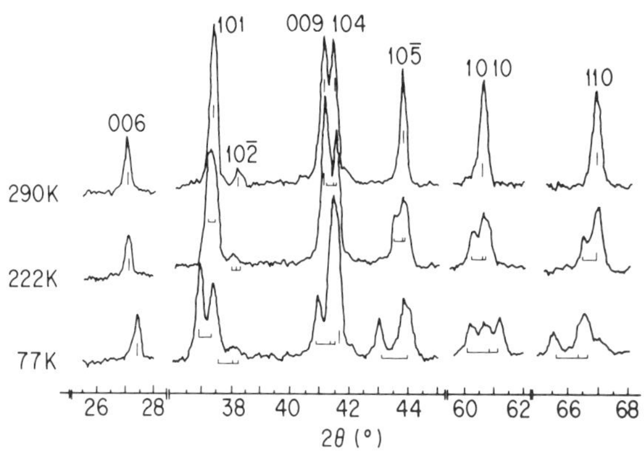

Fig. 48. Several X-ray diffraction peaks of the high-, intermediate- and low-temperature phases of stoichiometric YFe_2O_4 by Fe-Kα radiation. (Nakagawa et al. 1979.)

TABLE 28
Lattice parameters of stoichiometric YFe_2O_4 (Nakagawa et al. 1979).

Parameter	Temperature (K)		
	290	222	77
a (Å)	3.513	3.534	3.574
b (Å)	3.513	3.513	3.540
c (A)	24.779	24.771	24.505
α (°)	90	90.23	90.43
β (°)	90	89.54	88.82
γ (°)	120	120.20	120.62
V (Å^3)	264.83	265.74	266.78

second transition becomes visible at lower temperatures. According to the specific heat measurement shown in fig. 50, the entropy increase at the higher transition is much larger than that at the lower. This is consistent with the magnitude of thermal hysteresis at each transition. A larger hysteresis at a lower transition means a small difference in the free energy of the two states, realized in the lowest and the intermediate temperature region.

It is interesting that the temperature dependence of the resistivity is not affected by the oxygen deficiency at temperatures higher than the transition, in contrast to that below. In the stoichiometric compound, the increase of the resistivity is about two orders of magnitude at each transition point.

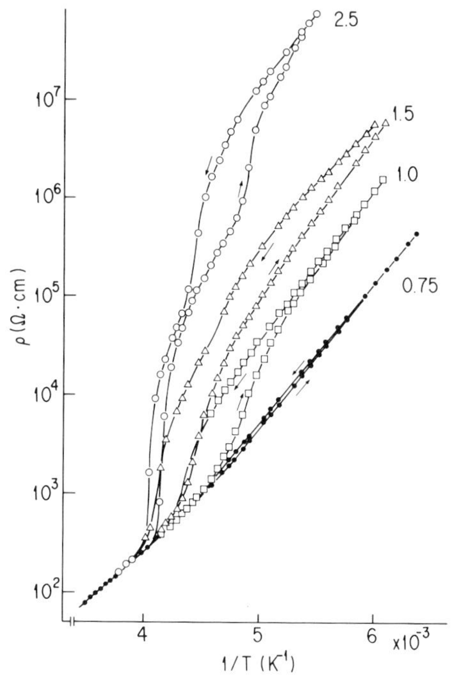

Fig. 49. The temperature dependence of the electrical resistivity of sintered YFe_2O_4 specimens with different stoichiometries. Arrows in the figure indicate the direction of temperature change during the measurement and the numbers indicate the ratio of H_2 to CO_2 of the atmosphere during annealing at 1200°C. The compositions of the specimens are estimated to be $x = 0.0$, 0.03, 0.04 and 0.05 from top to bottom, where x is the amount of the oxygen deficiency in the compound YFe_2O_{4-x}. (Tanaka et al. 1982.)

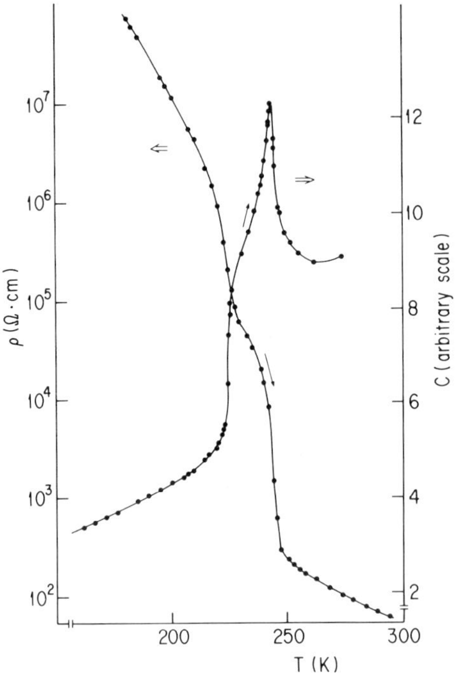

Fig. 50. The temperature dependence of the specific heat measured by the AC method and the electrical resistivity of stoichiometric YFe_2O_4. (Tanaka et al. 1982)

All these data resemble transitions in Fe_3O_4 or Ti_4O_7 (Matsui et al. 1977, Schlenker and Marezio 1980) and suggest strongly that the transition in stoichiometric YFe_2O_4 is a Verwey transition.

One can have a doubt on the absolute value of the data shown in fig. 49, since experiments were carried out on sintered polycrystals. Resistivity of a single crystal is shown in fig. 51. The specimen is not stoichiometric but the composition is estimated to be $x \sim 0.04$. The crystal may not be homogeneous. According to fig. 51, errors in the coordinate of fig. 49 are not so large. Resistivity of the order of 1 Ω cm at room temperature is about three orders of magnitude higher than that of Fe_3O_4 or Ti_4O_7.

Anisotropy in the conductivity is evident: at 300 K, that is lower along the c-axis than the a-axis by two orders of magnitude. The activation energy is 0.37 eV below 240 K and 0.20 eV above 250 K along the a-axis and 0.27 eV along the c-axis. Note that there is no trace of a higher transition in the electrical conductivity along the c-axis. As for the lower transition, we can say nothing at present since no single crystal is stoichiometric enough to exhibit it.

Not only the stoichiometry but also compression or substitution of the element R affects the transition very much. A sharp two-step transition as is shown in figs. 47 and 49 is observed only in the Y-compound. There are reasons to believe that Ho- and Er-compounds will exhibit the transition if they are stoichiometric (see next subsection). Unfortunately, these compounds exist only in an oxygen-deficient form (to $x = 0.09$ in Ho- and $x = 0.015$ in Er-compounds) when synthe-

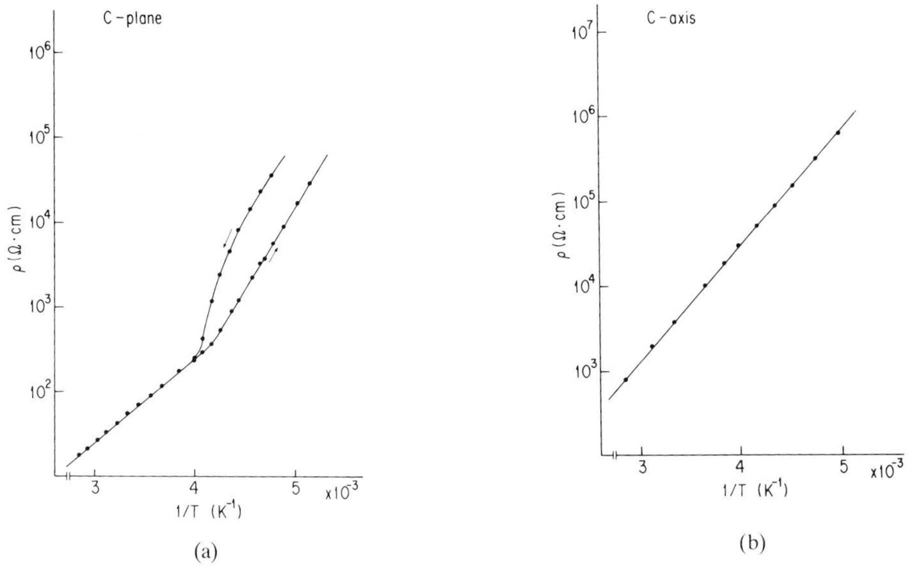

Fig. 51. The temperature dependence of the electrical resistivity of a YFe_2O_4 single crystal. (a) within the *c*-plane; (b) parallel to the *c*-axis. The composition of the crystal is estimated as $YFe_2O_{3.96}$, although it may not be homogeneous. (Tanaka et al. 1982.)

sized at 1200°C. Only the higher step of transition is observed in $ErFe_2O_4$ and no transition appears in $HoFe_2O_4$.

By a partial substitution of Y by Lu, the low-temperature phase is very easily suppressed. Only 2% Lu is enough to wipe out the lower transition at least in a temperature region above 77 K. The higher transition disappears by a 20% substitution. On the other hand, replacement of Y by Dy shifts both transitions to higher temperatures (fig. 52, Kishi et al. 1983). A higher transition temperature

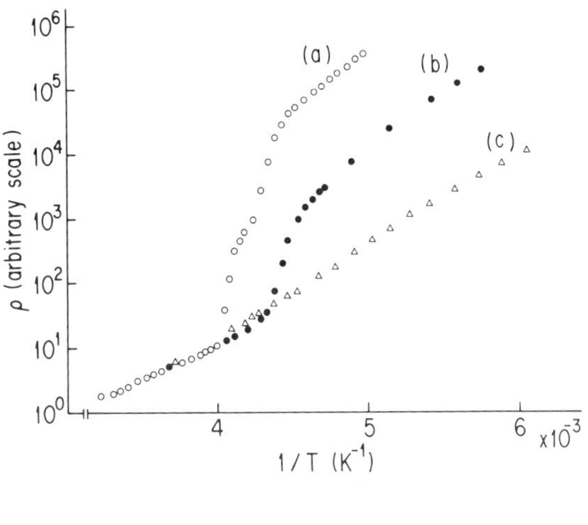

Fig. 52. Examples of the temperature dependence of the resistivity with no, one- or a two-step transition.
(a) $Y_{0.95}Dy_{0.05}Fe_2O_{3.979}$;
(b) $Y_{0.95}Lu_{0.05}Fe_2O_{3.992}$;
(c) $Y_{0.80}Lu_{0.20}Fe_2O_{3.993}$.
(Kishi et al. 1983.)

of 250 K during heating is observed in $Y_{0.95}Dy_{0.05}Fe_2O_{3.98}$. The transition disappears in crystals with more Dy. This is attributed to the oxygen deficiency, inevitably introduced in those crystals. Note that only indirect effects are expected to the lattice of Fe by the substitution of Y by Lu or Dy, as far as stoichiometric crystals are concerned. Nothing but the atomic distance changes in the Fe lattice, where transport of electrons as well as ordering of ions is considered to occur.

The transition is much influenced also by compression. Figure 53 shows the temperature dependence of electrical resistivity during heating, under hydrostatic pressure. Both transitions shift to lower temperature and smear out by compression (Siratori et al. 1990). Thermal hysteresis of the transition becomes larger, too. Pressure dependence of the transition temperature is very large: −3 K/kbar for the higher transition and −10 K/kbar for the lower at atmospheric pressure. This magnitude of the pressure coefficients is comparable to that of V_2O_3.

According to Mössbauer spectroscopy (Tanaka et al. 1979) and neutron diffraction (Funahashi et al. 1984) of stoichiometric YFe_2O_4, the spins order antiferromagnetically below the higher transition, whereas the crystal is paramagnetic above: the higher transition is the Néel point. This fact is in contrast to the spins in Fe_3O_4 or Ti_4O_7. In the former case, the spins of the Fe ions order at much higher temperature and the freedom of spins plays no role in the Verwey transition. In the latter, magnetic susceptibility decreases to a low value at the higher transition and remains unchanged at lower temperatures. It is argued that Ti^{3+} ions form diamagnetic pairs accompanied by lattice distortion, a bipolaron, at the higher transition point and the bipolarons order at the lower transition. We can compare the transition in YFe_2O_4 to that in V_2O_3 and the transition in Ti_4O_7 to VO_2.

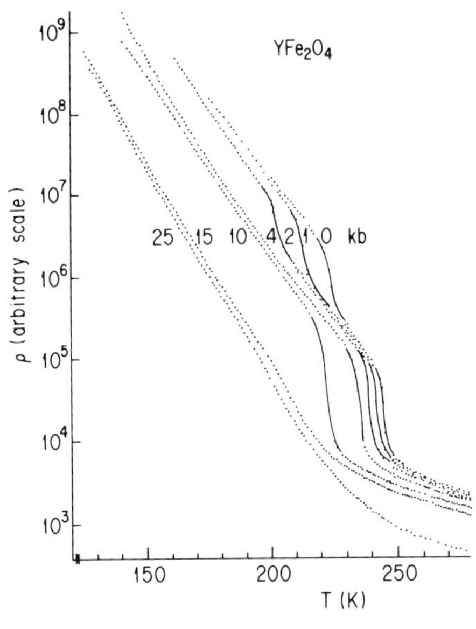

Fig. 53. The temperature dependence of the electrical resistivity of stoichiometric YFe_2O_4 under hydrostatic pressure. (Siratori et al. 1990.)

Mössbauer spectroscopy has given further very interesting information on the metal–nonmetal transition in YFe_2O_4 (Tanaka et al. 1984a). As is shown in fig. 54, the spectrum of stoichiometric YFe_2O_4 (indicated by S in the figure) at room temperature has a component that is absent in the spectrum of the non-stoichiometric compound, indicated by N. The difference between these two spectra, indicated by D, is composed of two peaks at 0.19 and 0.94 mm/s. Note that if this is a pair split by an electric quadrupole interaction, the isomer shift corresponds to $Fe^{2.5+}$. No such absorption is observed in the Tm-, Yb- or Lu-compounds but a trace is found in $HoFe_2O_4$ and $ErFe_2O_4$. Apparently, these absorption peaks are precursors of the Verwey transition, due to some special state of the Fe ions that is responsible for the transition. One possible explanation is that this state is a bipolaron, although we have found no direct evidence. Trace of this absorption observed in Ho- and Er-compounds suggests the existence of the transition, if we can synthesize them in the stoichiometric form.

5.2. *Hopping frequency of charge carriers*

Mössbauer spectroscopy provides microscopic information on the electronic state of atoms. The method applied here to ^{57}Fe nuclei gives direct evidence that Fe has two charged states in RFe_2O_4. At the same time, the analysis of the profile of the spectrum by a stochastic theory of the motional narrowing of resonances gives a frequency of the fluctuation of these state as a function of temperature.

The Mössbauer effect is a resonant absorption of γ-rays by nuclei, 14.4 keV in the case of ^{57}Fe. In this case, the nuclear spin, I, of the ground and the excited states are $\frac{1}{2}$ and $\frac{3}{2}$, respectively. The levels of a nucleus, and hence the frequency of the γ-ray, is determined by effective magnetic field and electrostatic interactions between the nucleus and surrounding electrons. Multipole expansion up to the second order is sufficient for the latter: the so-called isomer shift for zeroth order and e^2qQ or E_Q term for the second-order interaction (nuclei have no electric dipole). In the paramagnetic region, there is no effective magnetic field if an external field is absent. Since electric energy does not depend on the sign of I_z, there are only two absorption peaks, corresponding to $I_z = \pm\frac{1}{2}$ and to $\pm\frac{3}{2}$ of the

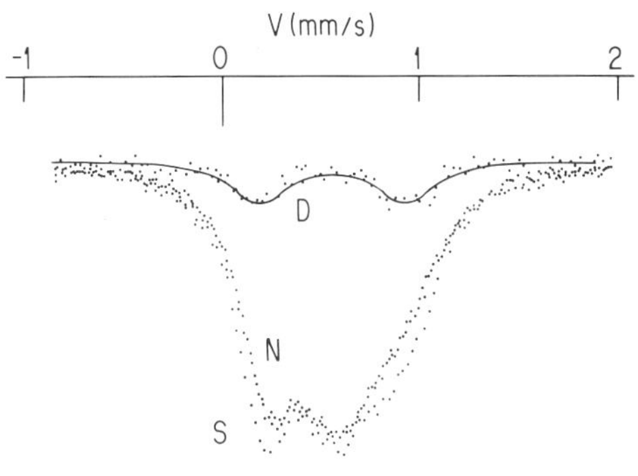

Fig. 54. Mössbauer spectra at 312 K of stoichiometric (S), non-stoichiometric (N) YFe_2O_4 and the difference (D = S − N) between them. Intensity of S and N are set equal at the tail. (Tanaka et al. 1984a.)

excited state, for one electronic state of a Fe atom. If Fe ions in the crystal have two charged states, the number of absorption peaks is $2 \times 2 = 4$. Experiments on RFe_2O_4 show three absorption peaks at room temperature (Narayanasamy et al. 1981, Tanaka et al. 1984a).

In the present case, however, the charged state of one Fe ion fluctuates from one (e.g., Fe^{2+}-like state) to another (Fe^{3+}-like). If the frequency of the fluctuation is too high to be recognized by the Mössbauer effect, one pair of absorptions will appear at an averaged position. When the fluctuation frequency is comparable to the difference of two resonance frequencies, the line profile provides information on the fluctuation frequency.

Figure 55 shows Mössbauer spectra of a powdered specimen of stoichiometric $LuFe_2O_4$ in the paramagnetic state (Tanaka et al. 1980). Three absorption peaks at around 300 K amalgamate into one peak at 400 K. This is an example of motional narrowing of resonance absorption and is well accounted for by a stochastic theory developed by Kubo and Anderson and applied to the Mössbauer effect by Blume (Tjon and Blume 1968, Blume 1968). Solid lines in the figure are profiles calculated by

$$W(\omega) \propto \mathrm{Re}[(p + 2w)/p^2 + \alpha^2 + 2pw)] \,, \tag{5}$$

where

$$p = -\mathrm{i}(\omega - \omega_0) + \tfrac{1}{2}\Gamma_0 \,, \qquad \alpha = \beta - \tfrac{1}{2}\mathrm{i}\gamma \,,$$

and

$$\omega_0 = \tfrac{1}{2}(\omega_1 + \omega_2) \,, \qquad \Gamma_0 = \tfrac{1}{2}(\Gamma_1 + \Gamma_2) \,,$$

$$\beta = \tfrac{1}{2}(\omega_1 - \omega_2) \,, \qquad \gamma = \tfrac{1}{2}(\Gamma_1 - \Gamma_2) \,,$$

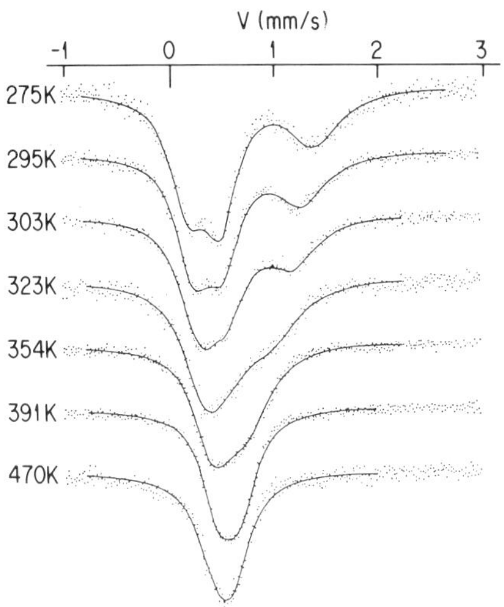

Fig. 55. Mössbauer spectra of $LuFe_2O_4$ in the paramagnetic region. Solid curves are calculated by the stochastic theory of motional narrowing, using parameters shown in figs. 56 and 57 (see text). (Tanaka et al. 1980.)

i is the unit of imaginary numbers, w is the fluctuation frequency and Re means the real part of the terms in brackets. This equation was obtained by Tjon and Blume (1968), and has been modified to include the difference of line width by the differently charged state.

The fundamental assumption of eq. (5) is that the principal axis of the electric field gradient at the ^{57}Fe nuclei remains unchanged regardless of the charged state, parallel to the c-axis, for example. In this case, transitions to $I_z = \pm\frac{1}{2}$ and that to $\pm\frac{3}{2}$ can be treated separately. The total absorption profile is a superposition of these two absorptions, each of which is expressed by eq. (5).

Seven parameters are determined to fit the calculated line profile to the experimental results. They are line positions, ω_1 and ω_2 for two transitions to $\pm\frac{1}{2}$ and $\pm\frac{3}{2}$, widths, Γ_1 and Γ_2, and the hopping frequency, w. The widths are assumed equal for two lines of the same charged state, as usual. As is seen in fig. 55, eq. (5) explains experimental results satisfactorily. The parameters used are shown in figs. 56 and 57 for six compounds from YFe_2O_4 to $LuFe_2O_4$. For Y-, Ho- and $ErFe_2O_4$, the specimens are non-stoichiometric and do not show the Verwey transition. Line positions are expressed by isomer shifts and quadrupole splitting. The isomer shift is an average of two line positions, the transition to $I_z = \pm\frac{1}{2}$ and to $\pm\frac{3}{2}$, whereas E_Q, the quadrupole splitting, is half the difference between them. It should be noted that no part of the spectrum in fig. 55 remained unexplained. This means that the specimen is homogeneous. In other words, the charge of all the Fe ions fluctuates in the same way. In the case of stoichiometric YFe_2O_4, such a homogeneity does not hold. Figure 58 shows this spectrum. A peak indicated by an arrow remains up to 450 K. The difference between the spectra of figs. 55 and 58 at 312 K was shown in fig. 54. As already stated in the preceding subsection, this part of the spectrum is a characteristic of materials in which Verwey-type transitions take place.

The isomer shift is a measure of the electronic charge density at the nucleus and, thus, the charged state of the atom. The existence of two different values of the isomer shift in fig. 56 is an indication of the existence of two charged states of the Fe ions, 2+ like and 3+ like. A point to be made here is their values. They are not typical for 2+ or 3+ ions of Fe but intermediate between them. Moreover, the lower value in the figure, which corresponds to a higher ionic charge, is decreased by about 0.10 mm/s in a compound containing 2+ cations other than Fe, such as $LuFeMgO_4$. These facts mean that Fe^{2+} and Fe^{3+} ions in RFe_2O_4 hybridize, although not into an equivalent state. Hereafter, we will denote this as [2+] or [3+] ions of Fe, instead of 2+ or 3+, to express explicitly the difference in the charged state. From the itinerant-electron point of view, this can be considered that there are charge density waves travelling in the crystal. If so, the distribution of a differently charged state at an instant may not be random. Coupling of charge density waves, or different ionic states, with the dynamics of Fe spins is an interesting problem, although conclusive evidence has not yet been disclosed. This point will be discussed briefly in the next section from the magnetic structure of compounds.

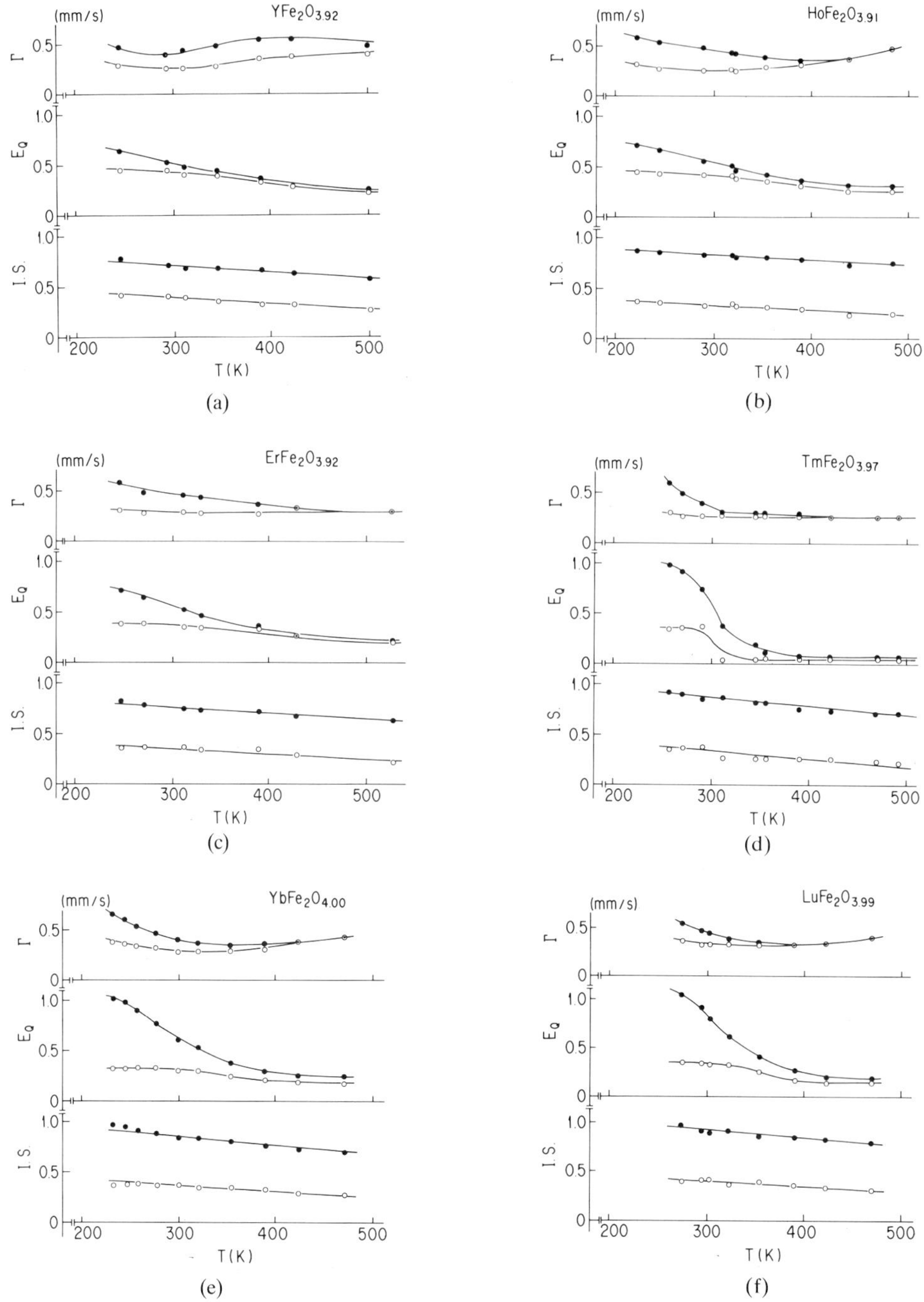

Fig. 56. The temperature dependence of the isomer shift, IS, quadrupole splitting (E_Q) and width (Γ) of RFe_2O_4. (R = Y (a); Ho (b); Er (c); Tm (d); Yb (e); Lu (f).) Specimens of Y-, Ho- and $ErFe_2O_4$ are nonstoichiometric, without Verwey transition. (Tanaka et al. 1980, 1984a.)

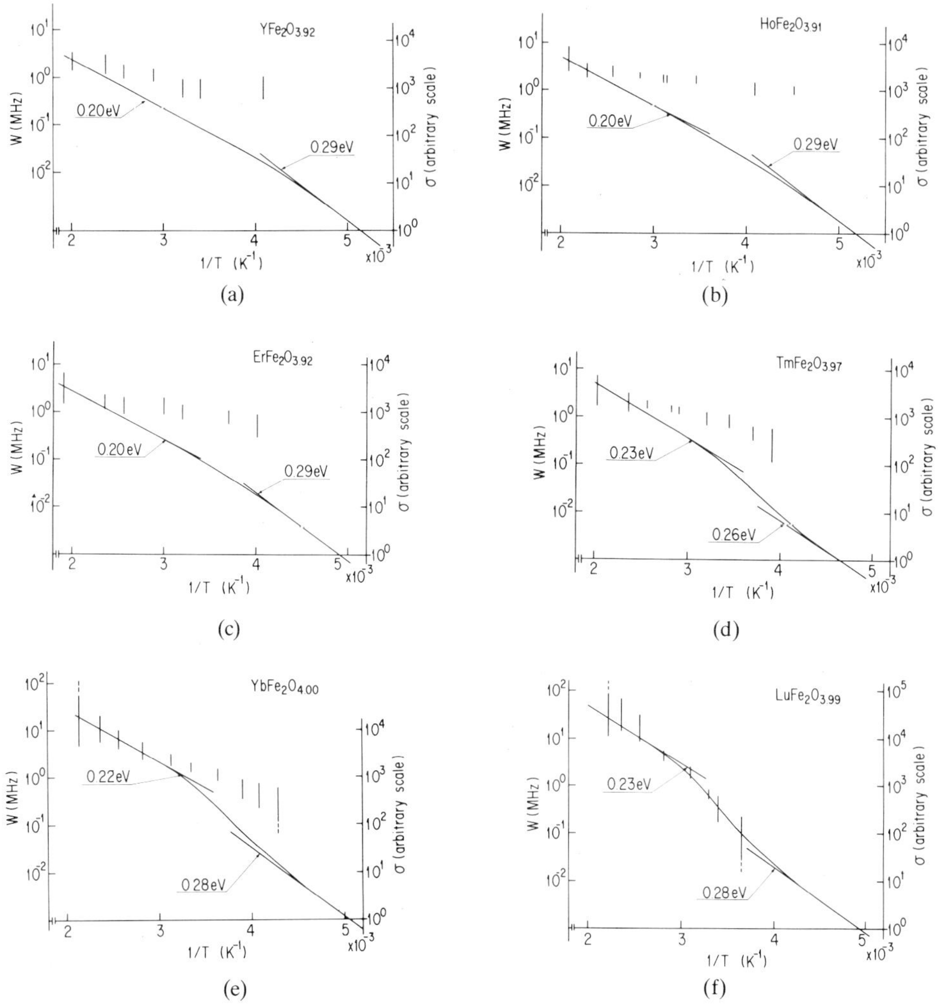

Fig. 57. The temperature dependence of the hopping frequency, w, in RFe_2O_4 (R = Y (a); Ho (b); Er (c); Tm (d); Yb (e); Lu (f)), determined by Mössbauer spectroscopy of the ^{57}Fe nuclei, as is indicated by vertical bars. Solid lines are the electrical conductivity of a polycrystalline specimen. (Tanaka et al. 1980, 1984a.)

The temperature dependence of the quadrupole splitting in fig. 56 is unusually large, but no explanation has been proposed.

The temperature dependence of the hopping frequency, w, is shown in fig. 57. The solid lines in the figures are the electrical conductivity, σ, of polycrystalline specimens. The coincidence in $LuFe_2O_4$ in the temperature dependence of w and σ is remarkable, but in other compounds that does not hold.

The frequency of the fluctuation of Fe ions between differently charged states

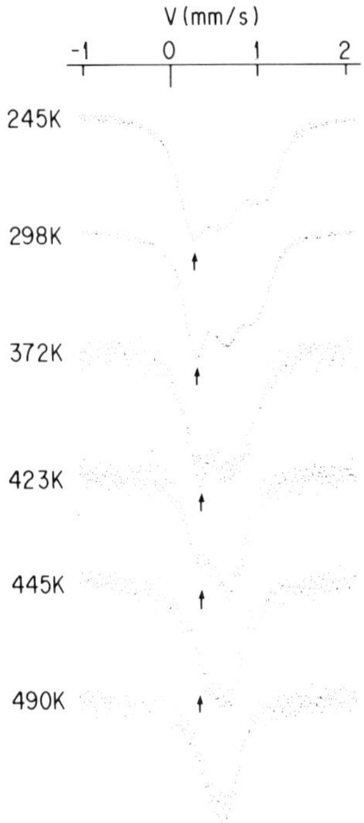

Fig. 58. The temperature dependence of the Mössbauer spectrum of stoichiometric YFe_2O_4. A peak indicated by arrows remains unexplained by the stochastic theory. (Tanaka et al. 1984a.)

means movement of charges between them and is correlated to the electrical conductivity by the Einstein relation,

$$\sigma = \frac{ne^2}{2k_B T} a^2 w \,, \tag{6}$$

where n is the carrier density, e is the electronic charge, k_B is the Boltzmann constant, T is the temperature and a is the hopping distance. The coincidence of the temperature dependence of w and σ shown in fig. 57 suggests that the carrier number is constant above 275 K in $LuFe_2O_4$ and the activation-type temperature dependence of σ is due to that of the mobility: the hopping conduction of electrons or small polarons. This interpretation is, however, questionable because w is too low or σ is too high. If we assume that the carrier number is equal to the number of Fe^{2+} (or Fe^{3+}), 1.1×10^{22} cm^{-3}, averaged hopping distance should be about 50 Å to account for the conductivity of about 10^{-1} $(\Omega\,\mathrm{cm})^{-1}$ by eq. (6). Apparently, this value is too large for the hopping conduction of electrons or holes.

The proportional relation between w and σ disappears in compounds other than $LuFe_2O_4$. By the decrease of the atomic number, w becomes relatively independent of temperature. This is another evidence suggesting that the correlation of a large number of charge carriers is necessary to account for the transport phenomena observed in these compounds.

We would like to make a point here. According to the temperature dependence of the electrical conductivity (fig. 57) RFe_2O_4 can be divided into two groups: Y (non-stoichiometric), Ho, Er and Tm, Yb, Lu. The change of the conductivity above and below the spin ordering temperature, at around 240 K, is different in these two groups. It is interesting that this grouping is the same as that by the appearance of the Verwey type transition or by the existence of the "hard component" in the Mössbauer spectrum shown in figs. 54 or 58. Since yttrium forms one group with Ho and Er, not the atomic number but the ionic radius of the R^{3+} ions can be considered as an important factor.

5.3. *Transport phenomena*

The conduction mechanism in YFe_2O_4 is discussed by Tsuda and collaborators on the basis of the temperature dependence of the electrical conductivity σ and the Seebeck coefficient S (Enomura et al. 1983, Sakai et al. 1985, 1986). The Seebeck coefficient of stoichiometric and non-stoichiometric YFe_2O_4 is shown in fig. 59. The sign of S is negative, except near the higher transition point in the stoichiometric specimen. As is shown in fig. 59, S in the paramagnetic region can be expressed well by a linear function of temperature. Tsuda and his collaborators analyzed this result by a theory of Cutler and Mott (1969) on the Seebeck effect due to the hopping motion of charge carriers, although their results, $S = aT + b$, does not have the full support of other theorists. Tsuda and his collaborators concluded that there are two carriers, electrons and holes, and estimated the state density at the Fermi level to be larger than 2.0 per formula unit per eV.

As for the region below the Néel temperature, the authors of these reports pointed out that resistivity of the non-stoichiometric specimen obeys

$$\rho = \rho_0 \exp[(T_0/T)^{1/3}] , \qquad (7)$$

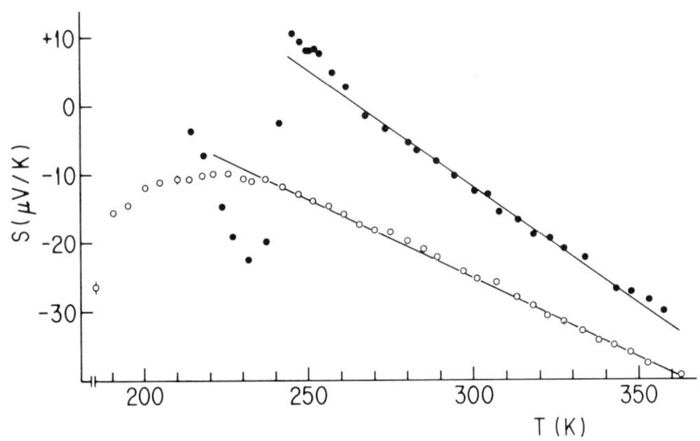

Fig. 59. The temperature dependence of the Seebeck coefficient in stoichiometric and non-stoichiometric YFe_2O_4. (Enomura et al. 1983.)

deduced for the variable range hopping model in a two-dimensional lattice (see fig. 60). Experiment gives

$$T_0 = (3.03 \pm 0.02) \times 10^7\ \mathrm{K}, \quad \text{for } x = 0.04,$$

$$T_0 = (3.27 \pm 0.02) \times 10^7\ \mathrm{K}, \quad \text{for } x = 0.05.$$

According to the theory (Mott and Davis 1979), T_0 is given by

$$T_0 = 27\alpha^2/\pi k_B N(\varepsilon_F), \tag{8}$$

where $N(\varepsilon)$ is the state density per unit area per unit energy and α is a factor in the envelope function of 3d electrons, $\exp(-\alpha r)$. If we assume that the state density has a rectangular shape with a width E_w and that each Fe atom contributes one state, the observed value of T_0 gives $E_w\alpha^2 = 5.68 \times 10^{17}$ eV cm^{-2} for the specimen with $x = 0.04$. If E_w is 1 eV, this equation gives an estimation of $\alpha = 7.5 \times 10^8$ cm^{-1} and $\alpha^{-1} = 0.13$ Å. Apparently this magnitude of α^{-1} is too small for the 3d atomic function. The authors of these reports argued that this is due to an inappropriate estimation of parameter values and that the variable range hopping mechanism itself is valid in non-stoichiometric YFe_2O_4.

The situation is a little different in the stoichiometric compound. Tsuda and his collaborators argued that the temperature dependence of the electrical resistivity is well explained by a semiconductor model where the energy gap, E_g, varies with temperature through the change of the degree of ordering,

$$\rho = \rho_0 \exp[E_g(T)/2k_B T]. \tag{9}$$

They calculated $E_g = E_0\,\Delta(T/T_c)$, by a molecular-field approximation, as

$$\Delta(T/T_c) = (T/2T_c)\log[(1 + \Delta(T/T_c))/(1 - \Delta(T/T_c))], \tag{10}$$

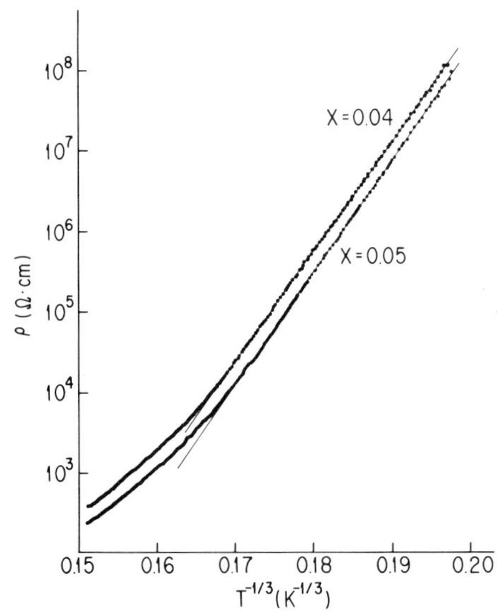

Fig. 60. The temperature dependence of the electrical resistivity of non-stoichiometric YFe_2O_{4-x} below room temperature, indicating the $T^{-1/3}$ law at low temperatures. (Sakai et al. 1986.)

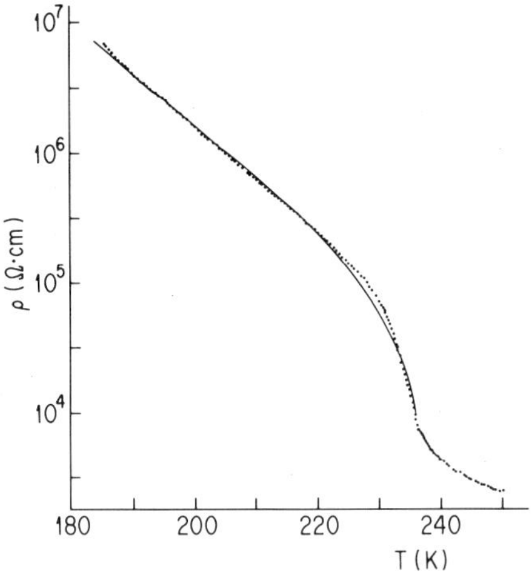

Fig. 61. The temperature dependence of the electrical resistivity of stoichiometric YFe_2O_4 below the transition point. A solid curve is calculated on the molecular-field approximation (see text). (Sakai et al. 1986.)

and determined T_c, E_0 and ρ_0 from a best fit. Their results are $T_c = 236$ K, $E_0 = 0.29$ eV and $\rho_0 = 7.8 \times 10^3$ Ω cm, respectively. Electrical resistivity calculated by eq. (10) with the above parameter values are shown in fig. 61 by a solid line. Equation (10) accounts well for experimental results, as indicated by the dots.

As is shown in fig. 59, the Seebeck coefficient below T_N depends strongly on the temperature in the stoichiometric compound. Moreover, it was observed in another stoichiometric specimen that S decreases down to -350 μV/K at 200 K, although the difference due to the specimen is small in the paramagnetic region (Sakai et al. 1985). This fact indicates that the Seebeck effect is highly structure sensitive below the transition, as is Fe_3O_4 (Kuipers and Brabers 1976). The Seebeck coefficient of the low-temperature phase of Fe_3O_4 was explained by the use of a theory of Austin and Mott (1969) on a two-carrier model ignoring the kinetic term, both carriers having a thermally activated mobility, and a band gap of about 0.1 eV. The activation energy of the mobility was also assumed to be of the order of 0.1 eV to account for the conductivity, with rather a strong dependence on the temperature; about 0.11 eV at 120 K and 0.04 eV at 80 K. In the case of the low-temperature phase of YFe_2O_4, Sakai et al. (1986) assumed coexistence of itinerant carriers and small polarons of hopping. Phenomenologically, a similarity is apparent in the character of the transport phenomena in the low-temperature phase of YFe_2O_4 and Fe_3O_4. They give examples to investigate the nature of carriers in a system of large correlation and strong electron–phonon coupling as well as details of the Verwey transition.

6. Magnetic properties

The main problem concerning the magnetic properties of this series of compounds is spin correlation of the transition metal ions which form triangular nets in the c-plane. As in the case of transport phenomena, magnetic coupling in the

c-plane is much stronger than that along the c-axis. This is an example of a two-dimensional spin system.

The magnetic moment of the rare earth ions is believed to be paramagnetic down to liquid-helium temperature, although a detailed study was performed only for $YbFe_2O_4$ (Kishi et al. 1982). Magnetization of a single crystal of $YbFe_2O_4$ at low temperatures was compared with that of non-stoichiometric YFe_2O_4 and it is concluded that Yb^{3+} ions are paramagnetic down to 4.2 K and their susceptibility obeys the Curie–Weiss law with an asymptotic Curie temperature of about −6 K. This corresponds to a molecular-field constant of about $-0.4\,mol/cm^3$. At the same time, there is a magnetic interaction between the Fe and Yb sublattices that corresponds to a molecular-field constant of $-1.5\,mol/cm^3$ for Yb^{3+}. In the following, nothing will be said about the magnetic properties of rare earths directly. However, there are several data indicating that the magnetic properties of the transition metal ions are affected by R, the rare earth element.

As for the transition metal elements, M and M′, we will confine ourselves to compounds containing Fe^{3+}, mainly because studies have been concentrated on them. The magnetism of $YMnO_3$ will not be referred to, although this is interesting because of the coexistence of magnetic ordering and ferroelectricity (Bertaut et al. 1963a,b, Bertaut and Mercier 1963, Kohn and Tasaki 1965).

6.1. *Magnetization*

6.1.1. *Induction of ferromagnetic net moment*

The first measurement of magnetization of this series of compounds was carried out on YFe_2O_4 by Sugihara et al. (1976). The specimen used was an oxygen-deficient non-stoichiometric powder. They found a peak in thermomagnetic curves at around 200 K, when the specimen was heated from 77 K after cooling without an external magnetic field. If the sample is cooled to 77 K in a magnetic field, thermoremanent magnetization is induced that decreases monotonically with increasing temperature and disappears at 200 K (see fig. 62). The magnitude of the induced moment is small, about $0.1\mu_B$ per formula unit when the cooling

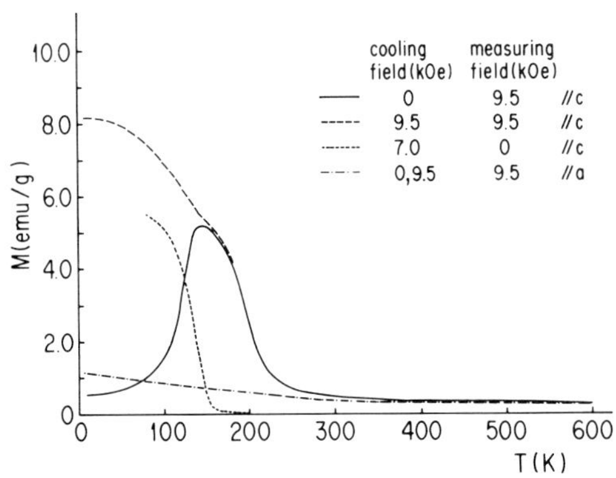

Fig. 62. The thermomagnetic curves of a single crystal of non-stoichiometric YFe_2O_4. The measurement was carried out with increasing temperature. (Sugihara et al. 1978.)

field is 7 kOe. They also measured the Mössbauer spectrum of the same specimen at 90 K and observed a sextet with an internal field of about 460 kOe. They concluded that YFe_2O_4 is an antiferromagnet with a Néel temperature of about 200 K, accompanied by weak parasitic moment.

The magnetic susceptibilities of compounds with other rare earth elements, Ho, Er, Tm, Yb, Lu as well as Y, were reported by Baynes et al. (1978). They found a peak near 200 K in all the compounds and assigned it as the Néel point. On the other hand, they reported weak ferromagnetic behavior only below 120–155 K, depending on the rare earth element. Thermomagnetic curves of $InFe_2O_4$ was reported later by Gérardin et al. (1981). It shows a very small peak at 200 K, an increase of magnetization with further decrease of temperature and a net moment induction only below 75 K (fig. 63). By Mössbauer spectroscopy they confirmed that an internal field appears below 200 K.

To elucidate the mechanism of the appearance of the moment, the experiment was extended to a single crystal (Sugihara et al. 1978). The main points of their results can be summarized as follows:

(i) Magnetic anisotropy is very large and spins at low temperatures can be assumed parallel to the *c*-axis. That is confirmed by the Mössbauer spectrum with incident γ-rays along that axis. According to the susceptibility measurement up to 650 K, the magnetic anisotropy energy is estimated to be about 80 K. The Curie constant is consistent with an idea that there are equal numbers of Fe^{2+} and Fe^{3+} ions with $g = 2.0$. The paramagnetic Curie temperature is about -400 K. This fact indicates that the main exchange coupling is strongly antiferromagnetic.

(ii) There is a marked after effect in the magnetization process, in a temperature range containing the peak in the thermomagnetic curves measured with increasing temperature (see fig. 62). Even at 77 K, magnetization increases after an application of an external magnetic field in proportion to the logarithm of time, i.e., relaxation is of the Jordan type.

(iii) The magnetic moment is induced by field cooling only along the *c*-axis: an induced moment is parallel to the spins. This is contradictory to ordinary parasitic ferromagnetism, where a small ferromagnetic moment appears by canting of

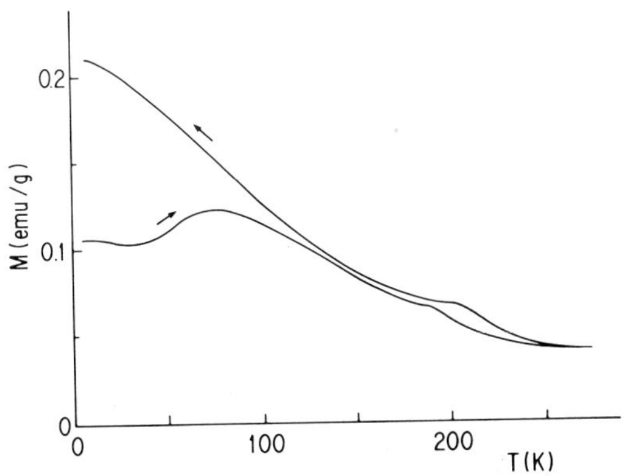

Fig. 63. The thermomagnetic curves of polycrystalline $InFe_2O_4$. In an external field of 1 kOe. (Gérardin et al. 1981.)

antiferromagnetic sublattices (Dzyaloshinski 1958, Moriya 1963). Sugihara et al. named this case parasitic ferrimagnetism.

These points were confirmed later by a measurement on a $LuFe_2O_4$ single crystal (Iida 1988) and believed to be common in RFe_2O_4. Thus, the problem in the magnetic properties of YFe_2O_4 made its appearance as a small moment induction by an external magnetic field parallel to the spin axis.

As was already described in sect. 5.1, the character is totally different in the stoichiometric YFe_2O_4 where the Verwey transition takes place. No ferromagnetic moment is induced in that case. On the other hand, the situation is similar in RFe_2O_4 with R different from Y. This fact suggests that the random distribution (without the long-range ordering) of the variable charged ions of Fe in an equivalent crystal site is necessary for the net moment induction. Sugihara et al. (1978) proposed a model which assumes a tendency of Fe^{3+} and Fe^{2+} to locate in different antiferromagnetic sublattices: a coupling of spin density waves and charge density waves. This model will be discussed in sect. 6.3.

Magnetic measurement is extended to higher fields, up to 100–290 kOe, for single crystals of the Y-, Yb- or Lu-compounds. (Nakagawa et al. 1980, Kishi et al. 1982, Iida et al. 1987, Iida 1988, Iida et al. 1989). The effect of field cooling is also examined. When an external magnetic field is applied in the c-plane, magnetization is proportional to the field at any temperature and no moment is induced by the field cooling. When the external field is parallel to the c-axis, a ferromagnetic moment appears. In the case of $LuFe_2O_4$ at 4.2 K, a major hysteresis loop can be observed only for samples cooled in an external field higher than 100 kOe. A coercive field at 4.2 K is about 110 kOe. By the application of fields stronger than 230 kOe, however, magnetization is almost saturated even in a sample cooled without an external field. In that case, Barkhausen jumps are observed several times in the magnetization curve (see fig. 64). The remanent magnetization is 45 emu/g or $2.8\mu_B$ per formula unit. This value cannot be considered "parasitic". It should be noted that the ordering of Fe^{2+} and Fe^{3+} ions in each antiferromagnetic sublattice will result in only $1\mu_B$ per formula unit. At 77 K, H_c decreases to about 35 kOe.

It is disclosed that the external magnetic field, necessary to saturate the crystal at low temperatures, depends on the rare earth element. In the case of $YbFe_2O_4$ or YFe_2O_4, hysteresis cannot be observed at 4.2 K in the magnetization curves up to 150 kOe. The magnetization curve is reversible. Magnetic anisotropy of Yb^{3+} ions may not be the cause of this difference, since YFe_2O_4 is still harder to saturate. The magnetization curves in the first quadrant are shown in fig. 64 for these three compounds. The large magnetization of $YbFe_2O_4$ at a finite external field is due to the paramagnetic Yb^{3+} ions.

Thermoremanent magnetization (TRM) is a function of the intensity of the cooling field as well as the temperature where the field is applied during cooling. Figure 65 shows the induced moment at 4.2 K as a function of the cooling field, along the c-axis of Lu-, Yb- and YFe_2O_4. The specimen was cooled down from room temperature. The curve of $LuFe_2O_4$ almost saturated at about 100 kOe, but TRM still increases at 300 kOe in the case of YFe_2O_4. The spin system is not

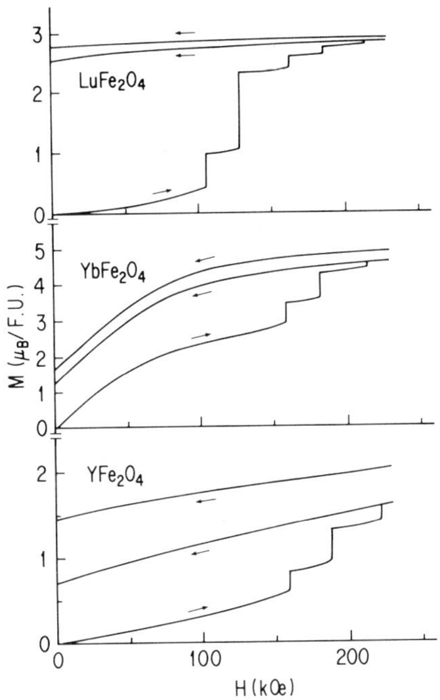

Fig. 64. The first quadrant of the hysteresis curves of Lu-, Yb- and YFe_2O_4 single crystals along the *c*-axis at 4.2 K. The uppermost curves are from the specimens cooled to 4.2 K in a field of 230 kOe. (Iida et al. 1989.)

homogeneous. This is shown clearly in the temperature dependence of the induced moment (see fig. 66). The moment induced by a higher cooling field disappears at low temperatures during the heating of the specimen and coincides at certain temperature with that induced by a lower cooling field. It is to be noted that the curve for $H_{cool} = 1$ kOe is identical to the temperature dependence of the saturation moment if that is normalized by the value at 4.2 K. Inhomogeneity of the spin system is confirmed by fig. 67 which indicates the dependence of the thermoremanent magnetization on the temperature where field cooling has begun. The experiment was carried out by observing the temperature dependence

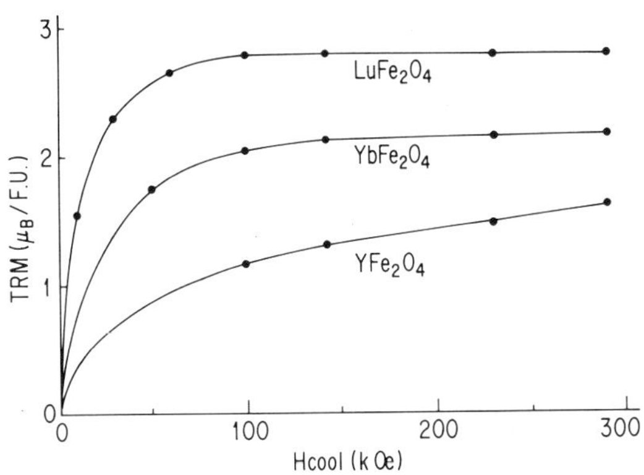

Fig. 65. The thermoremanent magnetization at 4.2 K along the *c*-axis of a single crystal of $LuFe_2O_4$, $YbFe_2O_4$ and YFe_2O_4, as a function of cooling field. (Iida ct al. 1989.)

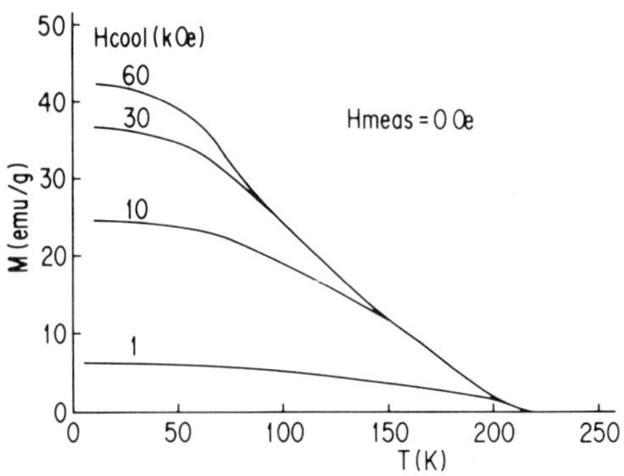

Fig. 66. The temperature dependence of the thermoremanent magnetization along the *c*-axis of $LuFe_2O_4$, for several values of the cooling field. (Iida 1988.)

of the induced moment, produced by the cooling down of the sample from room temperature to 4.2 K in 140 kOe, in successive heating and cooling cycles. Magnetization decreases with heating but remains almost constant during the subsequent cooling process and does not regain the initial value. The decrease of the magnetization by one heating–cooling cycle shows the moment fixed between T_m and T'_m, the maximum temperature in the preceding and the present cycle, respectively. These facts suggest that the induction of the macroscopic moment is due to the freezing of the local moment, and the freezing temperature is different from place to place, i.e., the specimen is magnetically inhomogeneous.

6.1.2. *Cases without* $Fe^{[2+]}$

To explain the appearance of a magnetic moment by the segregation of differently charged Fe ions to each antiferromagnetic sublattice, coexistence of two kinds of Fe ions is necessary in a crystallographically equivalent site. If the 2+ ions are other than Fe, exchange of ions cannot be expected below room

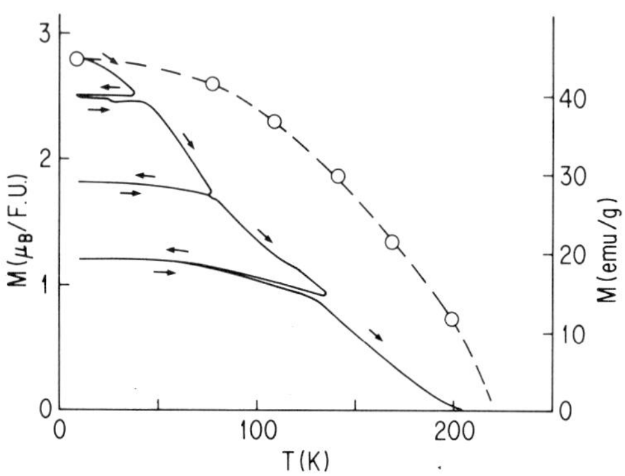

Fig. 67. The temperature dependence of the saturation magnetization (open circles and a broken line) and the thermoremanent magnetization (solid lines) along the *c*-axis of $LuFe_2O_4$. The latter is induced by cooling from 250 K down to 4.2 K in an external magnetic field of 140 kOe and measured without field by successive heating–cooling cycles with increasing T_m, the maximum temperature of the cycle. (Iida et al. 1987.)

temperature, where antiferromagnetic spin ordering appears. This point was examined for two cases, replacement of $Fe^{[2+]}$ by Mn^{2+} or Co^{2+} and an experiment on a single T layer composed of only Fe^{3+}.

Thermomagnetic curves of $YFeMnO_4$ powder are shown in fig. 68 (Siratori et al. 1983). Magnetization measured during heating, after cooling without magnetic field, shows a broad maximum at around 40 K, whereas that with decreasing temperature in an external field of 11 kOe increases monotonically to 2.0 emu/g = $0.09\mu_B$/formula unit. Two curves coincide above the peak. Thermoremanent magnetization is induced by field cooling, which disappears at about 40 K, the peak temperature. These features are similar to that of RFe_2O_4. Differences are the lower temperature (by a factor of about 5) of the peak and the smaller magnetization (by a factor of about 3) compared with YFe_2O_4. A similar feature was discovered by Iida (1988) through magnetization measurements on a single crystal of $LuFeCoO_4$ up to 140 kOe. In an external field of 1 kOe, a peak appears at 70 K in the thermomagnetic curve during heating after the cooling of the specimen without a magnetic field. At this field strength, thermoremanent magnetization at 4.2 K is about 0.15 emu/g = $0.01\mu_B$/formula unit. It should be noted that the magnetic moment induced along the *c*-axis is one order of magnitude larger than that within the *c*-plane. The latter is estimated to be a little larger than the magnitude expected from suspected experimental errors, in the orientation of the crystal, etc. To saturate the thermoremanent magnetization at 4.2 K, a cooling field of 100 kOe is sufficient. This is similar to $LuFe_2O_4$. Although hysteresis disappears between 40 and 50 K, magnetization curves are not linear up to 200 K when a field up to 140 kOe is applied along the *c*-axis.

T layers composed of only Fe^{3+} ions were investigated in $Lu_2Fe_3O_7$. As described in sect. 4, there are isolated T layers in this crystal. The charge of the ions in that layer is supposed to be 3+. A Mössbauer study by Tanaka et al. (1983), as well as that by Bahgat (1984), disclosed that the spectrum is composed of two parts; one is very similar to the spectrum of $LuFe_2O_4$ and orders at 260 K, and the other is a doublet down to 60 K, with an isomer shift which corresponds to Fe^{3+}. Tanaka et al. attributed the former to double T layers similar to that in

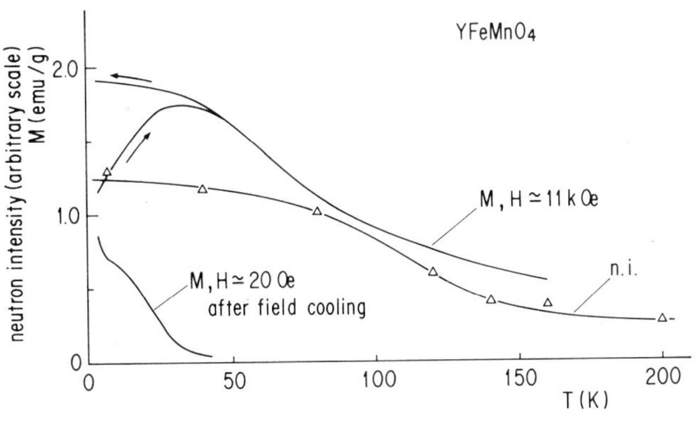

Fig. 68. Thermomagnetic curves of polycrystalline $YFeMnO_4$. (Siratori et al. 1983.)

$LuFe_2O_4$ and the latter to isolated T layers, concluding that the isolated layer, paramagnetic down to 60 K, is occupied by Fe^{3+} ions only. This is one of the evidences indicating the two-dimensional character, i.e. much stronger spin coupling in the *c*-plane compared with that along the *c*-axis, in this system. This conclusion of Tanaka et al. is confirmed by the magnetic measurements of Sugihara et al. (1985). They found a small increase below about 70 K in the temperature dependence of thermoremanent magnetization (fig. 69), and assigned it as a moment induced in the isolated T layers.

By these experiments, it was disclosed that the coexistence of differently charged ions of Fe is not indispensable to the appearance of, so called, parasitic ferrimagnetism. At the same time, "amplification" of the moment is demonstrated as well as of the ordering temperature by their coexistence. The magnetic properties of $RFeM'O_4$ described in this subsection resemble the properties of spin glasses or mictomagnets

6.1.3. *Field-heating effect*

So far, remanent magnetization is induced by the cooling of the specimen in a magnetic field (thermoremanent magnetization) or applying a high magnetic field at a constant temperature (isothermal remanent magnetization). These methods are not unusual. In this series of compounds, another rather strange method of moment induction was reported on $LuFe_2O_4$: the magnetic moment is induced by heating, instead of cooling, the sample in a small external magnetic field (Iida et al. 1986). This was called the field heating effect. An example is shown in fig. 70. The magnetization measured in a field of 4.5 kOe during heating after cooling without an external field (lower curve in fig. 70a) increases from 120 K and becomes larger than the value for isothermal magnetization indicated by the open circles in the figure. The same tendency is observed in a field-cooled specimen too (upper curves). When the external field is 105 Oe, the thermomagnetic curve shows two peaks (fig. 70b). The two curves (measured by heating and cooling processes) and the open circles (isothermal magnetization) coincide at the high-

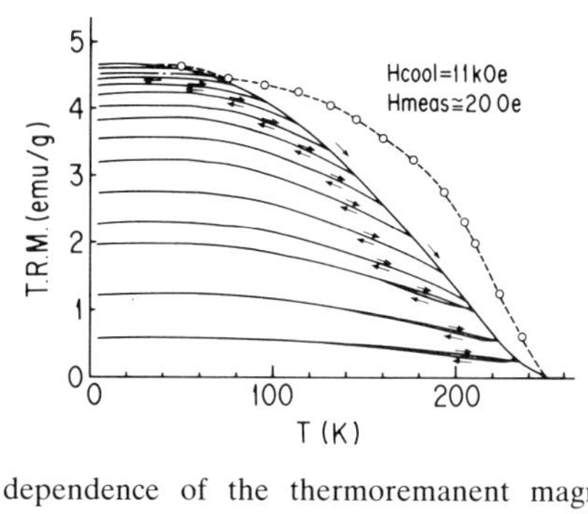

Fig. 69. The temperature dependence of the thermoremanent magnetization of polycrystalline $Lu_2Fe_3O_7$, induced by cooling from 320 K down to 4.2 K in a field of 11 kOe and measured without field by successive heating–cooling cycles. (Sugihara et al. 1985.)

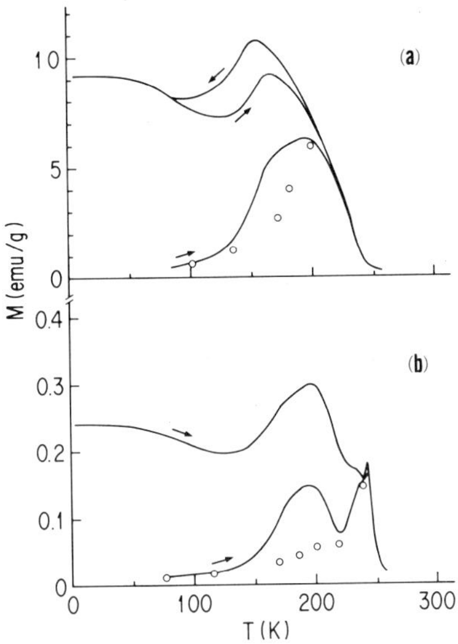

Fig. 70. The thermomagnetic curves of polycrystalline $LuFe_2O_4$ showing the field heating effect. (a) in an external field of 4.5 kOe; (b) in an external field of 105 Oe. In both cases, the upper curves are for a field-cooled specimen ($H_{cool} = H_{meas}$), whereas lower curves are for a specimen without field-cooling. Open circles are isothermal magnetization. (Iida et al. 1986.)

temperature peak. This fact indicates that spins are in thermal equilibrium at temperatures higher than the high-temperature peak, and there is an extraordinary magnetization process which becomes ineffective below this temperature. Still another magnetization process exists between 120 and 200 K and results in the second peak in the thermomagnetic curves. This process makes its appearance below 220 K by a finite relaxation time. An appreciable after-effect is observed in the magnetization process between 120 and 220 K. The point is that the temperature dependence of the relaxation time is unusual, being short at lower temperatures since magnetization is induced during heating. Although there is no direct evidence, Iida et al. assumed some phase transition as the origin of this effect.

The field-heating effect has been observed most clearly in $LuFe_2O_4$ but is supposed to exist in other compounds, too, at least in $YbFe_2O_4$. It is crucially dependent on the process of sample preparation, although no difference has been detected in composition, contamination of other elements or phases, lattice constants, etc., between specimens with and without the field-heating effect. The dynamics of the "melting" of frozen spins in these compounds are problems to be investigated.

6.2. *Spin correlation and magnetic structure*

6.2.1. *Magnetic Bragg lines along the c*-axis*

The first report of a neutron diffraction study of this series of materials appeared in 1979 (Akimitsu et al. 1979) and clearly manifested a two-dimensional character of the spin system. An experiment on a non-stoichiometric single crystal of YFe_2O_4 revealed that there are magnetic Bragg lines, instead of Bragg spots,

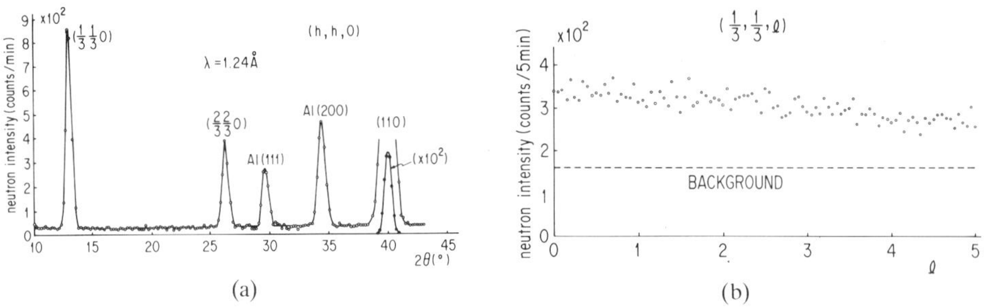

Fig. 71. Neutron diffraction patterns at 77 K. (a) along the $(h, h, 0)$; (b) along the $(\frac{1}{3}, \frac{1}{3}, l)$, of a non-stoichiometric single crystal of YFe_2O_4. (Akimitsu et al. 1979.)

parallel to the c^*-axis. Figure 71 is a reproduction of their results of $[h, h, 0]$ and $[\frac{1}{3}, \frac{1}{3}, l]$ scannings at 77 K. In the c^*-plane, there appear rather sharp peaks at $[\frac{1}{3}, \frac{1}{3}, 0]$ and $[\frac{2}{3}, \frac{2}{3}, 0]$. No change is observed in the intensity of $[1, 1, 0]$ nuclear diffraction by lowering the temperature from 300 down to 77 K, corresponding to the absence of the ferromagnetic moment. On the contrary, scanning along the c^*-axis at the magnetic peak shows no structure but a slight decrease with increasing l that is accounted for by the magnetic form factor of Fe ions. The following points can be concluded from these results:

(i) Along the c-axis, there is no spin correlation at 77 K. Subsequent experiments assured this down to 4.2 K.

(ii) In spite of (i), there is an ordering or marked development of spin correlation in the c-planes. The ordering scheme is such that edges of the magnetic unit cell are three times as large as the atomic unit cell, as shown in fig. 72. The spin state is repeated by each three (1, 1) lines.

As will be discussed later, this "triple period" magnetic structure is a characteristic of a triangular lattice with antiferromagnetic nearest-neighbor interaction. When the anisotropy of the spin direction is small, the period of three atomic lines in the spin structure is realized by a rotation of spins by 120°. If anisotropy is very large, as in the present case, the spin direction is restricted along, say, $\pm c$ and 120° rotation is impossible. At high temperatures, a sinusoidal modulation of the spin density, the period of which is three atomic lines, is expected and confirmed in such cases. At low temperatures, however, the thermal average of each spin moment approaches its full value, since thermal fluctuations vanish. To

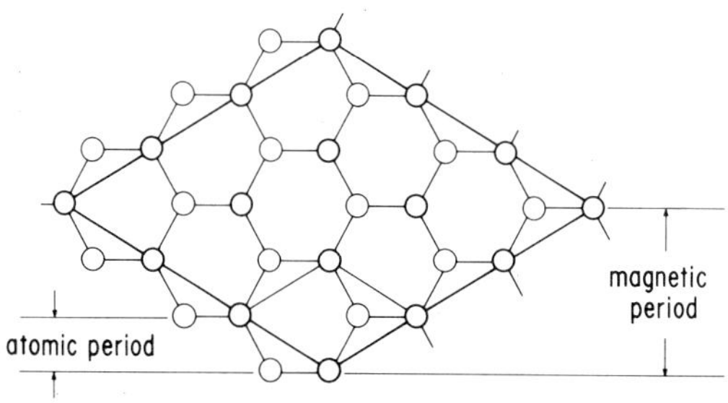

Fig. 72. The atomic and magnetic unit cell of RFe_2O_4 in a c-plane.

maintain the three atomic line period, a ferrimagnetic structure such as [+, +, −] should be realized. This point will be discussed later in connection with moment induction.

It should be noted here that the magnetic Bragg lines are observed in YFe_2O_4 only when the specimen is non-stoichiometric. In a crystal which exhibits a Verwey transition, the Bragg spots appear below the transition, indicating a three-dimensional spin ordering, though the magnetic structure is so complicated that it has not been determined yet. One reason for the lack of success is that no single crystal of the stoichiometric YFe_2O_4 has been available. Experiments on a polycrystalline specimen will be referred to in the next subsection.

6.2.2. *Neutron diffraction on polycrystals*

When the specimen is a polycrystal, diffraction in the form of a Bragg line gives a peculiar saw-tooth form, as is shown in fig. 73. In many cases of low-dimensional spin systems, short-range correlation of spins develop from temperatures higher than the ordering point. Here, too, saw-tooth diffraction peaks of $[\frac{1}{3}, \frac{1}{3}, l]$, etc., persist up to higher temperatures than the remanent magnetization or the splitting in Mössbauer spectrum. (An example of $YFeMnO_4$ is shown in fig. 82.) Development of the spin correlation is thoroughly investigated in stoichiometric YFe_2O_4 by Funahashi et al. (1984). They showed that the profile of the saw-tooth diffraction is explained well by assuming that the intensity of the Bragg line along the c^*-axis is constant, except for the decrease due to the magnetic form factor of Fe ions, and its profile in a direction perpendicular to the c^*-axis is expressed by a Lorentzian function. Examples are shown in fig. 74. The width of the Lorentzian function corresponds to the inverse of the correlation length κ and the peak intensity of the diffraction to staggered susceptibility χ. They obey an exponential law,

$$\kappa = K(T - T'_N)^{\nu}, \tag{11}$$

$$\chi = X(T - T'_N)^{-\gamma}, \tag{12}$$

as is shown in fig. 75. T'_N in eqs. (11) and (12) is 233.0 ± 0.5 K. As the transition

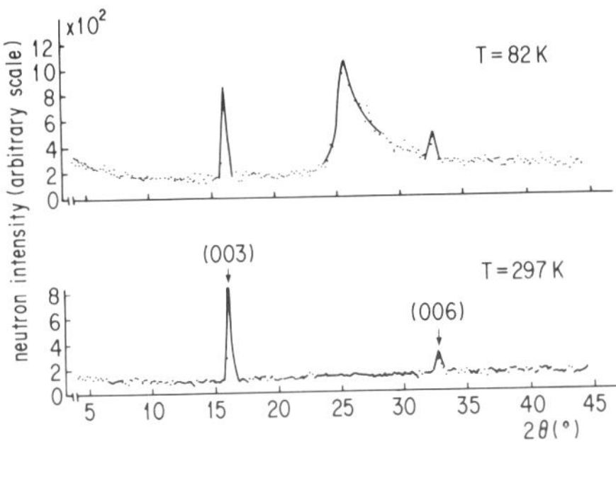

Fig. 73. Neutron diffraction patterns of polycrystals of non-stoichiometric YFe_2O_4 at 297 and 82 K. A saw-tooth diffraction profile at low temperature indicates a Bragg line. (Funahashi et al. 1984.)

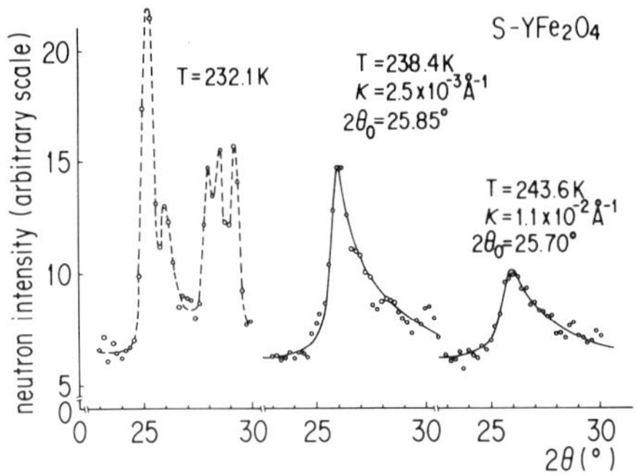

Fig. 74. Examples of neutron diffraction of polycrystals of stoichiometric YFe_2O_4 at around $(\frac{1}{3}, \frac{1}{3}, 0)$. Solid lines of saw-tooth shape are calculated profiles (see text). Below 234.5 K, the Bragg line disappears and Bragg spots indicate a three-dimensional spin ordering. (Funahashi et al. 1984.)

is of the first order, which is not equal to the apparent transition point, $T_N = 234.5$ K. The critical exponents are determined as

$$\nu = 1.5 \pm 0.05 , \qquad \gamma = 3.1 \pm 0.1 .$$

These values are larger than those rigorously evaluated for a two-dimensional ferromagnetic lattice of Ising spins (Onsager 1944).

As is shown in fig. 74, the saw-tooth form disappears below T_N and several peaks of ordinary shape due to Bragg spots instead of Bragg lines are observed. It is certain that the spins order three-dimensionally below the Verwey transition point.

It is interesting to note that the analysis of the line shape mentioned above cannot be applied to non-stoichiometric YFe_2O_4 at low temperatures. The saw-tooth profile of the diffraction peak cannot be interpreted well by an assumption of a Lorentzian or Gaussian function for the intensity profile perpendicular to the c^*-axis but is intermediate between them. The spin system is not

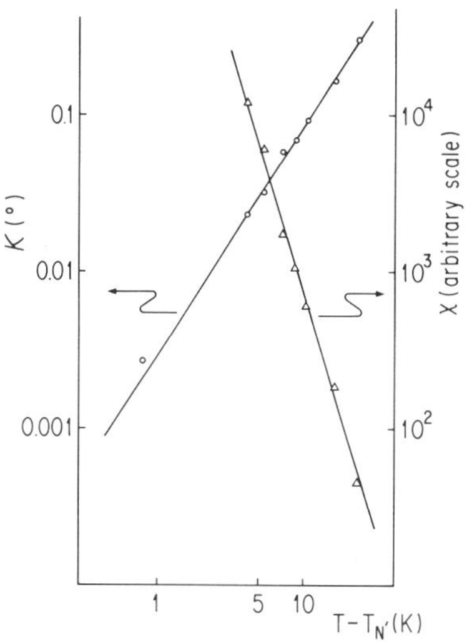

Fig. 75. A log–log plot of κ, the inverse of the correlation length, and χ, the staggered susceptibility at $(\frac{1}{3}, \frac{1}{3})$, against $T - T_N'$. T_N' is assumed 233.0 K, which is 1.5 K below the apparent transition point. (Funahashi et al. 1984.)

homogeneous. This is in contrast to the Mössbauer spectrum at room temperature described in the preceding section.

The situation in $RFeM'O_4$ and $R_2Fe_3O_7$ is similar to non-stoichiometric YFe_2O_4. The saw-tooth shaped magnetic diffraction assures a two-dimensional spin correlation, but the profile cannot be explained by a Lorentzian or Gaussian shape of intensity. An interesting point is that there is no anomaly in $Lu_2Fe_3O_7$ near 60 K in the temperature dependence of the intensity of the saw-tooth shaped magnetic diffraction. The correlation of spins in the isolated T layer develops at a much higher temperature than the "ordering" point.

The spin correlation of Fe^{3+} ions was also investigated in $LuFeMgO_4$ by Wiedenmann et al. (1983). Since Mg is non-magnetic in contrast to Mn or Co, an exchange coupled network is cut at that point. No magnetic Bragg peak was found down to 2 K in the neutron diffraction pattern. Wiedenmann et al. reported, however, an increase of diffuse scattering below 250 K, which was attributed to spin correlation. The amplitude of this scattering develops with decreasing temperature down to 30 K. The magnetic part of the scattering, separated from the nuclear part by subtraction of the neutron intensity at 300 K from that at 30 K, is shown in fig. 76. The main peak at $Q = (4\pi/\lambda)\sin\theta = 1.25\ \text{Å}^{-1}$ is asymmetric, although not so much as the peak in fig. 74. Wiedenmann et al. analyzed this scattering profile and concluded that the spin correlation develops mainly in a c-plane. The correlation of spins on adjacent c-planes is detected only for the nearest-neighbor pair and that is negative (antiferromagnetic). Spins of Fe^{3+} lie parallel to the c-axis. The Mössbauer spectrum of this material is a doublet down to 50 K, but a sextet at 5 K. The profile of each absorption line is asymmetric in the latter, indicating that the internal field of the Fe ions is not identical but distributes. They argued that $LuFeMgO_4$ at low temperatures is a spin glass due to spin freezing.

6.2.3. *Interlayer correlation in $LuFe_2O_4$*

Recently, Iida measured the neutron diffraction of a single crystal of $LuFe_2O_4$ (Iida et al. 1988). In an experiment at 14 K, he observed a weak line at $[0, 1, l]$,

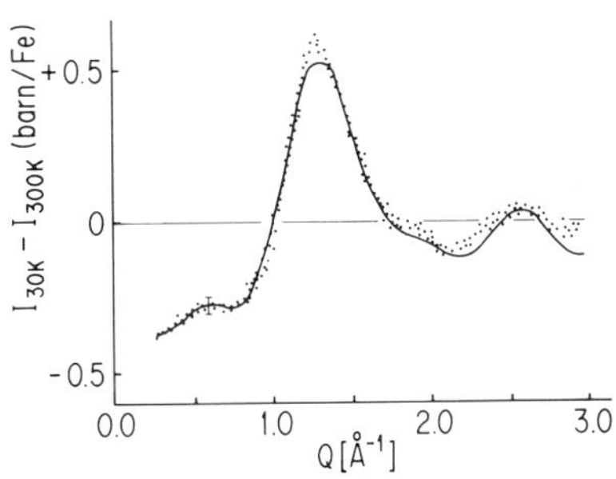

Fig. 76. The difference of neutron scattering of polycrystalline $LuFeMgO_4$ between 30 and 300 K. The solid line is calculated from a two-spin correlation up to the eighth neighbor in the same c-plane and the nearest-neighbor in the next c-plane. (Wiedenmann et al. 1983.)

besides Bragg lines at $(\frac{1}{3}, \frac{1}{3}, l)$, etc., reported earlier. The profile of $[0, 1, l]$ scattering along the c^*-axis is reproduced in fig. 77. This diffraction is concerned with the interlayer correlation of the ferrimagnetic moment noted in sect. 6.2.1. According to the analysis of Iida, the main feature of $[0, 1, l]$ profile is explained by a ferromagnetic correlation of net (ferrimagnetic) moments in two adjacent T layers. This is contradictory to the conclusion of Wiedenmann et al. (1983) on $LuFeMgO_4$. Moreover, the fine structure of the profile can be accounted for by a weak antiferromagnetic correlation between the two blocks, each composed of two T layers, separated by a U block containing rare earth ions. A solid line in fig. 77 is calculated by the use of the parameters, $q = -0.11$ and $s = 0.52$. Here, q and s are order parameters of inter- and intra-block spin correlation, respectively.

6.2.4. *Information from Mössbauer spectroscopy*

In contrast to the neutron diffraction which reflects the spatial correlation of spins, Mössbauer spectroscopy reveals information on the time correlation of one spin. If the system is not in thermal equilibrium, these two correlations are not identical. In addition, information can be drawn from the Mössbauer spectra on the electronic state including the charged state of the Fe atoms from the isomer shift, the electric quadrupole splitting and Zeeman splitting due to the internal field. When the specimen is a single crystal, the spin direction can also be determined from intensities of each component of the sextet.

Two points will be made here on the magnetic structure of RFe_2O_4 and $RFeM'O_4$.

6.2.4.1. *Three sublattices of Fe^{3+} in $YbFe_2O_4$.*

Mössbauer study at low temperatures has been carried out extensively in an external magnetic field (H_{meas}) on a single crystal of $YbFe_2O_4$ cooled with or without an external field (H_{cool}) (Tanaka et al. 1988, 1989). They drew a conclusion that there are three, not two, sublattices of Fe^{3+} spins at low temperatures, two of which align parallel and the other antiparallel to the two.

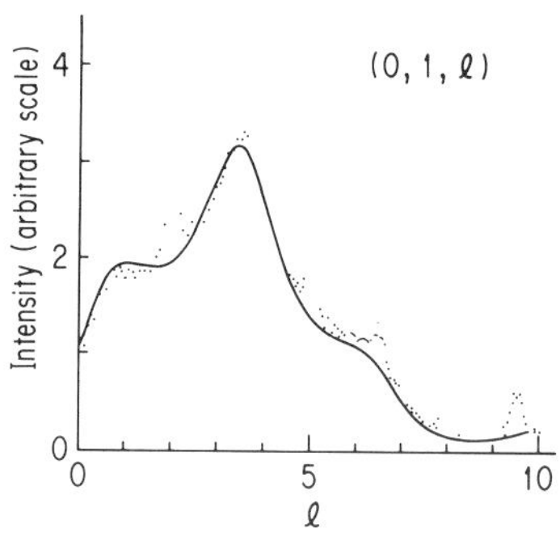

Fig. 77. Magnetic neutron diffraction along the $[0, 1, l]$ of a single crystal of $LuFe_2O_4$: the difference of the intensity at 14 K and that at 280 K. The solid line is a calculated profile by the use of $s = 0.52$ and $q = -0.11$. Here, s is the correlation parameter for adjacent T layers and q is that for T layers separated by a U block. (Iida et al. 1988.)

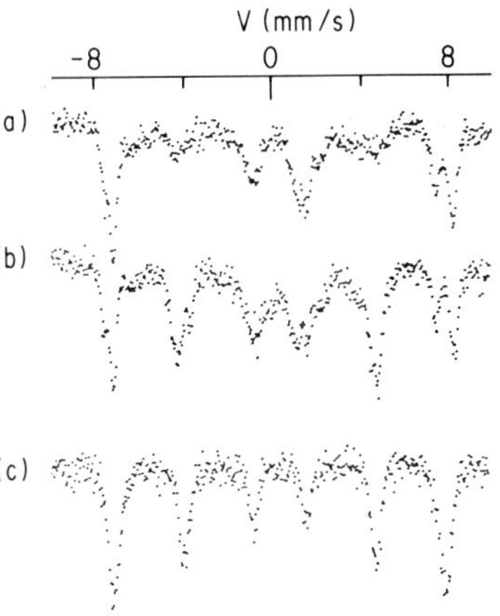

Fig. 78. Mössbauer spectra of $YbFe_2O_4$ and $LuFeCoO_4$ at 4.2 K. (a) Single crystal of $YbFe_2O_4$, incident γ-ray is parallel to the c-axis. (b) Polycrystal of $YbFe_2O_4$. (c) Polycrystal of $LuFeCoO_4$. (Tanaka et al. 1989.)

In their experiment, incident γ-rays were set parallel to the c-axis. Examples of the spectra are shown in figs. 78 and 79. There are six groups of absorption, each composed of several lines. The lines in the lowest (first) and the highest energy (sixth) group at around -8 and 8 mm/s, respectively, are sharp indicating that they are due to $Fe^{[3+]}$ ions. The second and fifth groups at around -5 and 5 mm/s are broad and assigned to [2+] ions. Since spins of Fe ions are parallel to the c-axis along which the γ-rays propagate, the sextet of each charged state, $Fe^{[3+]}$ or $Fe^{[2+]}$, lacks in intermediate (the second and the fifth) components, those of $Fe^{[3+]}$ will appear at around ± 5 mm/s.

Let us concentrate first to the sixth group. This group of absorption is composed of two lines, the intensity ratio of which is 1 : 2. This is naturally explained by the ferrimagnetic structure described in sect. 6.2.1. By the applica-

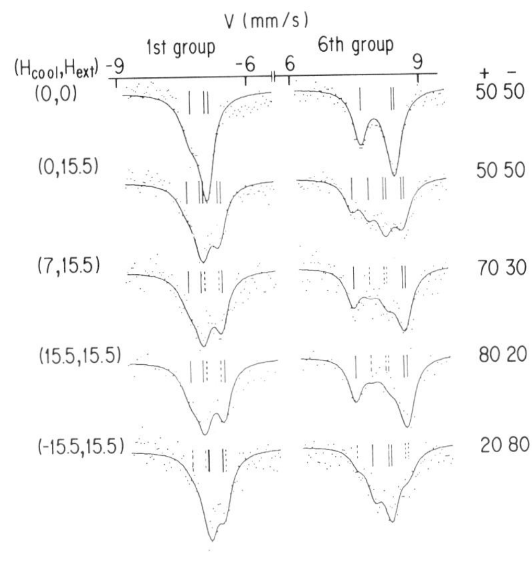

Fig. 79. The first and sixth absorption group of the Mössbauer spectra of a $YbFe_2O_4$ single crystal at 77 K as a function of the cooling and the measuring magnetic field indicated on the left. Numbers on the right are the estimated volume fraction of + and − domains (in %). γ-rays are incident along the c- or the spin axis. (Tanaka et al. 1989.)

tion of an external magnetic field (H_{meas}) parallel to the c-axis during the measurement, each line is split into two components with equal intensity, provided that the magnetic field during cooling is absent: four components are seen in the (0, 15.5) spectrum in fig. 79. Here, the condition of the measurement is indicated by a set of (H_{cool}, H_{meas}), each in units of kOe. The splitting of the lines means that there are two domains with opposite spin directions. The application of a magnetic field during cooling (H_{cool}) does not affect the position of the lines but does affect their intensity. As is shown in fig. 79, a positive H_{cool} intensifies the outer two components and a negative H_{cool} intensifies the inner [see the (15.5, 15.5) and (−15.5, 15.5) spectra]. Apparently this change corresponds to the net moment induction: the volume of the above-mentioned two domains becomes different by the field cooling. The ratio of the volume is estimated as 80:20 for $H_{cool} = 15.5$ kOe. This value is consistent with an induced moment of about 60% of the saturation value, shown in fig. 65. Within a domain, the effective field of the stronger line increases with H_{meas} parallel to H_{cool} whereas that of the weaker one decreases. The spin direction of these two lines should be opposite in a domain, consistent again with the ferrimagnetic structure.

The dependence of the first group on H_{meas} and H_{cool} is different. As is seen in fig. 79, this group of absorptions is also composed of two lines, the intensity ratio of which is 2:1, and the lines are split by H_{meas}. If these lines correspond to those of the sixth group directly, positive (parallel to H_{meas}) H_{cool} should intensify a component of the stronger line with lower (higher in the absolute value) velocity and that of the weaker line with higher velocity, resulting in a narrowing of the group. This is not realized. To explain the profiles of both the first and the sixth group of absorption intensity, $Fe^{[3+]}$ should be divided into three sublattices, P, C and M (as denoted by Tanaka et al. 1988, 1989) and the lines due to P and M coincide in the first group whereas those of P and C do so in the sixth group. The spin direction of P and C sublattices is parallel and that of M is opposite to them. The splitting scheme of each line is shown in fig. 80. Note the difference of both the isomer shift and the quadrupole splitting of each of the lines. The temperature dependence of the internal field is also different for these three sublattices. With

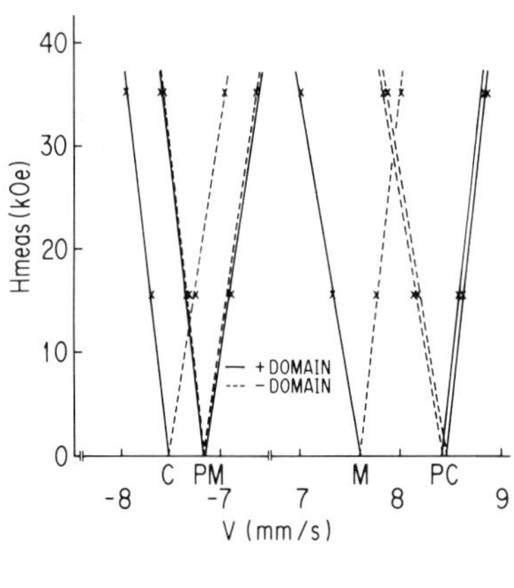

Fig. 80. The position of the first and sixth absorption lines as a function of H_{meas}, the external magnetic field during measurement. (Tanaka et al. 1989.)

increasing temperature above 100 K, the internal field of P decreases much faster than that of other two.

An important point here is thc H_{meas} dependence of the effective field. It is natural to consider that components intensified by H_{cool} correspond to a domain, the net moment of which is parallel to H_{cool}. As is shown in fig. 80, the effective field of this component of the P and C lines increases with externally applied H_{meas} whereas that of the M line decreases. The derivatives (slope of lines in fig. 80) are nearly those expected from γ, the magneto-mechanical ratio, of the ^{57}Fe nucleus. Since the hyperfine field at the nucleus of the transition metal element is ordinarily antiparallel to the electronic moment, an increase of the internal field by H_{meas} suggests that the direction of electronic moment is inverse to H_{meas}: P and C are minority sublattices, whose moments are opposite to the net moment. Note that the experiment is restricted to $Fe^{[3+]}$ and no information is available on the $Fe^{[2+]}$, at present.

The profile of the Mössbauer spectra of $LuFe_2O_4$ is similar to that of $YbFe_2O_4$. The magnetic structure in these two compounds is believed to be the same. On the other hand, the spectra of non-stoichiometric YFe_2O_4 and $InFe_2O_4$ are a little different, suggesting, at least, a microscopic difference in the spin structure as seen from Mössbauer spectroscopy. The dependence of physical properties on the R element appears here, as was mentioned before.

Here, information has been obtained only on the $Fe^{[3+]}$ ions. Absorption lines due to $Fe^{[2+]}$ are too broad to be analyzed in the same manner. Further investigation is needed on this point.

6.2.4.2. *Strength and direction of the internal field in RFeM'O_4*. As was stated in the last subsection, the magnetic properties are changed by the substitution of other divalent ions for $Fe^{[2+]}$ ions. The Mössbauer spectra of $YFeMnO_4$ powder below 155 K is shown in fig. 81 (Tanaka et al. 1984b). Below 120 K, there appears

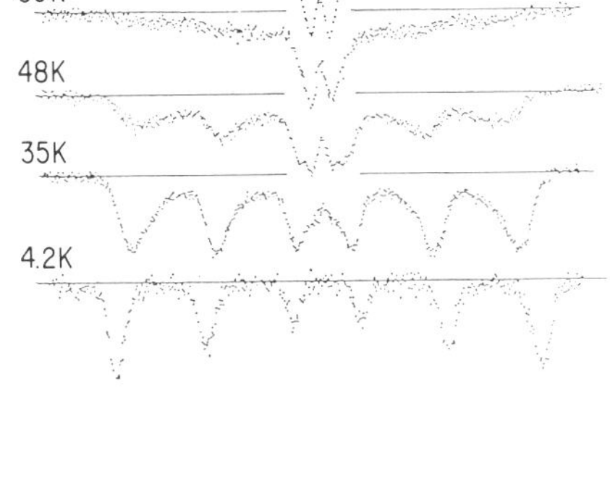

Fig. 81. Mössbauer spectra of polycrystalline $YFeMnO_4$ at low temperatures. (Tanaka et al. 1984b.)

a broad absorption besides the paramagnetic doublet, which develops with decreasing temperature and becomes a sextet below 55 K. The coexistence of the doublet and sextet continues to a much lower temperature. Figure 82 shows the temperature dependence of the intensity of the sextet, together with that of the peak intensity of the $(\frac{1}{3}, \frac{1}{3}, 0)$ neutron diffraction and the thermoremanent magnetization. The dependences are not identical.

At 4.2 K, the doublet disappears and the spectrum of $RFeM'O_4$ is composed of a symmetric sextet, each line of which is also symmetric. When $Fe^{[2+]}$ is substituted by non-magnetic ions such as Mg^{2+}, the situation is different (see fig. 83). The outer absorption lines have an asymmetric saw-tooth profile and are broader than the inner ones, in the spectrum of $LuFeMgO_4$ at 4.2 K. This is an indication of the distribution of the internal field, the effect of which is stronger for the outer lines than for the inner ones. The asymmetric distribution of the internal field can be accounted for by the thermal fluctuation of the Fe spins, if the exchange field acting on Fe ions is not identical but distributes because the ions interact with a different number of neighboring Fe atoms. Note that the thermal average of the electronic spin is a non-linear function of the exchange field. A similar saw-tooth line shape also appears at 35 K in the spectrum of $YFeMnO_4$ (see fig. 82). By lowering of the temperature, all Fe spins approach nearly the same saturation value despite the distributed exchange field, resulting in a symmetric line profile as is the case of $YFeMnO_4$ or $LuFeCoO_4$ at 4.2 K. The widths of the outer absorption lines in the spectrum of $YFeMnO_4$ at 4.2 K are still a little broader than that of the inner lines, still suggesting a distributed internal field.

The symmetry of the total spectrum gives information on the spin direction. When the Zeeman energy is dominant, the position of the second and the fifth lines relative to the first and the sixth lines are determined by the electric quadrupole coupling as

$$(v_1 - v_2) - (v_5 - v_6) = e^2qQ\,(3\cos^2\theta - 1)/2 \equiv 4E_Q\,, \tag{13}$$

where v_i is the resonance frequency of ith absorption measured by the relative velocity (in mm/s) of absorber to emitter, θ is the angle between the direction of

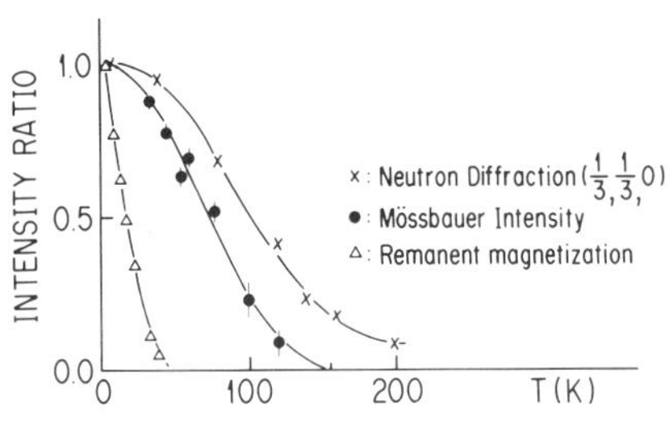

Fig. 82. The temperature dependence of the magnetic neutron diffraction peak at $(\frac{1}{3}, \frac{1}{3}, 0)$, the fraction of sextet intensity in Mössbauer absorption and thermoremanent magnetization, normalized by the value at 4.2 K. (Tanaka et al. 1984b.)

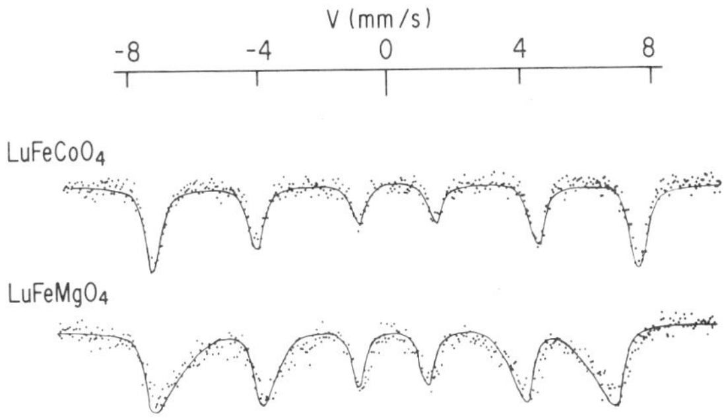

Fig. 83. Mössbauer spectra of polycrystalline $LuFeCoO_4$ and $LuFeMgO_4$ at 4.2 K. (Tanaka et al. 1984b.)

the internal field and the principal axis of the electric field gradient. Symmetric spectra of $YFeMnO_4$ and $LuFeCoO_4$ at 4.2 K indicate either $E_Q = 0$ or an averaged value of $\cos^2\theta$ equal to $\frac{1}{3}$. The former may not be the case, since paramagnetic spectra of these compounds do show quadrupole splitting. As for the latter, there are two possibilities: $\cos^2\theta$ is $\frac{1}{3}$ (or $\theta = 55°$) for all Fe spins, or the spin axis distributes to make the averaged value of $\cos^2\theta$ to be $\frac{1}{3}$. Tanaka et al. (1984b) succeeded in explaining the profile by the assumption that there is no special direction but that the Fe spin axis distributes homogeneously. Note that 55° from the *c*-axis has no significance in this crystal (cf. sect. 4.3). This conclusion is confirmed by an experiment on a single crystal of $LuFeCoO_4$ (Iwasaki 1988), with a little modification of the distribution. They argued that the distribution is not completely homogeneous but the distribution function has a broad maximum near $\theta = 30°$. This is in contrast to RFe_2O_4 or $RFeM'O_4$ (M′ means here a non-magnetic 2+ ion such as Mg or Zn) where the Fe spins lie along the *c*-axis.

Two conclusions can be drawn from above experiments.

(i) The easy axis of the Fe^{3+} ions in this crystal is the *c*-axis. When the Fe^{3+} compose a triangular lattice with other magnetic ions, however, spins cannot be parallel to the axis if the anisotropy of the other ions is not so strong.

(ii) The magnetic anisotopy of Fe^{2+} is large enough to align all spins parallel to the *c*-axis, whereas the Mn^{2+}, Co^{2+} or Cu^{2+} ions are not so anisotropic.

It is natural to consider that point (i) is due to the exchange energy, since the Fe^{3+} spins lie along the easy axis in the non-magnetic Mg-compound. As for point (ii), note that this is consistent with the atomic state described in sect. 4.4. The electronic state of Fe^{2+} is $(3d)^6$ and one electron is added to the S-state half shell of $(3d)^5$ Fe^{3+}. This electron will enter the ground orbital state of $l_z = \pm 1$, making the ion very anisotropic. The next electron in Co^{2+} will cancel the orbital moment and make the ion not so anisotropic in this crystal.

6.3. *Antiferromagnetism in $RFeM'O_4$*

As is described in this section so far, the magnetic properties of this series show several interesting characteristics. Although they have not been fully understood, they can be accounted for qualitatively by a two-dimensional coupling on a triangular lattice, a strong anisotropy of $Fe^{[2+]}$ ions and the coexistence of different kinds of ions in a crystallographically equivalent site.

Bragg lines parallel to the c^*-axis in the neutron diffraction and the paramagnetic-like doublet of isolated T layers in the Mössbauer spectra of $Lu_2Fe_3O_7$, at temperatures where spins on double T layers order, are a clear manifestation of a two-dimensionality of the spin correlation in these compounds. It is very likely that the rare earth elements in the U blocks couple weakly with transition metal ions on both sides of them and that the chain of the exchange interactions along the c-axis are cut at the U block.

A two-dimensional spin coupling is consistent with the anisotropy in the transport phenomena mentioned in the preceding section and can be understood as a consequence of the crystal structure.

6.3.1. *Antiferromagnetism on a triangular lattice*

The magnetic structure with a unit cell having three times larger edges in the c-plane than the atomic cell is a characteristic of an antiferromagnet on a triangular lattice.

Figure 84 shows a part of a double T layer with a "unit cell" and reciprocal vectors. There are two atoms in a unit cell. We will denote them as 1 and 2:

$$1:\quad \text{atoms at } (0,0) + l\boldsymbol{a} + m\boldsymbol{b}\,, \tag{14}$$

$$2:\quad \text{atoms at } (\tfrac{2}{3}, \tfrac{1}{3}) + l\boldsymbol{a} + m\boldsymbol{b}\,, \tag{15}$$

where l and m are integers. Each of the atoms, 1 and 2, compose a T layer with a different z-coordinate. The exchange energy is expressed as,

$$E = -\sum_{ij}\sum_{rr_0} J_{ij}(\boldsymbol{r})\, \boldsymbol{S}_i(\boldsymbol{r}_0)\, \boldsymbol{S}_j(\boldsymbol{r}_0 + \boldsymbol{r})\,, \tag{16}$$

where i and j denote the kind of atoms, 1 or 2 above, and $\boldsymbol{r}$ and $\boldsymbol{r}_0$ are cell positions. By the Fourier transformations

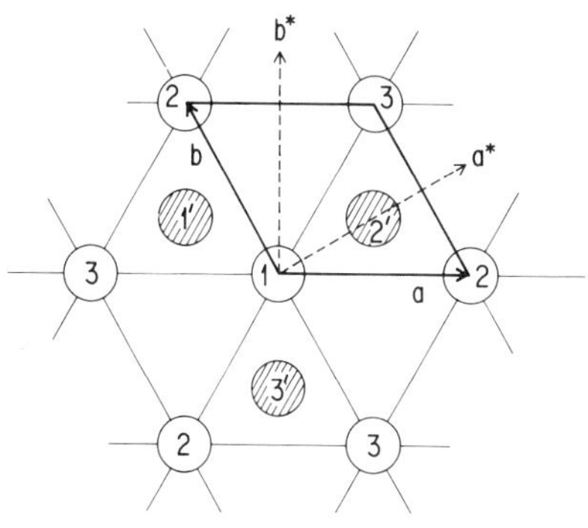

Fig. 84. The unit cell of a double triangular lattice, reciprocal vectors, $\boldsymbol{a}^*$ and $\boldsymbol{b}^*$, and three sublattices in a magnetic unit cell three times larger than the atomic cell.

$$S_i(\boldsymbol{q}) = \frac{1}{\sqrt{N}} \sum_{\boldsymbol{r}} S_i(\boldsymbol{r}) \exp[-\mathrm{i}\boldsymbol{q}\cdot\boldsymbol{r}] , \tag{17}$$

and

$$J_{ij}(\boldsymbol{q}) = \sum_{\boldsymbol{r}} J_{ij}(\boldsymbol{r}) \exp[-\mathrm{i}\boldsymbol{q}\cdot\boldsymbol{r}'] , \tag{18}$$

eq. (16) can be expressed as

$$E = -\sum_{ijq} J_{ij}(\boldsymbol{q})\, S_i(\boldsymbol{q})\, S_j(-\boldsymbol{q}) , \tag{19}$$

where a wave vector $\boldsymbol{q}$ is expressed as

$$\boldsymbol{q} = 2\pi(h\boldsymbol{a}^* + k\boldsymbol{b}^*) , \tag{20}$$

by the use of reciprocal vectors $\boldsymbol{a}^*$ and $\boldsymbol{b}^*$ (shown in fig. 84), and

$$\boldsymbol{r}' = \boldsymbol{r} + (1 - \delta_{ij})(\tfrac{2}{3}\boldsymbol{a} + \tfrac{1}{3}\boldsymbol{b}) ,$$
$$\boldsymbol{r} = l\boldsymbol{a} + m\boldsymbol{b} . \tag{21}$$

If there is no restriction on $S(\boldsymbol{q})$, an eigenstate $\{S_i(\boldsymbol{q})\}$ corresponding to the largest eigenvalue of the matrix,

$$J(\boldsymbol{q}) = \begin{pmatrix} J_{11}(\boldsymbol{q}) & J_{12}(\boldsymbol{q}) \\ J_{21}(\boldsymbol{q}) & J_{22}(\boldsymbol{q}) \end{pmatrix} , \tag{22}$$

will be realized.

If exchange coupling can be neglected except for nearest-neighbor spins in a T layer,

$$J_{11}(\boldsymbol{q}) = J_{22}(\boldsymbol{q}) = 2J\{\cos(2\pi h) + \cos(2\pi k) + \cos(2\pi(h+k))\}$$
$$\equiv 2Jy(h, k) , \tag{23}$$

and

$$J_{12} = J_{21} = 0 . \tag{24}$$

The eigenvalue of the matrix of eq. (22) takes a maximum at $h = k = \frac{1}{3}$, provided that J is positive, or the exchange interaction is antiferromagnetic. Bragg lines at $(\frac{1}{3}, \frac{1}{3}, l)$ prove that this is the case. Sublattices in the $(\frac{1}{3}, \frac{1}{3})$ structure are also shown in fig. 84. It is to be noted that, besides the repetition of the sublattices along the [1, 1] direction by every three atomic lines, a three-fold symmetry exists with a symmetry axis perpendicular to the plane of the cation sites.

Coupling between adjacent T layers modify eq. (24). If only the coupling between nearest-neighbor cations on an adjacent T layer, J', is important,

$$J_{12}(\boldsymbol{q}) = J^*_{21}(\boldsymbol{q})$$
$$= J'\{\exp[\tfrac{2}{3}\pi\mathrm{i}(2h+k)] + \exp[\tfrac{2}{3}\pi\mathrm{i}(-h-2k)] + \exp[\tfrac{2}{3}\pi\mathrm{i}(-h+k)]\} . \tag{25}$$

Eigenvalues of the matrix of eq. (22) are

$$E(\boldsymbol{q}) = Jy \pm J'\sqrt{3+y}\,. \tag{26}$$

If exchange coupling in a T layer, J, is ferromagnetic or $|J'|$, absolute value of exchange interaction between ions on adjacent T layers, is larger than $3J$, eq. (26) has a maximum at $h = k = 0$ and the magnetic unit cell will be the same as the atomic cell in which spins on each T layer align parallel. If $|J'|$ is not so large, however, the triple period magnetic structure is rather stable in a hexagonal lattice against interlayer interaction (Sakakibara 1984, Matsubara and Inawashiro 1987). At any rate, we can assume that the main exchange interaction is an antiferromagnetic coupling between the nearest-neighbor ions in a T layer as far as static spin structure is concerned. On the coupling of ions on an adjacent T layers, there are only a few direct experiments: on $LuFeMgO_4$ (Wiedenmann 1983) and $LuFe_2O_4$ (Iida 1988, Iida et al. 1988). At present, no quantitative estimate is available.

6.3.2. *Frustration and net moment induction*

Antiferromagnetism on a triangular net is one of the problems of recent interest in the physics of spin systems. The main point is the frustration, i.e., existence of infinitely degenerate ground states. A typical example is $CsCoCl_3$ (Mekata 1977, Mekata et al. 1986).

As is shown in the preceding subsection, the magnetic cell with triply larger edges in the c-plane than the atomic cell has the lowest energy as an antiferromagnet on a triangular lattice. If the direction of the spins is restricted to $\pm c$ by a large magnetic anisotropy, the lowest energy state at 0 K should be $[+,+,-]$ (or $[+,-,-]$). Here, $+$ or $-$ indicates the spin direction, along the c-axis, of the transition metal ions on the successive $(1,1)$ lines in a c-plane. (In the case of $[+,+,+]$ or $[-,-,-]$ structure, the magnetic cell is the same as the atomic cell. This is not the present problem.) In this structure, however, the molecular field on the $[+]$ ($[-]$) sublattice is zero if all the sublattice magnetization is the same, because the effect of the two other sublattices cancels each other. Then, there is no reason to have a $[+]$ ($[-]$) sublattice. Thus, the ground state cannot be unique but frustrates. This is the starting point of the problem of antiferromagnetism on a triangular lattice. At finite temperature, the sublattice magnetization need not be its full length and structures such as $[+,0,-]$, etc., are possible. In the case of $CsCoCl_3$, two successive phase transitions were reported: from a paramagnetic to a $[+,0,-]$ structure with a disordered sublattice $[0]$ and from a $[+,0,-]$ to a $[+,+,-]$ (or to a $[+,-,-]$) structure. The latter transition was explained by a weak ferromagnetic interaction between the next-nearest-neighbor cations (Mekata 1977). Note that the net moment is zero in the former, whereas it can be finite in the latter. The induction of the net magnetic moment by magnetic field cooling can be explained by such a ferrimagnetic structure.

In the case of $RFeM'O_4$ where M′ is Mn, Co or Cu, the spin direction is not restricted to $\pm c$, because the magnetic anisotropy of the constituent ions is not so large. If the magnetic anisotropy is absent, the spins rotate by 120° from one ion to the adjacent ion in the $[\frac{1}{3},\frac{1}{3}]$ scheme, composing a commensurate screw spin structure. Although the spin structure is determined to be unique in this case, there is a freedom in the spin direction: the energy of the spin system does not depend on the plane in which spins rotate as much as on the phase of the rotation, or the direction of the spin at the origin. In this sense, the system is also frustrated.

The magnetic anisotropy modifies the spin structure. If the easy axis of spins is the *c*-axis, planes of their rotation will contain the *c*-axis and the rotation angle will be modified a little to produce a net moment along the *c*-axis. On the latter, however, the anistropy energy is not so effective since it is averaged for the three sublattices of differently directed spins. If the anisotropy is treated as a perturbation to the exchange energy, the dependence of the total energy on the phase angle appears as the sixth order of the perturbation. The direction of the net moment can be either along the *c*-axis or within the *c*-plane, depending on the phase of rotation (see fig. 85). The magnitude of the induced moment will be dependent on the magnetic anisotropy, smaller for smaller anisotropy.

The magnetic measurement on a single crystal of $LuFeCoO_4$ (Iwasaki 1988) showed that the thermoremanent magnetization induced by a field cooling along the *c*-axis is about one order of magnitude smaller than that of $LuFe_2O_4$ in its saturated value, and that induced within the *c*-plane is one more order smaller. The former fact suggests that the anisotropy energy is about one order of magnitude smaller than the exchange energy. At the same time, the Mössbauer spectroscopy described in sect. 6.2.4.2 disclosed that the direction of the Fe^{3+} spins at 4.2 K is not limited to some directions but distributes in the solid angle, although the distribution is not rigorously homogeneous. This is an indication that the spin system is not thermally equilibrated at low temperatures.

6.3.3. *Response to disturbance and phase transformation in frustrated systems*

At low temperatures, frustrated spin systems are easily deformed by disturbances which split degenerate ground states by changing the magnetic interactions. If the disturbance has long-range character, deformation will be cooperative

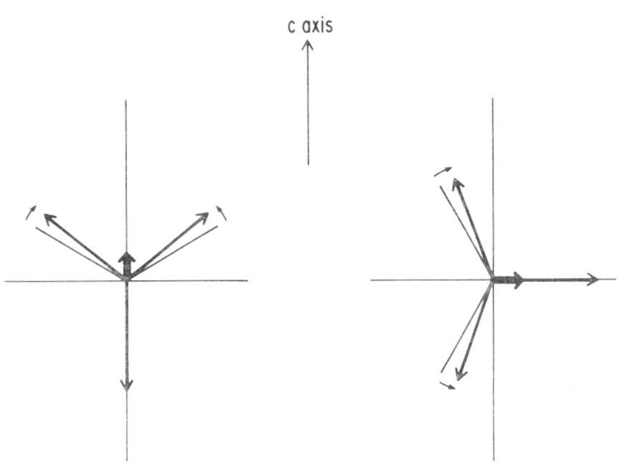

Fig. 85. The appearance of a small net moment by uniaxial anisotropy in a commensurate screw spin structure of pitch 3.

and there will be a phase change, which is of the first order in many cases. If the disturbance is short range and the relaxation time of the system is too long, a quenched glassy state will be realized. It seems that something is necessary to keep the system frustrated down to 0 K.

The case of stoichiometric YFe_2O_4 can be considered as an example of the long-range disturbance in the system. Note that the distribution of the same number of differently charged Fe ions on a triangular lattice can be compared with that of spins of a + and − *c*-direction, although a Coulomb interaction is of long range. This is the same situation pointed out by Anderson (1956) on the Verwey transition of Fe_3O_4. He made a point there that the lattice of the 16d site of the spinel structure forms a frustrated system if only the nearest-neighbor interaction is taken into account.

The three-sublattice ferrimagnetic structure, of $Fe^{[3+]}$ in $YbFe_2O_4$, mentioned in sect. 6.2.4.1 might be another example. If the three-fold symmetry of the triangular lattice is maintained, the easy axis of the magnetic anisotropy should be parallel to the symmetry axis and there should be only two, + and − spin, sublattices. In this case, the disturbance seems short range and the deformation is not macroscopic, since no deformation of the lattice has been observed in a diffraction experiment. Nevertheless, the disturbance may not be of atomic scale, or microscopic in that meaning, since well-defined three sublattices can be observed in neutron diffraction and in Mössbauer spectra. In the case of non-stoichiometric YFe_2O_4, absorption lines in the Mössbauer spectra are broad and fine structure as in Yb- or $LuFe_2O_4$ cannot be observed. There is no reason to consider that the situation is completely different for Y- and Yb- or Lu-compounds, but only the coherence length of the deformation of the triangular lattice supposed to be different. The coherence length might be longer in $LuFe_2O_4$ but shorter in non-stoichiometric YFe_2O_4.

In the case of $CsCoCl_3$, the strongest exchange coupling is antiferromagnetic along the *c*-axis. Components of the "two-dimensional" system are not individual spins but antiferromagnetic chains of spins parallel to the *c*-axis. The interactions between the elements of a two-dimensional system is averaged along the *c*-axis and the disturbances are cancelled to maintain the system frustrated down to the lower transition. At the same time, there is no possibility of the appearance of a ferromagnetic net moment. This might be the cause of the difference between $CsCoCl_3$ and RFe_2O_4.

It is natural to suppose that the origin of the disturbance in RFe_2O_4 is the distribution of two kinds of Fe ions in a T layer, not only in the case of a Verwey transition in the stoichiometric YFe_2O_4 but also in the spin structure of $YbFe_2O_4$. Indeed, dependence of the internal field at ^{57}Fe on the external field during measurement suggests that $\frac{2}{3}$ of $Fe^{[3+]}$ ions compose a minority spin sublattice whereas $\frac{1}{3}$ of $Fe^{[3+]}$ belong to a majority sublattice in the ferrimagnetic spin structure (cf. sect. 6.2.4.1). If this supposition is true, all the $Fe^{[2+]}$ ions should be located on the majority sublattice to produce a saturation moment of about $3\mu_B$ per formula unit: full coupling between charge and spin fluctuation.

6.3.4. *Relaxation time of the system and thermoremanent magnetization*

As already mentioned, there are several data indicating that spins of $RFeM'O_4$ are not in thermal equilibrium at low temperatures. The existence of Bragg lines down to the lowest temperature is one of them. If the system is thermally equilibrated, development of spin correlation in a *c*-plane increases the total coupling energy between two *c*-planes and it will overcome the thermal fluctuation at some finite temperature, resulting in a three-dimensional ordering. According to the experiment (Iida 1988), the half width of the $[\frac{1}{3}, \frac{1}{3}, l]$ Bragg line of $LuFe_2O_4$ remains constant below 200 K, at a little larger value than the nuclear peak. The development of the spin correlation within a *c*-plane is halted at 200 K and is maintained constant below this temperature. It is interesting to note that another example of the appearance of Bragg lines was reported on $Rb_2(Co, Mg)F_4$ by Ikeda et al. (1978) only when the sample was quenched through the Néel point.

Neutron diffraction gives information on the correlation of different spins at one time. If the system is in thermal equilibrium, this is equivalent to the correlation of one spin at different time. A long-range spacial correlation corresponds to a large autocorrelation. Mössbauer spectroscopy offers data on the latter through the internal field at the nuclei. In the case of ^{57}Fe, autocorrelation in the region of 10–100 ns can be measured precisely by this method, as is shown in the last section. In a low-dimensional spin system as in the present case, development of the long-range correlation is suppressed compared with the development of the short-range correlation because of the characteristic of the lattice. It is usual in such a case that the increase of the neutron diffraction intensity starts at a higher temperature than the increase of the internal field measured by Mössbauer spectroscopy. On the contrary, autocorrelation of a spin develops without spacial correlation at low temperatures where the relaxation time becomes too long and the system is not in thermal equilibrium.

Application of an external magnetic field changes the state of the spin system, a magnetic moment is induced, provided that the duration of the field is longer than the relaxation time of the system. If the relaxation time is increased to, say, 10^4 s or more keeping the system unchanged, the induced moment will remain unchanged even if the external field is reduced to zero. Since, ordinarily, the relaxation time increases with decreasing temperature, which is realized by cooling of the specimen: thermoremanent magnetization appears. Thus, the temperature dependence of the induction of thermoremanent magnetization is a measure of the temperature dependence of the relaxation time of the spins.

The experimental results are shown in fig. 67. Apparently, the induced thermoremanent magnetization is reduced part by part in the sample, by increasing temperature. It means that the relaxation time of the spins changes part by part. Additional information on the distribution is given by fig. 66, the effect of the cooling field on the temperature dependence of the induced thermoremanent magnetization. Compare, for example, the cases of $H_{cool} = 1$ kOe and $H_{cool} =$ 10 kOe. A higher cooling field induces more thermoremanent magnetization, but the difference disappears at a certain temperature below the Néel point. The temperature dependence of the thermoremanent magnetization induced by low field is proportional to the temperature dependence of saturation magnetization, persisting up to T_N.

These facts can be explained by an assumption that the specimen is divided into rather small domains, spins in one domain order coherently, with an appreciable distribution of size. Note that the domain wall energy will be small because of the frustration of the triangular lattice (see, e.g., Fazekas and Anderson 1974). Each domain has its moment because of the ferrimagnetic spin structure. It is natural to consider that a larger domain is harder to reverse, i.e., the relaxation time of the net moment of a domain is determined by the volume of the domain and is long when the domain is large.* Since the Zeeman energy is larger for a larger domain, the net moment induced by smaller external field should be due to the alignment of those of larger domains which is fixed at higher temperature. This is a natural explanation of fig. 66. At the same time, the temperature dependence of the thermoremanent magnetization induced by small H_{cool} will be similar to that of saturation magnetization since that is due to the large domain.

The difference of the thermoremanent magnetization (fig. 65) or the coercive field (fig. 64), due to the rare earth element, is attributed to the difference in the size of the domain or the coherence length of the spin ordering. That is consistent with a narrower Mössbauer absorption in $LuFe_2O_4$ than in non-stoichiometric YFe_2O_4 at low temperatures. This dependence on the rare earths should be correlated to the difference of electronic properties described in the last section.

Larger anisotropy will make spin flipping harder. Because of its strong anisotropy, the thermoremanent magnetization in RFe_2O_4 will appear from higher temperature than in $RFeM'O_4$ or single T layers in $Lu_2Fe_3O_7$ that are composed of only Fe^{3+} ions, even if there were no difference in the exchange interaction. This is consistent with the experimental observation described so far.

Acknowledgements

The present authors would like to express their sincere thanks to colleagues for their cooperation in the experiment, stimulating discussions and information of their data prior to publication. They include Dr. K. Kitayama, Dr. T. Sugihara and Dr. T. Sekine and Mr. T. Mohri in the synthesis and the phase diagram research of the compounds, Dr. S. Kimura, Dr. I. Shindo, Mr. S. Takekawa and Dr. J. Iida in the single crystal growth, Dr. K. Kato, the late Dr. I. Kawada and Dr. Y. Matsui for the structural analysis by X-ray or electron diffraction, Prof. Y. Nakagawa, Prof. M. Tanaka, Prof. J. Akimitsu, Prof. N. Tsuda, Dr. S.

*It is to be noted that the relaxation time considered here is not of single spin but of domain as a whole. Reversible temperature dependence of thermoremanent magnetization of $LuFe_2O_4$ or $Lu_2Fe_3O_7$ (figs. 67 and 69) in a cooling and a subsequent heating process suggests that spins in one "frozen" domain are in thermal equilibrium under a condition that the net moment is fixed.

Funahashi, Dr. T. Sugihara, Dr. J. Iida and Prof. N. Môri for the study of the physical properties. Organization of the study should be much appreciated in the chemistry side by Prof. Emeritus T. Katsura of the Tokyo Institute of Technology and in the physics side by Dr. N. Tsuda when he was at the National Institute for Research in Inorganic Materials. The authors also express their thanks to Prof. M. Mekata for critical reading of the manuscript (sect. 6) and to Prof. O. Evrard for sending copies of papers published by his group.

References

Akimitsu, J., Y. Inada, K. Siratori, I. Shindo and N. Kimizuka, 1979, Solid State Commun. **32**, 1065.

Anderson, P.W., 1956, Phys. Rev. **102**, 1008.

Austin, I.G., and N.F. Mott, 1969, Adv. Phys. **18**, 41.

Bahgat, A.A., 1984, private communication.

Bartholomew, R.F., and D.F. Frankl, 1969, Phys. Rev. **187**, 828.

Baynes, P., B. Viswanathan, A. Narayanasamy, N. Kimizuka, N.K. Appandairajan and T. Nagarajan, 1978, Proc. Nucl. Phys. & Solid State Phys. Symp. (India) C **21**, 592.

Bergerhoff, V.G., and H. Kasper, 1968, Acta Crystallogr. B **24**, 388.

Bertaut, F., and J. Mareschal, 1963, C.R. Acad. Sci. Paris **257**, 867.

Bertaut, F., and M. Mercier, 1963, Phys. Lett. **5**, 27.

Bertaut, F., F. Forrat and P. Fang, 1963a, C.R. Acad. Sci. Paris **256**, 1958.

Bertaut, F., R. Pauthenet and M. Mercier, 1963b, Phys. Lett. **7**, 110.

Bevan, D.J.M., and E. Summerville, 1979, Mixed Rare Earth Oxides, in: Handbook on the Physics and Chemistry of Rare Earths, Vol. 3, eds K.A. Gschneidner Jr and L. Eyring (North-Holland, Amsterdam) p. 401.

Blasse, G., G.J. Dirksen, N. Kimizuka and T. Mohri, 1986, Mater. Res. Bull. **21**, 1057.

Blume, M., 1968, Phys. Rev. **174**, 351.

Brauer, G., 1966, Structural and Solid State Chemistry of Rare Earth Oxides, in: Progress in the Science and Technology of the Rare Earths, Vol. 2, ed. L. Eyring (Pergamon Press, Oxford) p. 312.

Cannard, P.J., and R.J.D. Tilley, 1988, J. Solid State Chem. **73**, 418.

Cutler, M., and N.F. Mott, 1969, Phys. Rev. **181**, 1336.

Darken, L.S., and R.W. Gurry, 1945, J. Am. Chem. Soc. **67**, 1398.

Darken, L.S., and R.W. Gurry, 1946, J. Am. Chem. Soc. **68**, 798.

Drofenik, M., D. Kolar and L. Golic, 1974, J. Less-Common Met. **37**, 281.

Dzyaloshinski, I., 1958, J. Phys. & Chem. Solids **4**, 241.

Enomura, A., S.Asai, Y. Ishiwata, T. Inabe, Y. Sakai, N. Tsuda, M. Tanaka and K. Siratori, 1983, J. Phys. Soc. Jpn. **52**, 4286.

Evrard, O., B. Malaman, F. Jeannot, N. Tannieres and J. Aubry, 1974a, C.R. Acad. Sci. Paris, Ser. C **278**, 413.

Evrard, O., B. Malaman, N. Tannieres, F. Jeannot and J. Aubry, 1974b, C.R. Acad. Sci. Paris, Ser. C **279**, 1021.

Eyring, L., 1979, The Binary Rare Earth Oxides, in: Handbook on the Physics and Chemistry of the Rare Earths, Vol. 3, eds K.A. Gschneidner Jr and L. Eyring (North-Holland, Amsterdam) p. 337.

Fazekas, P., and P.W. Anderson, 1974, Philos. Mag. **20**, 423.

Funahashi, S., J. Akimitsu, K. Siratori, N. Kimizuka, M. Tanaka and H. Fujishita, 1984, J. Phys. Soc. Jpn. **53**, 2688.

Funahashi, S., Y. Morii and H.R. Child, 1987, J. Appl. Phys. **61**, 4114.

Gérardin, R., A. Alebouyeh, F. Jeannot, A. Courtois, B. Malaman and O. Evrard, 1980, Mater. Res. Bull. **15**, 647.

Gérardin, R., A. Alebouyeh, B. Malaman and O. Evrard, 1981, Rev. Chim. Miner. **18**, 625.

Gérardin, R., I. Nodari, H. Aqachmar and O. Evrard, 1982, C.R. Acad. Sci. Paris, Ser. II **295**, 863.

Goldschmidt, V.M., 1926, Skrifter Norske Videnskaps-Akad., Oslo, I. Mat.-Naturv. KI., No. 8.

Hägg, G., 1943, Arch. Kemi. Miner. Geol. B **16**, 1.

Holley Jr, C.E., E.J. Huber Jr and F.B. Baker, 1968, The Enthalpies, Entropies and Gibbs energies of Formation of the Rare Earth Oxides, in: Progress in the Science and Technology of the Rare Earths, Vol. 3, ed. L. Eyring (Pergamon Press, Oxford) p. 343.

Iida, J., 1988, Ph.D. Thesis (Department of Physics, Tohoku University).

Iida, J., Y. Nakagawa and N. Kimizuka, 1986, J. Phys. Soc. Jpn. **55**, 1434.

Iida, J., Y. Nakagawa, S. Takekawa and N. Kimizuka, 1987, J. Phys. Soc. Jpn. **56**, 3746.

Iida, J., Y. Nakagawa, S. Funahashi, S. Takekawa and N. Kimizuka, 1988, J. Phys. Colloq. (France) **49**, C8-1497.

Iida, J., S. Kakugawa, G. Kido, Y. Nakagawa, S. Takekawa and N. Kimizuka, 1989, Physica B **155**, 307.

Ikeda, H., M.T. Hutchings and M. Suzuki, 1978, J. Phys. & Chem. Solids **11**, L359.

Iwasaki, H., 1988, Thesis for Master's Degree (Department of Physics, Ochanomizu University).

Kasper, H., 1967, Z. Anorg. Allg. Chem. **349**, 113.

Kato, K., I. Kawada, N. Kimizuka and T. Katsura, 1975, Z. Kristallogr. **141**, 314.

Kato, K., I. Kawada, N. Kimizuka, I. Shindo and T. Katsura, 1976, Z. Kristallogr. **143**, 278.

Katsura, T., K. Kitayama, T. Sugihara and N. Kimizuka, 1975, Bull. Chem. Soc. Jpn. **48**, 1809.

Katsura, T., T. Sekine, K. Kitayama, T. Sugihara and N. Kimizuka, 1978, J. Solid State Chem. **23**, 43.

Khattak, C.P., and E.F.Y. Wang, 1979, Perovskites and Garnets, in: Handbook on the Physics and Chemistry of Rare Earths, Vol. 3, eds K.A. Gschneidner Jr and L. Eyring (North-Holland, Amsterdam) p. 525.

Kimizuka, N., and T. Katsura, 1974, Bull. Chem. Soc. Jpn. **47**, 1801.

Kimizuka, N., and T. Katsura, 1975a, J. Solid State Chem. **13**, 176.

Kimizuka, N., and T. Katsura, 1975b, J. Solid State Chem. **15**, 151.

Kimizuka, N., and T. Mohri, 1985, J. Solid State Chem. **60**, 382.

Kimizuka, N., and T. Mohri, 1989, J. Solid State Chem. **78**, 98.

Kimizuka, N., and E. Takayama, 1981, J. Solid State Chem. **40**, 109.

Kimizuka, N., and E. Takayama, 1982a, J. Solid State Chem. **41**, 166.

Kimizuka, N., and E. Takayama, 1982b, J. Solid State Chem. **42**, 22.

Kimizuka, N., and E. Takayama, 1982c, J. Solid State Chem. **43**, 278.

Kimizuka, N., and E. Takayama, 1984, J. Solid State Chem. **53**, 217.

Kimizuka, N., A. Takenaka, Y. Sasada and T. Katsura, 1974a, Solid State Commun. **15**, 1199.

Kimizuka, N., A. Takenaka, Y. Sasada and T. Katsura, 1974b, Solid State Commun. **15**, 1321.

Kimizuka, N., K. Kato, I. Shindo, I. Kawada and T. Katsura, 1976, Acta Crystallogr. B **32**, 1620.

Kimizuka, N., A. Yamamoto, H. Ohasi, T. Sugihara and T. Sekine, 1983, J. Solid State Chem. **49**, 65.

Kimizuka, N., T. Mohri, Y. Matsui and K. Siratori, 1988, J. Solid State Chem. **74**, 98.

Kishi, M., S. Miura, Y. Nakagawa, N. Kimizuka, I. Shindo and K. Siratori, 1982, J. Phys. Soc. Jpn. **51**, 2801.

Kishi, M., Y. Nakagawa, M. Tanaka, N. Kimizuka and I. Shindo, 1983, J. Magn. & Magn. Mater. **31–34**, 807.

Kitayama, K., and T. Katsura, 1976, Bull. Chem. Soc. Jpn. **49**, 998.

Kitayama, K., K. Nojiri, T. Sugihara and T. Katsura, 1985, J. Solid State Chem. **56**, 1.

Kohn, K., and A. Tasaki, 1965, J. Phys. Soc. Jpn. **20**, 1273.

Kuipers, A.J., and V.A.M. Brabers, 1976, Phys. Rev. B **14**, 1401.

Malaman, B., O. Evrard, N. Tannieres, J. Aubry, A. Courtois and J. Protas, 1975, Acta Crystallogr. B **31**, 1310.

Malaman, B., O. Evrard, N. Tannieres, A. Courtois and J. Protas, 1976, Acta Crystallogr. B **32**, 749.

Marezio, M., J.P. Remeika and P.D. Dernier, 1966, Mater. Res. Bull. **1**, 247.

Marezio, M., J.P. Remeika and P.D. Dernier, 1970, Acta Crystallogr. B **26**, 2008.

Matsubara, F., and S. Inawashiro, 1987, J. Phys. Soc. Jpn. **56**, 4087.

Matsui, M., S. Todo and S. Chikazumi, 1977, J. Phys. Soc. Jpn. **42**, 1517.

Matsui, Y., 1980, J. Appl. Crystallogr. **13**, 395.
Matsui, Y., 1988, unpublished data.
Matsui, Y., K. Kato, N. Kimizuka and S. Horiuchi, 1979, Acta Crystallogr. B **35**, 561.
McCarthy, G.J., and R.D. Fischer, 1972, J. Solid State Chem. **4**, 340.
Mekata, M., 1977, J. Phys. Soc. Jpn. **42**, 76.
Mekata, M., Y. Ajiro and K. Adachi, 1986, J. Magn. & Magn. Mater. **54–57**, 1267.
Mizusaki, J., T. Sasamoto, W.R. Cannon and H.K. Bowen, 1982, J. Am. Ceram. Soc. **65**, 363.
Moriya, T., 1963, Weak Ferromagnetism, in: Magnetism I, eds G.T. Rado and H. Suhl (Academic Press, New York) ch. 3, p. 85.
Moruzzi, V.L., and M.W. Shafer, 1960, J. Am. Ceram. Soc. **43**, 367.
Mott, N.F., and E.A. Davis, 1979, Electronic Processes in Non-Crystalline Materials, 2nd Ed. (Clarendon Press, Oxford).
Muan, A., and E.F. Osborn, 1965, Phase Equilibria among Oxides in Steelmaking (Addison-Wesley, Reading, MA) p. 1.
Nafziger, R.H., G.C. Ulmer and E. Woermann, 1971, Gaseous Buffering for the Control of Oxygen Fugacity at one Atmosphere, in: Research Techniques for High Pressure and High Temperature, ed. G.C. Ulmer (Springer, New York) p. 9.
Nakagawa, Y., M. Inazumi, N. Kimizuka and K. Siratori, 1979, J. Phys. Soc. Jpn. **47**, 1369.
Nakagawa, Y., M. Kishi, H. Hiroyoshi, N. Kimizuka and K. Siratori, 1980, Proc. Int. Conf. Ferrites, Kyoto, eds H. Watanabe, S. Iida and M. Sugimoto (Center for Academic Publications Japan, Tokyo) p. 115.
Nakamura, T., 1981, J. Solid State Chem. **38**, 229.
Nakamura, T., G. Petzow and L.J. Gonckler, 1979, Mater. Res. Bull. **14**, 649.
Narayanasamy, A., T. Nagarajan, P. Muthukumarasamy and N. Kimizuka, 1981, Phys. Status Solidi A **66**, 377.
Onsager, L., 1944, Phys. Rev. **65**, 117.
Patterson, A.L., and J.S. Kasper, 1967, Close Packing, in: International Tables for X-ray Crystallography, Vol. 2, eds J.S. Kasper and L. Lonsdale, 2nd Ed. (Kynoch, Birmingham, England) p. 342.
Pauling, L., 1960, The Nature of the Chemical Bond, 3rd Ed. (Cornell University Press, Ithaka, New York) p. 404.
Piekarczyk, W., W. Weppner and A. Rabenau, 1978, Mater. Res. Bull. **13**, 1077.
Roth, R.S., 1964, Phase Equilibria Studies in Mixed Systems of Rare Earth and other Oxides, in: Progress in the Science and Technology of the Rare Earth Oxides, Vol. 1, ed. L. Eyring (Pergamon Press, Oxford) p. 167.
Sakai, Y., K. Kaneda, N. Tsuda, M. Tanaka and K. Siratori, 1985, J. Phys. Soc. Jpn. **54**, 4089.
Sakai, Y., K. Kaneda, N. Tsuda and M. Tanaka, 1986, J. Phys. Soc. Jpn. **55**, 3181.
Sakakibara, T., 1984, J. Phys. Soc. Jpn. **53**, 3607.
Sato, M., 1971, Electrochemical Measurements and Control of Oxygen Fugacity and other Gaseous Fugacities with Solid Electrolyte Systems, in: Research Techniques for High Pressure and High Temperature, ed. G.C. Ulmer (Springer, New York) p. 43.
Schlenker, C., and M. Marezio, 1980, Philos Mag. B **42**, 453.
Schmitz-DuMont, O., and H. Kasper, 1965, Z. Anorg. Allg. Chem. **341**, 252.
Schneider, S.J., and R.S. Roth, 1960, J. Res. N.B.S. A **64**, 317.
Schneider, S.J., R.S. Roth and J.L. Waring, 1961, J. Res. N.B.S. A **65**, 345.
Scmaltzried, H., and A. Navrotsky, 1975, Festkörper Thermodynamik (Verlag Chemie, Weinheim) p. 119.
Sekine, T., and T. Katsura, 1976, J. Solid State Chem. **17**, 49.
Shannon, R.D., 1976, Acta Crystallogr. A **32**, 751.
Shannon, R.D., and C.T. Prewitt, 1969, Acta Crystallogr. B **25**, 925.
Shindo, I., N. Kimizuka and S. Kimura, 1976, Mater. Res. Bull. **11**, 637.
Siratori, K., S. Miura, S. Funahashi, J. Akimitsu, N. Kimizuka and S. Takekawa, 1983, J. Magn. & Magn. Mater. **31–34**, 799.
Siratori, K., N. Môri, H. Takahashi, G. Oomi, J. Iida, M. Tanaka, M, Kishi, Y. Nakagawa and N. Kimizuka, 1990, J. Phys. Soc. Jpn. **59**, 631.
Smiltens, J., 1957, J. Am. Chem. Soc. **79**, 4877.
Sreedharan, O.M., and M.S. Chandrasekharaiah, 1986, J. Mater. Sci. **21**, 2581.
Sugihara, T., 1978, Ph.D. Thesis (Department of Chemistry, Tokyo Institute of Technology).
Sugihara, T., N. Kimizuka and T. Katsura, 1975, Bull. Chem. Soc. Jpn. **48**, 1806.
Sugihara, T., M. Abe and T. Katsura, 1976, J. Phys. Soc. Jpn. **40**, 1211.

Sugihara, T., K. Siratori, I. Shindo and T. Katsura, 1978, J. Phys. Soc. Jpn. **45**, 1191.

Sugihara, T., K. Siratori, N. Kimizuka, J. Iida, H. Hiroyoshi and Y. Nakagawa, 1985, J. Phys. Soc. Jpn. **54**, 1139.

Takayama, E., and N. Kimizuka, 1979, J. Electrochem. Soc. **126**, 2012.

Tanaka, M., M. Kato, N. Kimizuka and K. Siratori, 1979, J. Phys. Soc. Jpn. **47**, 1737.

Tanaka, M., J. Akimitsu, I. Shindo, N. Kimizuka and K. Siratori, 1980, Proc. Int. Conf. Ferrites, Kyoto, eds H. Watanabe, S. Iida and M. Sugimoto (Center for Academic Publications Japan, Tokyo) p. 119.

Tanaka, M., J. Akimitsu, Y. Inada, N. Kimizuka, I. Shindo and K. Siratori, 1982, Solid State Commun. **44**, 687.

Tanaka, M., N. Kimizuka, J. Akimitsu, S. Funahashi and K. Siratori, 1983, J. Magn. & Magn. Mater. **31–34**, 769.

Tanaka, M., K. Siratori and N. Kimizuka, 1984a, J. Phys. Soc. Jpn. **53**, 760.

Tanaka, M., K. Siratori and N. Kimizuka, 1984b, J. Phys. Soc. Jpn. **53**, 4113.

Tanaka, M., H. Iwasaki and K. Siratori, 1988, Hyperfine Interactions **41**, 455.

Tanaka, M., H. Iwasaki, K. Siratori and I. Shindo, 1989, J. Phys. Soc. Jpn. **58**, 1433.

Tjon, J.A., and M. Blume, 1968, Phys. Rev. **165**, 456.

Tretyakov, Yu.D., V.V. Sorokin, A.R. Kaul and A.P. Erastova, 1976a, J. Solid State Chem. **18**, 253.

Tretyakov, Yu.D., V.V. Sorokin and A.P. Erastova, 1976b, J. Solid State Chem. **18**, 263.

Tretyakov, Yu.D., V.V. Sorokin and A.P. Erastova, 1976c, J. Solid State Chem. **18**, 271.

Tretyakov, Yu.D., A.R. Kaul and V.K. Portnoy, 1977, High Temp. Science **9**, 61.

Van Hook, J.H., 1961, J. Am. Ceram. Soc. **44**, 208.

Verwey, E.J., and P.W. Haayman, 1941, Physica **8**, 979.

Warshaw, I., and R. Roy, 1959, J. Am. Ceram. Soc. **42**, 434.

Wells, A.F., 1975a, Structural Inorganic Chemistry, 4th Ed. (Clarendon Press, Oxford) p. 132.

Wells, A.F., 1975b, Structural Inorganic Chemistry, 4th Ed. (Clarendon Press, Oxford) p. 490.

Went, J.J., G.W. Rathenau, E.W. Gorter and G.W. Oosterhout, 1952, Philips Tech. Rev. **13**, 194.

Wiedenmann, A., W. Gunsser, J. Rossat-Mignod and M.O. Evrard, 1983, J. Magn. & Magn. Mater. **31–34**, 1442.

Wood, J.S., 1972, Stereochemical and Electronic Structural Aspect of Five-coordination, in: Progress in Inorganic Chemistry, Vol. 16, ed. S.J. Lippard (Wiley, New York) p. 227.

Wyckoff, R.W.G., 1964, Crystal Structures, Vol. 1, 2nd Ed. (Interscience, New York) p. 111.

Yakel, H.L., W.C. Koehler, E. Bertaut and E.F. Forrat, 1963, Acta Crystallogr. **16**, 957.

Zhdanov, G.S., 1945, C.R. Acad. Sci. USSR **48**, 39.

Zoumalan, S., O. Evrard, N. Tannieres, F. Jeannot and J. Aubry, 1975, Rev. Chim. Miner. **12**, 577.

Handbook on the Physics and Chemistry of Rare Earths, Vol. 13
edited by K.A. Gschneidner Jr. and L. Eyring

Chapter 91

ELEMENTAL ANALYSIS BY ATOMIC EMISSION AND MASS SPECTROMETRY WITH INDUCTIVELY COUPLED PLASMAS

R.S. HOUK
Ames Laboratory-US Department of Energy, Department of Chemistry, Iowa State University, Ames, IO 50011, USA

Contents

1. Introduction

In recent years, atomic spectroscopy has become by far the most common technique for trace elemental determinations. In the last 15 years, atomic spectroscopy itself has undergone remarkable changes brought about primarily by the inductively coupled plasma (ICP) as an excitation source for atomic emission spectrometry (AES). Atomic absorption spectrometry remains a viable technique but, as usually practiced, is limited to determination of one element at a time.

Now, many elements are determined routinely at the same time by ICP–AES, as evidenced by the fact that several thousand of these instruments are in routine use throughout the world. In 1980, just when ICP–AES seemed secure in its niche as the all-around champion for elemental analysis, the ICP was shown to be an excellent ionization source for mass spectrometry (MS) (Houk et al. 1980). In just a few years, ICP–MS has already exerted a considerable impact on analytical spectroscopy, particularly for determination of rare earth elements, for which it is clearly superior to any other technique.

This chapter describes the instrumentation and basic analytical capabilities of ICP–AES and ICP–MS. Some illustrative examples of analyses involving rare earths as either analyte or matrix are also described. The chosen cases reflect some of the major areas of interest but are not meant to represent all the work in the literature. Complete reviews of publications in ICP–AES are provided, in even years, in the Emission Spectrometry section of the review issue of *Analytical Chemistry* (Keliher et al. 1986). Applications for analysis of specific sample types, e.g., waters, metals, etc., are also reviewed therein in odd years. Analytical uses of ICPs are surveyed regularly in the *ICP Information Newsletter*, the *Journal of Analytical Atomic Spectrometry*, and the *Journal of Analytical Spectroscopy*. ICP–AES is also described in detail in three recent books (Boumans 1987, Montaser and Golightly 1987, Thompson and Walsh 1983), and a fourth monograph deals with applications of ICP–MS (Date and Gray 1989). Numerous reviews on ICPs (Fassel 1978), ICP–AES (Barnes 1978), and ICP–MS (Douglas and Houk 1985, Houk 1986, Gray 1985, Houk and Thompson 1988) are also available.

2. Properties of ICPs

2.1. *Sample introduction*

The ICP serves as an energy source to atomize, ionize and excite atoms from the sample. The physical nature of the sample determines the means used to introduce it into the ICP. Solutions are by far the most common form, and these are generally nebulized and injected as aerosols in a flowing stream of argon. The pneumatic nebulizer shown in fig. 1 is a common way to produce aerosols from solutions. The spray chamber removes many of the larger droplets, which would dry into large solid particles that would not be vaporized and atomized properly. A nebulizer such as that depicted in fig. 1 typically consumes sample at a rate of 1–3 ml min^{-1}. One to three percent of the consumed sample is actually injected into the plasma. Samples can be changed in one or two minutes.

Dissolution of many rare earth compounds and alloys and many geological matrices containing rare earths is difficult or time consuming. Although dissolution procedures have been described for most matrices, several alternate ways to introduce samples into an ICP without prior dissolution are also of interest here. For example, the sample can be ground into a fine powder (diam. 1 μm or less),

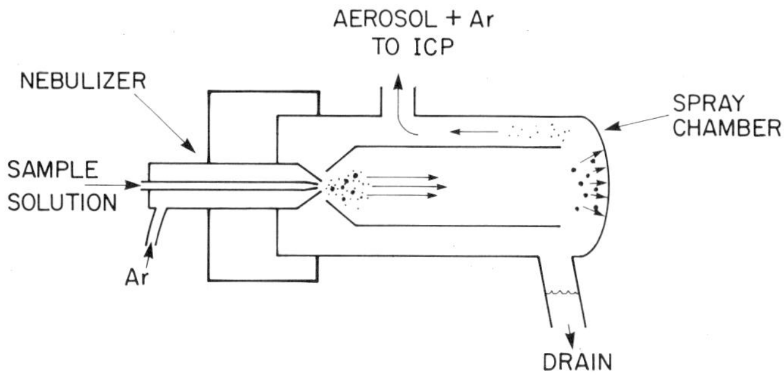

Fig. 1. Concentric pneumatic nebulizer and spray chamber for introducing solution into an ICP.

suspended as an aqueous slurry, and nebulized with a special nebulizer that resists plugging. Solids can be inserted directly into the axial channel in a graphite cup or tungsten wire loop, or they can be dispersed into mobile particulates (e.g., by an arc, spark, laser, or electrothermal device) and the particulates swept into the ICP. Despite the variety of sample-introduction techniques that have been devised, it is generally agreed that considerable room still exists for improvement in this area, particularly in terms of efficiency and precision (Browner and Boorn 1984).

2.2. *Plasma configuration*

There are several varieties of ICPs. The description here is limited to the most common ones but still serves to illustrate the general principles behind all the different types.

The ICP is generated and sustained in a plasma torch similar to the one shown in fig. 2. The torch is an assembly of concentric quartz tubes that define the gas flow patterns in the plasma. Argon is by far the most commonly used gas, and it is supplied to the plasma in several separate flows. The outer gas (usually 8–20 $\ell\ \mathrm{min}^{-1}$) is injected tangentially and spirals from the annular space between the outer and middle tubes. This gas flows in a vortex-like pattern into the torch region enclosed by the load coil. Electrons in the plasma interact with time-varying fields induced by the RF current flowing through the load coil. For example, an axially oriented magnetic field induced by the current flow causes the electrons to flow in closed, circular paths inside the plasma. As the plasma is at atmospheric pressure, the mean free path is very short ($\sim 10^{-3}$ cm) and the charged particles collide frequently with neutral gas particles. These collisions sustain the ionization and heat the gas, causing formation of an extended, stable plasma. Naturally, the inductive coupling process requires efficient impedance matching between the RF generator and the plasma. Typically, the power is 1–2 kW at a frequency of 27 MHz. It has been estimated that approximately 75% of this power actually is coupled into the plasma (de Galan and van der Plas 1987). Because of the skin depth effect, nearly all the power is coupled into the

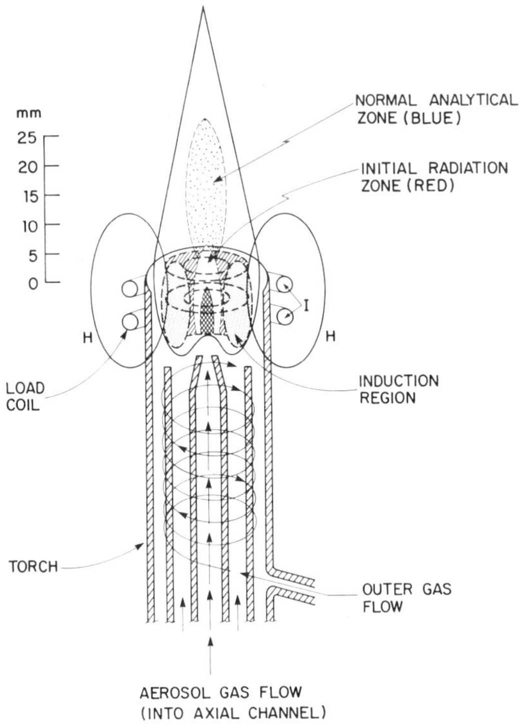

Fig. 2. Schematic diagram of ICP, torch, gas flows, and induced magnetic field. The shaded zones are observed when a nebulized sample containing Y is introduced via the aerosol gas flow.

outer region of the plasma where the argon flow is tangential, as described above. This zone is often called the induction region for this reason.

The argon flow containing the sample (e.g., from the nebulizer) travels through the inner tube into the center of the plasma. This region is often called the axial channel, and is enclosed and heated by the induction region. Most of the power is coupled into the plasma in a region physically distinct from that where the sample is located. Thus, the composition of the sample can change over a substantial range without perturbing the basic electrical processes that sustain the plasma. The confinement of the analyte in the axial channel facilitates collection of photons or ions. Furthermore, the ICP is not contaminated by electrode material, like arcs and sparks, because there are no electrodes and the load coil is not in physical contact with the plasma. Viewed from the side, as shown in fig. 2, the ICP looks like a wide flame with a bright white outer section split by the relatively dark axial channel. Viewed from the open end of the torch, the ICP looks like a luminous doughnut, with the bright induction region enclosing the darker axial channel.

2.3. *Conditions in the axial channel*

Gas kinetic temperatures in the axial channel reach approximately 5000 K (Barnes 1978, Hasegawa and Haraguchi 1987). Although not as hot as the induction region, the axial channel is still hot enough to atomize and ionize injected sample constituents. Furthermore, the analyte particles reside there for a fairly long time (~4 ms), which allows adequate transfer of energy to the analytes to convert them into free atoms and ions. The fate of analyte aerosol particles in the ICP can be visualized through the following sequence of reactions for the hypothetical salt MX dissolved in small, aqueous aerosol droplets:

desolvation

$$\mathrm{MX(aq, aerosol)} \rightarrow \mathrm{MX(s, aerosol)}\ ; \tag{1}$$

vaporization of dry particles

$$\mathrm{MS(s, aerosol)} \rightarrow \mathrm{MX(g)}\ ; \tag{2}$$

atomization

$$\mathrm{MX(g)} \rightarrow \mathrm{M(g)} + \mathrm{X(g)}\ ; \tag{3}$$

ionization

$$\mathrm{M} \rightarrow \mathrm{M}^{+} + \mathrm{e}\,, \qquad K_{\mathrm{ion}} = (n_{\mathrm{M}^{+}}/n_{\mathrm{m}})n_{\mathrm{e}}\ ; \tag{4}$$

excitation

$$\mathrm{M} \rightarrow \mathrm{M}^{*} \rightarrow \mathrm{M} + h\nu\ ; \tag{5}$$

$$\mathrm{M}^{+} \rightarrow (\mathrm{M}^{+})^{*} \rightarrow \mathrm{M}^{+} + h\nu'. \tag{6}$$

Here the asterisks denote excited electronic states of the ions or neutral atoms. These reactions lie farther to the right as temperature increases. Some of these steps can actually be seen to occur in the axial channel when a concentrated solution of yttrium is introduced (fig. 2). The initial radiation zone low in the plasma corresponds to emission from both excited Y atoms and YO molecules. As these species travel further up the axial channel, atomization becomes more extensive and ionization also occurs, so the red emission is supplanted by strong blue emisssion from singly charged, positive Y ions. These events can also be seen from vertical profiles of emission from various species (fig. 3). Many rare earths show a similar spatial pattern of emission, i.e., oxide or hydroxide bands and neutral atom lines are emited low in the initial radiation zone, ion lines are seen higher in the normal analytical zone of the plasma. The latter region is observed almost exclusively in emission spectrometric measurements.

The extent of ionization [eq. (4)] can be calculated from the Saha equation

$$\log K_{\mathrm{ion}} = \tfrac{3}{2}\log T - \frac{5040(\mathrm{IE})}{T} + \log\left(\frac{Z_{\mathrm{M}^{+}}}{Z_{\mathrm{M}}}\right) + 15.684\,, \tag{7}$$

where T is the temperature (K), IE the ionization energy (eV), and Z the electronic partition function. Calculated degrees of ionization under conditions prevalent in the normal analytical zone are shown in fig. 4. Most elements are ionized to a significant extent. Rare earths are almost completely ionized. Because of their low second ionization energies, the light rare earths also yield

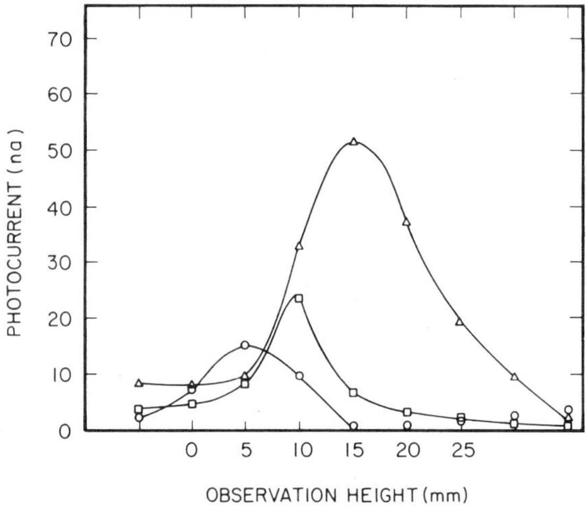

Fig. 3. Dependence of emission signals from various species on observation height for a solution of Y (500 mg L^{-1}) (○): YO; (□) Y I; (△) Y II. Reproduced with permission (Reeves et al. 1980).

some doubly charged ions, but these should number less than 10% of the singly charged ions, as illustrated by the values in parentheses in fig. 4. Most elements have second ionization energies comparable to or greater than that for lead (15 eV) and do not yield significant numbers of doubly charged ions. Thus, for mass spectrometric purposes, the ICP provides a high yield of singly charged ions and little else, so that the mass spectra of ions extracted from the ICP would be expected to be simple. To a first approximation, this is observed experimentally, with some qualifications to be described in detail later.

$$\left(\frac{M^{+}}{M^{+}+M}\right)\times 100\%$$

H 0.1																	He
Li 100	Be 75											B 58	C 5	N 0.1	O 0.1	F 9×10^{-4}	Ne 6×10^{-6}
Na 100	Mg 98											Al 98	Si 85	P 33	S 14	Cl 0.9	Ar 0.04
K 100	Ca 99(1)	Sc 100	Ti 99	V 99	Cr 98	Mn 95	Fe 96	Co 93	Ni 91	Cu 90	Zn 75	Ga 98	Ge 90	As 52	Se 33	Br 5	Kr 0.6
Rb 100	Sr 96(4)	Y 98	Zr 99	Nb 98	Mo 98	Tc	Ru 96	Rh 94	Pd 93	Ag 93	Cd 85	In 99	Sn 96	Sb 78	Te 66	I 29	Xe 8.5
Cs 100	Ba 91(9)	La 90(10)	Hf 98	Ta 95	W 94	Re 93	Os 78	Ir	Pt 62	Au 51	Hg 38	Tl 100	Pb 97(0.01)	Bi 92	Po	At	Rn
Fr	Ra	Ac															

Ce 98(2)	Pr 90(10)	Nd 99*	Pm	Sm 97(3)	Eu 100*	Gd 93(7)	Tb 99*	Dy 100*	Ho	Er 99*	Tm 91(9)	Yb 92(8)	Lu
Th 100*	Pa	U 100*	Np	Pu	Am	Cm	Bk	Cf	Es	Fm	Md	No	Lw

$$\left(\frac{M^{+2}}{M^{+2}+M^{+}+M}\right)\times 100\%$$

Fig. 4. Calculated values for degree of ionization of M^{+} and M^{+2} at $T = 7500$ K, $n_e = 1\times10^{15}$ cm^{-3}. Elements marked by asterisks yield significant M^{+2}, but partition functions were not available. Reproduced with permission (Houk 1986).

3. Analytical measurements by atomic emission spectrometry

3.1. *Instrumentation*

The two types of spectrometers used most commonly to disperse and detect radiation from ICPs are shown in fig. 5. With either spectrometer, an image of the plasma is focused through a lens onto the entrance slit. With the spectrometer shown on the left of fig. 5, the grating is fixed, and several spectral lines are monitored simultaneously with a collection of separate detectors. Commercial instruments have from 10 to 60 channels.

The diagram on the right of fig. 5 depicts a scanning device that isolates a single spectral window at a time. The grating is rotated to change the wavelength observed. Alternatively, a range of wavelengths can be observed simultaneously if some sort of multichannel detector such as a photodiode array or charge coupled device is used in place of the photomultiplier.

Many facilites have both types of instruments collecting radiation from the same plasma; the scanning device facilitates line selection and determination of unusual elements, while the polychromator is used for routine determination of a fixed set of elements. Two recent books provide several very nice discussions of optical instrumentation for ICP–AES (Boumans 1987, Montaser and Golightly 1987).

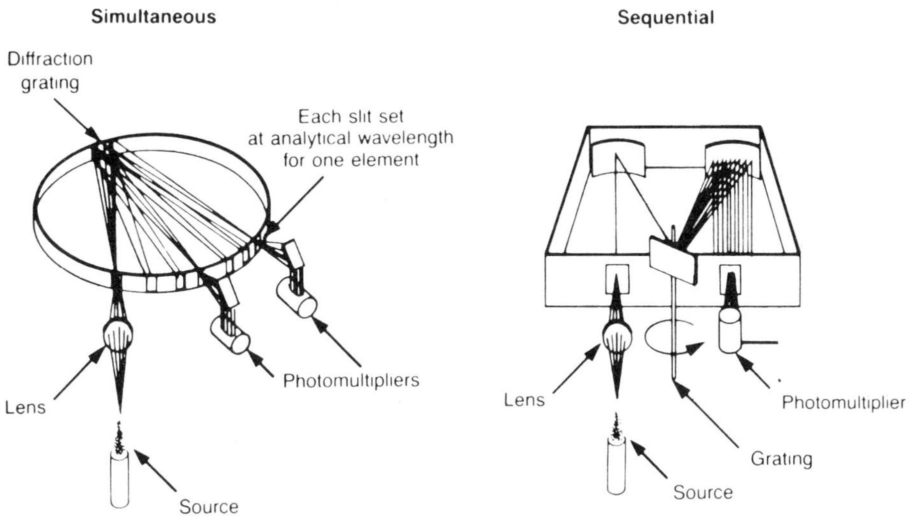

Fig. 5. Dispersive spectrometers for measuring emission from an ICP. Only two of the detectors are shown for the simultaneous instrument. Naturally, the grating is actually inside the spectrometer enclosure for the sequential instrument. Reproduced with permission (Rubinson 1987, Sobel and Dahlquist 1981).

3.2. *Spectral emission from the ICP*

The spectral characteristics of the ICP depend on the position viewed by the optical system. In the discussion below, the spectra shown are typical of those obtained from the normal analytical zone (fig. 2). Several concepts are also illustrated by vertical profiles of emission of a given spectral line. The position scale for these profiles is consistent with that shown for the plasma regions in fig. 2 and is also typical of that seen for most analytical ICPs under normal operating conditions.

Background spectra from the ICP are shown in fig. 6. A broad continuum whose intensity is wavelength-dependent underlies all the spectral features. The wavelength range between 190 and 350 nm is nearly free of background lines except for the OH bands. This open wavelength interval is a consequence of the choice of argon as the plasma medium. The energy level structure of argon is such that it emits only a few lines in this region, which is where many of the best analyte lines occur. Above 400 nm, numerous lines from argon and hydrogen are

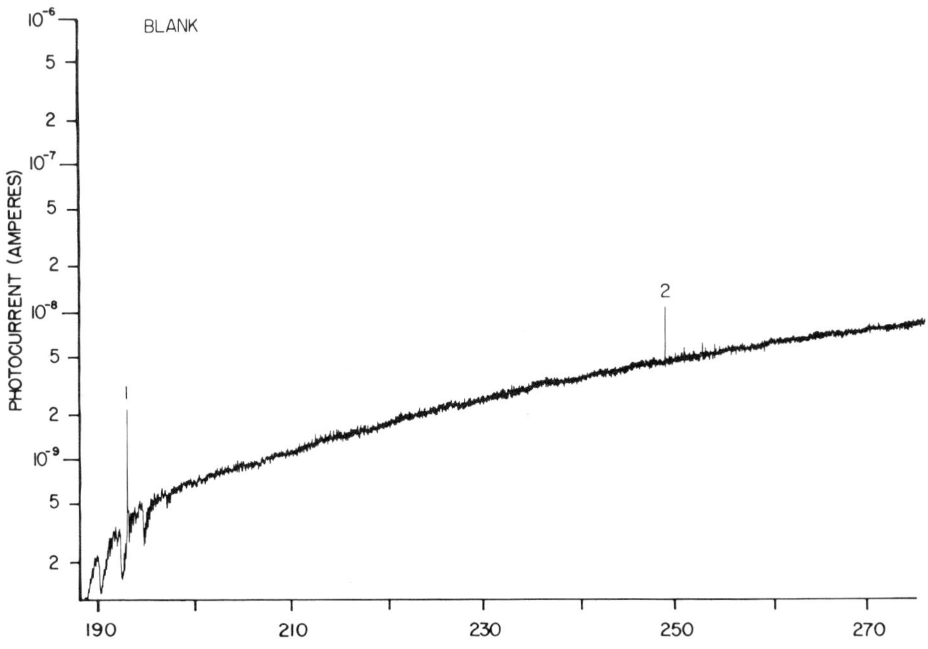

1	C	I	193.090
2	C	I	247.856

Fig. 6. Wavelength scans of background spectra obtained during nebulization of deionized water. Note logarithmic intensity scales. Wavelengths are listed for selected lines. Reproduced with permission (Winge et al. 1985).

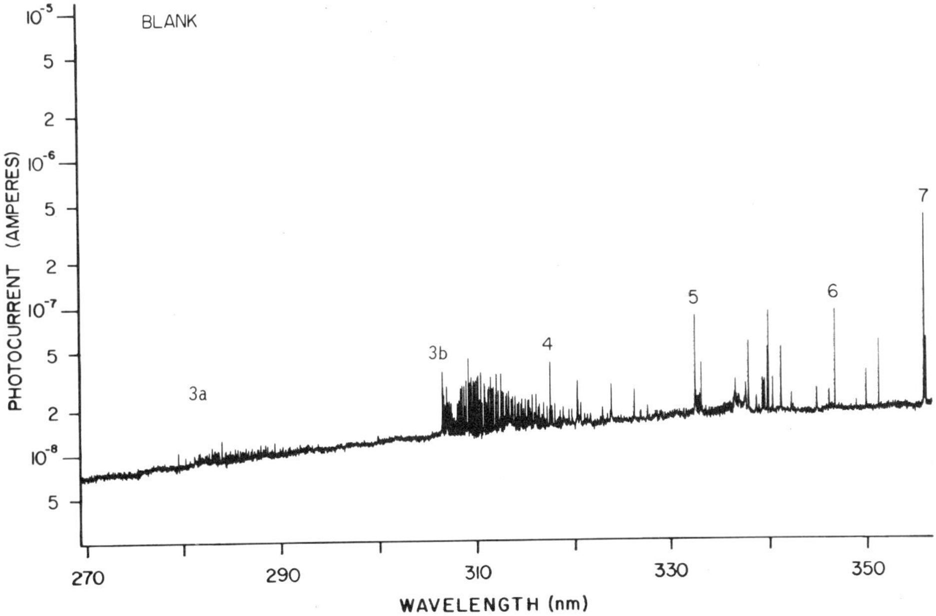

3a	OH	(bh)	281.13	5	Ar	I	331.935
3b	OH	(bh)	306.36	6	Ar	I	346.108
4	Ar	I	317.296	7	Ar	I	355.431

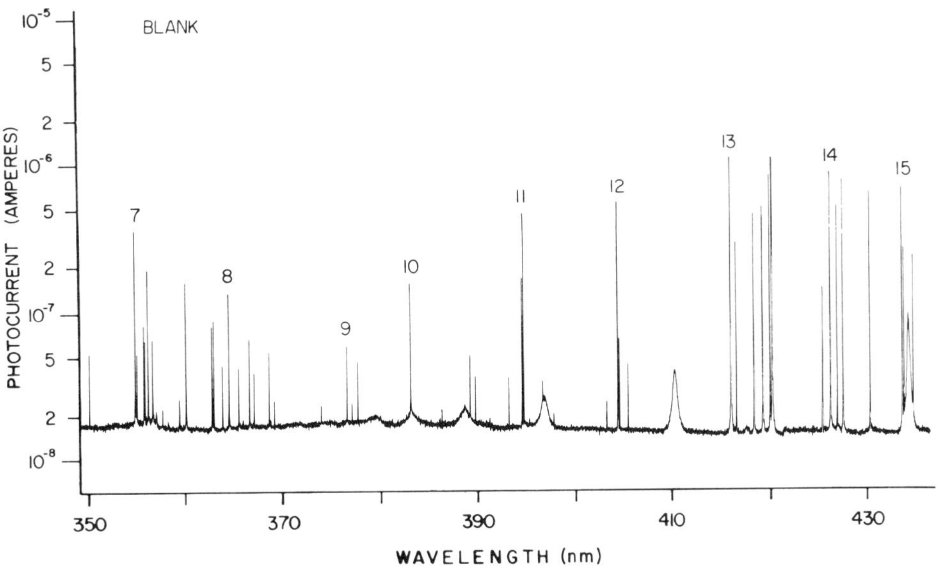

7	Ar	I	355.431	10	Ar	I	383.468	13	Ar	I	415.859
8	Ar	I	364.983	11	Ar	I	394.898	14	Ar	I	425.936
9	Ar	I	377.037	12	Ar	I	404.442	15	Ar	I	433.356

Fig. 6. (continued).

observed. This region at higher wavelength is used mainly for determination of the alkali metals, Ca, Sr, and Ba.

Portions of the spectra emitted by two lanthanides are shown in figs. 7 and 8. Analyte lines can be identified from these scans by comparing them with the background spectra in fig. 6. This comparison shows that nearly all the lines shown in figs. 7 and 8 are from the analyte elements. The spectrum of La is not overly complex, whereas that for Sm consists of many lines. Thus, the spectra emitted by some rare earths are indeed complex, but this is not particularly so for all of them. The actinides also emit very complex spectra, as shown for U in fig. 9. Most of these lines are from singly charged, positive ions.

Vertical profiles of emission intensity for ion lines are shown schematically in fig. 10. To a first approximation, all ion lines exhibit maximum intensity at the same vertical position in the ICP. Thus, for multi-element analysis, the optical region observed need not be changed on an element-by-element basis as long as ion lines are used. Neutral atom lines of high excitation energy (>5 eV) are most intense at the same position as the ion lines. In contrast, neutral atom lines of low to medium excitation energy peak lower in the plasma at various positions

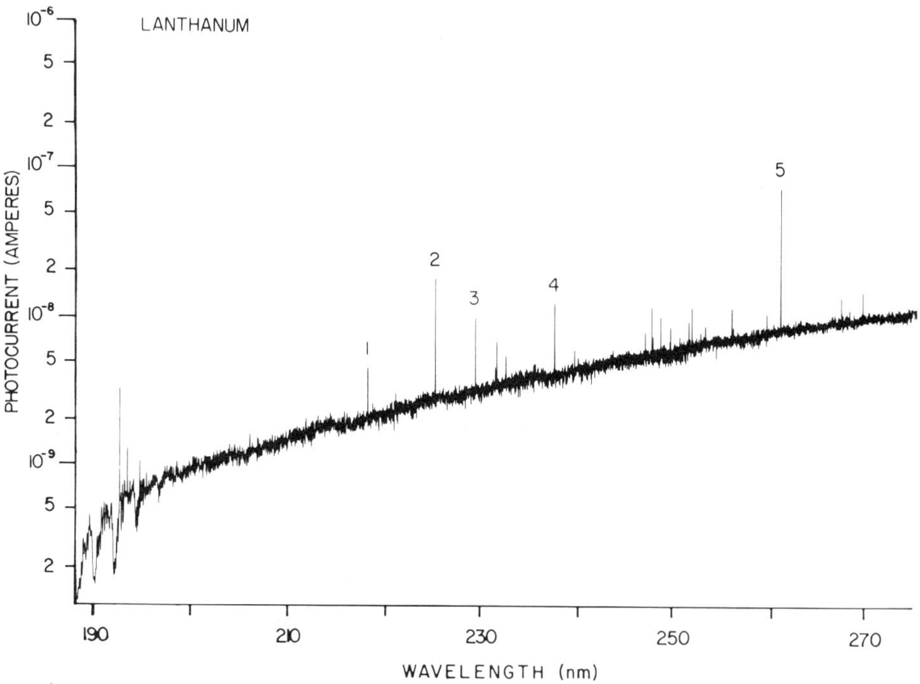

1	La	II	218.787	4	La	III	237.938
2	La	II	225.676	5	La	II	261.034
3	La	III	229.778				

Fig. 7. Wavelength scans showing emission from La (10 mg ℓ^{-1}). Reproduced with permission (Winge et al. 1985).

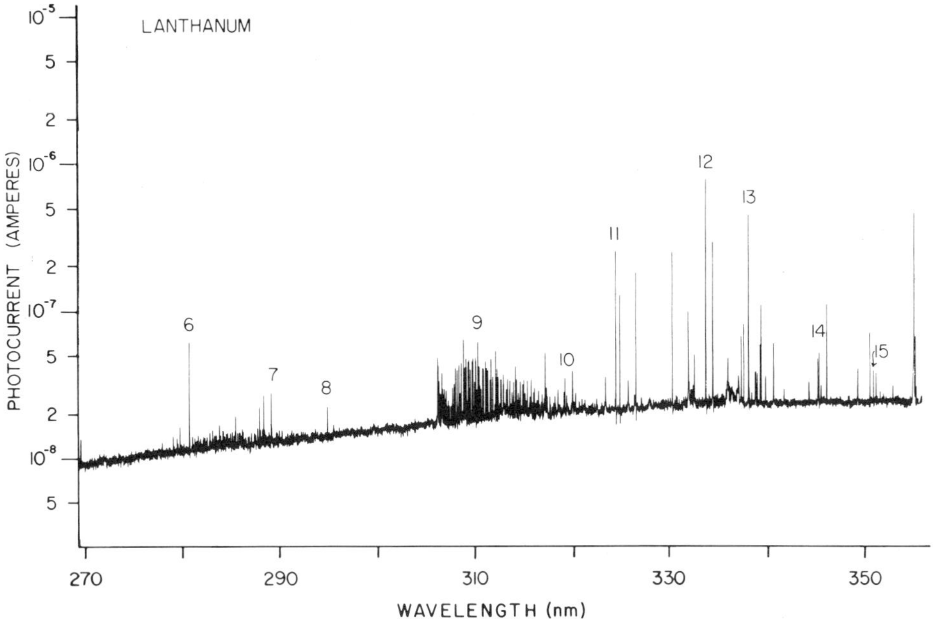

6	La	II	280.839	10	La	II	319.302	14	La	II	345.317
7	La	II	289.307	11	La	II	324.513	15	La	II	351.000
8	La	II	295.050	12	La	II	333.749				
9	La	II	310.459	13	La	II	338.091				

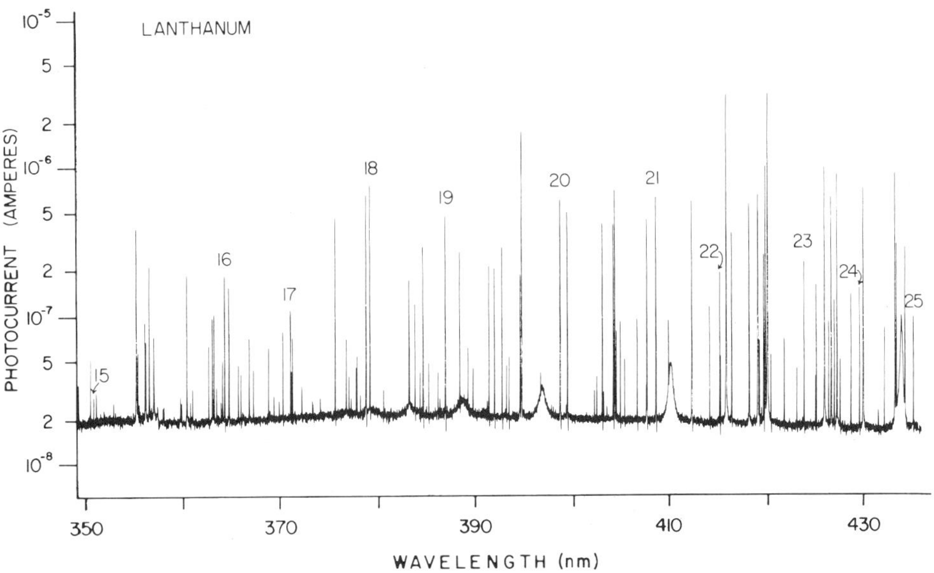

15	La	II	351.000	19	La	II	387.164	23	La	II	423.838
16	La	II	364.542	20	La	II	398.852	24	La	II	429.605
17	La	II	371.354	21	La	II	408.672	25	La	II	435.440
18	La	II	379.478	22	La	II	415.197				

Fig. 7. (continued).

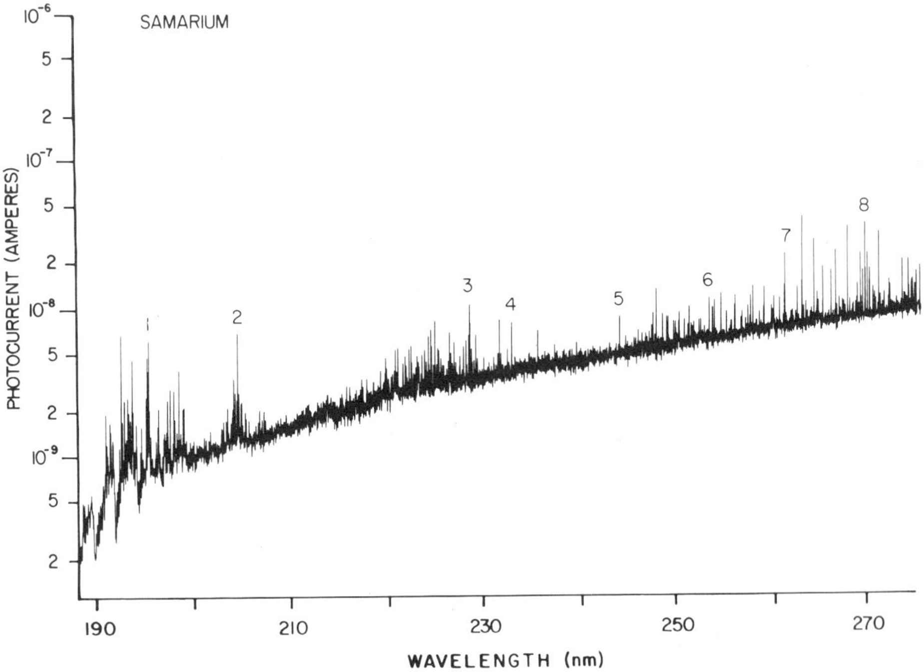

1		~195.9	4	Sm	233.129	7	Sm	I	261.007
2		~205.0	5	Sm	244.155	8	Sm		269.334
3	Sm	228.875	6	Sm	253.239				

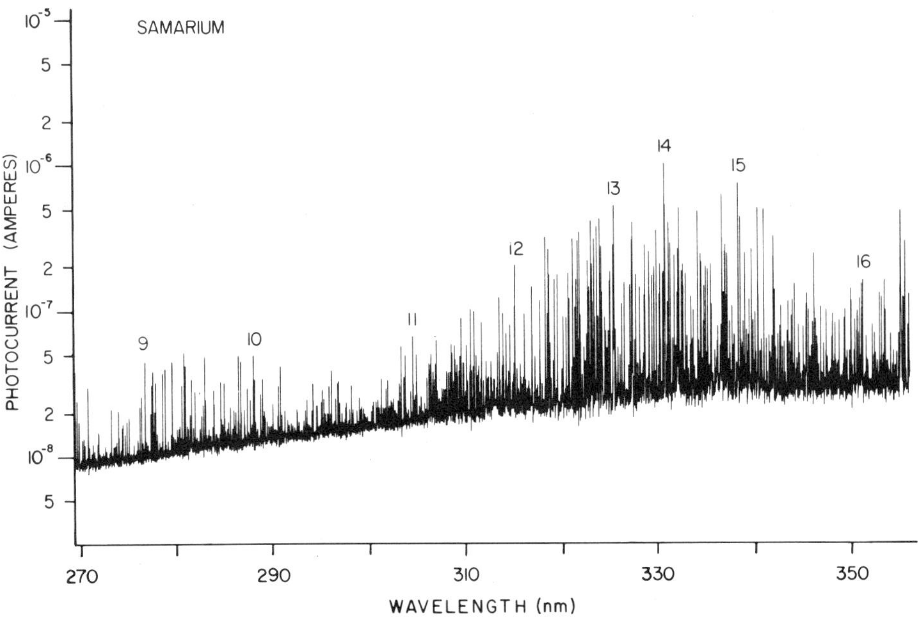

9	Sm	II	276.785	12	Sm	II	315.252	15	Sm	II	338.240
10	Sm	I	288.168	13	Sm	II	325.438	16	Sm	II	351.123
11	Sm	II	304.693	14	Sm	II	330.639				

Fig. 8. Wavelength scans showing emission from Sm (100 mg ℓ^{-1}). Reproduced with permission (Winge et al. 1985).

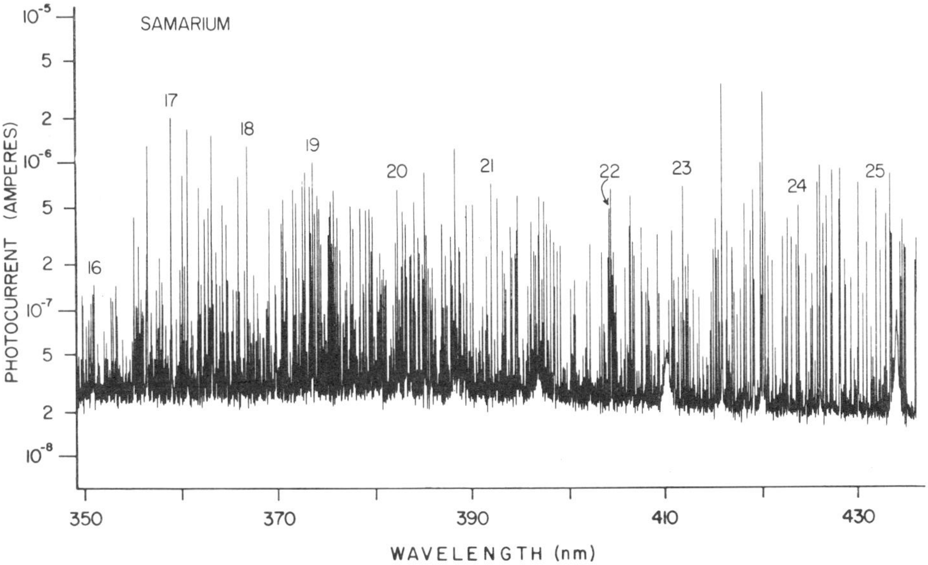

16	Sm	II	351.123	20	Sm	II	382.620	24	Sm	II	423.674
17	Sm	II	359.260	21	Sm	II	392.240	25	Sm	II	431.894
18	Sm	II	367.084	22	Sm	II	404.290				
19	Sm	II	373.912	23	Sm	II	411.855				
			373.920								

Fig. 8. (continued).

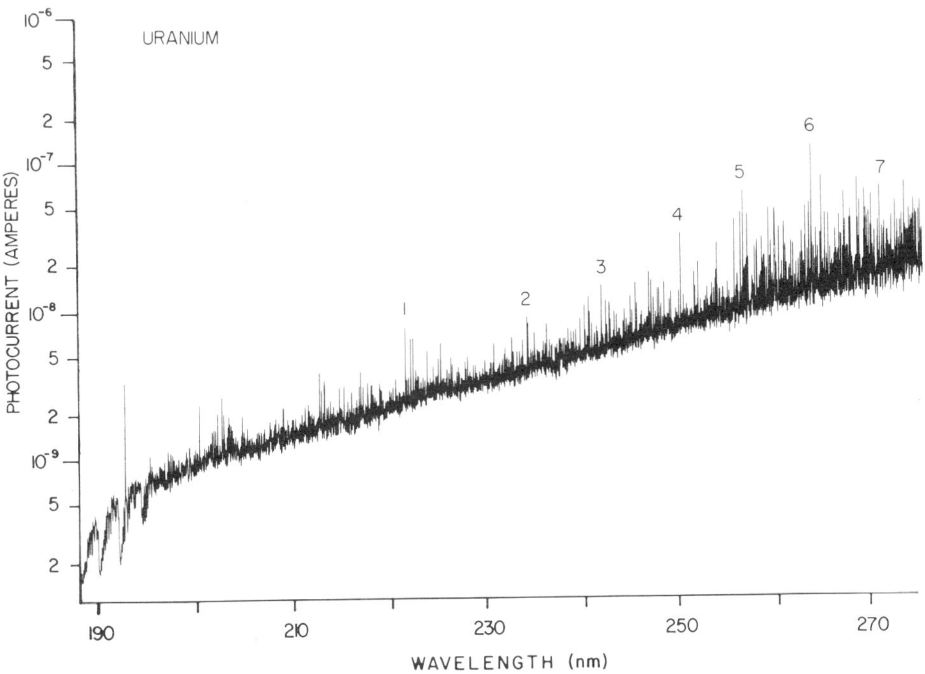

1	U		221.928	4	U	II	250.086	7	U	II	270.695
2	U		234.357	5	U	II	256.541				
3	U	II	241.957	6	U	II	263.553				

Fig. 9. Wavelength scans showing emission from U (100 mg ℓ^{-1}). Reproduced with permission (Winge et al. 1985).

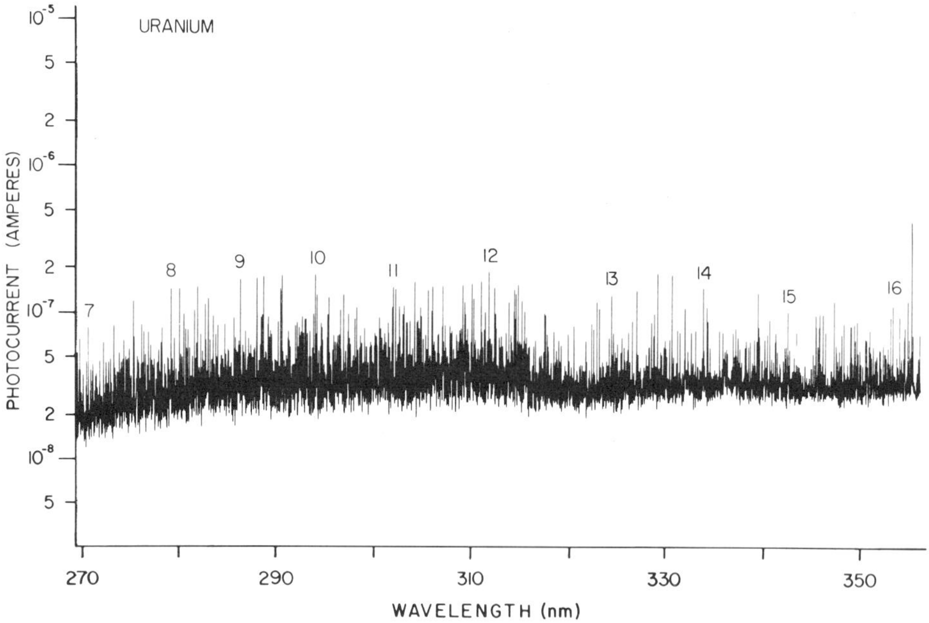

7	U	II	270.695	11	U	II	302.221	15	U	II	342.169
8	U	II	279.394	12	U	II	311.935	16	U	II	353.357
9	U	II	286.568	13	U	II	324.417				
10	U	II	294.192	14	U	II	333.779				

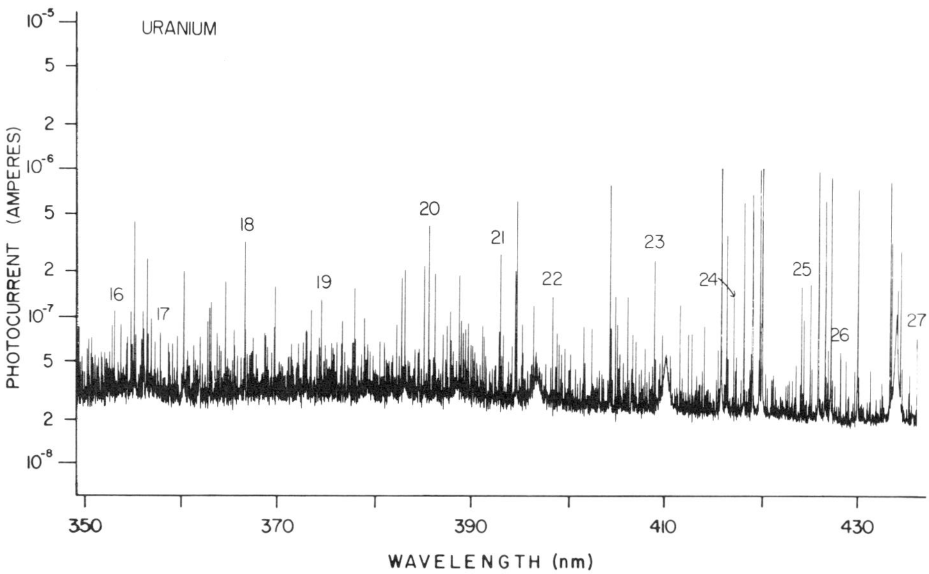

16	U	II	353.357	20	U	II	385.958	24	U	II	417.159
17	U	II	358.184	21	U	II	393.203	25	U	II	424.167
18	U	II	367.007	22	U	II	398.580	26	U	II	428.203
19	U	II	374.868	23	U	II	409.014	27	U	II	436.226

Fig. 9. (continued).

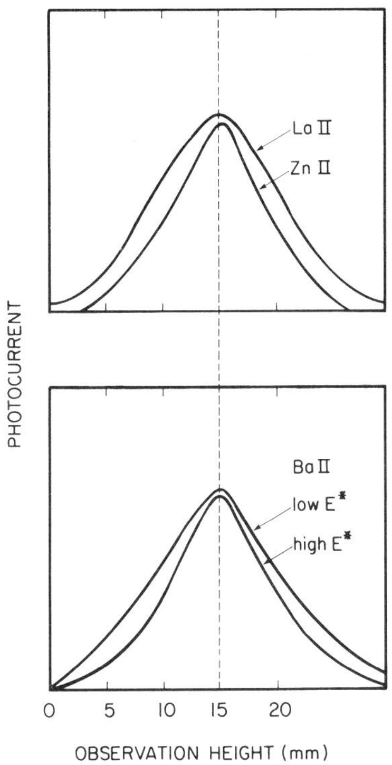

Fig. 10. Vertical profiles of emission intensity for ion lines of various elements. Profiles are shown for two Ba II lines of differing excitation energy (E^*). Adapted and reproduced with permission (Blades and Horlick 1981a, Furuta and Horlick 1982).

(Blades and Horlick 1981a,b, Furuta and Horlick 1982). These latter lines are, therefore, used less often, although there are some elements (e.g., alkali metals) which do not have practical alternative lines. Nevertheless, there is no other excitation source for which a single set of operating conditions is so nearly optimum for such a wide range of elements.

3.3. *Detection limits and interferences*

The first test of an analytical method is whether it can sense the analyte at concentrations likely to be present in the samples. The detection limit represents the lowest concentration of an element that can definitely be certified to be present in a sample. The preferred criterion is to express the detection limit as the solution concentration that yields a net signal equal to three times the standard deviation of the background at the wavelength monitored. Although a variety of instruments are used to detect radiation from an ICP, the overall magnitude of the detection limits does not vary greatly from instrument to instrument for the medium resolution spectrometers used in most commercial facilities. A selected set of values for rare earth elements and some common analytes is given in table 1 to illustrate typical performance. Detection limits vary substantially from element to element within the range 0.2–70 $\mu g\ \ell^{-1}$. Naturally, analyte concentrations near the detection limit cannot be measured with much confidence, and the lowest

TABLE 1
Typical detection limits for ICP–AES. Reproduced with permission (Winge et al. 1985).

Element	Line (nm)	Detection limit ($\mu g\ \ell^{-1}$)	Element	Line (nm)	Detection limit ($\mu g\ \ell^{-1}$)
Al I	309.271	20	Nd II	401.225	50
Ba II	455.403	1		430.358	70
Ca II	393.366	0.2	Sm II	359.260	40
Cd II	214.438	2		442.434	50
Co II	238.892	6	Eu II	381.967	3
Fe II	238.204	5		412.970	4
Hf II	277.336	15	Gd II	342.247	14
Mg II	279.553	0.2		336.223	20
Mo II	202.030	8	Tb II	350.917	20
Nb II	309.418	40		384.873	60
Pb II	220.353	40	Dy I	353.170	10
Ta II	226.230	25		364.540	20
Zr II	343.823	7	Ho II	345.600	6
Sc II	361.384	1.5		339.898	13
	357.253	2	Er II	337.271	10
Y II	371.030	4		349.910	17
	324.228	5	Tm II	313.126	5
La II	333.749	10		346.220	8
	379.478	10	Yb II	328.937	2
Ce II	413.765	50		369.419	3
	413.380	50	Lu II	261.542	1
Pr II	390.844	40		291.139	6
	414.311	40	Th II	283.730	60
				283.231	70
			U II	385.958	200
				367.007	300

quantifiable concentration is usually taken to be approximately five times the detection limit. The detection limits shown for U in table 1 are poor because so many lines are emitted that no single line is particularly intense.

Detection limits are by no means the only criterion by which to judge the value of an analytical method for a particular application. Ideally, the analyte signal should respond only to the concentration of analyte and should not be affected by other substances in the sample solution. In many other excitation sources such as flames or arcs, addition of matrix elements from the sample changes either the excitation capabilities of the source or the number densities of various analyte species. Thus, the emission intensities can depend on both the analyte concentration and the identity and concentration of concomitants, which represents an undesirable interference.

A general picture of the effect of concomitant on analyte ion emission in the ICP is shown in the spatial profiles in fig. 11. Sodium is chosen as a typical concomitant. Addition of sodium enhances the analyte intensity low in the plasma, whereas a slight suppression is observed higher in the normal analytical

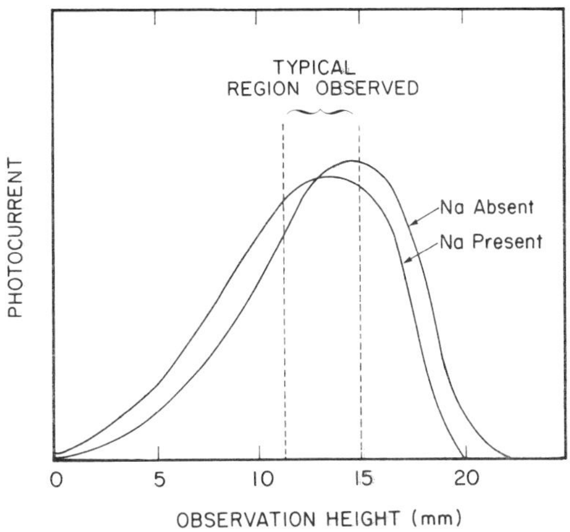

Fig. 11. Typical vertical profiles of emission from an analyte ion line in the absence and presence of Na concomitant (~500 mg ℓ^{-1}). Adapted and reproduced with permission (Blades and Horlick 1981b).

zone when sodium is present. Analytical signals are typically measured from an extended vertical slice of the plasma seen through a slit ~0.4 cm high. Thus, the emission signals observed from analytes are usually spatially integrated over the region represented by the dotted lines in fig. 11. If this observed section straddles the cross-over point of the two spatial profiles, as shown in the figure, the enhancement low in the plasma is cancelled by the suppression seen at higher observation positions (Blades and Horlick 1981b). The lack of severe interference shown in these curves represents typical behavior observed in actual analytical practice for many analytes and matrices. Thus, the intensity of analyte ion lines generally does not vary strongly if the matrix concentration changes within a reasonable range, typically up to ~1% in solution (Larson et al. 1975). As pointed out by Thompson and Ramsey (1985) and Thompson (1987), the analyst should not assume that matrix interferences are absent in ICP–AES, as they can be significant in some cases, e.g., Ca at ~1% levels.

The ability of the ICP to excite spectra from many elements simultaneously, which is desirable for multi-element analysis, also means that the probability of line overlap between spectral lines of different elements is high. Figures 8 and 9 illustrate this problem for Sm and U, two elements with very line-rich spectra. In ICP–AES, interferences of this nature are more serious problems than the chemical-type effects described above.

There are three basic solutions to spectral overlap problems. The first approach is to select analyte lines very carefully to minimize potential overlap problems. Compilations of wavelengths and intensities of spectral lines are invaluable for this purpose (Winge 1985, Boumans 1980, 1984). A recent series of papers addresses line selection specifically for the lanthanides (Boumans et al. 1988).

The second approach to correct for line overlap or spectral background shifts involves estimating the background contribution and subtracting it from the gross intensity observed. The difference is the net intensity of the analyte line, which should be related to analyte concentration. Some ways how this is done are illustrated in fig. 12. Most commercial instruments can implement such correction procedures. Obviously, this approach has limitations in cases where a matrix line falls directly underneath an analyte line or where the matrix line is much stronger than the analyte line and the difference of the two intensities is small and uncertain.

In general, the additional intensity measurements to define the background (e.g., along the sloping profile in fig. 12c) also lengthen the time required for a determination. Nevertheless, correction procedures for spectral overlap problems are common and, in many cases, improve the accuracy of the determined concentrations substantially. Some illustrative examples of the effect of background corrections on determined analyte concentrations are shown in table 2. Results are presented for several of the more abundant rare earths and ytterbium, one of the less abundant ones. All the determined values are higher if no background correction is done. For the set of lines labelled "weak interference", the extent of overlap is not substantial, and omitting the correction procedure yields results that do not differ much from the corrected values. In contrast, for a different set of lines from the same elements in the same sample, the extent of interference is substantial, and lack of correction induces serious errors in the determined concentrations. Clearly, the need for spectral overlap corrections should be evaluated on a line-by-line basis for the particular sample matrices to be analyzed.

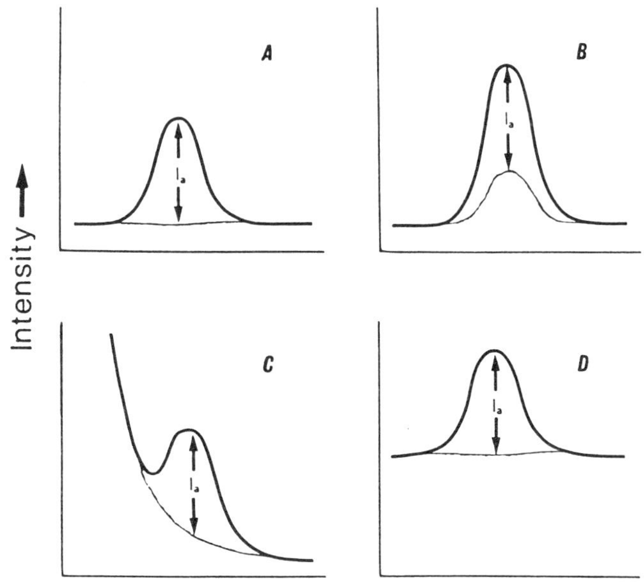

Fig. 12. Schematic wavelength scans showing gross signal (bold line), background (fine line), and net analyte signal (I_a): a, analyte peak on continuum background. b, analyte peak atop background or matrix line. c, analyte peak along wing of strong interferent line. d, analyte peak atop elevated background caused by matrix. Reproduced with permission (Thompson 1987).

TABLE 2

Analysis of monazite sand by ICP–AES showing effect of background or overlap correction (see, e.g., fig. 12) on determined concentrations. Reproduced with permission of Spectrochimica Acta B and the Publisher, Pergamon Press, Oxford (Boumans et al. 1988).

Weak interference				Strong interference			
	Line	Concentration (mg ℓ^{-1})			Line	Concentration (mg ℓ^{-1})	
Element	(nm)	Meas.	Corrected	Element	(nm)	Meas.	Corrected
Ce II	413.765	45.24	45.07	Ce II	413.380	46.05	44.80
La II	333.749	33.68	33.65	La II	412.323	34.70	33.27
Nd II	406.109	18.15	17.98	Nd II	401.225	19.85	17.54
Pr II	414.313	6.68	6.34	Pr II	390.843	12.49	6.62
Sm II	446.734	2.53	2.24	Sm II	359.260	2.94	2.27
Yb II	328.937	0.023	0.019	Yb II	222.446	0.028	0.018

With either of the common instruments shown in fig. 5, the resolution is typically 10–20 pm. At these resolution values, the instrument function is the major component of the observed spectral linewidth. The actual physical widths of the spectral lines are substantially narrower (2–5 pm), so the third potential solution to spectral interferences is to use a spectrometer with sufficient resolution to separate most of the overlapping lines. Therefore, several types of high-resolution instruments have seen recent use with ICPs. These include grating monochromators with long focal lengths (e.g., 3 m), echelle spectrometers, and Fourier transform interferometers (Edelson 1987, Faires et al. 1984). An example of a spectrum from one of these instruments is shown in fig. 13. The two lines depicted differ in wavelength by 30 pm and are completely resolved. Most of the lines in the complex spectra of elements like Sm and U (figs. 8 and 9) can indeed be resolved with such instruments. A recent series of papers by Boumans and co-workers also shows that detection limits can often be improved by use of high spectral resolution, particularly in the presence of line-rich matrix elements (Boumans et al. 1988).

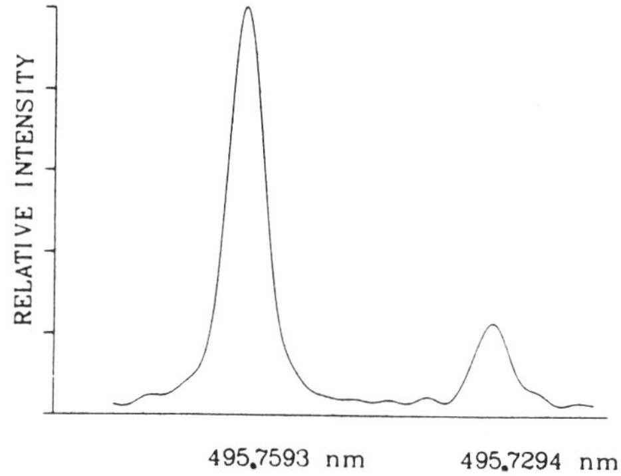

Fig. 13. Two iron lines observed in high resolution with a Fourier transform spectrometer. Reproduced with permission (Faires et al. 1984).

4. Typical applications of ICP atomic emission spectrometry

4.1. *Determination of lanthanides in geological matrices*

Geochemical studies often require determination of all the lanthanides (and often yttrium as well) in various geological matrices such as rocks, sediments, coal, etc. In most geological matrices, the light rare earths (e.g., La and Ce) are much more abundant than the heavier rare earths (e.g., Tb and Tm). Therefore, the analytical technique for this application should have multi-element capabilities, large dynamic range, and limited susceptibility to interferences. This problem was not easily fulfilled before the advent of ICP–AES, which provided the first analytical technique with the potential to cover the whole suite of rare earths in a reasonable time.

Several workers have investigated determination of rare earths in geological matrices. A paper by Crock and Lichte (1982) illustrates the capabilities of ICP–AES for these studies. The samples are dissolved (generally by a fusion procedure). Generally, the rare earths are separated from the resulting solution by ion exchange. They are also pre-concentrated to facilitate determination of the less abundant rare earths. Analytical results obtained by ICP–AES are plotted, in fig. 14, as a function of the accepted concentrations. In general, the points fall close to the line of slope 45°, indicating good agreement between determined and accepted values. The confidence limits (95% level), which represent a measure of the precision of the method, are between 0.4 to 10% of the mean determined values, with an average confidence limit of about 2%. Calibration of the instrument response is necessary when the plasma is started, but only occasionally thereafter. The actual spectroscopic measurement typically takes a few minutes for each sample, as compared to several hours for the sample dissolution and separation (Crock and Lichte 1982).

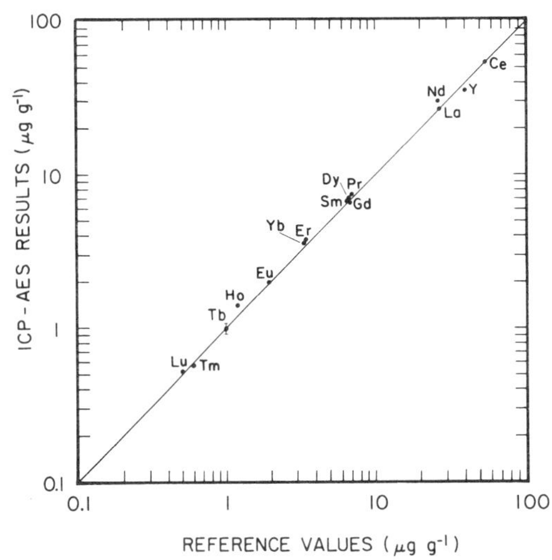

Fig. 14. Determination of rare earths in standard basalt BCR-1 (US Geological Survey). The solid line of slope 45° is not fitted to the points. Reproduced with permission (Crock and Lichte 1982).

4.2. *Determination of rare earth impurities in rare earth reagents*

There are numerous applications in materials sciences and electronics that require individual rare earth(s) in as pure a form as possible. Preparation of highly pure, individual rare earths is a challenging task because of their chemical similarity and their tendency to occur together naturally. The purer the reagent, the lower the levels of contaminants, and the more difficult the analysis becomes because the impurity must be measured in the presence of a large exess of the major rare earth. In these cases, dilution of the sample and spectral interferences from the major constituent can limit the detectability for impurities.

One approach to the interference problem was demonstrated by Yoshida and Haraguchi (1984). They used cation exchange liquid chromatography to separate different rare earths with selective monitoring of emission lines for each by ICP–AES. The chromatographic apparatus was simply inserted before the nebulizer. The plasma then atomized, ionized, and excited the rare earth in each chromatographic band so that an element-selective chromatogram could be generated for each impurity. Such chromatograms are illustrated for a cerium oxide sample in fig. 15. The large peaks at retention times of 30 min represent emission from cerium at the same wavelengths as the analyte lines. Normally, these cerium lines would interfere with the analyte lines. However, with the chromatographic separation before the emission measurement, the trace analyte

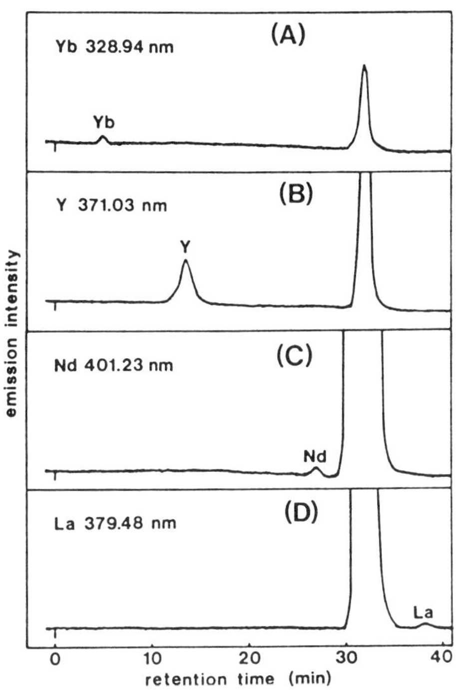

Fig. 15. Cation-exchange chromatography with detection by ICP–AES. Element-selective chromatograms for rare earth impurities in Ce_2O_3. Cerium elutes at 30 min. Reproduced with permission (Yoshida and Haraguchi 1984).

TABLE 3
Determination of impurities in rare earth reagents by cation exchange separations with on-line detection by ICP–AES. Reproduced with permission (Yoshida and Haraguchi 1984).

Impurity	Reagent and impurity concentrations ($\mu g\ g^{-1}$)				
	Y_2O_3	La_2O_3	Nd_2O_3	$EuCl_3$	$GdCl_3$
Y	–	10	23	70	39
La		–	98		23
Cl		42	39		42
Pr		50	112		
Nd			–		
Sm			250	38	
Eu			6	–	120
Gd	14		20		–
Tb				20	40
Dy	66	13	15	14	
Ho	30		37		
Er					
Tm					
Yb			8.2	2	15
Lu	0.6			0.3	2

elements are also separated in time and are easily monitored without interference from cerium. Analytical results for various "pure" rare earth reagents are given in table 3 to illustrate the general applicability of this technique. Each of the chromatograms shown in this paper took 30–40 min, so analysis time is sacrificed to some extent by use of such a separation.

4.3. *Monitoring of actinide isotopes*

Emission spectrometry is generally thought of as a technique for elemental determinations, with isotope ratio determinations being the domain of mass spectrometry. In the manufacture, handling, and analysis of nuclear materials, there is considerable need for techniques to determine both elemental and isotopic concentrations of U, Pu, and other actinides in as near an on-line, process control fashion as possible. These elements emit many lines when excited in an ICP, as shown in fig. 9. Edelson and Fassel (1981) point out that some of these lines exhibit isotopic shifts of sufficient magnitude to be separated by a high-resolution monochromator. An example is shown in fig. 16. Separate lines from each isotope are clearly detectable. In these experiments, the fact that actinides are the heaviest elements actually helps resolution of isotopic lines, because the Doppler width of a line decreases as atomic weight increases. The plasma is operated inside a secure containment facility to prevent excretion of actinides into the environment. Since the analytical information is carried by photons, the optical instrumentation for the actual measurement is completely isolated from the radioactive source, so deposition of actinides in the spec-

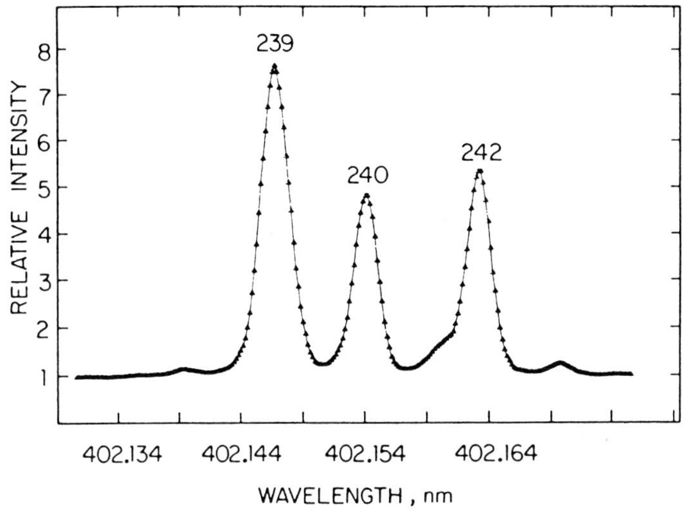

Fig. 16. Isotopic structure of Pu II 402.145 nm emission line. Each isotope was present at approximately 0.2 g ℓ^{-1}. Reproduced with permission (Edelson 1987).

troscopic apparatus is not a problem (Edelson 1987). Several other containment facilities for monitoring actinides and actinide isotopes have been installed in nuclear processing plants for various applications, e.g., for monitoring product purity and for ensuring that fissionable materials are not diverted surreptitiously from a process stream.

5. Analytical measurements by mass spectrometry

5.1. *Instrumentation*

As indicated above (e.g., fig. 4), the ICP is an efficient source of singly charged, positive ions. The ICP operates at atmospheric pressure, whereas mass spectrometry requires transport of ions under high-vacuum, collisionless conditions. Therefore, the ions from the ICP are extracted into a vacuum system before they are mass analyzed. At present, ions are extracted using the interface shown in fig. 17. In most ICP–MS instruments, the ICP is operated horizontally. A metal cone with a sampling orifice (~1 mm diameter) is immersed into the ICP, usually in the center of the axial channel. Gas flows from the ICP through the sampling orifice into the first stage of the vacuum chamber, which is evacuated by a mechanical pump to a pressure of ~1 torr. Ions are entrained in this gas flow. A supersonic jet forms downstream from the sampling orifice. The central section of the jet flows through a second orifice in another metal cone, referred to as the skimmer. Downstream from the skimmer, the ions next encounter an ion lens, i.e., a collection of metal cylinders, grids, and/or apertures biased at voltages

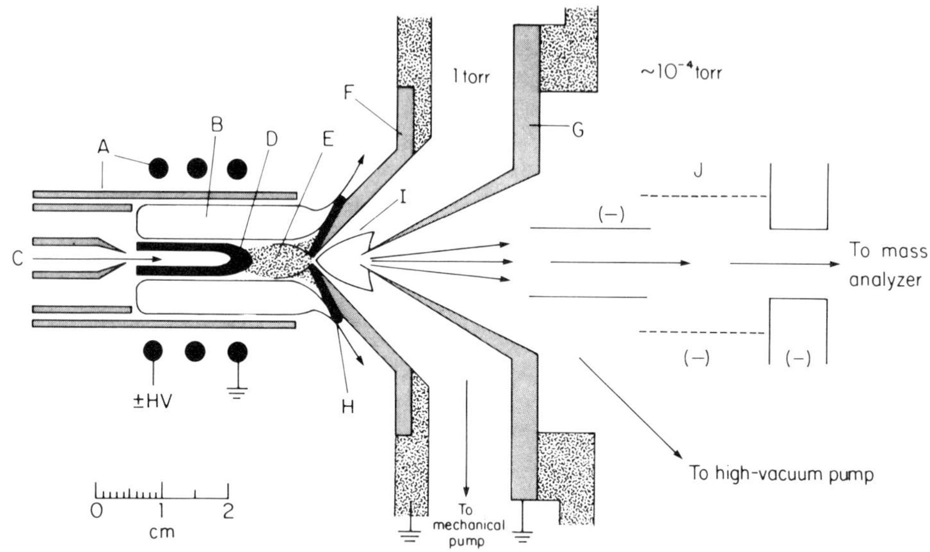

Fig. 17. ICP and sampling interface for ICP–MS. A solution of Y is injected into the plasma with resulting red emission from excited YO and Y and blue emission from Y^+. A, torch and load coil; B, induction region; C, aerosol gas flow; D, initial radiation zone; E, normal analytical zone (see fig. 2); F, sampler; G, skimmer; H, boundary layer of ICP gas deflected outside sampler; I, supersonic jet; J, ion lens. Reproduced with permission (Houk 1986).

chosen to focus and transmit ions. The lens keeps ions in a beam, while neutral gas from the ICP is allowed to flow into the pumping system for removal.

After the ions have been extracted from the plasma, they are conducted into a quadrupole mass analyzer (fig. 18). This is an array of four metal rods biased at carefully regulated DC and RF voltages. The voltages are selected so that ions of a given mass-to-charge ratio (m/z) have a stable path through the rod structure while other ions are deflected into the rods and are thus removed from the beam (Dawson 1976). The selected ions leave the mass analyzer and strike a detector, usually a Channeltron electron multiplier (Kurz 1979) operated in a pulse counting mode. In this fashion, discrete pulses from individual ions can be observed. The multiplier is offset and the optical axis is screened with a baffle to prevent detection of photons from the ICP.

The m/z value transmitted by a quadrupole mass analyzer can be scanned or switched very rapidly. The most common ways to acquire data with one of these devices are illustrated in fig. 19. The spectrum can simply be scanned continuously once, as shown in fig. 19a. In the peak-hopping mode (fig. 19b), the mass analyzer transmits a certain m/z value for a chosen time and then quickly switches to another value. This peak hopping is illustrated for three separate m/z positions spanning each of two peaks. This peak-hopping process is generally done repetitively with each mass peak monitored many times. The background can also be determined by hopping to an m/z position devoid of ions.

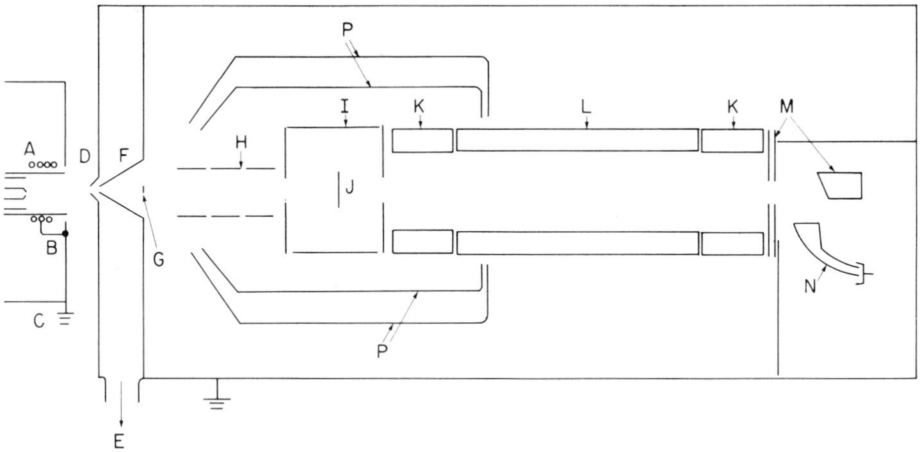

Fig. 18. Diagram of an ICP–MS device. A, torch and load coil; B, grounded center tap; C, grounded shielding box; D, sampler; E, line to rotary vacuum pump; F, skimmer; G, grounded metal disk; H, ion lens; I, Bessel box with photon stop in center (J); K, RF-only quadrupole rods; L, quadrupole mass analyzer; M, exit ion lens and ion deflector; N, Channeltron electron multiplier; P, cryogenic baffles for evacuation. Reproduced with permission (Houk and Thompson 1988).

In the multichannel scanning mode (fig. 19c), the mass analyzer scans continuously over the mass region of interest. The scans are done repetitively (typically 100–200 times in 1–2 min) and individual scans are added for a signal averaging effect. With either scan mode, at any instant in time, only ions of one m/z value are transmitted through the rods; the other ions are lost. Also, the widths of the peaks shown in fig. 19 are realistic for a quadrupole mass analyzer. Unit mass

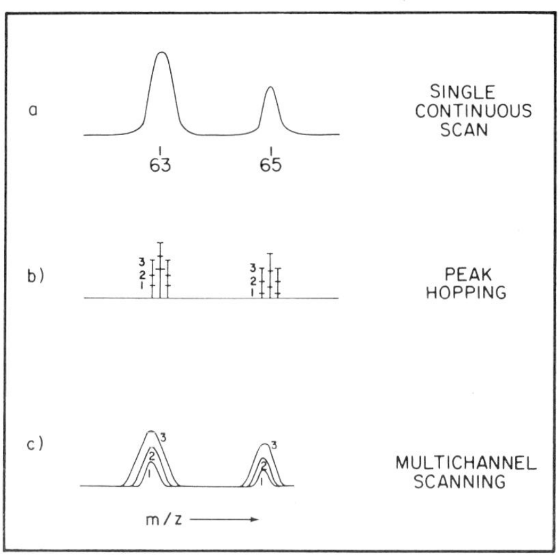

Fig. 19. Scanning and data acquisition modes. Three successive cycles (indicated by 1, 2, 3) are shown for peak hopping and multichannel scanning. In peak hopping, each m/z position is typically monitored for ~0.1 s. In multichannel scanning, each scan takes typically 0.01 s per m/z unit in the window observed.

resolution is typical, and chemically different ions at the same nominal mass generally can not be resolved using a quadrupole mass analyzer. Recent work at VG Elemental has demonstrated that a double focusing MS can be used with an ICP to provide a spectral resolution of ~9000, which is sufficient to resolve many cases of overlap between atomic analyte ions (e.g., $^{32}S^+$) and nearby background ions (e.g., $^{16}O_2^+$) (Bradshaw et al. 1989).

5.2. *Mass spectra, detection limits, and interferences*

Typical background spectra observed during introduction of de-ionized water or nitric acid (typically 1%) are shown in fig. 20. There are major background ions at m/z values corresponding to some analytes such as Ca (Ar^+, $m/z = 40$), S (O_2^+, $m/z = 32$), Fe (ArO^+, $m/z = 56$) and Se (Ar_2^+, $m/z = 76$–80). Other common acids such as HCl, $HClO_4$ and H_2SO_4 should be avoided because they yield additional troublesome peaks such as ClO^+, $ArCl^+$, and SO^+. Some of the classical sample dissolution procedures have been revised for ICP–MS to eliminate $HClO_4$ from the final solution injected into the instrument. Generally, aqueous samples yield no background peaks above $m/z = 80$, except for the occasional occurrence of weak peaks from Kr or Xe in argon. Some literature generates the impression that elements below $m/z = 80$ cannot be determined at all by ICP–MS because of polyatomic ion interferences like those described above. It is more accurate to state that the light elements are not done as well as the heavy ones. As mentioned previously, many of the polyatomic ion interferences for light ions can be resolved if a double focusing MS is used (Bradshaw et al. 1989).

Representative spectra from a lanthanide and an actinide are shown in figs. 21 and 22. The most abundant analyte peaks are from monatomic ions (M^+), and these are observed at sensitivities of 10^5–10^6 count s^{-1} per mg ℓ^{-1} in solution. Ion count rates as low as 2 counts s^{-1} can be distinguished from the background, so the detection limits for most elements are of the order of 10–100 ng ℓ^{-1}. At present, these powers of detection are superior to those obtainable with any other common multi-element technique. Atomic absorption spectrometry with electrothermal vaporization does provide detection limits in a similar range but is generally used only for single-element determinations.

Rare earths tend to yield the most complex spectra of any element in ICP–MS in that there can be significant amounts of M^+, M^{+2}, MO^+ and MOH^+ under the same conditions. Both the absolute and relative abundances of these species depend on the operating conditions chosen (Olivares and Houk 1985, Horlick et al. 1985, Zhu and Browner 1987, Gray and Williams 1987, Longerich et al. 1987, Vaughan et al. 1987). Generally, operating conditions can be identified such that the M^+ peak is by far the major analyte ion observed from all the elements. However, even a small residual amount (e.g., 1%) of oxide ion from the major matrix element(s) can obscure peaks from trace analytes. An example is shown in fig. 23; the oxide ions FeO^+, MnO^+, and $MnOH^+$ overlap with some analyte peaks from Ga and Ge (Date et al. 1987). Fortunately, it is easy to predict where

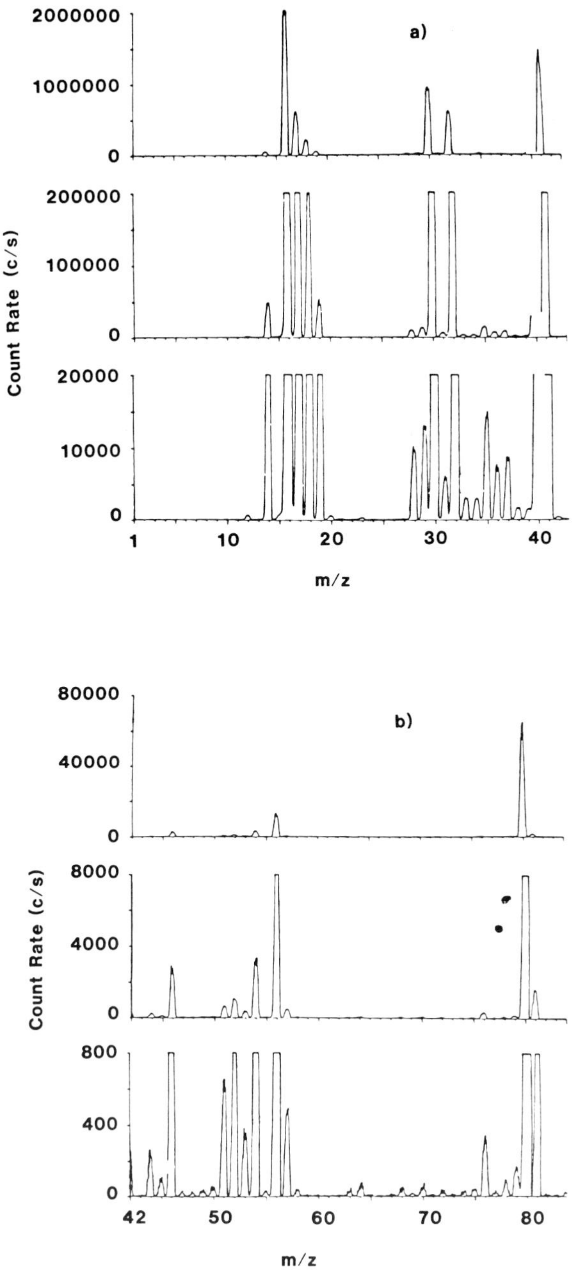

Fig. 20. Background spectra obtained by ICP–MS for 5% HNO_3 in de-ionized water. $^{40}Ar^+$ is not shown because of its very high count rate. Other background ions include N^+ ($m/z = 14$), O^+ (16), H_2O^+ (18), NO^+ (30), O_2^+ (32), ArH^+ (41), ArN^+ (54), ArO^+ (56) and Ar_2^+ (76–80). Some Cl^+ (35, 37) and ClO^+ (51, 53) are also observed from impurities in the acid. Reproduced with permission (Tan and Horlick 1986).

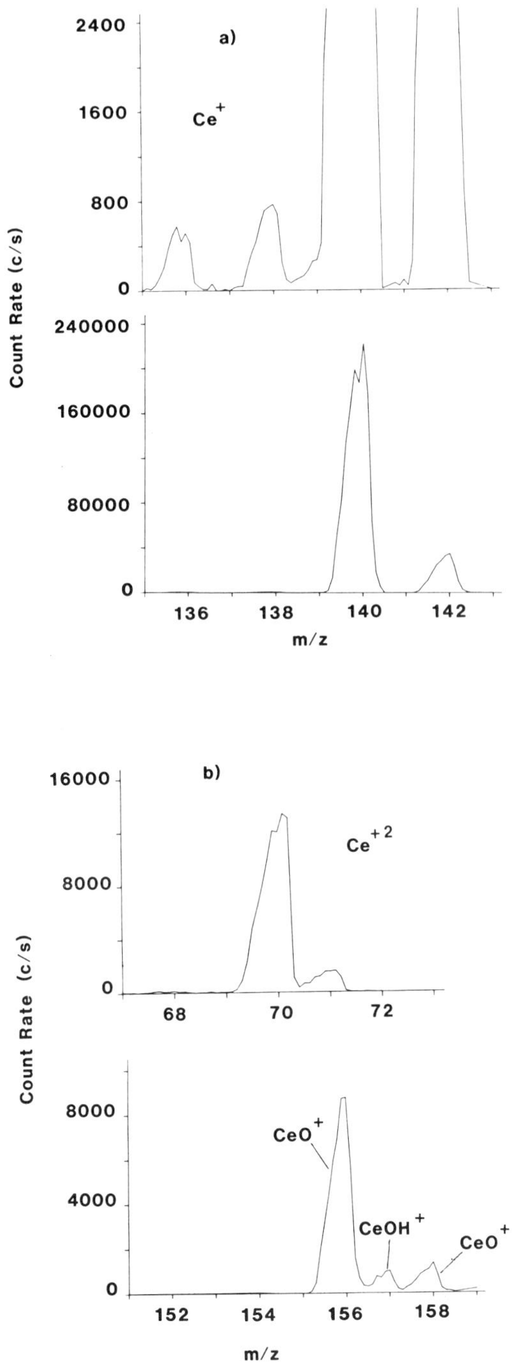

Fig. 21. ICP mass spectra showing Ce^+ (a) and other Ce species (b). The Ce concentration was 400 μg ℓ^{-1}. Note different vertical scales.

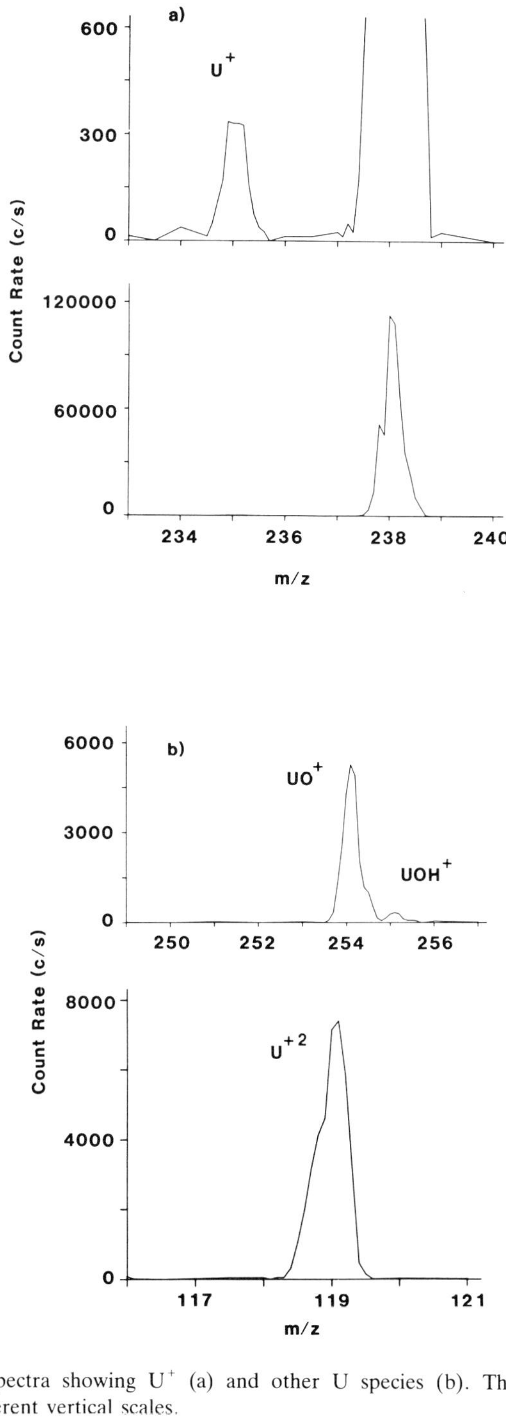

Fig. 22. ICP mass spectra showing U^+ (a) and other U species (b). The U concentration was 400 μg ℓ^{-1}. Note different vertical scales.

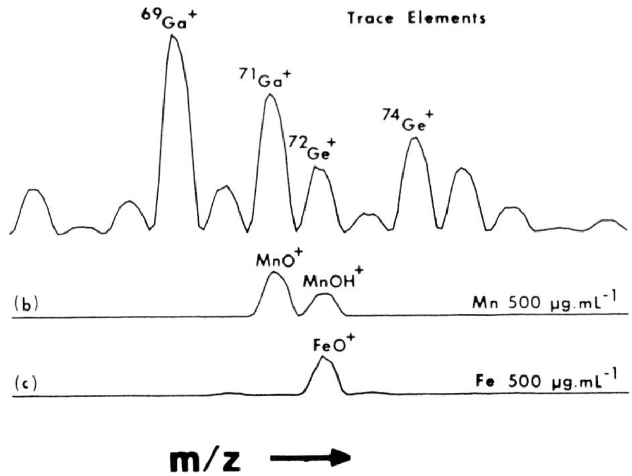

Fig. 23. Spectra obtained for standard solutions containing trace elements (1 mg ℓ^{-1}, top) and matrix elements (Mn and Fe) showing oxide and hydroxide ion interferences. Reproduced with permission (Date et al. 1987).

the oxide peaks occur. In many cases, accurate corrections can be done for overlap problems due to matrix oxide ions, particularly if either the analyte or the interferent have more than one isotope. Lichte et al. (1987) have described a clever correction method based on standardization of oxide abundances relative to that for ThO^+. Although other polyatomic ions from the matrix element, such as ArM^+, are at even lower levels than the oxides, they can be troublesome in some samples as well. Still, for the lanthanides and actinides, problems with spectral interferences are less frequent, less severe, and more easily correctable for ICP–MS than for ICP–AES. For example, one of the major projected applications of ICP–MS is for analysis of uranium and other actinides.

In the absence of spectral overlap, analyte ion signals in ICP–MS depend more extensively on the identity and concentration of concomitants than in ICP–AES. With present ICP–MS devices, analyte signal generally decreases as the total solute level of the sample increases, as indicated in fig. 24 (Olivares and Houk 1986, Thompson and Houk 1987). In addition, the magnitude of the interference depends on the matrix element, the analyte element, and the operating conditions (Beauchemin et al. 1987, Tan and Horlick 1987). At the time of writing, the physical reasons for this effect are not yet fully understood (Tan and Horlick 1987, Gillson et al. 1988, Crain et al. 1988), but there are several solutions to this problem. The first approach is to simply dilute the solution so that the total solute level is too low to cause extensive interference. A total solute level of 0.1% takes care of much of the interference in many cases. For comparative purposes, 1% solutions can usually be tolerated in ICP–AES. In ICP–MS, the sampling orifice begins to plug if the solute level is much above 0.5%, and the nebulizer and/or

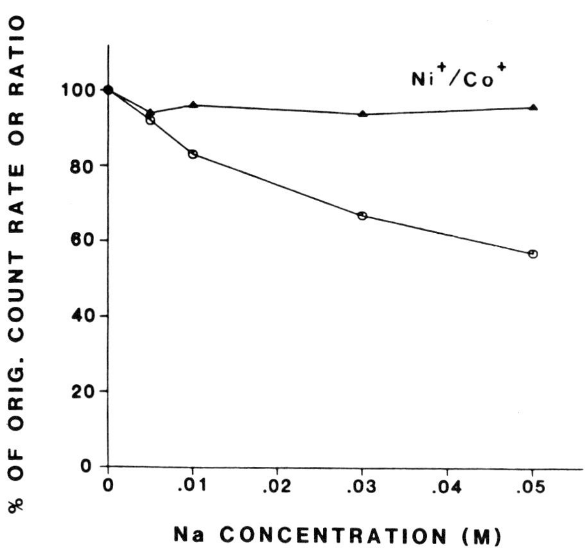

Fig. 24. Effect of Na concomitant in ICP–MS. Co^+ signal (○) is suppressed as Na concentration increases, but the signal ratio (Ni^+ signal/Co^+ signal) (▲) does not change greatly with Na concentration. Reproduced with permission (Thompson and Houk 1987).

torch tip plug anyway at solution concentrations much above 1%, so the additional dilution is often not a severe handicap.

The second approach to circumventing matrix effects in ICP–MS is to compensate for the interference. An example is shown in fig. 24. The Co^+ signal decreases as the Na concentration increases, but the ratio of Ni^+ signal to Co^+ signal is not greatly affected by the sodium concentration. Thus, internal standardization can correct for this matrix effect, although the analyst can not assume this is so and must verify that the signals for analyte and internal standard respond similarly to the concomitants expected (Thompson and Houk 1987, Beauchemin et al. 1987). Internal standardization can also be used to compensate for drift of the sensitivity with time; with present commercial ICP–MS instruments, the stability of the analyte signal is typically 3–5% RSD over a few hours, which is somewhat worse than seen for ICP–AES (usually 1% or better).

The spectra in figs. 21 and 22 also show the stable isotope distributions for the rare earths chosen. In general, isotope ratios can be determined by ICP–MS with a precision in the range 0.2–0.6% relative standard deviation. A solution concentration of at least 1 $\mu g\ \ell^{-1}$ is required. Although the precision is not as good and more sample is required relative to classical ionization techniques such as thermal ionization MS, the sample throughput rate for ICP–MS can be very fast (up to approximately one sample per minute). In addition, concentrations of elements with at least two significant isotopes can be determined by isotope dilution, which can also compensate for the matrix effects and drift described in the preceding section.

6. Applications of ICP mass spectrometry

6.1. *Determination of rare earths in dissolved geological matrices*

This application has already been described in the previous section on ICP–AES. Relative to ICP–AES, ICP–MS yields simpler spectra and better detection limits for rare earths. Representative detection limits typical of the performance of most devices under multi-element operating conditions are given in table 4. The detection limits follow the trend of isotopic abundances, i.e., detection limits are best for the mono-isotopic elements and poorest for Gd and Nd where less abundant isotopes are monitored.

A mass spectrum of rare earths in a digested rock is depicted in fig. 25. The various rare earths are readily identified by their isotopic distributions. Even though La and Ce are much more abundant than the heavier rare earths, in this study it is straightforward to select at least one useful m/z value for each rare earth without prohibitive interference. Determined values for rare earth concentrations in several standard reference materials are presented in table 5. The agreement with the reference values is, generally, very good, even in those cases where corrections for overlap from oxide ions were necessary. The work presented in table 5 was performed by the same group that provided the ICP–AES results shown in fig. 14. The whole rare earth group could be determined by ICP–MS without a separation and pre-concentration step.

6.2. *Determination of rare earths in oils and solids*

Hutton (1986) reports analytical results for the determination of numerous elements in an oil standard reference material. Some caution is necessary for determination of elements at low m/z values because of additional background ions observed when organic solvents are nebulized. However, at the high m/z values corresponding to rare earth ions, the spectrum is not much different from that observed during nebulization of aqueous solutions. Determined concentrations are shown in table 6. The concentrations of these elements are not certified in this particular reference material; this situation may well change with the excellent capabilities of ICP–MS for rare earth determinations.

TABLE 4

Typical detection limits for rare earths by ICP–MS. Reproduced with permission (Lichte et al. 1987).

Isotope	Abundance (%) of isotope used	Detection limit* ($\mu g\ \ell^{-1}$)	Isotope	Abundance (%) of isotope used	Detection limit* ($\mu g\ \ell^{-1}$)
^{139}La	99.9	0.01	^{159}Tb	100	0.01
^{140}Ce	88.5	0.01	^{163}Dy	24.9	0.03
^{141}Pr	100	0.01	^{165}Ho	100	0.01
^{146}Nd	17.3	0.05	^{168}Er	27.1	0.03
^{152}Sm	26.6	0.03	^{169}Tm	100	0.01
^{151}Eu	47.9	0.015	^{174}Yb	31.9	0.03
^{157}Gd	15.7	0.05	^{175}Lu	97.4	0.01

*Represents total elemental concentration.

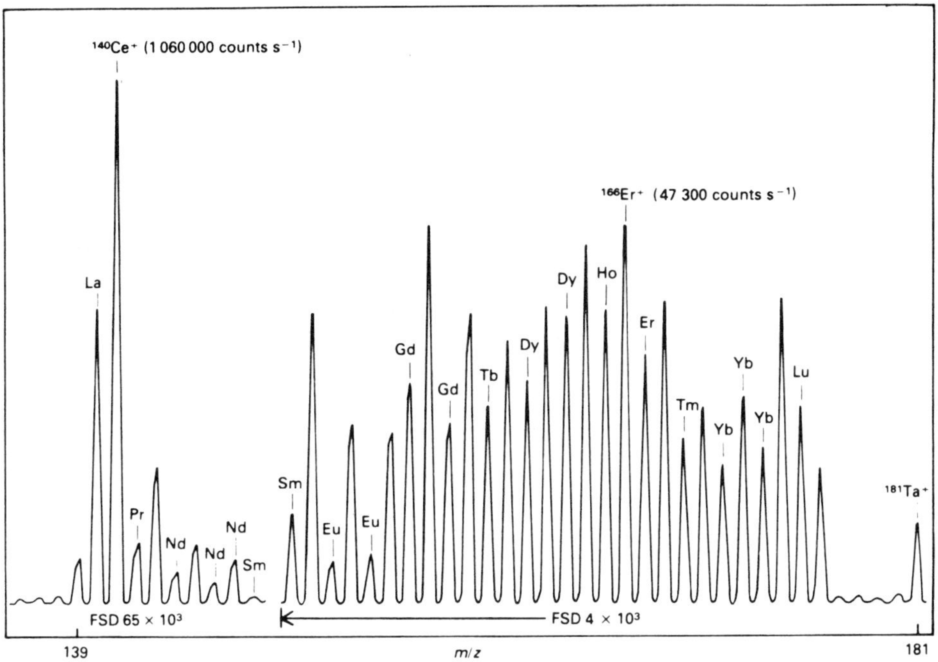

Fig. 25. ICP–MS spectrum for rare earths in a dissolved syenite standard reference material (SY-3). Ce is present at ~20 mg ℓ^{-1}. In this sample, Lu is the least abundant rare earth and is easily observed at ~80 μg ℓ^{-1} (8 μg g^{-1} in the original sample). Reproduced with permission (Date and Hutchinson 1987).

TABLE 5

Determination of rare earths in geological standard reference materials by ICP–MS. Reproduced with permission (Lichte et al. 1987).

Element	Basalt BCR-1		Coal NBS 1632a	
	ICP–MS value (μg g^{-1})	Ref. value(s)	ICP–MS value (μg g^{-1})	Ref. value
La	26.7	26.6, 25.3	14.7	14.5
Ce	54.4	53.8, 54.0	30.6	28.5
Pr	7.0	7.29	3.74	3.3
Nd	28.9	29.7, 28.6	13.4	13.0
Sm**	6.8	6.7, 6.7	2.62	2.6
Eu*	1.99	1.98, 1.95	0.53	0.49
Gd**	7.1	6.9, 6.63	2.01	2.4
Tb**	1.08	1.0	0.31	0.3
Dy**	6.59	6.72, 6.3	1.84	2.1
Ho*	1.29	1.4	0.36	0.38
Er*	3.74	3.8, 3.61	1.01	0.91
Tm*	0.53	0.57	0.15	0.4
Yb*	3.55	3.7, 3.39	0.98	0.90
Lu**	0.54	0.524, 0.54	0.13	0.15

*Result corrected for overlap of one oxide peak.

**Result corrected for overlap of two oxide and/or hydroxide peaks.

TABLE 6
Determination of rare earths in fuel oil (NBS 1634a) by ICP–MS. Reproduced with permission (Hutton 1986).

Element	Determined concentration ($\mu g\ g^{-1}$)	Element	Determined concentration ($\mu g\ g^{-1}$)
La	1.29	Tb	0.01
Ce	0.56	Dy	0.04
Pr	0.22	Ho	0.006
Nd	0.09	Er	0.02
Sm	0.08	Tm	0.005
Eu	0.02	Yb	0.04
Gd	0.05	Lu	0.007

As stated in the section on sample introduction (section 2.1), dissolution of materials containing rare earths is often difficult and/or time-consuming. Therefore, some research has addressed various ways of circumventing sample dissolution by introducing solids directly into ICPs. Mass spectra for rare earth analytes are shown for arc nebulization of an ore (fig. 26) and laser ablation of a ceramic (fig. 27). Because of the rapid scanning possible with the mass spectrometer, virtually all the elements can be identified and determined semi-quantitatively (i.e., to within a factor of two to five) quickly by ICP–MS. For these studies, ICP–MS is particularly attractive because of the high and fairly uniform sensitivity for most elements, the simple, readily interpreted spectra, and the ability to follow a transient sample introduction pulse. The combination of a solid sampling

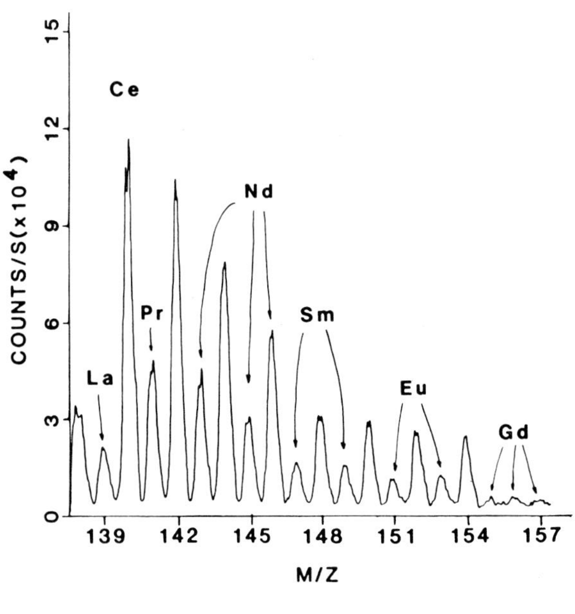

Fig. 26. Mass spectrum of Lemki Pass ore (No. 58098) nebulized by an arc discharge. Nd is present at 1.68%, Gd at 0.08% in the original sample. Reproduced with permission (Jiang and Houk 1987).

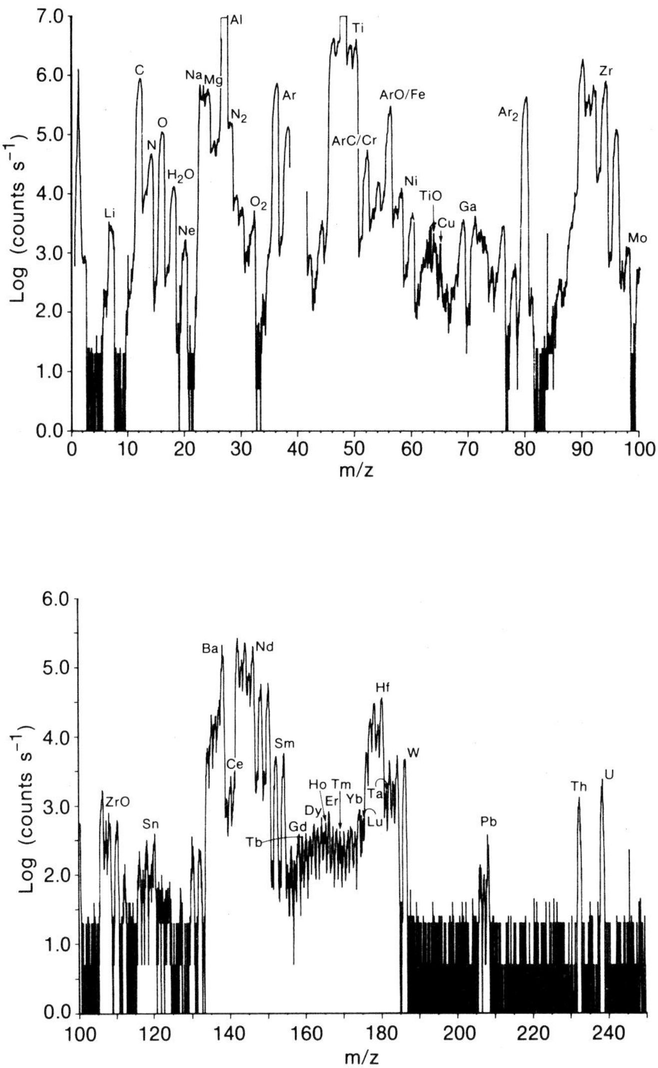

Fig. 27. Mass spectra of Al_2O_3/TiC/ZrO_2 ceramic obtained by ICP–MS with laser ablation of dry solid. Zirconium is present at ~1%, Sm is at ~50 $\mu g\ g^{-1}$, and Tm is observable at ~1 $\mu g\ g^{-1}$. Note low levels of oxide ions (e.g., ZrO^+/Zr^+ ~10^{-3}) when sample is introduced without nebulized water. Courtesy of P. Arrowsmith, IBM Research Center.

technique with ICP–MS would be very valuable if a variety of solids could be analyzed without standards that closely mimic the sample matrix, as preparation of matched, homogeneous solid standards is very difficult and expensive. This capability has not yet been demonstrated but is an important grail for future research in this area.

7. Conclusion

ICP–AES has already exerted a major impact on analytical spectroscopy, as more multi-element determinations are presently done by this technique than by any other. ICP–MS could exert a similar impact in the near future, particularly for applications requiring detectability as the main figure of merit. There are some weaknesses and room for improvement in both ICP techniques, however. The main problems in ICP–AES for rare earths are spectral interferences and detection limits that are not quite good enough for many applications. In ICP–MS, suppression of oxide ions, improvements in precision, and alleviation of matrix interference effects are desirable. Nevertheless, both techniques represent substantial improvements upon what was available previously for multi-element determinations, and both are likely to be used for some time to come.

Acknowledgement

Ames Laboratory is operated for the US Department of Energy by Iowa State University under Contract No. W-7405-Eng-82. This research was supported by the Director for Energy Research, Office of Basic Energy Sciences. The author gratefully acknowledges the assistance of Jeffrey Crain.

References

Barnes, R.M., 1978, CRC Crit. Rev. Anal. Chem. **7**, 203.

Beauchemin, D., J.W. McLaren and S.S. Berman, 1987, Spectrochim. Acta B **42**, 467.

Blades, M.W., and G. Horlick, 1981a, Spectrochim. Acta B **36**, 861.

Blades, M.W., and G. Horlick, 1981b, Spectrochim. Acta B **36**, 881.

Boumans, P.W.J.M., 1980, Line Coincidence Tables for Inductively Coupled Plasmas Atomic Emission Spectrometry, Vol. 1 (Pergamon Press, Oxford).

Boumans, P.W.J.M., 1984, Line coincidence tables for Inductively Coupled Plasmas Atomic Emission Spectrometry, Vol. 1 (Pergamon Press, Oxford).

Boumans, P.W.J.M., ed., 1987, Inductively Coupled Plasmas Emission Spectrometry, part 1 (Wiley, New York).

Boumans, P.W.J.M., J.A. Tielrooy and F.J.M.J. Maessen, 1988, Spectrochim. Acta B **43**, 173.

Bradshaw, N., D. Gregson and P.D. Blair, 1989, Pittsburgh Conf. on Analytical Chemical Application Spectroscopy (Atlanta, Georgia) paper No. 1274.

Browner, R.F., and A.W. Boorn, 1984, Anal. Chem. **56**, 786A, 875A.

Crain, J.S., R.S. Houk and F.G. Smith, 1988, Spectrochim. Acta B **43**, 1355.

Crock, J.G., and F.E. Lichte, 1982, Anal. Chem. **54**, 1329.

Date, A.R., and A.L. Gray, eds, 1989, Applications of Inductively Coupled Plasma Mass Spectrometry (Blackie, London).

Date, A.R., and D. Hutchison, 1987, J. Anal. Atomic Spectrom. **2**, 269.

Date, A.R., Y.Y. Cheung and M.E. Stuart, 1987, Spectrochim. Acta B **42**, 3.

Dawson, P.H., ed., 1976, Quadrupole Mass Spectrometry and Its Applications (Elsevier, Amsterdam).

de Galan, L., and P.S.C. van der Plas, 1987, Low-Gas-Flow Torches for ICP Spectrometry, in: Inductively Coupled Plasmas in Analytical Atomic Spectrometry, eds A. Montaser and D.W. Golightly (VCH Publishers, New York) ch. 14.

Douglas, D.J., and R.S. Houk, 1985, Prog. Anal. Atomic Spectrosc. **8**, 1.

Edelson, M.C., 1987, High-Resolution Plasma Spectroscopy, in: Inductively Coupled Plasmas in Analytical Atomic Spectrometry, eds A. Montaser and D.W. Golightly (VCH Publishers, New York) ch. 7.

Edelson, M.C., and V.A. Fassel, 1981, Anal. Chem. **53**, 2345.

Faires, L.M., B.A. Palmer, R. Engleman Jr and T.M. Niemczyk, 1984, Spectrochim. Acta B **39**, 819.

Fassel, V.A., 1978, Science **202**, 183.

Furuta, N., and G. Horlick, 1982, Spectrochim. Acta B **37**, 53.

Gillson, G.R., D.J. Douglas, J.E. Fulford, K.W. Halligan and S.D. Tanner, 1988, Anal. Chem. **60**.

Gray, A.L., 1985, Spectrochim. Acta B **40**, 1525.

Gray, A.L., and J.G. Williams, 1987, J. Anal. Atomic Spectrom. **2**, 599.

Hasegawa, T., and H. Haraguchi, 1987, Fundamental Properties of Inductively Coupled Plasmas, in: Inductively Coupled Plasmas in Analytical Atomic Spectrometry, eds A. Montaser and D.W. Golightly (VCH Publishers, New York) ch. 8.

Horlick, G., S.H. Tan, M.-A. Vaughan and C.A. Rose, 1985, Spectrochim. Acta B **40**, 1555.

Houk, R.S., 1986, Anal. Chem. **58**, 97A.

Houk, R.S., and J.J. Thompson, 1988, Mass Spectrom. Rev. **7**, 425.

Houk, R.S., V.A. Fassel, G.D. Flesch, H.J. Svec, A.L. Gray and C.E. Taylor, 1980, Anal. Chem. **52**, 2283.

Hutton, R.C., 1986, J. Anal. Atomic Spectrom. **1**, 259.

Jiang, S.-J., and R.S. Houk, 1987, Spectrochim. Acta B **42**, 93.

Keliher, P.N., W.J. Boyko, R.H. Clifford, J.L. Snyder and S.F. Zhu, 1986, Anal. Chem. **58**, 335.

Kurz, E.A., 1979, Am. Lab. **11**(13), 67.

Larson, G.F., V.A. Fassel, R.H. Scott and R.N. Kniseley, 1975, Anal. Chem. **47**, 238.

Lichte, F.A., A.L. Meier and J.G. Crock, 1987, Anal. Chem. **59**, 1150.

Longerich, H.P., B.J. Fryer, D.F. Strong and C.J. Kantipuly, 1987, Spectrochim. Acta B **42**, 75.

Montaser, A., and D.W. Golightly, 1987, Inductively Coupled Plasmas in Analytical Atomic Spectrometry (VCH Publishers, New York).

Olivares, J.A., and R.S. Houk, 1985, Anal. Chem. **57**, 2674.

Olivares, J.A., and R.S. Houk, 1986, Anal. Chem. **58**, 20.

Reeves, R.O., S. Nikdel and J.D. Winefordner, 1980, Appl. Spectrosc. **34**, 477.

Rubinson, K.A., 1987, Chemcal Analyses (Little, Brown and Co., Boston) p. 728.

Sobel, H.R., and R. Dahlquist, 1981, Am. Lab. **13**, 152.

Tan, S.H., and G. Horlick, 1986, Appl. Spectrosc. **40**, 445.

Tan, S.H., and G. Horlick, 1987, J. Anal. Atom. Spectrom. **2**, 745.

Thompson, J.J., and R.S. Houk, 1987, Appl. Spectrosc. **41**, 801.

Thompson, M., 1987, Analytical Performance of Inductively Coupled Plasma-Atomic Emission Spectrometry, eds A. Montaser and D.W. Golightly, in: Inductively Coupled Plasmas in Analytical Atomic Spectrometry (VCH Publishers, New York) ch. 5.

Thompson, M., and M.H. Ramsey, 1985, Analyst **110**, 1413.

Thompson, M., and J.N. Walsh, 1983, A Handbook of Inductively Coupled Plasma Spectrometry (Blackie, London).

Vaughan, M.-A., G. Horlick and S.H. Tan, 1987, J. Anal. Atom. Spectrom. **2**, 765.

Winge, R.K., V.A. Fassel, V.J. Peterson and M.A. Floyd, 1985, Inductively Coupled Plasma-Atomic Emmision Spectroscopy. An Atlas of Spectral Information (Elsevier, Amsterdam).

Yoshida, K., and H. Haraguchi, 1984, Anal. Chem. **56**, 2580.

Zhu, G., and R.F. Browner, 1987, Appl. Spectrosc. **41**, 349.

Handbook on the Physics and Chemistry of Rare Earths, Vol. 13
edited by K.A. Gschneidner Jr. and L. Eyring

Chapter 92

RARE EARTH ELEMENTS IN BIOLOGICAL SYSTEMS

Patrick H. BROWN
Department of Pomology, University of California – Davis, Davis, CA 95616, USA

Anne H. RATHJEN
CSIRO Division of Horticulture, Urrbrae, 5064, South Australia,

Robin D. GRAHAM
Waite Agricultural Research Institute, University of Adelaide, Glen Osmond, 5064, South Australia

Derek E. TRIBE
Crawford Fund for International Agricultural Research, 1 Leonard Street, Parkville, 3052, Victoria, Australia

---

Contents

1. Introduction

Originally named because of their similarity to the "earths" (i.e. magnesia, lime, etc.) the rare earth elements (REE) are not at all rare but as a group are the fifteenth most abundant component of the earth's crust. The first such element discovered was named yttria in 1794, and by 1904 all except promethium (Pm) had been identified. It was not until 1948 that this final rare earth was isolated as a radioactive byproduct of uranium fission as it does not occur naturally in a stable form.

This discovery completed the sequence of 14 elements occurring between the atomic numbers 58 and 71 inclusive. The group did not fit into the original periodic tables and was included as the separate "lanthanide series". Lanthanum itself, together with two other group 3A elements, yttrium and scandium (and sometimes thorium), is often included with the lanthanide series elements in discussions of the REE. This is because they frequently occur together in rare earth minerals, having similarities in ionic radii and chemical activity.

The REE are all remarkably similar in their chemical and physical characteristics, indeed so much so that several lanthanides always occur together in the one mineral. Their separation, which was for many years one of the most difficult problems in inorganic chemistry, involves complex chelation and column absorption techniques which depend on only slight differences in equilibrium constants between the elements.

The REE are frequently subdivided into three groups (table 1); the light [lanthanum (La) to europium (Eu)], middle [samarium (Sm) to holmium (Ho)] and heavy earths [gadolinium (Gd) to lutetium (Lu)] (Topp 1965).

TABLE 1
Classification of the lanthanide elements into light, middle and heavy earths (after Topp 1965).

Element	Symbol	Atomic number	Descriptive classification
Lanthanum	La	57	
Cerium	Ce	58	
Praseodymium	Pr	59	
Neodymium	Nd	60	--------**Light earths** (La–Eu)
Promethium	Pm	61	
Samarium	Sm	62	
Europium	Eu	63	
Gadolinium	Gd	64	
Terbium	Tb	65	-------**Middle earths** (Sm–Ho)
Dysprosium	Dy	66	
Holmium	Ho	67	
Erbium	Er	68	--------- **Heavy earths** (Gd–Lu)
Thulium	Tm	69	
Ytterbium	Yb	70	
Lutetium	Lu	71	

1.1. *Abundance and distribution of the REE*

Data on the abundance of the REE in igneous rocks are presented in table 2 (Moeller 1963). The most plentiful rare earths are cerium (Ce) and yttrium (Y) occurring at an abundance similar to tin (Sn) and cobalt (Co). Neodymium (Nd) and lanthanum (La) occur in similar quantities to lead (Pb) and all the other REE occur at a level of abundance somewhat above silver (Ag) and cadmium (Cd).

Table 2 indicates that the REE are not rare as such, many being more abundant than some established trace elements. They are widely distributed but are generally found in only small amounts (Gschneidner 1977).

There are three groups of minerals in which the REE are found. Minerals in the first group contain major quantities of lanthanides. All of these are associated with crystallizations from magmatic mother liquors of pegmatic character (Topp 1965). Important examples are the minerals monazite and xenotime. The second group includes minerals with the REE as minor constituents. Many calcium minerals, such as apatite, are members of this group. Minerals in the third group contain the REE in the bipositive state in small isolated distributions. These are not used as sources of rare earths.

It is well-known that rare earths with even atomic numbers are more abundant than those with adjacent odd atomic numbers (Markert 1987). It is also true that the quantity of elements with even atomic numbers decreases approximately linearly with increasing atomic number.

TABLE 2
Abundances of elements in igneous rocks of the crust of the earth (after Moeller 1963).

Symbol	Atomic number	Abundance (mg kg^{-1})	Symbol	Atomic number	Abundance (mg kg^{-1})
Sc	21	5	B	5	<3
Y	39	28.1	N	7	46.3
La	57	18.3	Co	27	23
Ce	58	46.1	Ni	28	60
Pr	59	5.53	Cu	29	70
Nd	60	23.9	Zn	30	132.0
Pm	61	~0	Ga	31	15
Sm	62	6.47	Ge	32	7
Eu	63	1.06	As	33	5
Gd	64	6.36	Br	35	1.62
Tb	65	0.91	Mo	42	2.5–15
Dy	66	4.47	Ag	47	0.1
Ho	67	1.15	Cd	48	0.15
Er	68	2.47	Sn	50	40
Tm	69	0.2	Sb	51	1
Yb	70	2.66	I	53	0.1
Lu	71	0.75	Pb	82	16
Be	4	6	Bi	83	0.2

The chemical similarities of the natural REE (which excludes Pm) results in their universal occurrence in all rare earth minerals. The rare earths have a valence of three, are strongly electropositive and have little tendency to hydrolyze. Their strong affinity for both oxygen and phosphate results in the tendency for rare earth minerals to occur as phosphate complexes. In table 3 the major rare earth minerals, their usual form and the abundance of the REE in them are listed (Moeller 1963). Table 4 indicates known rare earth reserves.

The REE content of the earth's crust averages about $154\,\mathrm{mg\,kg^{-1}}$ and in soil $100–200\,\mathrm{mg\,kg^{-1}}$.

1.2. *Chemical and physical characteristics*

The chemical and physical characteristics of the REE have been adequately reviewed in earlier editions of the *Handbook on the Physics and Chemistry of Rare Earths* and elsewhere (Cockerill et al. 1973, Gschneidner 1981, 1988, Topp 1965, Weiss 1974) and only brief mention will be made here of the major properties that will be relevant in later sections.

The ground-state electronic configuration of La is the xenon core plus $5d^1$ and $6s^2$. After La, subsequent electrons are added to the lower-energy 4f orbitals rather than filling the 5d orbitals. This phenomenon is important in understanding the REE as many properties result from the shielding of the inner 4f orbitals (Mikkelson 1976). The neutral atoms contain three outer electrons which distribute themselves between the 6s, 5d and 4f subshells. These electrons are removed to the negative ions in the rare earth trivalent compounds and do not take part in electronic configuration of the rare earth ions.

The REE are generally isolated in the +3 valence state. Cerium, however, may exist in a stable tetrapositive state and Pr and Tb are known to form higher valence oxides. Europium ions (Eu^{2+}), Sm^{2+} and Yb^{2+} can also be isolated although they are strong reducing agents.

TABLE 3
Important rare earth-containing minerals (after Moeller 1963).

Name	Generalized composition	Location of significant deposits
Monazite	49–74% Ce earths; 1–4% Y earths	China; Western Australia; India; Brazil; Zaire; South Africa; USA (Florida, North and South Carolian, Idaho); USSR
Bastnasite	65–70% Ce earths; <1% Y earths	China; USA (California, New Mexico); USSR
Euxenite*	13–35% Y earths; 2–8% Ce earths	Australia; USA (Idaho)
Xenotime	54–65% Y earths; ~0.1% Ce earths	Western Australia; Norway; Brazil
Gadolinite	35–48% Y earths; 2–17% Ce earths	Sweden; Norway; USA (Texas and Colorado)

*Name used when $(Nb, Ta)_2O_5/TiO_2 = 1:4$ or more; when 1 : 3 or less, it is called polycrase.

TABLE 4
World rare earth reserves (Jackson 1985).

Location	Reserves (tonnes)
North America	
United States	4 900 000
Canada	182 000 +
Total	*5 100 000*
South America	
Brazil	20 000
Total	*20 000*
Europe	
USSR	450 000
Finland, Norway, Sweden	50 000 +
Total	*500 000*
Africa	
Burundi	1 100
Egypt	100 000
Kenya	12 500
Malagasy Republic	50 000
Malawi	297 000
Republic of South Africa	357 000 +
Total	*820 000*
Asia	
Australia	184 000
China	36 000 000
India	2 200 000
Malaysia	30 000
Republic of Korea	45 000
Sri Lanka	13 000
Thailand	1 000 +
Total	*38 500 000*

All REE have the same configuration of outer electrons and, as these electrons predominantly determine chemical properties, the rare earths possess similar chemical and physical attributes. The REE exhibit diamagnetism due to their completed subshells and successive filling of the 4f subshell gives rise to the characteristic paramagnetic properties of the middle and heavy rare earths.

1.2.1. *Ionic size*

An unusual and significant property of the REE is the so-called "lanthanide contraction", a term referring to the reduction in atomic radius that occurs with increasing atomic number. This arises from insufficient shielding of the nuclear charge with each additional 4f electron. The largest contraction occurs between La and Ce, each successive decrease being smaller than the previous (Wolniak et al. 1980). These sequential reductions in ionic radius are exhibited both by the rare earth ions in various charge states and by the rare earth metals themselves (Topp 1965).

One of the most striking effects of the lanthanide contraction is displayed in the behaviour of yttrium, which, although it has a much smaller nuclear charge ($Z = 39$), has a radius similar to holmium ($Z = 67$). For this reason yttrium is frequently isolated among the heavy earths and often included in discussions of the REE.

Perhaps the most significant biological effect of the lanthanide contraction results from the fact that the REE ions have an ionic radius similar to the calcium ion Ca^{2+}. Within the lanthanides the radius of the tripositive ions ranges from 8.5×10^{-2} to 1.15×10^{-1} nm. Calcium has an ionic radius of 9.9×10^{-2} nm. The extra positive charge on the REE, as compared to Ca^{2+}, will, in most cases, contribute to the stability of complexes formed by lanthanides (Mikkelson 1976). This results in the inhibition of many Ca^{2+}-dependent enzymes by the REE (see Polya et al. 1987). It is this similarity to and potential to replace and compete with Ca^{2+} that has been the emphasis of most research into rare earths in biological systems.

1.2.2. *Bonding characteristics*

An investigation of the bonding characteristics of Ca and the rare earths demonstrates further the similarities between the two. In contrast to the transition metals, neither Ca nor the REE show significant covalent bonding or crystal-field stabilization (Mikkelson 1976). This is because, unlike the transition metals, the outer orbitals of Ca are of very high energy and thus unfavourable to covalent bonding. In the lanthanides, the extra d electron tends to increase the covalent character of the bonding, which, however is still predominately ionic. These similarities between Ca and the REE have been put forward to account for many of the interactions between the two in both biological and nonbiological systems.

2. The role of REE in biological systems

2.1. *Introduction*

The use of REE in biological systems can be subdivided into three categories. First, a large amount of work has been devoted to the use of rare earths, in particular La, as a probe in systems of biological interest (Reuben 1979), particularly for Ca in biomembranes (Mikkelson 1976). Second, many studies have utilized REE as markers to trace the movement and deposition of elements in tissues. These include the use of La to monitor the movement of Ca and water within isolated membranes, whole cells and whole organisms both animals and plants [for a review see Mikkelson (1976) and Weiss (1974)]. Finally, lanthanides have been used in investigating the role of Ca^{2+} in muscle and nerve activity. This has been done by using La^{3+} and other REE as competitive inhibitors of Ca^{2+}.

Much of the early work using rare earths as monitors of Ca^{2+} in cells was based on their close physical and chemical similarities and was aided by some useful REE characteristics. Examples of these are their paramagnetism, useful NMR spectra and the electron dense properties which make the REE visible under the electron microscope.

The work performed on animal systems utilizing the REE has been largely devoted to developing an understanding of the role of calcium in the cell. In plants, bacteria and some fungi, other studies have been carried out on aspects of the REE, including effects on disease resistance, enzyme activity, nitrogen fixation, interactions with hormones and growth responses.

2.2. *Influence of the REE on calcium metabolism*

La^{3+}, by virtue of an ionic radius similar to and a valence higher than Ca^{2+}, was predicted by Lettvin et al. (1964) to bind at superficially located Ca^{2+} adsorption sites in a less reversible manner than does Ca^{2+} itself. This work was closely followed by a report on the displacement of Ca^{2+} from membranes in the presence of La^{3+} (Doggenweiler and Frenk 1965). A differential binding of La^{3+} by the constituents making up the outer strata of the unit membrane – the lipid polar groups and the nonpolar components – was demonstrated. Lanthanum ions were found to displace Ca^{2+} from a monolayer of phosphatidyl serine and phosphatidyl ethanolamine but not from phosphatidyl choline.

Doggenweiler and Frenk (1965) concluded that La^{3+} acts as a "super calcium", i.e., La^{3+} is some 20 times as effective as Ca^{2+} in its effect on the Na conductance curve (Takata et al. 1966). Similarly, the REE terbium (Tb^{3+}) has been shown to compete with Ca^{2+} for binding sites both on proteins (Gross et al. 1981) and at cell and vesicle membranes (Deschenes et al. 1981). Most of the evidence for the competition of rare earths for external Ca^{2+} binding sites has, however, been determined using La^{3+} (Mikkelson 1976, Neldhuis 1982, Revel and Karnowsky 1967, Weiss 1974).

Many enzymes and other proteins have been demonstrated to be subject to inhibition by La^{3+}. An example of this is the inhibition of spinach ferredoxin by micromolar concentrations of $LaCl_3$ (Surek et al. 1987). Interestingly, it required *milli*molar concentrations of $MgCl_2$ to achieve the same level of restriction. The isolation of two Ca^{2+}-dependent protein kinases from silver beet leaves allowed the determination of the inhibition of these enzymes by various rare earth ions, all at a concentration of 1 mM (Polya et al. 1987). This ranged from 66% (La(acetate)$_3$) to 89% ($CeCl_3$), 91% ($Gd(NO_3)_3$), 93% ($Ho(NO_3)_3$) and finally a 97% inhibition of enzyme I by $Sm(NO_3)_3$. Results were fairly similar for enzyme II, being 97, 99, 99, 98 and 98%, respectively. It is interesting to note that lanthanum inhibited enzyme II much more dramatically than enzyme I. This could be due to a greater number of Ca^{2+} sites on the second enzyme, as in all cases the percentage inhibition was increased. Similar results have also been obtained with Ca^{2+}-dependent protein kinases isolated from wheat germ (Polya and Micucci 1985).

In almost all cases La^{3+} replaces Ca^{2+} on protein binding sites. In the horseradish peroxidase, however, the reverse is the case, with the enzyme having a *lower* affinity for lanthanum than calcium ions (Morishima et al. 1986). Extrapolation of the effect of competition of lanthanum ions for calcium binding sites is therefore fairly difficult without intimate knowledge of the individual characteristics of the enzyme concerned.

Recently, Ca^{2+} has been identified as occurring in two distinct bound phases in the cell. Most evidence suggests La^{3+} competes only for phase-1 (easily exchangeable) Ca^{2+}. Langer and Frank (1972), for instance, found that 0.5 mM La^{3+} in cultured rat heart cells displaces 0.52 mM Ca^{2+}, all from phase-1 Ca^{2+} sites.

The effect of La^{3+} on Ca binding has been summarized by Sanborn and Langer (1970). They have concluded the following for work with muscle and nerve membranes:

(i) La^{3+} effects a displacement of tissue-bound Ca^{2+};
(ii) The La^{3+} sensitive Ca^{2+} is kineticaly homogeneous and is identifiable as contractile-dependent Ca^{2+};
(iii) The site of the La^{3+}–Ca^{2+} interaction is superficial rather than internal;
(iv) La^{3+} in general and in concentrations up to 40 μM does not effect the total displacement of phase 2 Ca^{2+};
(v) A significant proportion of phase-2 Ca^{2+} may be internal;
(vi) Contractability or nerve activity can be maintained or at least partially sustained by an internal Ca^{2+} source.

The influx and efflux of Ca^{2+} from animal systems is also known to be influenced by the presence of REE in bathing solution (Freeman and Daniel 1973, Mikkelson 1976, Weiss 1974). Results suggest that La^{3+} can displace Ca^{2+} from extracellular binding sites and can inhibit efflux of extracellular and part of the intracellular Ca^{2+}. Lanthanum ions do not appear to seal the membrane entirely from Ca^{2+} movement, possibly because of the existence of both exposed and internal Ca^{2+} binding sites (Burton and Fernandez 1973, Freeman and Daniel 1973, Gögeleir et al. 1981, Langer and Frank 1972, Revel and Karnowsky 1967). The H^+/Ca^{2+} antiports in oat roots and *Beta vulgaris* storage tissue, however, are also subject to inhibition by La (Blumwald and Poole 1986, Shumaker and Sze 1986), this perhaps causing the observed influences on Ca^{2+} influx and efflux. The restriction of H^+/Ca^{2+} antiports has been attributed to the occupation of calcium binding sites by La^{3+}.

Evidence of rare earths inhibiting the passage of Ca^{2+} through cells is also available for Pr^{3+} (Hunt and Jawaharlal 1980) and Tb^{3+} (Deschenes et al. 1981, Gross et al. 1981). Several divalent and trivalent cations have been ranked in order of their effectiveness in inhibiting Na^+-induced Ca^{2+} uptake (Sarkardi et al. 1977). This was carried out in the presence of 40 mM Na^+ and 20 μM Ca^{2+} and results in the finding that La^{3+} offers the greatest inhibition, followed by $Nd^{3+} > Tm^{3+} = Y^{3+} > Cd^{2+} \gg Sr^{2+} > Ba^{2+} = Mn^{2+} \gg Mg^{2+}$. The effectiveness of these ions was found to be related to their ionic radius. Ca^{2+} has also been found to be capable of displacing Tb^{3+} from vesicle membranes at quite low concentrations (Gross et al. 1981). The degree of lanthanide inhibition of calcium may be related to ionic concentration. Low concentrations ($<1.6 \times 10^{-7}$ M) stimulate Ca^{2+} exchange whereas higher concentrations may cause 80–90% inhibition (Trosper and Philipson 1983).

2.3. *REE in animal systems*

Of all the trivalent rare earths, La has long been considered to have chemical properties most like the alkaline earths (Weiss 1974). Most of the work carried out on rare earths in animal systems is based upon the use of this rare earth ion as a substitute or antagonist for Ca^{2+} in a variety of cellular and subcellular reactions. The large volume of literature devoted to the subject reflects the complex and varied functions of Ca^{2+} in membrane systems and coupling reactions.

The importance of Ca in living tissue was recognized more than a century ago by Ringer (1883) when he determined its effect in maintaining heat activity and later its part in the cellular adhesion process. These and subsequent demonstrations of the actions of Ca have been reviewed by Mikkelson (1976) who lists the following effects in which Ca^{2+} plays a role:

(i) Cell communication via junctional membranes;
(ii) Actinomyosin contraction;
(iii) Activation of phosphorylase kinase in muscle;
(iv) Hormone–cell interactions;
(v) Determination of cellular permeability;
(vi) Stabilization of cellular membranes.

Calcium in cellular membranes influences permeability and is thus involved in muscular and nerve transmissions. It also has more general effects on monovalent ion transport and has been described as a trigger in the initiation of numerous cellular events.

2.3.1. *Binding sites*

The rare earths appear to bind specifically to anionic groups in the membrane matrix, possibly to the negatively charged groups on the mucopolysaccharides or to a mucopolysaccharide/protein complex (Burton and Fernandez 1973, Freeman and Daniel 1973, Langer and Frank 1972). REE may also bind to the polar head, phosphate head groups or the lipid bilayer (Batra 1982, Gögeleir et al. 1981).

There is some evidence that La^{3+} or other lanthanides may bind to and inhibit the Ca^{2+}/Na^{+} transport apparatus (Trosper and Philipson 1983), either by a direct conformational effect on the Ca^{2+} pump mechanism or by influencing the metabolic activity of the transporting ATPase (Batra 1982, Hunt and Jones 1982, Parker and Barritt 1981, Sarkardi et al. 1977). This could be achieved through a combination either with transport proteins or with phospholipids of the plasma membrane (Parker and Barritt 1981).

Lanthanide cations also restrict the Ca^{2+}/Mg^{2+}-dependent ATPase enzymes with the extent of the inhibition increasing with ionic radius (Agarwal and Kalra 1983, Batra 1982, Hunt and Jones 1982, Sarkardi et al. 1977).

2.3.2. *Effect on specific mammalian functions*

The influence of REE on aspects of Ca^{2+} metabolism has resulted in much work being carried out on their effects on mammalian nerve, heart and muscle tissues (for reviews see Maier-Maercker 1980, Weiss 1974; and also Hatae 1982, Neldhuis 1982, Takata et al. 1966, Trosper and Philipson 1983, Wolniak et al. 1980).

Lanthanum has been observed to modify cell permeability to Na^+, K^+ and Ca^{2+} ions (Hatae 1982, Martinez-Palomo et al. 1973, Takata et al. 1966, Trosper and Philipson 1983), perhaps explaining the dramatic but reversible uncoupling of excitation and contraction in the presence of REE in mammalian (Langer and Frank 1972, Martinez-Palomo et al. 1973, Takata et al. 1966, Trosper and Philipson 1983) and frog heart tissue (Hatae 1982, Mines 1910) as well as in nerve (Takata et al. 1966) and muscle tissue (Freeman and Daniel 1973). This effect is caused by a modification of excitability and transmembrane action potential. Takata et al. (1966) suggests these results "*support the notion of Ca^{2+} ions (and Ca^{2+}-like substances [REE ions]) acting like gates, occluders or competitors for sodium and potassium channels, whatever their character. Adsorbed La^{3+} ions might also alter the distribution of the charged particles within the membrane without changing the overall potential difference between inside and out*".

Amongst other specific effects of REE in animal systems is the inhibition of cell growth by the inhibition of spindle fibre orientations during cell division (Neldhuis 1982, Wolniak et al. 1980). For more detailed information on this and other effects of rare earths in animals the reader is referred to reviews by Mikkelson (1976) and Weiss (1974).

2.3.3. *Active levels of REE in cells*

Much conflicting evidence exists as to active levels of REE in the cell. Quoted values vary from $<2 \times 10^{-7}$ M up to 2 M solution levels. Many of the arguments can be resolved, however, through the realization of differences in the experimental pH, REE source and form and the type of buffer used.

Electron micrographs indicate that colloidal La^{3+} does not penetrate the cell membrane. The ionic form, however, can cross this barrier (Freeman and Daniel 1973). At pH 4–5, $LaCl_3$ is in the dissociated ionic form and in mice the distribution of La^{3+} depends on whether it is ionized or non-ionized, the latter complex being far more mobile and diffusible (Laszlo et al. 1952).

The many conflicting results are further complicated by the effect of particle size and the presence of free phosphate groups (in buffers) to which La^{3+} is strongly attracted. Unfortunately, relatively little is known of the state of rare earths in solution at different pH levels and in various mixes. Most of the information available on the distribution of the REE is based on their visibility under the electron microscope. This may, of course, vary according to particle aggregation or repulsion.

One of the major points of contention amongst researchers is whether or not the REE can pass through the plasma membrane in plants and animals. Almost half the references reviewed favoured the rare earths, and La^{3+} in particular, passing through the membrane to influence cellular metabolism whilst most of the remainder opposed this view. Discrepancies are due to the effects of pH, buffer composition and various other factors on the physical and chemical characteristics of the REE in solution.

2.3.4. *Toxicity*

According to Hodge and Sterner (1943) the REE are only slightly toxic and the extensive review by Haley (1979) similarly concluded that REE have "a very low acute toxicity". Lethal dose rates for various species and routes of administration have been reported by Haley (1965).

The tolerance of animals to oral doses of the stable REE is presumably due largely to the low rate of REE absorption from the intestinal tract. For this reason many of these elements have been used successfully as markers to measure the rates of passage of food residues through the alimentary tract and to estimate the apparent digestibilities of diets (Luckey et al. 1975).

Recent verbal reports from China (personal communications to D.E. Tribe) claim that the supplementation of rations with small amounts of soluble mixed extract of rare earths produced increased growth rates in chickens and pigs, and increased wool growth in sheep. However, no REE has yet been shown to be an essential nutrient for any species and an extensive intestigation by Hutcheson et al. (1975) showed no significant effects on survival, growth or reproduction when the oxides of Sc, La, Sm, Eu, Dy, Tb, Tm, and Yb were fed to mice for three generations.

2.4. *REE in plant and bacterial systems*

As with animal systems a large amount of work has been devoted to the delineation of the involvement of the rare earths in extracellular and extra membrane pathways. These elements have been implicated in a wide range of physiological processes in plants, bacteria and nematodes associated with plants. Many of these, but by no means all, are associated with REE/Ca^{2+} interactions.

2.4.1. *Uptake of REE in plants*

The uptake of REE from soil and solution has been investigatead in a wide range of plants. Concentrations in plant tissue are, on the whole, very low (Pant 1981, Williams 1967) ranging from $<30\ \mu g\ kg^{-1}$ La in dry matter of *Yucca flaccida* (Wolterbeek and Van Die 1980) [and even less for the uptake of the radionuclides ^{144}Ce and ^{147}PR by oat plants (Cummings and Bankert 1971)] to levels of $2.3\ g\ kg^{-1}$ in dry matter of hickory (Robinson 1943). In dry root matter a value of $24–29\ mg\ kg^{-1}$ is approximately the average although the final concentration is influenced by soil chelates present (Knaus and El-Fawaris 1980, Lawton et al. 1981). Non-chelated dysprosium ions (Dy^{2+}) are sorbed by roots of red alder up to a concentration of $29\ mg\ kg^{-1}$ in dry matter while sorption of Dy–DPTA has been found to be less by a factor of eight (Knaus and El-Fawaris 1980).

Alternatively, it has also been suggested that soil chelate systems may enhance the uptake of REE (Garten et al. 1981, Laul et al. 1979).

Uptake may be facilitated by increased levels of nitrogen and potassium fertilizers and decreased by addition of phosphate (Cummings and Bankert 1971, Lawton et al. 1981). The rate of uptake appears to be dependent on the REE ionic radii (Laul et al. 1979) with a decrease in absorption occurring with each successive reduction in ionic radius.

2.4.2. *Transport of REE in plants*

Work using La as a marker for water movement demonstrated that La^{3+} was largely distributed in the inner cellulosic walls (Oparka and Gates 1981). The primary endodermis of the spruce root (*Picea abies*), however, has been demonstrated to prevent the apoplastic transport of lanthanum into the stele (Jorns 1987). This indicates the presence of a casparian strip which is impermeable to La^{3+}. In *Yucca flaccida*, however, La^{3+} penetrated to the phloem cells, demonstrating clearly that it must have crossed the cell membrane at least at the endodermis (Wolterbeek and Van Die 1980).

Transport across the casparian strip may be dependent on the extent of development of the casparian strip (Maier-Maercker 1980), as well as by soil REE levels and pH (Lawton et al. 1981, Oross and Thomson 1982). At a pH of 7.8, small and infrequent deposits of La^{3+} were found in cap cell walls of the desert plant *Cynodon* while at pH 7.2, La^{3+} was found throughout the walls (Oross and Thomson 1982). At a lower pH it would seem that no barrier exists to La^{3+} movement, lanthanum precipitates being found throughout apoplastic and symplastic spaces.

Other factors affecting REE distribution in plants are concentrations in solution culture, species differences and humidity with high humidity favouring La^{3+} uptake. High levels of La^{3+} ions in plant tissue resulted in deposition of the element in many membrane bound compartments as well as a wide distribution in the cytoplasm (Peterson and Hull 1981b).

2.4.3. *Depositon in tissues*

Evidence for REE deposition in plant tissues can be divided into two groups – those which report REE as being restricted to extracellular or apoplastic locations (Campbell et al. 1974, Campbell and Thomson 1977, Hanzely and Harmet 1982, Harmet 1979, Johnen 1978, Lawton et al. 1981, Maier-Maercker 1980, Oparka and Gates 1981, Platt-Aloia and Thomson 1982, Taylor and Hall 1978, Thomson et al. 1973) and secondly those which describe intracellular deposition (DuPont and Leonard 1977, Oross and Thomson 1982, Peterson and Hull 1981a,b, Robards and Robb 1974, Robinson 1943, Toriyama and Jaffe 1972, Van Steveninck et al. 1976, Wainwright 1977, Wilson and Robards 1980, Wolterbeek and Van Die 1980). As with studies on animal systems, inconsistency in REE source, pH at which experiments are conducted and incomplete or unsatisfactory detection methods have led to this considerable conflict. For the purpose of this review the two groups shall be discussed separately.

Among the studies suggesting that REE do not cross cell membranes are many indicating rare earth accumulation at or in the outer surface of cytoplasmic membranes (Campbell et al. 1974, Hanzely and Harmet 1982, Harmet 1979, Oparka and Gates 1981, Taylor and Hall 1978, Thomson et al. 1973). Lanthanum fed to branches was found to distribute through the cell wall, form local associations with outer leaf membranes (Thomson et al. 1973) and to bind tightly at external Ca^{2+} sites on the cell surface (Campbell et al. 1974, Campbell and Thomson 1977, Harmet 1979). This, sometimes heavy, staining was not removable with a $CaCl_2$ wash (Taylor and Hall 1978). Similarly, experiments with the chelated form of the rare earth europium [Eu $(TTA)_3$] showed no significant entry into wheat root hairs although penetration into bacterial cells in the rhizosphere did occur (Johnen 1978), while marked deposition of La^{3+} has been observed to occur in the cellulosic and interfibrillar region of cell walls with small pockets accumulating in suberin layers and plasmodesmata (Maier-Maercker 1980, Oparka and Gates 1981). In the outer epidermal cell walls of *Avena sativa*, electron dense deposits of La^{3+} were found to be mostly confined to the exterior surfaces of the plasmalemma. In inner epidermal cells the accumulation of La^{3+} was most pronounced in the area of the middle lamellae at Ca^{2+}-containing sites (Hanzely and Harmet 1982). Evidence that the casparian strip and hypocotyl of some plants may prevent REE movement is presented by several authors (Bahadur et al. 1978, DuPont and Leonard 1977, Lawton et al. 1981, Nagahashi et al. 1974, Peterson et al. 1986, Williamson 1979). A cytoplasmic barrier may also be present in ungerminated seeds (Platt-Aloia and Thomson 1982).

Observations of REE penetrating barley plasmalemma have been made by Robards and Robb (1974). They found that entry occurred in conjunction with cell plate formation during cell division. Clarkson (1965) and Clarkson and Sanderson (1969) showed that La^{3+} entered cells at levels high enough to affect root elongation and cell division. DuPont and Leonard (1977) also describe penetration of cell membranes by La^{3+} into vesicles in the cytoplasm of meristematic cells.

Lanthanum has been detected in the tannin vacuoles of *Mimosa pudica* (Toriyama and Jaffe 1972), in hickory leaves (Robinson 1943), lupin leaf cytoplasm (Wainwright 1977), *Yucca flaccida* phloem exudate (Wolterbeek and Van Die 1980), in cytoplasm and vacuoles of young *Allium cepa* roots (Clarkson 1965, Clarkson and Sanderson 1969, Wilson and Robards 1980) and in maize (Peterson and Hull 1981a), *Cynodon* sp. (Oross and Thomson 1982), *Zea mays*, *Hordeum vulgare*, *Salicornia virginica* and *Spartina alterniflora* cytoplasm, vacuoles, vesicles and endoplasmic reticulum in root and leaf cells (Peterson and Hull 1981b, Peterson et al. 1986).

Penetration through the cell membrane appears to be influenced by plant age, with young tissue being more permeable (Clarkson 1965, Clarkson and Sanderson 1969, Robards and Robb 1974, Van Steveninck et al. 1976, Wilson and Robards 1980); high concentrations (Peterson and Hull 1981a,b, Wilson and Robards 1980); time – especially if more than 2–4 hours (DuPont and Leonard 1977, Peterson and Hull 1981b); and pH, low pH favouring penetration (DuPont and Leonard 1977, Oross and Thomson 1982, Wilson and Robards 1980). The ability of La^{3+} to cross the membrane may be dependent on molecular groupings on cell walls and internal pH buffering (Van Steveninck et al. 1976, Wilson and Robards 1980). At low pH it appears that no barrier to La^{3+} transport exists at the casparian strip (Oross and Thomson 1982). Lanthanum penetration may also depend on its form. In an investigation using electron-opaque tracers, $La(NO_3)_3$ did not penetrate the cytoplasm, unlike colloidal $La(OH)_3$ which was found in cytoplasmic residues in endodermal cells and ultimately within the stele. The charged La^{3+} supplied as $La(NO_3)_3$ never reached the stele (Robards and Robb 1974). At pH 7.7, 80% of La is charged and 20% is colloidal. Thus, reducing the pH increases the colloidal proportions and may increase the amount entering into the cytoplasm. If the REE do indeed pass into the cytoplasm, the consequences for many Ca^{2+}-dependent enzymes are likely to be quite marked.

Based on the evidence that La^{3+} and other REE do not enter the cytoplasm, many workers have used them to delineate the extent of apoplastic movement of ions and water (Campbell et al. 1974, DuPont and Leonard 1977, Fineran and Gilbertson 1980, Lawton et al. 1981, Maier-Maercker 1980, Oparka and Gates 1981, Oross and Thomson 1982, Thomson et al. 1973, Van Steveninck et al. 1976, Wilson and Robards 1980). In view, however, of the conflicting reports of REE penetrating into the cytoplasm, this technique must be questioned although the unique electron dense characteristics of the rare earths still make them useful tools in this type of work. Evidence suggests that La within cells may not aggregate to a sufficient degree to appear clearly in electron microscope sections. This may explain the failure of many workers to detect its presence. It is unlikely that any evidential differences were a result of the nutrient supply, as ions were almost invariably balanced in each experiment.

3. Physiology of REE in plants

3.1. *REE and membrane stability*

3.1.1. *Membrane stabilization*

Another major effect of the rare earths occurs, as with Ca^{2+}, as a result of their effects on membrane stability (Mikkelson 1976, Weiss 1974).

Membrane stabilization takes place when an agent such as La^{3+} or Ca^{2+} causes proteins and protein/lipid complexes to become less fluid. Some fluidity is essential to maintain optimum functioning and destabilization may result in an improved membrane selectivity and overall function. Destabilization, however, also contributes to cell "leakiness". La^{3+} may thus contribute to many aspects of ion transport and membrane binding characteristics through its effects on cellular membranes.

Early evidence demonstrated that La^{3+} attaches to polyanionic polysaccharides in the cell membrane, resulting in disruption of cellular transport systems (Robards and Robb 1974). La^{3+} ions were later shown also to influence the binding of certain hormones and solutes to cell membranes (Poovaiah 1979).

Stimulation of binding was observed for several ionic species, the order of effectiveness – $La^{3+} > Ca^{2+} > Mg^{2+} > K^+$, $Na^+ > NH_4^+$ – being not dissimilar to the Hofmeister series of stability (Poovaiah and Leopold 1976b), with La^{3+} being a more potent stabilizing agent than Ca^{2+}. La^{3+} and other rare earths may, therefore, restrict leakiness by altering membrane characteristics, notably membrane fluidity (Edwards 1983, Poovaiah 1979, Poovaiah and Leopold 1976a, Williamson 1979). Indeed, only 1 mM is required to inhibit the leakage of betacyanin from beetroot discs in comparison to the 10 mM of Mn^{2+}, Ca^{2+}, Sr^{2+} and Mg^{2+} (Cooke et al. 1986).

One important result of the stabilization of cell membranes is the observed reduction in disease resistance in barley to *Erysiphe graminis* when exposed to La^{3+}, detailed later (Edwards 1983).

3.1.2. *Effects on transport and accumulation of ions*

The rare earths have important effects on membrane stability and interact strongly with Ca. It is therefore not surprising to find that they also strongly affect ionic interactions within the cell.

An early report of an interaction between the uptake of lanthanides and other ionic species was that of Cummings and Bankert (1971) who demonstrated an increase in Ce and Pr uptake in the presence of Na and K. This may have been due to direct competition or perhaps soil interactions prior to uptake but is more probably due to increased plant growth. Competition between REE and Ca^{2+} at uptake sites has been described earlier (Bahadur et al. 1978, Campbell and Thomson 1977, Clarke and Hennessy 1981, Hanzely and Harmet 1982, Harmet 1979, Leonard et al. 1975, Mikkelson 1976, Toriyama and Jaffe 1972, Ueki 1979, Weiss 1974) and will not be discussed further except where it affects the uptake of other ions.

Evidence from studies of seismonasty in *Mimosa* species suggests that La^{3+} disrupts K^+ efflux from motor cells and, therefore, the contractile protein activity involved in leaf movement (Toriyama and Jaffe 1972). Another example of interference with K^+ transport is presented by Leonard et al. (1975) who demonstrated that the short term influx of Rb-labeled K^+ into corn root segments was disrupted by $La(NO_3)_3$ (0.025 mM La^{3+}) in a manner suggesting competitive inhibition. There was no such effect on Cl^- or phosphate uptake. Active transport of K^+ and the production of starch (an event requiring K^+) are inhibited by La^{3+} in the desert plant *Lupinus arizonicus*, probably by altering the K^+ pump system (Wainwright 1977). Maier-Maercker (1980) suggests, however, that the effect of La^{3+} on K^+ movement is indirect; La^{3+} does not serve as an analogoue for potassium ions and there is no stoichiometric relationship between the two ions. This is in contrast to the earlier reports described. More recent evidence suggests that by blocking Ca^{2+} binding sites La^{3+} may have a direct effect on the H^+/K^+ antiport system in *Avena* coleoptile segments (Hanzely and Harmet 1982).

Generalized observations that the rare earths disrupt monovalent ion transport have been provided by several writers (Brogardh and Johnsson 1975, Campbell and Thomson 1977, Clarke and Hennessy 1981, Hall et al. 1977, Harmet 1979,

Poovaiah and Leopold 1976a). Explanations of these effects include the influence on water permeability of cells (Brogardh and Johnsson 1975, Hall et al. 1977), the binding of REE to Ca^{2+} sites involved in regulating monovalent ion transport (Campbell and Thomson 1977, Harmet 1979) and through generalized effects on membrane stability and leakiness (Poovaiah and Leopold 1976a).

Whatever the nature of the response, REE clearly influence the flux of ions into cells. These fluxes in turn may be expected to influence many plant processes.

3.2. *Physiological manifestations*

The REE influence a wide range of plant processes through their effects on monovalent cation transport and membrane structure. Before discussing these, brief mention will be made of the effects of REE on plant exudates involved in nematode hatching.

Hatching of the parasitic nematode *Globodera rostochiensis* has been found to be restricted by low concentrations of $LaCl_3$ (Atkinson and Ballantyne 1979). Fifty percent (50%) inhibition occurred at 83–100 μM La^{3+}, this appearing to be caused by a prevention of the binding and action of essential stimulators in the root exudate of host potato plants. Contradictory evidence (Clarke and Hennessy 1981) suggests that La^{3+} is, in fact, a moderately effective stimulator of hatching and than any apparent inhibition occurs only with high concentrations of $LaCl_3$. This is due to a complexation of La^{3+} with the hatching stimulus in potato root exudate.

3.2.1. *Influence on leaf and plant movement*

The movement of plants in response to changes in the position of the sun, humidity or physical stimulation may also be influenced by rare earths. Such movements may be indirectly affected by altering the influx and efflux of ions and the subsequent uptake or loss of water from “motor” cells.

Phototaxis in the alga *Chlamydomonas reinhardtii* is reversibly inhibited by La^{3+}, the addition of these ions causing the restriction of Ca uptake on the side near light (Nultsch 1979). Lanthanum ions also appear to influence seismonasty in the sensitive plant *Mimosa pudica* L. Toriyama and Jaffe (1972) propose that the cations interfere with the activity of the tannin vacuole, which in turn is responsible for leaf movements. It is suggested that this interference occurs through the prevention of K^+ mobility, restricting its availability to contractile proteins. Later reports refute the role of these proteins but nevertheless demonstrate an effect of La^{3+} on leaf movement. Campbell et al. (1974) observed that the presence of La^{3+} prevents the backflow of K^+ that usually occurs when Ca^{2+} is present in the cell membrane. Thus the closing of leaflets in *Mimosa pudica* L. is inhibited by the effect of La^{3+} on membrane permeability to K^+. The effect of La^{3+} on the mobility of K^+ ions has also been recognized as the cause of La^{3+} inhibition of sun tracking in *Lupinus arizonicus* L. (Wainwright 1977). This is due to the two-fold inhibition of K^+ transport and K^+-dependent starch granule formation.

3.2.2. *Influence on stomatal movements*

Divalent cations have long been known to influence the opening and closing of stomata. In investigations of the effect of Ca^{2+}, Mg^{2+} and La^{3+} on the transpiratory movements of stomata of *Avena savita*, 2.5 mM La^{3+} gave the same degree of inhibition as 20 mM Ca^{2+} and 40 mM Mg^{2+} (Brogardh and Johnsson 1975). These effects are proposed to occur through the following processes (Brogardh and Johnsson 1975, Maier-Maercker 1980):

(i) The interference of the ions with ionic processes central to stomatal regulation;

(ii) The reduction in water permeability of guard cells caused by the ions;

(iii) The action of La^{3+} as an extra-membrane osmotic agent in drawing water out of guard cell vacuoles.

This last proposal would depend on the membrane being impermeable to La^{3+} when present in sufficiently high concentrations to affect osmotic potential.

Lanthanum ions have also been shown to act as stimulators of the opening of adaxial stomata in *Commelina communis* L. (De Silva et al. 1986). This is proposed to be due to their action as Ca channel blockers. The closure of abaxial stomata due to abscisic acid is also inhibited by La^{3+} (De Silva et al. 1985). It may be, however, that different mechanisms of La^{3+} action exist and are dependent on whether the stomata close in response to light or to humidity. Hall et al. (1977) have demonstrated that 2.5 mM La^{3+} strongly inhibits stomatal opening in response to increased humidity, but has little or no effect on stomatal response to light. Lanthanum may also act by restricting the entry of potassium ions into guard cells, but its ineffectiveness at inhibiting K^+ influx in reponse to increased light suggest other mechanisms may be operating. These require further investigation.

In general terms, however, La^{3+} has been observed to reduce leaf water conductance and photosynthesis. This reduction in photosynthesis may be unrelated to stomatal effects, as Tb^{3+} is known to inhibit chlorophyll *a* fluorescence in vitro by binding with CF_1, the chloroplast coupling factor protein, and transferring excitation energy from the protein ligands (Mills and Hind 1978). It is not known whether this would operate in the intact system.

3.2.3. *Effects on bacteria; nitrogen fixation and disease*

As early as 1948, La was recognized as being capable of altering bacterial and viral activity when Björkman demonstrated that La produces an effect similar to ultraviolet radiation on the influenza virus (Björkman and Horsfall 1948). Wurm (1951) showed an inhibition of growth and metabolism of *Streptococcus faecalis* due to the presence of La. These interactions probably occur through the inhibition of Ca uptake, but may also be due to specific effects on nucleic acids and proteins within the cell and nucleus (Laszlo et al. 1952).

More recently La has been demonstrated to inhibit plaque bacteria by competitively binding available phosphate (Wåler and Rölla 1983). A range of rare earths (La^{3+}, Nd^{3+}, Sm^{3+}, Eu^{3+}, Tb^{3+}, Dy^{3+}) at concentrations of 100–200 μM restrict the active transport of Ca^{2+} in *Myobacterium phlei* by disrupting the availability of respiratory linked substrates and ATP (Agarwal and Kalra 1983). The extent of this inhibition increases with ionic radius.

Other investigations suggest that not only do the rare earths such as Eu and La penetrate substantially into rhizosphere microorganisms (Johnen 1978) but that they may also considerably stimulate peptolytic, amylolytic and aminolytic activity in some media (Naguib et al. 1979). Differences in effectiveness can be related to differences in the chelation of the elements by the soil.

Two important effects of REE on microbial systems include nitrogen fixation and disease responses. One group working on nitrogen fixation by a range of nonsymbiotic organisms has reported substantial increases in the presence of La and other REE (Bahadur and Jyotishmati 1976, Bahadur and Tripathi 1976, Bahadur et al. 1978, Ranganayaki et al. 1981). Concurrent with this is a considerable increase in carbon consumption and an increase in the amount of N_2 fixed (mg N) per gram carbon consumed (Bahadur et al. 1978, Ranganayaki et al. 1981). Maximum stimulation has been found to occur with 300 μM La although not all species respond equally. Addition of Mo further enhances the stimulatory effects of the REE (Bahadur et al. 1978).

Relatively little work has been performed on the effects of REE on disease (Bahadur and Tripathi 1976, Edwards 1983, Velasco et al. 1979). Some reports indicate an increase in disease incidence (Bahadur and Jyotishmati 1976, Bahadur and Tripathi 1976, Velasco et al. 1979), or decreased resistance to disease (Edwards 1983), while a Chinese paper suggests that REE help to control disease (Wang and Chen 1985). Coconut trees growing throughout the Philippines may suffer from a disease called cadang-cadang. Its occurrence has been found to coincide with areas high in thallium and REE (Velasco et al. 1979). Earlier experiments demonstrated a relationship between rare earth levels and disease incidence. Although it is proposed that the presence of rare earths at high concentrations brings about a shortage of one or more essential trace elements resulting in the disease symptoms shown (Velasco et al. 1977), cadang-cadang has recently been demonstrated to be a result of infection by the coconut cadang-cadang viroid (CCCV) by Haseloff et al. (1982). It is possible, however, that the high concentration of rare earths contributes to the sensitivity of coconut palms to the viroid.

Another reference to an REE/disease interaction suggests that La increases the susceptibility of barley plants to the fungus *Erysiphe graminis* (Edwards 1983). It is proposed that this heightened sensitivity is a result of the stabilization of cell membranes by La^{3+}, thus preventing the hypersensitive tissue collapse and stimulating growth of secondary fungal haustoria. It is possible that REE/ hormone interactions are involved, and this point will be discussed later. Abscisic acid is, however, definitely involved in the stimulation of rishitin and lubimin (fungitoxic sesquiterpenes) accumulation in potato tubers, an effect which is prevented by the application of La (Jackson 1985).

The only paper found proposing an increase in disease resistance due to the application of rare earths is that of Wang and Chen (1985). They found that applying a REE paste to the bark of rubber tree helped control the incidence of "brown bast disease".

3.2.4. *Effects of REE on plant growth*

Early reports indicated that the rare earths were inhibitory to plant growth (Pickard 1970, Van Steveninck et al. 1976). Pickard (1970), working with La^{3+}, Pr^{3+} and Nd^{3+}, found that the elongation of oat coleoptile sections was inhibited. Application of LaCl resulted in increased turgidity of segments, perhaps because La decreased permeability of cells to K^+ resulting in a tendency for water to enter the cells. Neodymium was more restrictive of oat coleoptile growth than La^{3+}, but Pr^{3+} did not cause any inhibition at all. This may have been due to differences in ionic radius or charge (Pickard 1970). Lanthanum chloride in low concentrations has been reported to restrict the suberization of potato leaf cuttings dipped in solution; 70% inhibition is obtained after 7 days in 10 μM $LaCl_3$ (Balamani et al. 1986). Colloidal La is also recorded as causing an almost complete inhibition of cell division and root elongation in the root tips of barley plants after 15 hours. The cause of this effect is still unclear (Van Steveninck et al. 1976).

More recent evidence, however, has demonstrated the stimulation of plant growth in the presence of REE. Application of Ce sulphate (up to 100 mg/kg) enhanced the root and shoot growth of the tropical plants *Phaseolus radiatus* and *Brassica pekinensis* and the root crop *Ipomea triloba* developed larger darker green leaves and fleshier roots. It is suggested that, at low levels, the rare earths act as micronutrients, with toxicity occurring at high concentrations (Velasco et al. 1979). No elemental analyses were performed in these experiments and the increase in growth may have been due to some other trace element.

Other studies indicate that La^{3+}, as well as divalent cations, causes rapid growth responses in *Avena* coleoptile segments. Application of La^{3+} produces a prolonged stimulatory effect on the rate of elongation at concentrations of 20 μM to 50 mM. This occurs, however, only after an initial depression (Hanzely and Harmet 1982, Harmet 1979). It has been suggested that La^{3+} stimulates elongation through its effect on Ca^{2+} binding sites, Ca^{2+} perhaps having a role in the regulation of monovalent cation transport through interaction with acidic phospholipids on the membrane (Helmet 1979).

Irreversible replacement of Ca^{2+} with La^{3+} may have a direct effect on the K^+/H^+ antiport, as it does on the Ca^{2+}/H^+ antiport (see section 2.2). Since it is known that H^+ extrusion is responsible for acidification of the cell wall, leading to softening and finally elongation of the coleoptile, La^{3+} might be expected to influence the growth of the *Avena* coleoptile (Hanzely and Harmet 1982). It is also possible that La^{3+} affects auxin binding sites and, thus, cell elongation. This point will be discussed later.

Metal ions such as La^{3+} and Ca^{2+} cause a synergistic stimulation in growth when applied as a foliar spray with the plant growth substance 1-Triacontanol (Welebir 1982). Significant responses using this technique have been recorded in wheat, barley and rye.

These conflicting reports on REE and interactions with plant growth force the conclusion that further work with the use of more consistent application and analysis techniques is required.

3.2.5. *Effects of REE/hormone interactions*

One of the most important means by which REE may influence plant growth, either positively or negatively, is through their effects on the binding of hormones. It has been suggested that hormones acting through an attachment to a site of action on a membrane may be subject to the effects of solutes (especially La^{3+}) that alter membrane characteristics (Poovaiah and Leopold 1976a). Lanthanum can be expected to be particularly significant in this respect because of its potency as a membrane stabilizer. Lanthanum has been demonstrated to increase greatly auxin binding to *Avena* coleoptile segments (Poovaiah and Leopold 1976b). The mechanism of action is proposed to be a modification of the stability of an anionic lipid protein complex in the plasmalemma (Harmet 1979).

The stimulation of the closure of abaxial stomata after the application of abscisic acid of *Commelina communis* was inhibited by La (De Silva et al. 1985). The action of cytokinin in stimulating the adventitious bud initiation in *Torenia* step segments was inhibited by 0.3 mM $LaCl_3$ (Tanimoto and Harada 1986). The consequences of this effect on crop yield is theoretically quite severe.

Other hormone responses affected by membrane-altering solutes include the binding of ethephon (Poovaiah 1979), the effectiveness of 1-Triacontanol (Welebir 1982), and ABA binding and kinetin effectiveness (Edwards 1983).

The potential of rare earths as hormone synergists suggest that La and other closely related REE will influence many aspects of plant physiology. Through their actions as membrane stabilizers they have also been demonstrated to be the most potent hormone effectors of any of the common cations whilst remaining non-toxic to cellular activity.

4. REE in agriculture

The literature on the use of rare earths in agriculture is predominantly from mainland China (Guo 1985a,b, 1987), with earlier reference by them to Russian work and one report from the USA. A small number of preliminary and so far unpublished studies have been carried out in Australia. The work done in China is extensive, involving numerous field experiments and pot studies of both physiological and yield-response types.

The work in China has been carried out intensively since 1972 but it is only recently that some of this literature has become available internationally [*New Frontiers in Rare Earth Science and Applications*, Vol. II; *Proceedings of the International Conference on Rare Earth Development and Applications*, Beijing, People's Republic of China, 1985 (Science Press, Beijing, China) and the *Proceedings of the First International Conference on the Metallurgy and Materials Science of Tungsten, Titanium, Rare Earths and Antimony*, 1988 (International

Academic Publishers)]. The increases in crop yield reported by workers from all parts of China range between 8–50% (table 5), with the common response being of the order of 8–15% (Guo 1985a). The magnitude of these increases appears to indicate that many of these crops would be deficient in rare earths under normal circumstances. Records of deficiency symptoms, however, have not been found and there is no evidence that rare earths are essential elements.

The yield increases reported in table 5 are from experiments which showed a yield improvement with the application of rare earths; those with a neutral or negative response are not included in the average yield response, although the percentage of the total number of experiments represented by the latter group is apparently not high. In wheat studies (Young 1935), for instance, 14% of experiments did not respond to rare earths whilst the remainder averaged an 8% yield increase (range 5–15%).

Beside yield responses, there are reports, including photographs and videos, of visual differences in plants caused by REE application (table 6). These include darker green foliage due to more chlorophyll (Gao et al. 1985, Lin et al. 1986, Ning 1985, Young 1935), enhanced rate of development (Lin et al. 1986, Ning 1985, Young 1935), greater production of roots (Wu et al. 1984), stronger tillering (Ning 1985, Young 1935), better fruit colour in apples, oranges and watermelons (Gao et al. 1985) and more polyphenols in tea (Ning 1985).

Vitamin C content was also raised in Delicious apples and Muscat Hamburg grapes, the protein and fat contents of soybean seeds were increased and the amount of nicotine in tobacco was amplified by the application of rare earths. Similarly, the activity of some enzymes and the total chlorophyll content have been found to be enhanced by an increased concentration of rare earths in plant tissue.

TABLE 5

Increase in the yield of a range of crops after the application of rare earths.

Crop	Average increase	Ref.
Apple	15–30%	Gao et al. (1985)
Day lily flower	20.36%	Ning (1985)
Grapes	7.8–22.5%	Gao et al. (1985)
Peanut	>10% (300 kg ha^{-1})	Ning (1985)
Rice	391.5 kg ha^{-1}	Ning (1985)
Rubber	14.86%	Wang and Chen (1985)
Soybean	>10% (150 kg ha^{-1})	Ning Jiaben (1985)
Sugarcane	4.5–7.5 t ha^{-1}	Ning (1985)
Tea	13.2%	Ning (1985)
Tobacco	34.2–49%	Ning (1985)
Watermelon	13%	Gao et al. (1985)
Wheat	8.0%	Young (1935)
Anabaena	43.2–60%	Wang et al. (1985)
Narcissus	20.8%	Lin et al. (1986)
Cotton	10–15%	Zhao and Yang (1988)

TABLE 6
Increase in quality components of different crops after rare earth application.

		Increase	Ref.*
Sugar content			
Grape		0.15–0.74%	[1]
Rubber		0.56 mg ml^{-1}	[2]
Sugar		450–300 kg ha^{-1}	[3]
Watermelon		1.16%	[1]
Chlorophyll			
Day lily flower		?	[3]
Grape		?	[1]
Wheat		30%	[4]
Narcissus		?	[5]
Enzymes			
Corn	Peroxidase and esterase activity	Density of gel bands	[6]
Day lily flower	Amylase and ascorbic acid oxidase activity	11.5%	[3]
Peanut	Dehydrogenase activity	20.02%	[3]
Rice	Dehydrogenase activity	30.81%	[3]
Miscellaneous			
Delicious apple	Peel anthocyanin	55.3–313.5%	[1]
	Vitamin C	$(2.7–6.7) \times 10^{-3}$ mg g^{-1}	
	Soluble solids	1.5–1.7%	
	Amino acids and polyphenols	?	[3]
Grapes	Vitamin C	$3.5–1.3 \times 10^{-3}$ mg g^{-1}	[1]
Tobacco	Nicotine	42.64%	[3]
	Smoke value	Down 25.46%	
Soybean	Protein	5.49%	[3]
	Fat content	8.5%	

*References: [1] Gao et al. (1985); [2] Wang and Chen (1985); [3] Ning (1985); [4] Young (1935); [5] Lin et al. (1986); [6] Takata et al. (1966).

Agronomic factors contributing to the yield of a crop also have been studied and found to be increased by the application of rare earths. The percentage germination of seeds of rape, wheat, turnip, cabbage and barley was enhanced after soaking in various concentrations of RE–nitrate for 8 hours (Wu et al. 1984). The increase was of the order of 19% for winter wheat. Similarly, the seedling, root and shoot growth of wheat, cucumber, soybean, mungbean and corn were aided by REE (Tang and Li 1983, Wu et al. 1984, 1985, Young 1935). An increase in seedling vigour would, of course, tend to result in an increase in yield. Studies by Wu et al. (1985) demonstrated that the addition of REE chlorides to pots in which soybean plants were growing resulted in an increase in shoot growth of 32%. This may have been due to the rise of 50% in nodulation observed on the plants. The magnitude of this latter increase, as with that of nodule fresh weight, suggests that the advancement in shoot growth may have been due to enhanced nitrogen fixation. These authors propose, however, that it is caused by a rise in phosphate uptake. This hypothesis is supported by various other workers who report an increase in the uptake of ^{32}P in tomato, cucumber, soybean, mungbean, tobacco and day lily flower in response to the application of rare earths (Ning 1985, Wu et al. 1984, 1985). Other reports (Leonard et al. 1975, Ueki 1979) confirm that La^{3+} stimulates P uptake, possibly by the release of phosphatases into the soil medium.

Other plant measurements reported to be increased by the addition of rare earths to the plant are dry weight and leaf area of grapes (Gao et al. 1985), sprouting of tea (Ning 1985), both fresh and dry weight of wheat (Xie and Chang 1985), height and number of effective stems of sugar cane (Ning 1985) and the dry rubber content produced by rubber trees (Wang and Chen 1985). The heading of wheat has also been reported to be advanced by 2 days (Xie and Chang 1985).

In a summary of Chinese research since 1972, involving "hundreds of scientists and technicians from more than 60 units", Zhang et al. (1988) have reported that more than 20 crops and trees respond to REE applications and it is currently thought that one the major reasons for the improved yields and product quality is that photosynthetic rates are improved following increased chlorophyll contents.

Numerous toxicity tests, including some using primates (see, e.g., Ji et al. 1988), have been reported and their results have "firmly testified that REE is, within normal range, convincingly safe". A special issue of the Journal of the Chinese Rare Earth Society, published in September, 1985, reported thirteen experimental studies of REE toxicity, from which it was concluded (Ji 1985) that "the possible harmful effect of trace rare earths used in agriculture on human health was almost non-existent".

Since field tests have established that, under Chinese conditions, REE applications return significant profits to farmers, their use is said to have increased remarkably in recent years. Between 1981 and 1987, the area on which REE was applied increased 260-fold to more than 1 million ha, mainly in the provinces of Heilongjiang, Jiangxi, Hunan, Henan and Anhui. The use of REE applications to crops grown in other provinces is also reported to be "stepping forward very rapidly." (Zhang et al. 1988).

There have been few reports from other countries on the benefits of REE in increasing plant production, apart from the earlier Russian papers previously mentioned. Young (1935) recorded a slight increase in the shoot growth of timothy when La was applied to the soil in a pot trial. Likewise in a small pot experiment, a 10% increase in yield of snail medic (*Medicago scutellata*) was found after the foliar application of Pr. While this increase was significant ($P = 0.05$), applications of La, Ce, Nd or a mixture of the four were without effect (Glynn and Graham, unpublished results). In a similar study with wheat, no response was found to La, Ce or a REE mixture from a Western Australian mine irrespective of the phosphate status of the soil (Guerin and Graham, unpublished results).

In order to determine the extent to which the Chinese results might be reproduced under Australian conditions, an Australian mineral extraction company (SX Holdings Ltd.) has encouraged a group of agricultural scientists in universities and government research agencies to undertake experiments to measure the response of farm crops (wheat, sugar, cotton and pastures) to applications of rare earths.

Preliminary experiments carried out at the University of Queensland (Weggler and Asher, personal communication) showed no significant effect on *P*-uptake by sorghum or soybean roots of low concentrations of La or Ce added to the plant culture solution. In a well-replicated field trial carried out at the Pastoral Research Institute in the Western District of Victoria, Australia, no responses were found to REE applications (by spray and seed dressing) to a summer fodder crop (*Brassica sp.*) (Quigley, personal communication). However, pot trials carried out in northern Queensland gave significant responses, where rare earths (Nongle, or a mixture of laboratory grade cerous and lanthanum nitrate) were applied to sugar cane (Reghenzani, personal communication). A cane sett dip increased significantly ($P < 0.01$), the speed of germination of slower germinating cultivars. Dry weights of cane tops at eight weeks were increased significantly ($P < 0.01$), by more than 20% by sett/root dips, but not be foliar spray. Further results are expected during the next year.

Zhu and Liu (1986, 1988) have reported the total and soluble REE contents of numerous Chinese soils. After correlation with field responses in crops to REE, they determined the following critical values of soluble REE contents, based on an unpublished analytical technique involving extraction with acetate solution (pH 4.8) and a spectrophotometric determination (table 7).

Generally speaking, the responsive soils are alkaline in pH but responses have also been reported from all agricultural regions in China and are not restricted to high pH soils. Liu (1988) lists the distribution of REE in various soil groups and the influence of parent material.

Crop responses to REE applications are reported to be most probable when soils contain less than 10 ppm of available, or soluble, rare earths, and plants grown in soils with more than 20 ppm soluble REE are unlikely to respond.

However, many Chinese publications (e.g., Zhang et al. 1988) indicate that responses to REE applications depend on a combination of many factors, including soil properties (such as pH, organic matter and mineral contents and parent rock), the methods, rates and timing of applications, crop conditions (such as type, variety and stage of growth), weather conditions, and the amounts, purity and pH of the water in which the REE are dissolved.

The most common methods of REE application are spraying, seed dressing, smearing or soaking, and for some crops a combination of these methods is

TABLE 7
Critical values of soluble REE contents.

<5.0 ppm REE	Very low
5–10 ppm REE	Low
11–15 ppm REE	Medium
16–20 ppm REE	High
>20 ppm REE	Very high

recommended. In some circumstances the REE are applied with fungicides as seed dressings or with herbicides as sprays. However, their incorporation with mineral fertilisers for subsequent application directly to the soil is reported to negate their effectiveness. Further Chinese studies are in progress in order to understand better the physiological effects of rare earths on plants and animals, and to determine how application technologies can best be simplified and made more effective.

The primary source of rare earths for agricultural use in China are soluble extracts of mineral ores, that were originally known as "Nongle" (=happy farmer) but have recently been re-named "Changle" (=happiness forever). These compounds have usually been extracted with nitric acid and are said to contain an equivalence of 38% RE oxide. As well as the rare earths they also contain a number of elements essential to plant growth (table 8). Trace element deficiencies are common in China, particularly those of B, Fe, Mn, Mo and Zn (Liu et al. 1983, 1984). Many of these soils are also low in REE (Zhu and Liu 1986) and it remains possible that some of the reported effects of rare earths may be due to other trace elements, especially Mo, present in Changle as contaminants. The

TABLE 8
Quantitative chemical analysis of Nongle/Changle (Kerven, unpublished).

	Percent found		
Analyte	Sample 1	Sample 2	Method of analysis
Ce	14.7	16.2	ICPAES**
La	9.7	8.9	ICPAES
NO_3	49.6	47.8	Calorimetric
Cl	4.41	5.64	Ag titration
H_2O*	17.0	17.4	NMR
Insoluble matter	0.16	0.05	Gravimetric
Ca	1.54	2.83	FAAS
Mg	0.033	0.065	FAAS
Mn	0.040	0.056	FAAS
Fe	<0.002	<0.002	FAAS
Cu	<0.002	<0.002	FAAS
Co	<0.002	<0.002	FAAS
Zn	0.0047	0.0185	FAAS
Mo	0.019	0.023	ICPAES
B	0.013	0.003	ICPAES
Ni	<0.002	<0.002	FAAS
P	<0.002	<0.002	ICPAES
S	<0.002	<0.002	ICPAES
Sr	<0.002 +	<0.002 +	ICPAES
Total	*97.21%*	*98.98%*	

*Water of crystallization. Sample dried in desiccator overnight prior to weighing out.

**All element analyzed by ICPAES were corrected for spectral interference by Ce and La.

lack of analyses, including source of rare earths, in the published papers leaves this question unresolved but it is unlikely that such contaminants could account for all the effects that have been reported. The nitrate commonly used to solubilize the rare earths is also unlikely to be significant at the rates of applications.

There are several reasons why the results published by the Chinese are difficult to evaluate. In addition to the inevitable ambiguities and lack of clarity that sometimes occur in translations, some Chinese journals do not require contributors to detail their experimental designs, methodologies, statistics and results. Furthermore, because the China National Non-Ferrous Metals Import and Export Corporation is seeking to market overseas the rare earth compounds (Changle) produced in China, together with the associated information, analytical and application technologies which are used in Chinese agriculture, there is an understandable reluctance both to publish comprehensively in the scientific literature or to reveal all the available information in discussion with scientists from other countries.

5. Summary

The REE bind in biological systems at superficially located Ca^{2+} sites in a manner less reversible than Ca^{2+}, thereby influencing biochemical reactions involving Ca^{2+}. They may, either, have little influence due to Ca^{2+} specificity or may act in opposition to Ca^{2+} binding. Ca^{2+}/REE interactions can result in a number of effects, including the displacement of Ca^{2+} from its binding sites, the inhibition of various Ca^{2+}-dependent enzymes and also may cause a modification of cell permeability to Na^+ and K^+. Other effects of the REE involve membrane stabilization, the blockage of K^+ and Ca^{2+} transport and the general disruption of monovalent ion transport. As a result of these influences the REE can cause various physiological effects, particularly the inhibition of phototaxis, seismonasty and stomatal opening and may initiate a general reduction in leaf water conductance. Rare earths may also influence bacterial systems resulting in increased nitrogen fixation and efficiency in nonsymbiotic bacteria and the stimulation of the growth of fungal haustoria by the prevention of hypersensitive cell collapse.

Growth increases may be initiated by various effects of rare earth application, particularly those of membrane stabilization, improved hormone effectiveness, growth responses in coleoptile segments, improved nitrogen fixation and efficiency and a reduction in water loss by the plant.

The extensive range of physiological and biochemical effects of the lanthanide elements holds out the possibility of significant effects on plant and animal growth and development. The results reported from China indicate that such effects occur under field conditions when a soluble extract of rare earths is appropriately applied to crops under a wide range of soil and climatic conditions.

A full understanding of the physiological basis of the Chinese results, and the extent to which they may be reproduced in other farming systems, can only be determined by further research. Such studies must include investigations into the availability and pH responses of rare earths in different soils and their ability to cross cellular membranes, together with soil/REE interactions, including chelation and soil micro-organism effects. Well-designed field studies are also needed in which various methods and rates of REE application to a range of crops are studied under various soil and climatic conditions. Finally, the physiological and agronomic responses of crops to REE applications need to be identified and measured.

Only when such studies have extended or modified the evidence reported from China will it be possible to determine how important rare earths are as crop nutrients and to predict the circumstances in which their agricultural application would be merited. In the meantime sufficient evidence exists to support the possibility that REE applications could improve crop yields.

Acknowledgements

The authors wish to acknowledge the support of Muswellbrook Energy and Minerals Ltd. (AHR) and the Hannaford Bequest (PHB). The Australian Department of Science and the Chinese National Non-Ferrous Metals Corporation, and the Australian Academy of Technological Sciences and Engineering and the State Science and Technology Commision of China are also thanked for the provision of funds allowing Dr. R.D. Graham and Prof. D.E. Tribe, respectively, to travel to China for investigation into the uses of rare earths in agriculture. Mr. G. Kerren of the University of Queensland carried out the analysis of Nongle. The literature survey presented was undertaken following approaches from Mr. Norton Jackson of Muswellbrook Energy and Minerals Ltd.

References

Agarwal, N., and V.K. Kalra, 1983, Biochim. Biophys. Acta **727**, 285.

Atkinson, H.J., and A.J. Ballantyne, 1979, Ann. Appl. Biol. **93**, 191.

Bahadur, K., and U. Jyotishmati, 1976, Zentralbl. Bakteriol. Parasitenkd. Infektionskr. Hyg. Zweite. Naturwiss. Abt. Mikrobiol. Landwirtsch. Technol. Umweltschutzes **131**, 157.

Bahadur, K., and P. Tripathi, 1976, Zentralbl. Bakteriol. Parasitenkd. Infektionskr. Hyg. Zweite. Naturwiss. Abt. Mikrobiol. Landwirtsch. Technol. Umweltschutzes **130**, 486.

Bahadur, K., S. Prakash and U. Jyotishmati, 1978, Zentralbl. Bakteriol. Parasitenkd. Infektionskr. Hyg. Zweite. Naturwiss. Abt. Mikrobiol. Landwirtsch. Technol. Umweltschutzes **133**, 632.

Balamani, V., K. Veluthambi and B.W. Poovaiah, 1986, Plant Physiol. (Bethesda) **80**, 856.

Batra, S., 1982, Acta Physiol. Scand. **114**, 447.

Bjorkman, S.E., and F.L. Horsfall, 1948, J. Exp. Med. **88**, 445.

Blumwald, E., and R.J. Poole, 1986, Plant. Physiol. (Bethesda) **80**, 727.

Brogardh, T., and A. Johnsson, 1975, Planta (Berl.) **124**, 99.

Burton, P.R., and H.L. Fernandez, 1973, J. Cell Sci. **12**, 567.

Campbell, N.A., and W.W. Thomson, 1977, Plant. Physiol. (Bethesda) **60**, 635.

Campbell, N.A., W.W. Thomson and K. Platt, 1974, J. Exp. Bot. **25**, 61.

Clarke, A.J., and J. Hennessy, 1981, Nematologica **27**, 190.

Clarkson, D.T., 1965, Ann. Bot. (London) **29**, 309.

Clarkson, D.T., and S. Sanderson, 1969, Planta (Berl.) **89**, 136.

Cockerill, A.F., G.L.O. Davies, R.C. Harden and D.M. Rackham, 1973, Chem. Rev. **73**, 553.

Cooke, A., A. Cookson and M.J. Earnshaw, 1986, New Phytol. **102**, 491.

Cummings, S.L., and L. Bankert, 1971, Radiol. Health Data Rep. **12**, 83.

De Silva, D.L.R., R.C. Cox, A.M. Hetherington and T.A. Mansfield, 1985, New Phytol. **101**, 555.

De Silva, D.L.R., R.C. Cox, A.M. Hetherington and T.A. Mansfield, 1986, New Phytol. **104**, 41.

Deschenes, R.J., H.G. Mautner and J.K. Marquis, 1981, Biochim. Biophys. Acta **649**, 515.

Doggenweiler, C.F., and S. Frenk, 1965, Proc. Natl. Acad. Sci. USA **53**, 425.

DuPont, F.M., and R.T. Leonard, 1977, Protoplasma **91**, 315.

Edwards, H.H., 1983, Phytopathol. Z. **107**, 22.

Fineran, B.A., and J.M. Gilbertson, 1980, Eur. J. Cell Biol. **23**, 66.

Freeman, D.J., and E.E. Daniel, 1973, Can. J. Physiol. Pharmacol. **51**, 900.

Gao, Deliang, Zhu Tongshun and Lu Zhengyuan, 1985, Studies on the Application of Rare Earth Elements to Fruit Trees and Watermelon Plants, in: Proc. Int. Conf. on Rare Earth Development and Applications, Beijing, People's Republic of China, Vol. 2, eds Xu Guangxian and Xiao Jimei (Science Press, Beijing) p. 1504.

Garten Jr, C.T., E.A. Bondietti and R.L. Walker, 1981, J. Environ. Qual. **10**, 207.

Glyn, A., and R.D. Graham, year unknown, unpublished results.

Gogeleir, H., H. De Smedt, W. Van Driessche and R. Borghgraef, 1981, Biochim. Biophys. Acta **640**, 185.

Gross, D.S., S.W. Rice and H. Simpkins, 1981, Biochim. Biophys. Acta **656**, 167.

Gschneidner Jr, K.A., 1977, U.S. Navy J. Underwater Acoustics **27**, 159.

Gschneidner Jr, K.A., 1981, Rare Earth Speciality Inorganic Chemicals, in: Speciality Inorganic Chemicals, ed. G. Thompson (Royal Society of Chemistry, London) pp. 403–443.

Gschneidner Jr, K.A., 1988, Physical Properties of Rare Earth Metals, in: CRC Handbook of Chemistry and Physics, ed. R.C. West, 69th Ed. (CRC Press, Boca Raton, FL) pp. B208–B215.

Guerin, S., and R.D. Graham, year unknown, unpublished results.

Guo, Bosheng, 1985a, Present and Future Situation of Rare Earth in China's Agronomy, in: Proc. Int. Conf. on Rare Earth Development and Applications, Beijing, People's Republic of China, Vol. 2, eds Xu Guangxian and Xiao Jimei (Science Press, Beijing, China) p. 1522.

Guo, Bosheng, 1985b, J. Chin. Rare Earth Soc. **3**, 93.

Guo, Bosheng, 1987, A new Application Field for Rare Earth-Agriculture, in: Rare Earth Horizons 1987 (Department Ind. Tech. and Commerce, Canberra, Australia) pp. 237–246.

Haley, T.J., 1965, J. Pharm. Sci. **54**, 663.

Haley, T.J., 1979, Toxicity, in: Handbook on the Physics and Chemistry of Rare Earths, eds K.A. Gschneidner Jr and L. Eyring (North-Holland, Amsterdam) pp. 553–585.

Hall, A.E., W.W. Thomson, C.W. Asbell, K. Platt-Aloia and R.T. Leonard, 1977, Physiol. Plant. **41**, 89.

Hanzely, L., and K.H. Harmet, 1982, Z. Pflanzenphysiol. **107**, 223.

Harmet, K.H., 1979, Plant Physiol. (Bethesda) **64**, 1094.

Haseloff, J., N.A. Mohamed and R.H. Symons, 1982, Nature (London) **299**, 316.

Hatae, J., 1982, Jpn. J. Physiol. **32**, 609.

Hodge, H.C., and J.H. Sterner, 1943, Am. Ind. Hyg. Assoc. Quart. **10**, 93.

Hunt, G.R.A., and K. Jawaharlal, 1980, Biochim. Biophys. Acta **601**, 678.

Hunt, G.R.A., and I.C. Jones, 1982, Biosci. Rep. **2**, 921.

Hutcheson, D.P., D.H. Gray, B. Venugopal and T.D. Luckey, 1975, J. Nutr. **105**, 670.

Jackson, N., 1985, Rare Earths Report on a Visit to the People's Republic of China by an Australian Delegation, 9–24 September, 1985 (Department of Science, Australia).

Ji, Yunjing, 1985, J. Chin. Rare Earth Soc. (Special Issue) **4**.

Ji, Yunjing, Li Janlin, Li Laiyu and Huang Zhaopei, 1988, The Study on the Toxicity, Absorption and Excretion of Mixture of Rare

Earths in Monkeys, in: Proc. 1st Int. Conf. Metallurgy and Materials Sci. of Tungsten, Titanium, Rare Earths and Antimony, ed. Fu Chong Yue (International Academic Publishers) p. 1296.
Johnen, B.G., 1978, Soil Biol. Biochem. **10**, 495.
Jorns, A.C., 1987, J. Plant Physiol. **129**, 493.
Knaus, R.M., and A.H. El-Fawaris, 1980, Environ Exp. Bot. **21**, 217.
Langer, G.A., and J.S. Frank, 1972, J. Cell Biol. **54**, 441.
Laszlo, D., D.M. Ekstein, R. Lewin and K.G. Stern, 1952, J. Natl. Cancer Inst. **13**, 559.
Laul, J.C., W.C. Ceimer and L.A. Rancitelli, 1979, Phys. Chem. Earth **11**, 819.
Lawton, J.R., A. Todd and D.K. Naidoo, 1981, New Phytol. **88**, 713.
Leonard, R.T., G. Nagahashi and W.W. Thomson, 1975, Plant Physiol. (Bethesda) **55**, 542.
Lettvin, J.Y., W.F. Pickard, W.S. McCulloch and W. Pitts, 1964, Nature (London) **202**, 1338.
Lin, Xijie, Fu Zian, Dai Jinxing and Liu Quancheng, 1986, J. Fujian Agric. Coll. **15**, 246.
Liu, Zheng, 1988, The Effects of Rare Earth Elements on Growth on Crops, in: Proc. Int. Symp. on Hardly Known Trace Elements, Budapest, 1988, in press.
Liu, Zheng, Zhu Qi-qing and Tang Li-hua, 1983, Soil Sci. **135**, 40.
Liu, Zheng, Zhu Qi-qing, Tang Li-hua, Xu Jung-xiang and Yen Chuliang, 1984, Micronutrient News **4**, 1.
Luckey, T.D., A. Kotb, J. Vogt and D.P. Hutcheson, 1975, J. Nutr. **105**, 660.
Maier-Maercker, V., 1980, Z. Pflanzenphysiol. **105**, 121–130.
Markert, B., 1987, Phytochemistry (Oxford) **26**, 3167.
Martinez-Palomo, A., D. Benitez and J. Alanis, 1973, J. Cell Biol. **58**, 1.
Mikkelson, R.B., 1976, Lanthanides as Calcium Probes in Membranes, in: Biological Membranes, Vol. 3, eds D. Chapman and D.F. Wallach (Academic Press, New York) pp. 153–190.
Mills, J.D., and G. Hind, 1978, Photochem. Photobiol. **28**, 67.
Mines, G.P., 1910, J. Physiol. (London) **40**, 327.
Moeller, T., 1963, The Chemistry of the Lanthanides (Chapman and Hall Ltd., London).
Morishima, I., M. Kurono and Y. Shiro, 1986, J. Biol. Chem. **261**, 9391.
Nagahashi, G., R.T. Thomson and R.T. Leonard, 1974, Science (Washington, DC) **183**, 670.
Naguib, M.I., M.K. Zeinat and M. Sobhy, 1979, Egypt. J. Bot. **22**, 225.
Neldhuis, J.D., 1982, Biochim. Biophys. Acta **720**, 211.
Ning, Jiaben, 1985, The Results of Applying Rare Earth Elements to Rice and other Crops, in: Proc. Int. Conf. on Rare Earth Development and Applications, Beijing, People's Republic of China, Vol. 2, eds Xu Guangxian and Xiao Jimei (Science Press, Beijing) p. 1518.
Nultsch, W., 1979, Arch. Microbiol. **123**, 93.
Oparka, K.J., and P. Gates, 1981, Planta (Berlin) **152**, 388.
Oross, J.W., and W.W. Thomson, 1982, Eur. J. Cell Biol. **28**, 257.
Pant, G., 1981, Indian J. For. **4**, 186.
Parker, J.C., and G.J. Barritt, 1981, Biochem. J. **200**, 109.
Peterson, T.A., and R.J. Hull, 1981a, J. Cell Biol. **91**, Suppl. 420a, No. 24055.
Peterson, T.A., and R.J. Hull, 1981b, Plant Physiol. (Bethesda) **67**, 157.
Peterson, T.A., E.S. Swanson and R.J. Hull, 1986, J. Exp. Bot. **37**, 807.
Pickard, B.G., 1970, Planta (Berlin) **91**, 314.
Platt-Aloia, K.A., and W.W. Thomson, 1982, Plant Physiol. (Bethesda) **67**, 159.
Polya, G.M., and V. Micucci, 1985, Plant Physiol. (Bethesda) **79**, 968.
Polya, G.M., E. Klucis and M. Haritou, 1987, Biochim. Biophys. Acta **931**, 68.
Poovaiah, B.W., 1979, J. Am. Soc. Hortic. Sci. **104**, 164.
Poovaiah, B.W., and A.C. Leopold, 1976a, Plant Physiol. (Bethesda) **58**, 182.
Poovaiah, B.W., and A.C. Leopold, 1976b, Plant Physiol. (Bethesda) **58**, 783.
Ranganayaki, S., K. Bahadur and C. Mohan, 1981, Z. Allg. Mikrobiol. **21**, 329.
Reuben, J., 1979, Bioinorganic Chemistry: Lanthanides as Probes in Systems of Biological Interest, in: Handbook on the Physics and Chemistry of Rare Earths, Vol. 4, eds K.A. Gschneidner Jr and L. Eyring (North-Holland, Amsterdam) ch. 39.
Revel, J.P., and M.J. Karnowsky, 1967, J. Cell Biol. **37**, C7.
Robards, A.W., and M.E. Robb, 1974, Planta (Berlin) **120**, 1.
Robinson, W.O., 1943, Soil Sci. **56**, 1.

Sanborn, W.G., and G.A. Langer, 1970, J. Gen. Physiol. **56**, 191.
Sarkardi, B., I. Szasz, A. Gerloczy and G. Gardos, 1977, Biochim. Biophys. Acta **464**, 93.
Shumaker, K.S., and H. Sze, 1986, J. Biol. Chem. **261**, 12172.
Surek, B., G. Kreimer, M. Melkonian and E. Latzko, 1987, Planta (Berlin) **171**, 565.
Takata, M., W.F. Pickard, J.Y. Lettvin and J.W. Moore, 1966, J. Gen. Physiol. **50**, 461.
Tang, Xike, and Li Guofeng, 1983, J. Chin. Rare Earth Soc. **1**, 56.
Tanimoto, S., and H. Harada, 1986, Plant Cell Physiol. **27**, 1.
Taylor, A.R.D., and J.L. Hall, 1978, Protoplasma **96**, 113.
Thomson, W.W., K. Platt and N.A. Campbell, 1973, Cytobios. **8**, 57.
Topp, N.E., 1965, The Chemistry of Rare Earth Elements (Elsevier Publishing Company, Amsterdam).
Toriyama, H., and M.J. Jaffe, 1972, Plant Physiol. (Bethesda) **49**, 72.
Tribe, D.E., year unknown, personal communication.
Trosper, T.L., and K.D. Philipson, 1983, Biochim. Biophys. Acta **731**, 63.
Ueki, K., 1979, Plant Cell Physiol. **20**, 789.
Van Steveninck, R.F.M., M.E. Van Steveninck and D. Chescoe, 1976, Protoplasma **90**, 89.
Velasco, J.R., L.E. Domingo, Z.N. Sierra and F.F. Coronado, 1977, Philipp. J. Coconut Stud. **2**, 1.
Velasco, J.R., L.E. Domingo, A.S. Lansangan and Z.N. Sierra, 1979, Philipp. J. Coconut Stud. **4**, 1.
Wainwright, C.M., 1977, Am. J. Bot. **64**, 1032.
Waler, S.M., and G. Rolla, 1983, Scand. J. Dent. Res. **91**, 260.
Wang, Guohong, and Chen Yucai, 1985, Study on the Application of Rare Earths to Rubber Trees, in: Proc. Int. Conf. on Rare Earth Development and Applications, Beijing, People's Republic of China, Vol. 2, eds Xu Guangxian and Xiao Jimei (Science Press, Beijing) p. 1510.
Wang, Lanxian, Xu Zhong, Wu Xiangyu and Tang Renhuan, 1985, Effects of Rare Earth Elements on Photosynthesis and Photofixation of Nitrogen in Anabaena Azotica, in: Proc. Int. Conf. on Rare Earth Development and Applications, Beijing, People's Republic of China, eds Xu Guangxian and Xiao Jimei (Science Press, Beijing) p. 1527.
Weiss, G.B., 1974, Ann. Rev. Pharmacol. **14**, 343.
Welebir, A.J., 1982, Plant Physiol. (Bethesda) **69**, Suppl. 4, 37, No. 204.
Williams, R.J.P., 1967, Endeavour **26**, 96.
Williamson, F.A., 1979, Planta (Berlin) **144**, 209.
Wilson, A.J., and A.W. Robards, 1980, Protoplasma **104**, 149.
Wolniak, S.M., P.K. Hepler and W.T. Jackson, 1980, J. Cell Biol. **87**, 23.
Wolterbeek, B., and J. Van Die, 1980, Acta Bot. Neerl. **29**, 307.
Wurm, M., 1951, J. Biol. Chem. **192**, 707.
Wu, Zhaoming, Tang Xike and Tsui Cheng, 1984, J. Chin. Rare Earth Soc. **2**, 75.
Wu, Zhaoming, Li Jiage, Xu Ji and Xin Shuying, 1985, The effect of Rare Earth Elements on Nodulation and Nitrogen Fixation of Soybean Plants, in: Proc. Int. Conf. on Rare Earth Development and Applications, Beijing, People's Republic of China, Vol. 2, eds Xu Guangxian and Xiao Jimei (Science Press, Beijing) p. 1515.
Xie, Huiguang, and Chang Qingzhong, 1985, Study on the Effect of Rare Earth Elements on the Yield of Wheat, in: New Frontiers in Rare Earth Science and Applications, Vol. II. Proc. Int. Conf. on Rare Earth Development and Applications, Beijing, People's Republic of China, eds Xu Guangxian and Xiao Jimei (Science Press, Beijing) p. 1505.
Young, R.S., 1935, Cornell Ag. Exp. Sta. Memoir 174.
Zhang, Yulin, Liu Zhongxiu and Zhang Jizhen, 1988, Proc. 1st Intl. Conf. Metallurgy and Materials Science of Tungsten, Titanium, Rare Earths and Antimony, ed. Fu Chongyue (International Academic Publishers).
Zhao, Qinghua, and Yang Huiyuan, 1988, Proc. 1st Intl. Conf. Metallurgy and Materials Sci. of Tungsten, Titanium, Rare Earths and Antimony, ed. Fu Chongyue (International Academic Publishers).
Zhu, Qi-qing, and Liu Zheng, 1986, Curr. Prog. Soil Res. People's Republic of China, p. 335.
Zhu, Qi-qing, and Liu Zheng, 1988, Proc. 1st Int. Conf. Metallurgy and Materials Science of Tungsten, Titanium, Rare Earths and Antimony, ed. Fu Chongyue (International Academic Publishers).
Zook, M.N., J.S. Rush and J.A. Kuc, 1987, Plant Physiol. (Bethesda) **84**, 520.

ERRATA

Vol. 1, ch. 6, McEwen

1. Page 487, Mackintosh, A.R., 1962: insert the word "Phys." between "1962" and "Rev."

Vol. 6, ch. 49, Rogl

1. Chaban et al. (1980) should be Chaban et al. (1979), and the page number 875 should be 873
   p. 387, top reference list (for the Gd–Fe–B system),
   p. 426, top reference list (for the Nd–Fe–B system),
   p. 444, first reference on the page (for the Sm–Fe–B system).

Vol. 10, ch. 65, Freeman et al.

1. Page 215, line 3 (bottom): the equation number is 7 instead of "17".

Vol. 11, Prologue, Svec

1. Page 23, 1954 entries, second and fifth entries: J.L. Dye instead of "J. Dye".
2. Page 27, 1966 entries, first entry: *with D.A. Csejka* . . . J. Phys. Chem. instead of "J. Chem. Phys.".
3. Page 27, 1966 entries, second entry: *with M.J. Pikal* . . . J. Phys. Chem. instead of "J. Chem. Phys.".
4. Page 29, 1974 entries, eleventh entry: *with D.L. Witte* . . . J.A. Rard instead of "A. Rard".
5. Page 31, 1977 entries, last entry: *with J.A. Rard, L.E. Shiers* . . . the page numbers should be 337–347 instead of "56–58".
6. Page 31, 1979 entries, second entry: *with Habenschuss* . . . II . . . the page numbers should be 3758–3763 instead of "2797–2806".
7. Page 31, 1981 entries: the following entry was omitted: *with J.A. Rard*, Isopiestic determination of the activity coefficients of some aqueous rare earth electrolyte solutions at 25°C. 5. $Dy(NO_3)_3$, $Ho(NO_3)_3$ and $Lu(NO_3)_3$. J. Chem. Eng. Data, **26**, 391–395.

Vol. 12, ch. 83, Szytuła and Leciejewicz

1. Page 133, entry 4.2 (second column): RCo_2Ge_2 instead of "RCo_2Ce_2".

SUBJECT INDEX